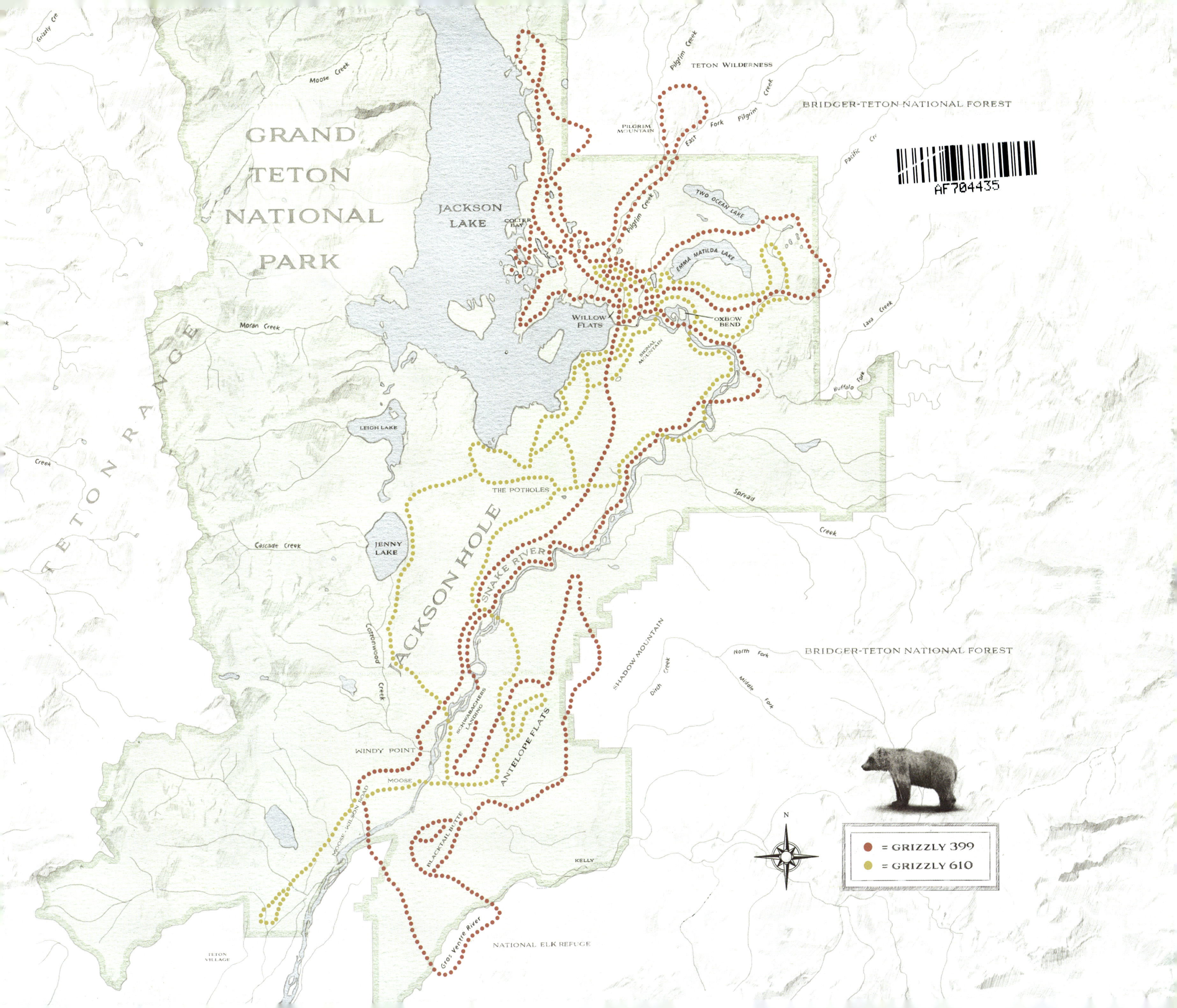
GRAND TETON NATIONAL PARK
JACKSON LAKE
COLTER BAY
TETON WILDERNESS
BRIDGER-TETON NATIONAL FOREST
PILGRIM MOUNTAIN
Pilgrim Creek
East Fork Pilgrim Creek
Pacific Cr
TWO OCEAN LAKE
EMMA MATILDA LAKE
WILLOW FLATS
OXBOW BEND
SIGNAL MOUNTAIN
Moose Creek
Moran Creek
Grizzly Cr
TETON RANGE
LEIGH LAKE
JENNY LAKE
Cascade Creek
Creek
Cottonwood Creek
THE POTHOLES
JACKSON HOLE
SNAKE RIVER
Lava Creek
Buffalo Fork
Spread Creek
SHADOW MOUNTAIN
Ditch Creek
North Fork
Middle Fork
BRIDGER-TETON NATIONAL FOREST
SCHWABACHERS LANDING
ANTELOPE FLATS
WINDY POINT
MOOSE
MOOSE-WILSON ROAD
BLACKTAIL BUTTE
KELLY
Gros Ventre River
NATIONAL ELK REFUGE
TETON VILLAGE
N
= GRIZZLY 399
= GRIZZLY 610
AF704435

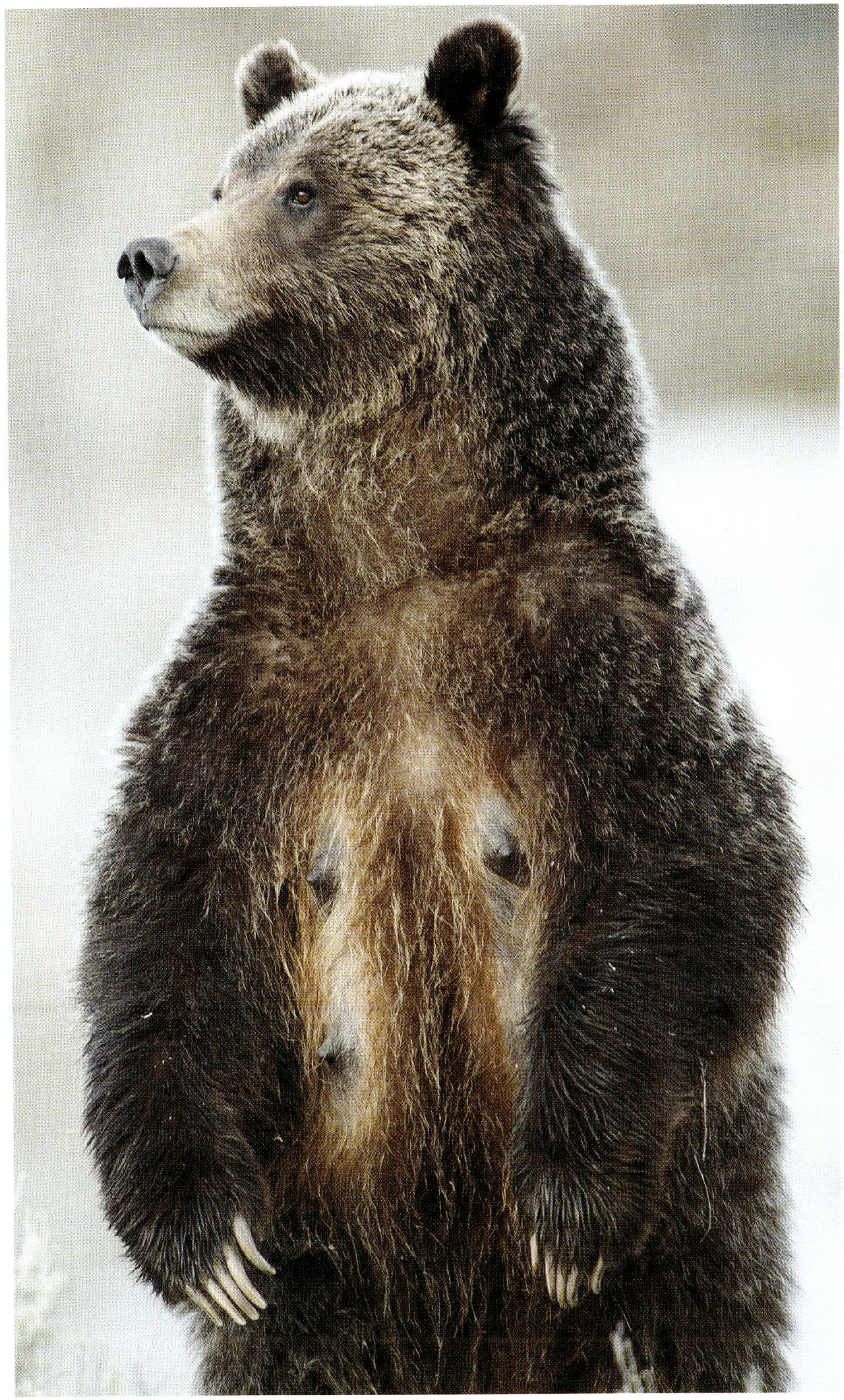

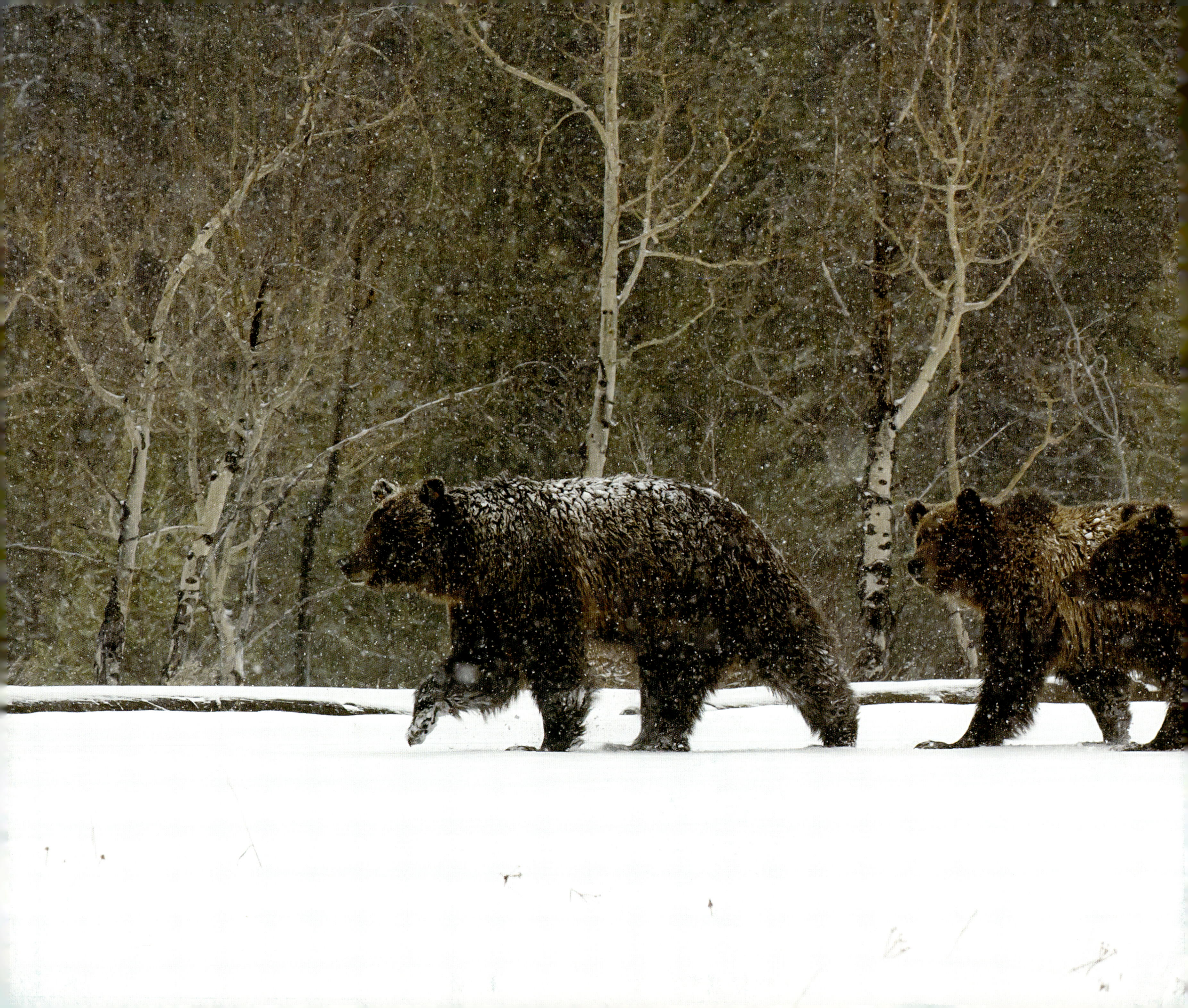

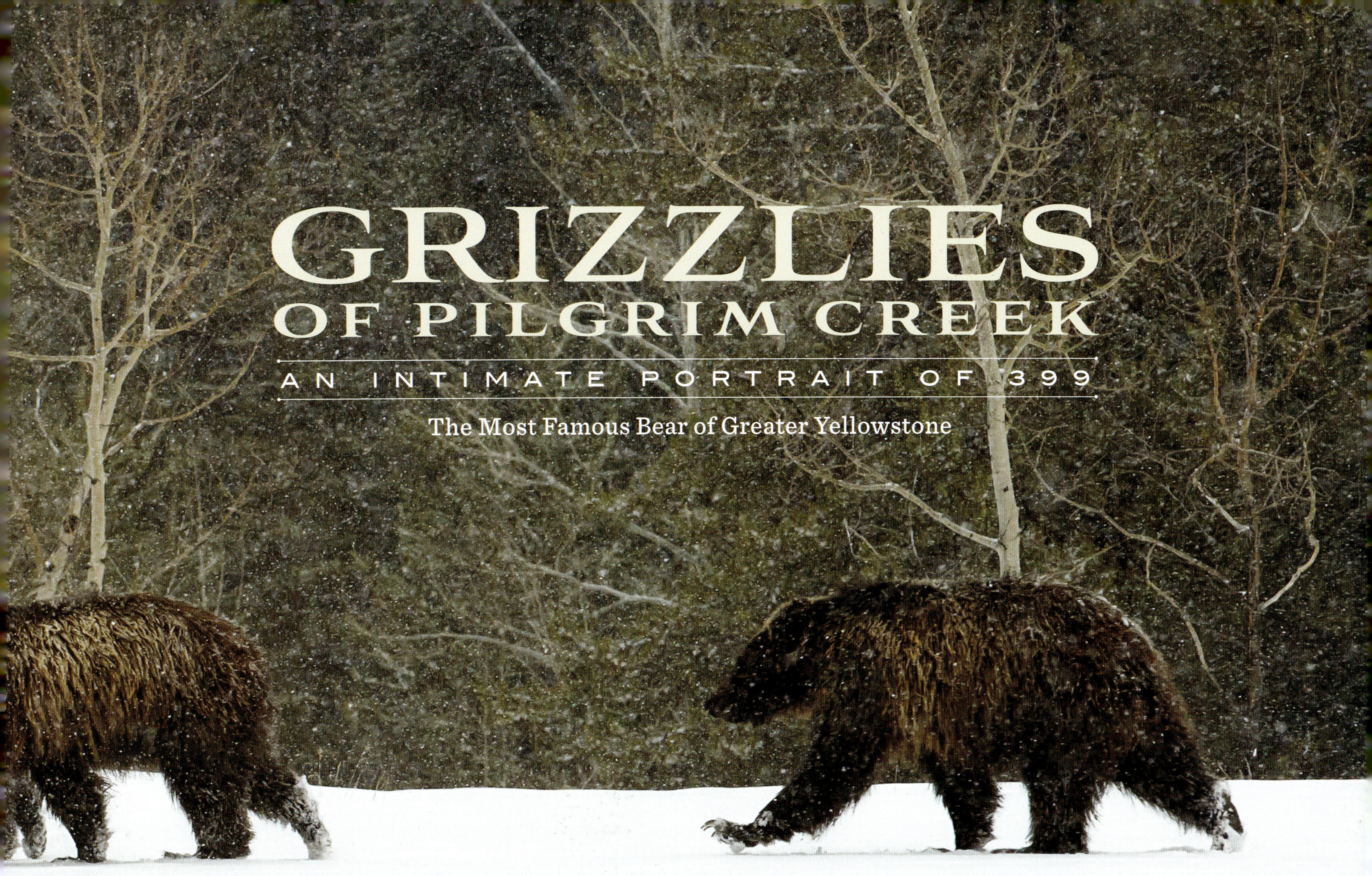

GRIZZLIES OF PILGRIM CREEK

AN INTIMATE PORTRAIT OF 399

The Most Famous Bear of Greater Yellowstone

Photography by THOMAS D. MANGELSEN

TODD WILKINSON | Foreword by TED TURNER

MANGELSEN IMAGES OF NATURE GALLERY

RIZZOLI NEW YORK

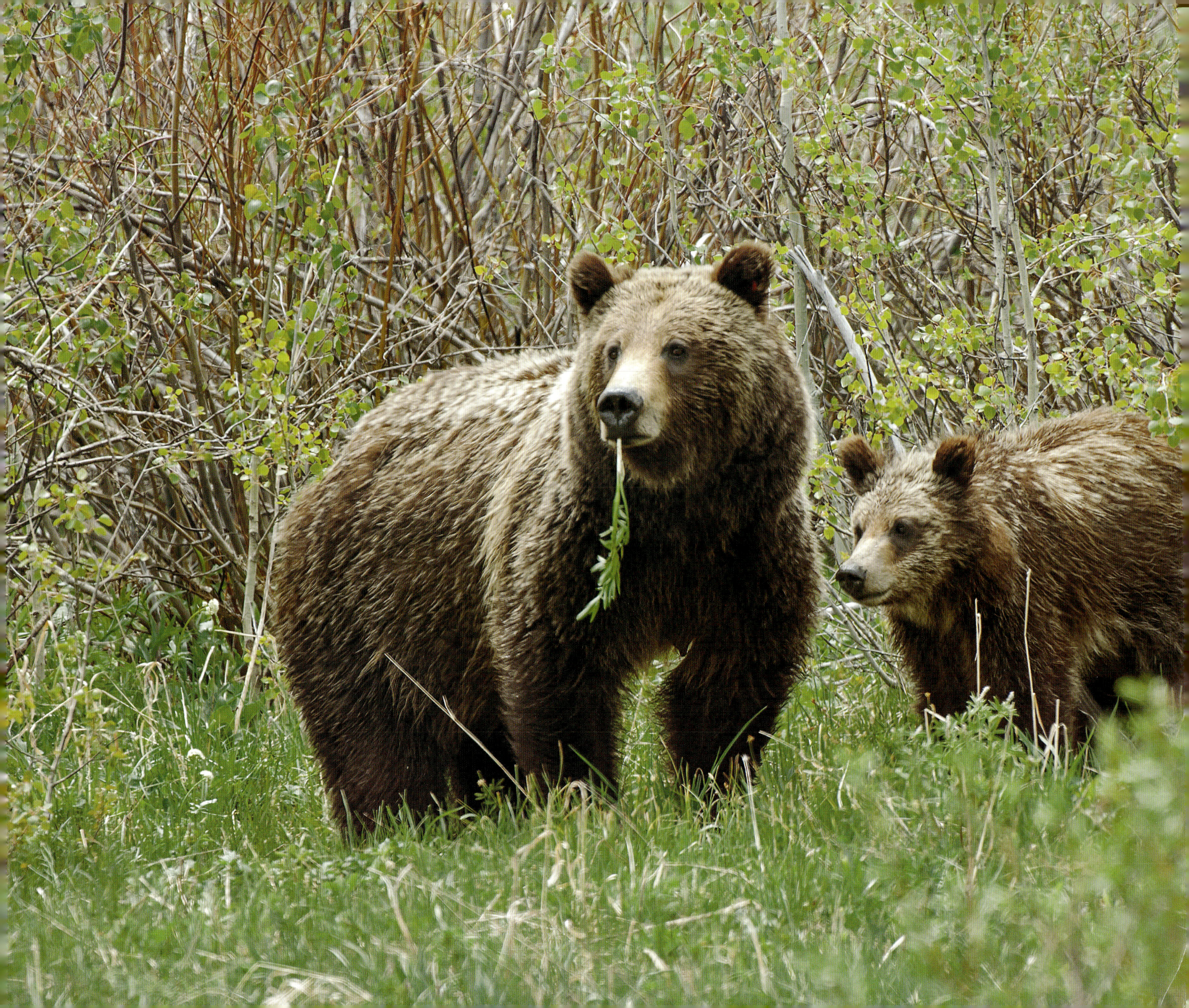

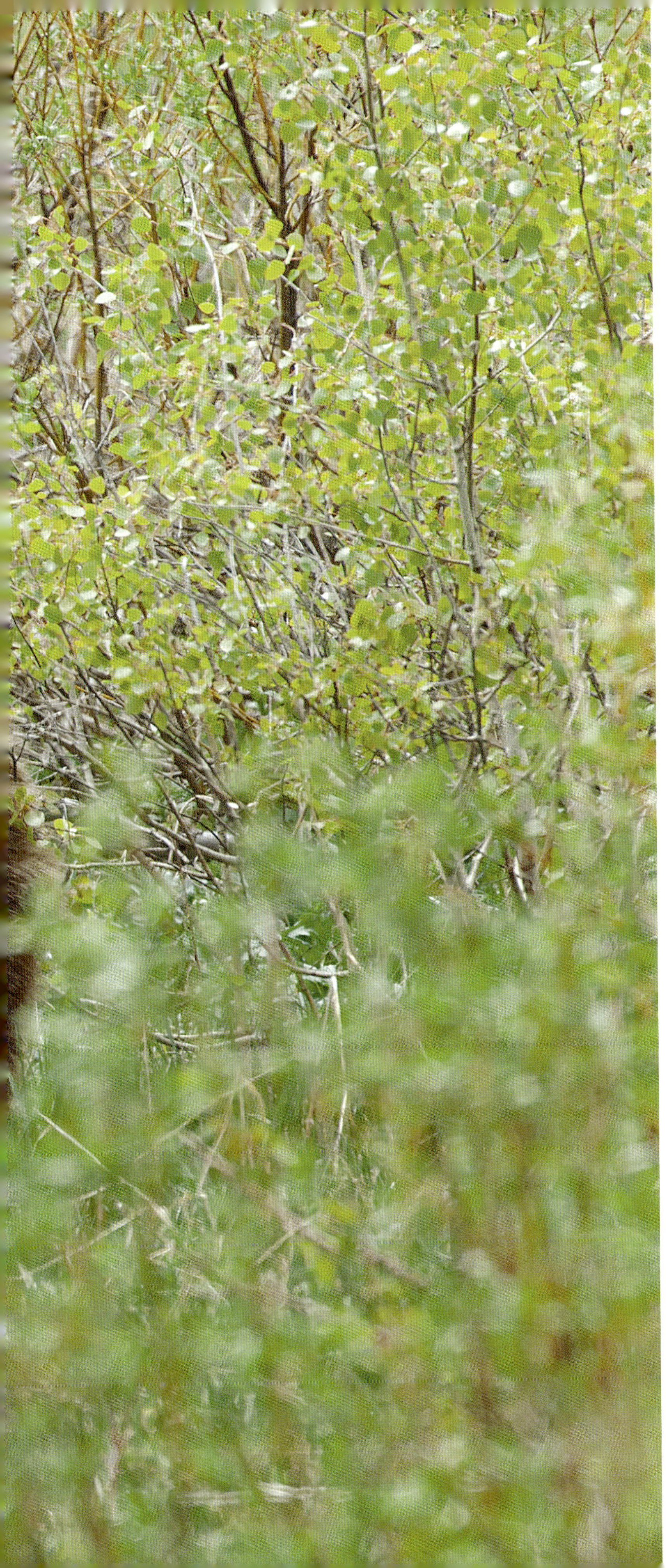

Contents

FOREWORD | TED TURNER

I HOWL FOR WILD WOLVES and I brake for wild grizzly bears and butterflies. As a citizen of Planet Earth, I am proud—and feel extremely fortunate—to call the Greater Yellowstone Ecosystem home. Having grizzlies present in our landscape is one of the many rare wonders of nature found here that sets it apart from other wild regions in the world.

What we have in Greater Yellowstone is a front-row seat, together as witnesses, to American wildlife conservation history in the making. Others are paying close attention to what happens here, hoping to learn lessons about how coexistence between humans and nature is supposed to work.

Bringing back the wolf to Yellowstone National Park was a momentous, unprecedented feat that many said could never be done in our lifetime. It stands out and is monumental. But truthfully, saving the Yellowstone grizzly—a species pushed to the brink of extinction, then recovering its numbers to the point where they are today—has been a challenge even more remarkable. I've heard it described as nothing short of a modern miracle.

It's an accomplishment shared by all people of Greater Yellowstone, all Westerners, and all Americans. It is, collectively, a gift from us to those we'll never know in the future, folks who will be extremely grateful to us for saying that losing the Great Bear was unacceptable. Yes, think about that: we came close to losing the grizzly, an animal that serves as an icon for the mother of all national parks. The federal Endangered Species Act, a law that is under siege in Congress, was an essential tool. Its existence is why we now have grizzlies thriving in Grand Teton National Park and greater Jackson Hole, Wyoming, the domain of the famous bear family that Thomas Mangelsen has been tracking and photographing for nearly a decade..

What can you say about the photographic excellence of Mangelsen that hasn't already been said? He has earned his reputation as one of America's greatest nature photographers ever and this book asserts an exclamation point at the end of the statement. The power of this book starts with the diary of jaw-dropping images he has amassed. They speak to his intimate knowledge of and passion for the grizzly known as 399 and her expanding tribe.

As for Todd Wilkinson, I am quite familiar with his work and got to know him well when he wrote the recent book about me, *Last Stand: Ted Turner's Quest to Save a Troubled Planet*. He is widely known as an American environmental journalist, though it is his knowledge of conservation in Greater Yellowstone, his home ecosystem, that sets him apart.

A few summers ago, my children and grandchildren and I embarked on an educational retreat in Yellowstone. One of the highlights was sharing the joy of observing grizzlies with other visitors from around the world. The grizzlies of Greater Yellowstone are an important part of our common heritage.

Maybe the most important thing to remember is that our job of ushering grizzlies forward safely and healthfully into the future is far from over. A work in progress, it demands that we redouble our commitment to insisting upon wildlife habitat protection and expanding attitudes of tolerance for grizzlies so that the handoff we make to the next generation is easier.

People often ask me, "Ted, what can I do to make a difference?" The answer is simple. Share your passion about grizzlies and other wildlife with friends and loved ones. For those who aren't familiar with bears, enlighten them by having them pick up a copy of this extraordinary volume. You don't have to venture very far into the pages before you become hooked for life as fans of a bear family led by the unforgettable 399.

Tom and Todd have done their part in bringing this amazing bear story to our attention. Now it's up to us.

Like mother, like daughter: 610, progeny of 399, moves through the dense riverine thicket of Oxbow Bend just prior to the brilliant eruption of spring colors.

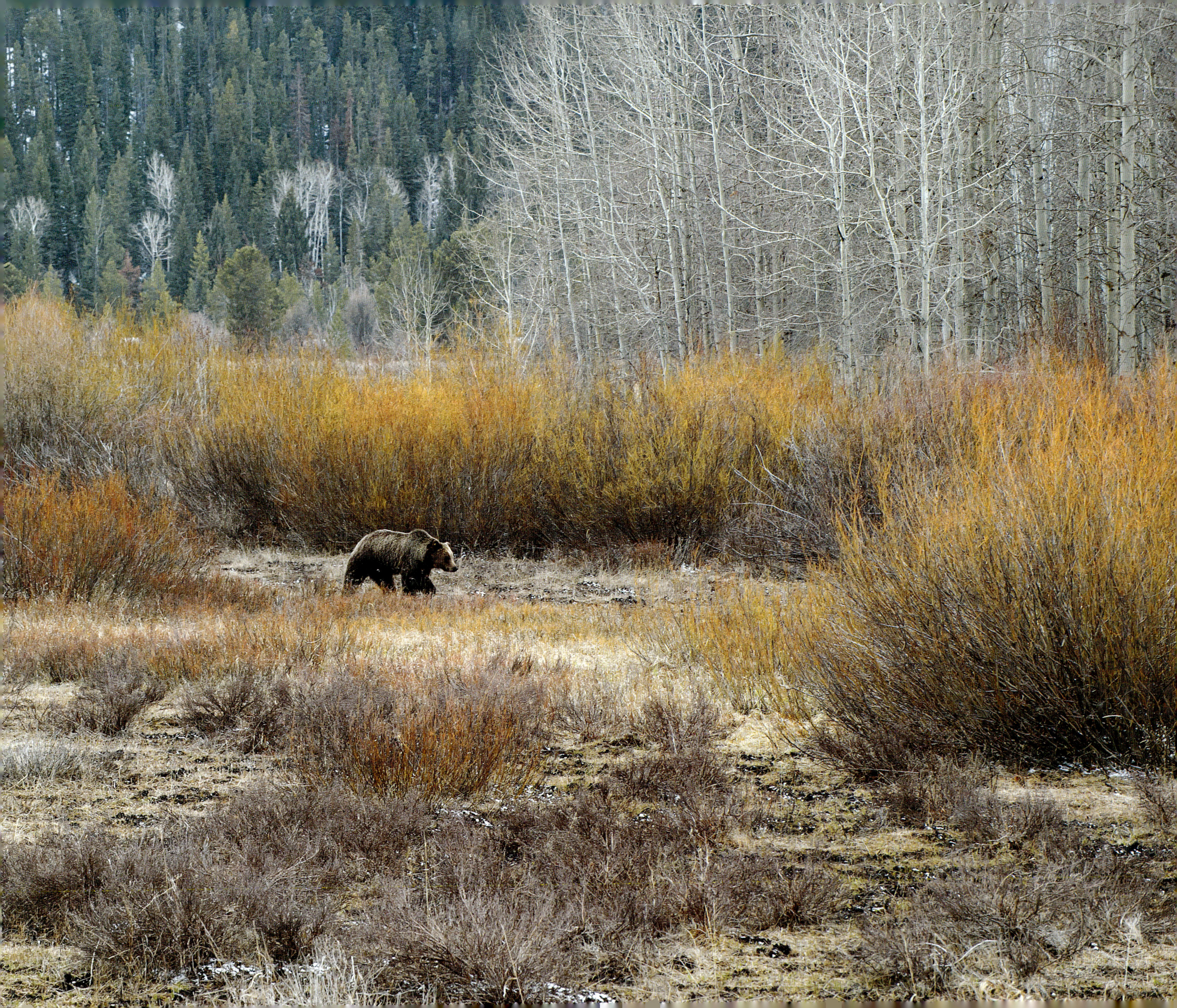

INTRODUCTION | TODD WILKINSON

THE MOST FAMOUS BEAR FAMILY IN THE WORLD

THOMAS MANGELSEN AND I, as celebrated wildlife photographer and humble environmental journalist, respectively, have been keeping our eyes peeled to the plight of grizzly bears in America for most of our adult lives. It has been playing out figuratively in our backyards, literally in Mangelsen's, and where survival is concerned it is a touchstone for our time.

The saga of Jackson Hole Grizzly Bear Number 399 is a real-life natural history cliffhanger. Will 399 and her offspring, which make up the most famous bear family in the world, live or die? Their fates paradoxically are both beyond our control and held firmly in our hands. As you contemplate their dramatic story, consider the following circumstantial events that also speak to fate.

It would be perfectly reasonable to believe that botanist Erwin Evert had little inkling he would be fatally mauled by a grizzly bear on the day he went for a walk east of Yellowstone National Park near Kitty Creek. On the day he died, he was seeking out his own version of paradise.

We could also reason that the late wildlife photographer Jim Cole didn't expect to be attacked by a grizzly in Hayden Valley, the second bear attack of his life. Neither did Dennis VanDenbos, a Wyoming schoolteacher out for a morning power stroll, realize that a legendary mother bear, 399, and her three cubs were in the bushes feeding upon an elk carcass not far from Jackson Lake Lodge when he fatefully passed by them and suffered a misadventure.

Nor did Stephen Westmoreland, a big-game hunter, ascertain that 399's adolescent daughter, 615, was dining upon a moose carcass before he startled her. Once the bear caught scent of him, she reared to inspect, prompting Westmoreland to lift his rifle and pull the trigger.

It would also be logical to conclude that field biologist Adam Stewart, in the prime of his young life, suspected accurately that it wasn't wise to be hiking alone in the Cub Creek drainage of the Bridger-Teton National Forest on that early day in September 2014.

These are some of the hair-raising incidents involving human encounters with grizzlies that you'll read about in the following pages. But, in fact, they are exceptions to the experiences millions of people have while traveling through what remains of grizzly country, people who don't often realize that they have a much higher probability of being struck by an errant strike of lightning or perishing in an auto wreck than meeting their end by a bear. What's notable here is that all of the above individuals were drawn to Greater Yellowstone because it is wild, their hunch of it confirmed indisputably by the fact that this place supports grizzlies.

On the other hand, if we are willing to ponder a different view of reality, the world as it appears from a bear's perspective, it is similarly sensible to assume that 760—the grandson of 399—could not possibly understand why he was trapped by wildlife managers in 2014 and banished from his home valley, handed what would amount to a controversial and tragic death sentence.

Clearly, a bear's life is not equal to a human's. But what is it that makes a grizzly expendable? How people and bears intersect, which is to say how we coexist, figures centrally into the calculus of whether grizzlies in Greater Yellowstone will persist robustly or be imperiled in the decades to come.

399's first famous trio of cubs—615, 610, and 587—huddle closely at Mama's side after nursing near Jackson Lake Junction. The fate of each bear would play out very differently in the years to follow.

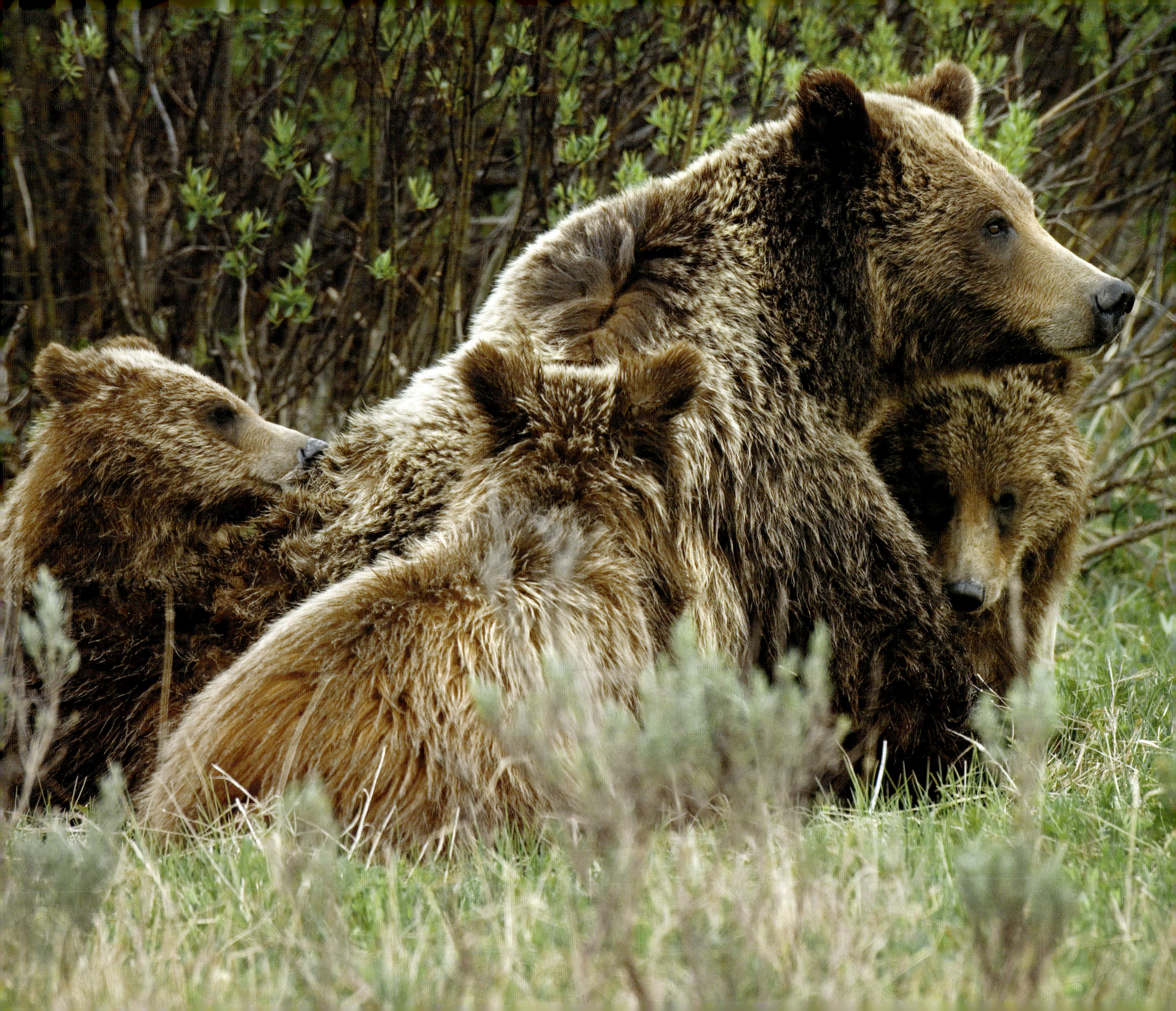

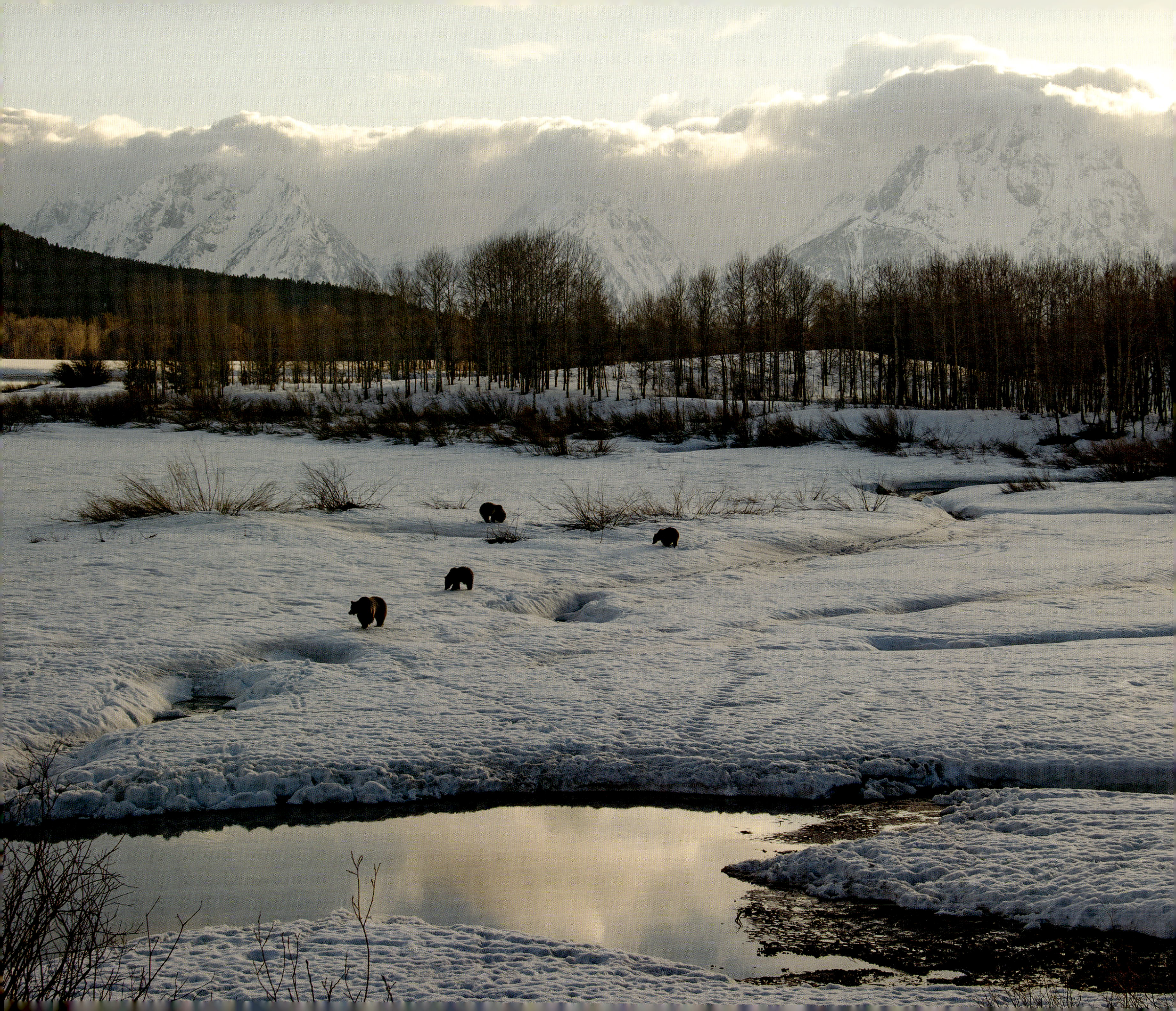

With Mangelsen's exquisite collection of photographs functioning as a visual journal of 399, you will see why her tale serves as a window into understanding the threats facing not only grizzlies but also all large wildlife predators on Earth.

In 1985, Dr. Stephen Herrero, today professor emeritus of environmental science and biology at the University of Calgary, wrote a book now considered a classic—*Bear Attacks: Their Causes and Avoidance*. It was published at a time when the grizzly population of Greater Yellowstone was crashing. Years before, grizzly bears had been eliminated from Grand Teton National Park, a setting that today lies at the heart of the 399 drama. Back when *Bear Attacks* appeared, which wasn't so long ago, industrial-strength logging, mining, and energy development were occurring in national forests surrounding the national parks of Greater Yellowstone, eviscerating bear habitat. Roads were being carved into backcountry, enabling poachers easy access to the last bear strongholds. Huge numbers of cattle and domestic sheep were being turned out onto millions of acres of federal grazing allotments in the heart of grizzly habitat. Bears that showed up in the middle of them often were shot on sight. Except in national parks, where they were protected, grizzlies were treated as species non grata.

In 1975, Congress had granted the bears emergency protection under a fairly new federal law, the Endangered Species Act. You can imagine the public outcry that made the bear's ambulance ride into triage happen: there was the possibility that the very emblems of America's first national park could vanish. Bears were dying at a faster rate than they were naturally being reproduced.

Hunting of grizzlies was outlawed (over opposition from some old western firebrands), poachers faced stiff monetary penalties and jail time if caught killing a bear, and public land managers were forced to modify heavy-handed methods of extracting natural resources. Alarmingly, the Greater Yellowstone grizzly population continued to fall in number to fewer than 200. But grizzlies hung on. Conservation intervention and triage saved them, barely, their turnaround today representing one of the greatest recoveries of a wildlife population ever.

In his tome, Herrero held up a mirror to human society asking us to reflect. "If a book titled *People Attacks* were written for bears, it could only depict our species as being typically bloodthirsty killers—aggressive, dangerous, often inflicting fatal injury to bears," he wrote. "Since the arrival of Europeans in North America, the grizzly bear has been shot, trapped, poisoned, and had expansive portions of its habitat converted to ranch land, housing, highways, and recreational playgrounds."

More than 50,000 grizzlies once roamed the United States largely west of the Mississippi River. Today, using the best guess of scientific researchers and following 40 years of aggressive stewardship via the Endangered Species Act, bear specialists say there may be almost 2,000 grizzlies in the Lower 48, roughly half residing in the 20-million-acre Greater Yellowstone Ecosystem. Those numbers, however, are in dispute.

There is a fierce national debate being waged over the issue of whether the grizzly population of Greater Yellowstone is secure enough that the same states that killed off grizzlies—Wyoming, Montana, and Idaho—can be trusted long term to steward them. Even more vociferously, there is growing disagreement over the question of whether grizzlies should be hunted as trophy animals for sport. It's analogous to the same bitter divides that have broken out over the hunting of elephants, tigers, lions, polar bears, and whales.

This foremost is a bear story, not a people story. 399 is a symbol of what's at stake.

399 and her three juvenile cubs search for beavers and muskrat in their lairs under the snow along the edge of Oxbow Bend.

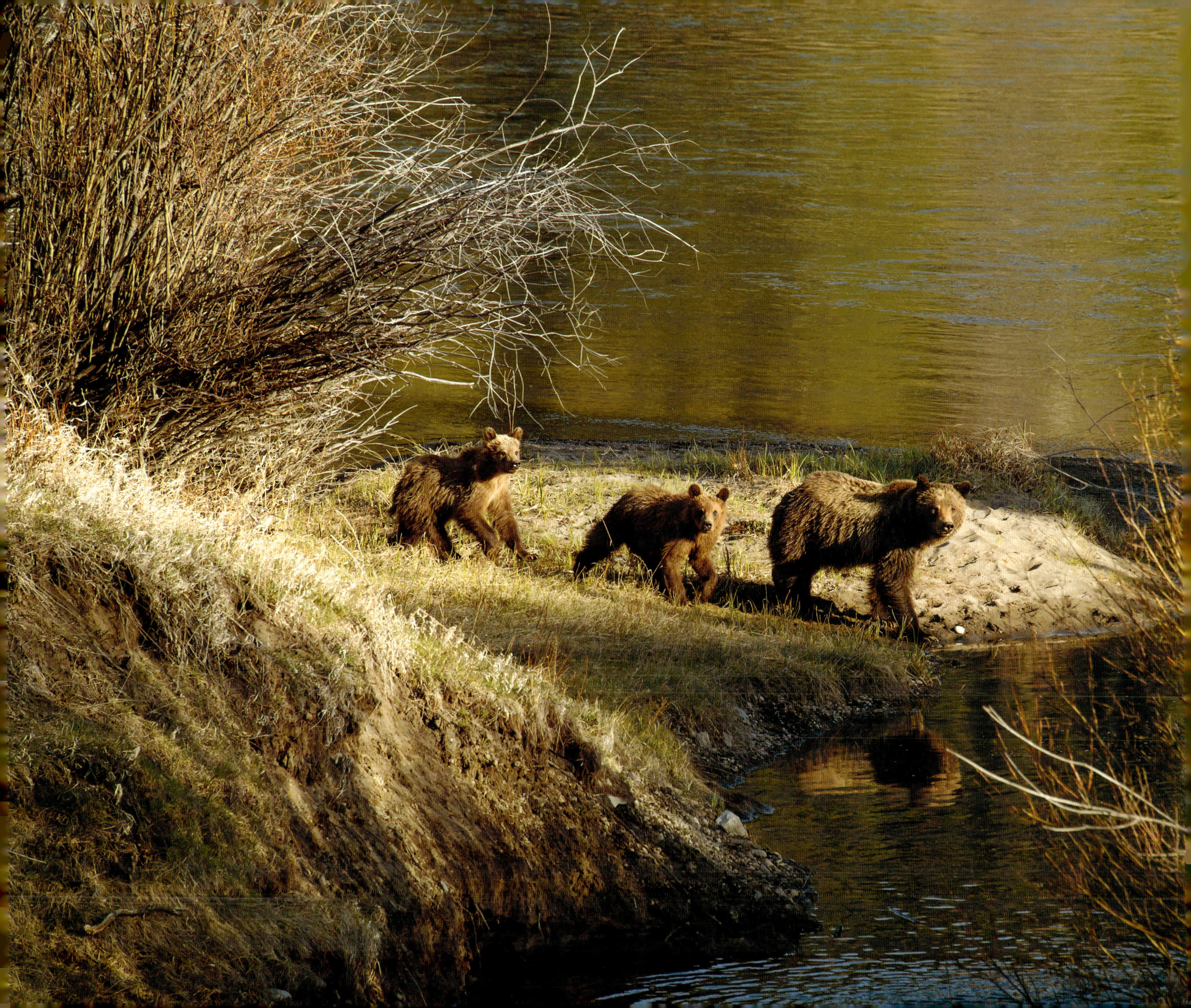

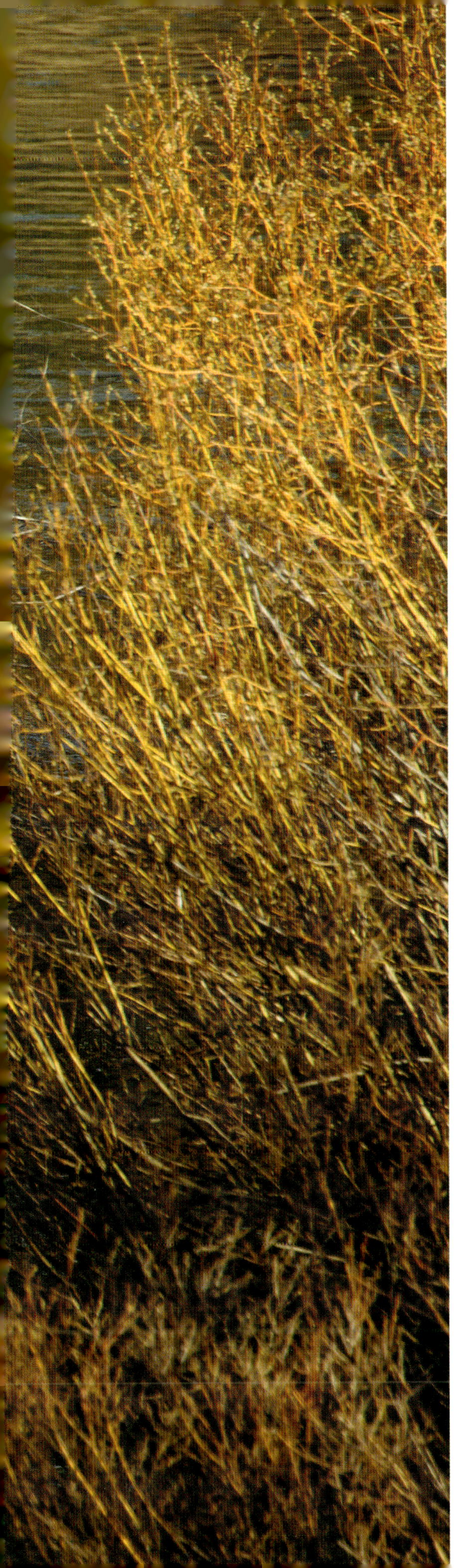

Bear Family Tree

Grizzly 399
(1996–)

- **Single Cub**
  - Unidentified (2004–2004)
- **First Set of Triplets**
  - 587 (Male) (2006–2013)
  - 615 (Female) (2006–2009)
  - 610 (Female) (2006–)
    - **First Set of Twins**
      - 760 (Male) (2011–2014)
      - Unidentified (Male) (2011–)
    - **Single Cub**
      - Unidentified (2014–2014)
    - **Second Set of Twins**
      - Unidentified (2015–)
      - Unidentified (2015–)
- **Second Set of Triplets**
  - Unidentified* (Male) (2011–)
  - Brownie (Male) (2011–2012)
  - Ash (Female) (2011–2012)
- **Third Set of Triplets**
  - Unidentified (2013–)
  - Unidentified (2013–)
  - Unidentified (2013–2014)

*Adopted by 610 in July 2011.

THE GRIZZLIES OF PILGRIM CREEK

TODD WILKINSON

THERE WAS TURBULENCE IN THE northern Rocky Mountains. May can be a very windy time for a pilot in a small Cessna 172 Skyhawk navigating the invisible updrafts hitting towers of granite and rooster tailing vertically as raging vaults of air. Passenger Thomas Mangelsen normally had a solid disposition, but the gusts pelting the southern Absarokas due east of Jackson Lake, Wyoming, made him queasy.

Mangelsen had taken to the air because for years the exact source of his subject had eluded him, looming as a still unresolved piece in an ursine mystery. He had spent every waking hour during many months observing *the matriarch*—been there to welcome her emergence from the high country with generations of offspring beside her, trailed them from a distance, babysat them, advocated for them, barked to members of his kind to stay back, enthralled visitors with tales of the bears' exploits, tried to guard them from hunters, sniffed the gut piles and carrion they vacated, examined their scat, and cried over their losses and reunions. In a way, he had helped make them legends.

Sometimes left with only paw prints as clues to decipher the course of the bears' direction, Mangelsen and his companion Sue Cedarholm had documented their peregrinations all over Greater Yellowstone: through the draws of Pacific, Ditch, Spread, and Cottonwood Creeks (among others); over the flanks of Signal Mountain and down to the shore of Colter Bay; across open lake to Elk Island and Donoho Point, then back again to land; across the Snake River into Oxbow Bend, using Cattleman's Bridge Road, and over to Schwabachers Landing (back and forth, again and again, sometimes over the perilously busy US Highway 89); around Emma Matilda and Two Ocean Lakes, as well as Cygnet, Christian, and Heron Ponds; through the tangle of Willow Flats; from the head of the Gros Ventre River Valley to the other side of Jackson Hole south of Teton Village; into the Potholes and across Antelope Flats toward Mormon Row; back up to Leeks Marina and over to the rumple of Shadow Mountain; through the national park barracks at Moose and the Murie Ranch; past the outskirts of Kelly, skirting the National Elk Refuge; and then westward to the other side of the Tetons in Idaho—where one member of 399's family was once relocated and then, with determination, made it back home.

The bears had wandered north into Yellowstone and way down to the dell of the Upper Green River 60 miles distant from Jackson Hole in the opposite direction, not heeding—not understanding, ever—the invisible artificial boundary lines of national park, national forest, National Elk Refuge, and private land. And often, they had looked curiously at the mob scenes of creatures on two legs. Heart-poundingly, they also moved through firing lines formed by humans shooting live ammunition. In all, their family territory encompassed a couple hundred square miles, and inside of it they had logged, collectively, enough mileage to travel from coast to coast if the crisscrosses of their perambulations were straightened and laid end to end. For the bears, the purpose wasn't complicated; it was really about a quest to find food and, along the way, cement their own sense of belonging to place. A wild grizzly, Mangelsen said, does not and cannot dwell patly inside a tiny box of space the way domestic creatures or stir-crazy captive zoo animals do. It is not in their nature.

Around the time autumn became winter, with the first bite of a deep freeze closing in, the matriarch 399 always headed for the terra firma of natal home, to a discreet hole in the ground somewhere not terribly far from the Continental Divide up along the muscly shoulders of Pilgrim Creek. It's where she gave birth to several broods of cubs. Maybe, too, Mangelsen mused, it's where she herself first crawled out into the bright daylight from a cozy dark burrow her mama might have excavated.

Mangelsen had done his own sleuthing driven by curiosity and an instinct of protectiveness. 399's origin was the point in a GPS coordinate marked on a topo map spread out across his lap. The pilot tried to ease his craft gently lower. In the rattling cabin, Mangelsen clutched his camera, squinting into the viewfinder as he sailed over a pine forest still buried in snow. Along a slope falling away from a ridge a few hundred feet beneath

399, with Ash and one of her darker cubs, digs for tubers and grazes on biscuitroot near Pilgrim Creek.

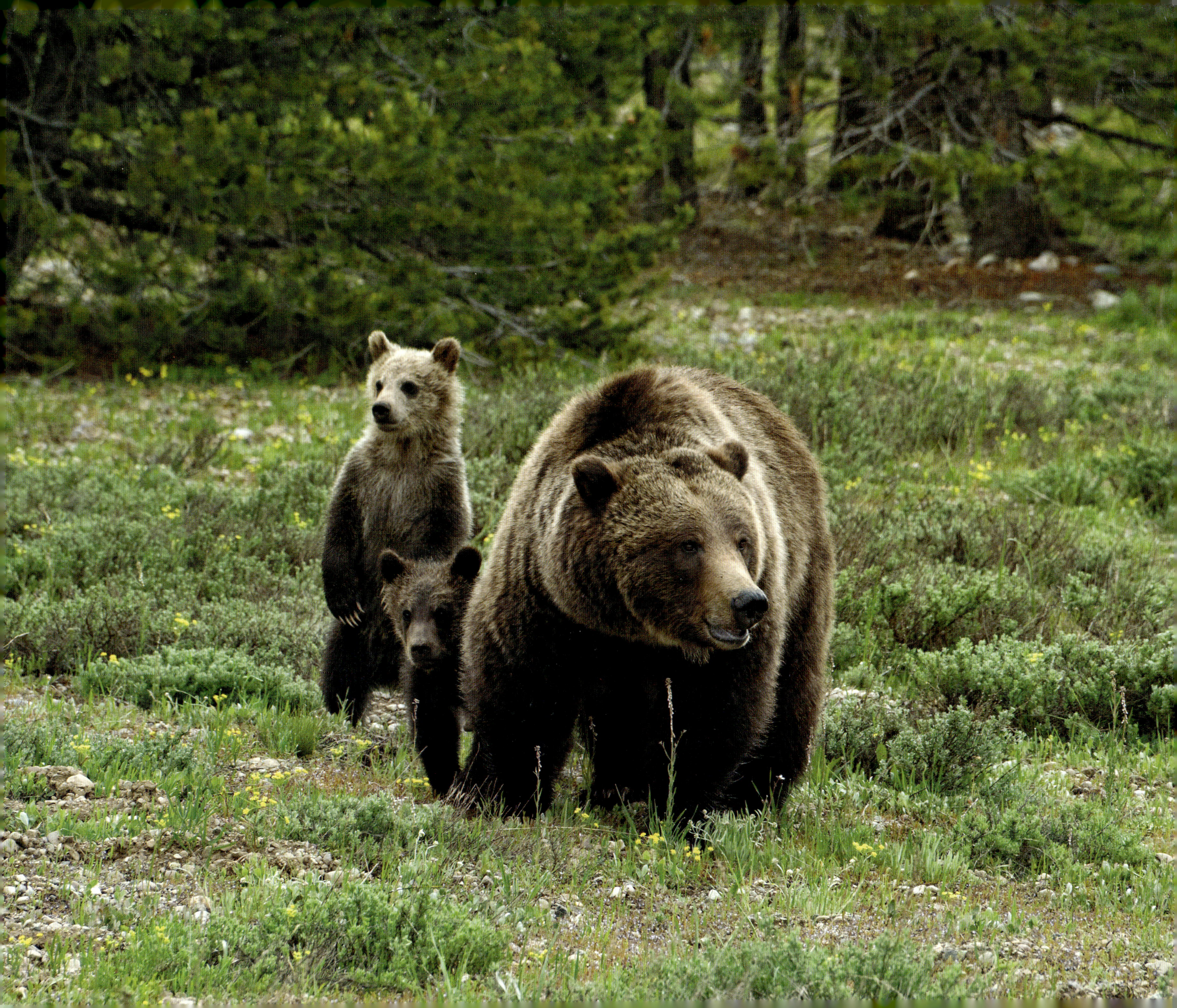

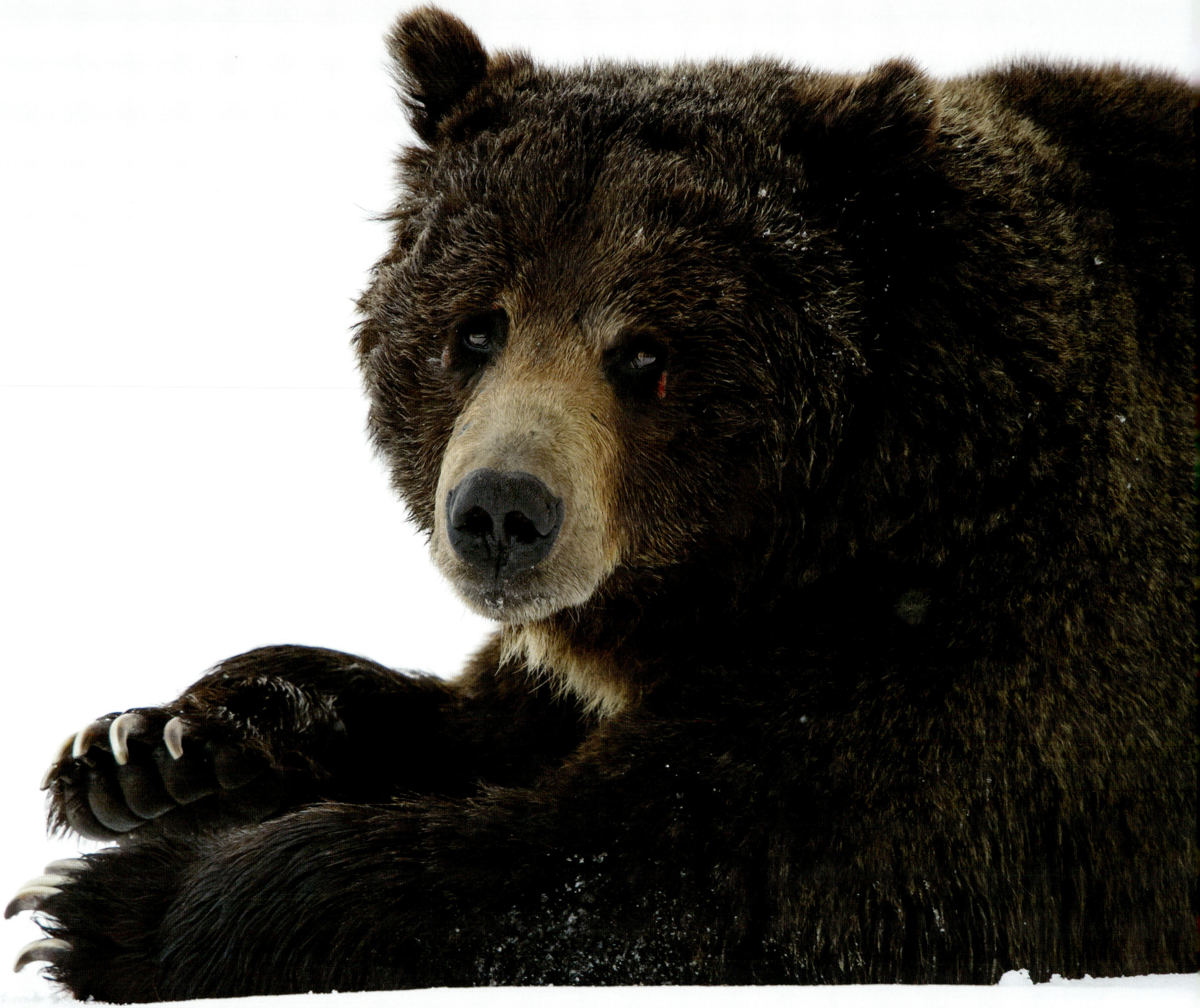

them, there it was. Abruptly. Inconspicuously. Not grand or majestic. Just smudges of mud tracked out of a tiny entrance at the base of a massive Douglas fir and out onto a white melting apron. That was it. "Aha," Mangelsen said to his friend Jim Laybourn, who was seated behind him in the plane. "Home sweet home."

GRIZZLY. For some of us, no matter where we call home, the Latin scientific name of this bear tells us all we need to know—all we are willing to consider because we've never been told to think any differently—about a notorious beast that has evolved for a million years dating to the Pleistocene epoch. *Ursus arctos horribilis.* Literal translation: the horrible bear.

Feeding our worst primitive fears, it is said that grizzlies can outsprint a horse over short distances; that their canid fangs, trademark thrasher claws, and formidable jaws are capable of crushing the femur of a one-ton bison; that these hulking brutes pack temperaments prone to easy agitation, unpredictability, and lethal aggression.

They've been maligned as American man-eaters. We love to make ourselves scared thinking about grizzlies, indulging the reptilian part of our brains, just contemplating the paralyzing prospect: Should you run when a grizzly approaches or collapse to the ground, curl into the fetal position, and play dead? Answer: It depends.

Did I mention that those frightening claws a grizzly flaunts on its front paws (which actually are useless for bears attempting to climb trees) function, in fact, as tools for digging plants, worms, and insects, and excavating ground squirrels from subterranean warrens—not as nightmarish talons similar to a velociraptor's? Our urban legends stalk the grizzly.

Native Americans in the arcs of their ancient oral traditions ascribed mystical powers to the Great Bear. And yet, according to anthropologists, the sacred bruin was probably on the wane in the open prairies of the Great Plains not long after indigenous peoples asserted themselves as expert horsemen, riding steeds that were brought to this continent by the Spanish in the early sixteenth century. That was the start of grizzlies losing 98 percent of their original home range.

Members of the Lewis and Clark Expedition dutifully took note of their first encounter with the shoulder-humped bruin as they were headed westward up the Missouri River in 1805. On April 29 of that year, they felled a curious grizzly with muskets near the confluence of the Missouri and Big Muddy Creek in northeastern Montana. A few days later, on May 5, a Sunday, Captain William Clark scrawled this written reflection (spelling and punctuation is his, not mine):

> *In the evening we saw a Brown or Grisley beare on a sand beech, I went out with one man Geo Drewyer & Killed the bear, which was verry large and a turrible looking animal, which we found verry hard to kill we Shot ten Balls into him before we killed him, & 5 of those Balls through his lights This animal is the largest of the carnivorous kind I ever saw we had nothing that could way him, I think his weight may be stated at 500 pounds . . . we had him skined and divided, the oile tried up & put in Kegs for use.*

In literature following in the explorers' wake, books such as Jack Olsen's chilling *Night of the Grizzly* about two different grizzly attacks in Glacier National Park on the same evening, and the science-challenged imaginations of Hollywood scriptwriters, grizzlies have been portrayed—you could argue, demonized—as bloodthirsty shadowy beings that, like fairy-tale wolves, are agents of nocturnal terror, anachronistic animals that have outlived their useful tenure of existence in a twenty-first-century world with the permanent trappings of civilization.

Point of fact: grizzlies are a unique North American offshoot of European brown bears and the large salmon-eating bruins (both *U. arctos*) found along the coasts of British Columbia, Alaska, and East Asia. Likely they crossed the Bering Land Bridge 60,000 years ago. A subspecies that roams our continental interior, mostly in mountains today, they are smaller in stature, a geomorphic response to their terrain; the harsh winters they endure push them into hibernation for up to half the year as food becomes scarce. Female grizzlies in Greater Yellowstone on average tip the scales at around 300 to 350 pounds, males 100 pounds more. Around Jackson Hole, Mangelsen has seen a giant elusive male that he estimates may surpass 700 pounds. If a trophy sport hunt is ever brought back to Wyoming, he has no doubts that big bears like that would be the first targeted and turned either into a floor rug or a mounted head on someone's wall. Nobody kills a grizzly, after all, to fill a human larder. The one-time consumption of such bears is instead purely for show, a sign of prestige.

For males, it's a hard-knock life. A massive grizzly bear marked with battle scars takes a rest in the pristine snow near Fishing Bridge along the Yellowstone River in Yellowstone National Park.

Frank Van Manen, a Dutchman by birth, offers a perspective. Van Manen is head of the federal Interagency Grizzly Bear Study Team, the foremost bear research unit in the world and bulwark for the longest continuous examination globally of a major large mammal population. He says the brown bears of Europe and grizzlies of Yellowstone are nearly identical genetically but qualitatively different in a key respect. Bears in Europe have survived by living far more cryptically, forced into being more elusive and striving to be unseen. Humans over millennia, relentlessly dispatching aggressive bears, have caused them to evolve behaviorally in this way. The grizzlies of Greater Yellowstone, he says, still exude a bold, wild spirit reflecting the big and brawny—and still relatively unsubdued—landscape around them. They have not been compelled, by constant human pressures and lethal harassment, to be only meek, stealthy prowlers. They reveal themselves to us in the light of day when we peer into each other's eyes. Americans may not fully appreciate how special this visual accessibility is, Van Manen says.

According to keen observers like Mangelsen, there could be profound consequences for bears going forward if grizzly hunts—having been gone now for more than 40 years—are brought back to the Lower 48. If the states of Wyoming, Montana, and Idaho persist in clinging to their antagonistic attitudes toward grizzlies, still held over from the frontier, they will undermine an accomplishment that makes Greater Yellowstone a beacon for bringing a species back from oblivion, with wisdom prevailing over intolerance. "Why should we make grizzlies think that no amount of distance between us and them is safe?" Mangelsen asks. "Why is it that some managers seem to believe we need to teach grizzlies that every time they see us, they had better run for their lives in the other direction?"

IN 2006, THE YEAR 399 TURNED 10, Mangelsen the photographer and naturalist was traveling far away from Jackson Hole photographing wildlife in far-flung corners of the world, expanding his image portfolio of other wild animals, as spring turned to summer. By late August, when colder nights in the high country were causing aspen trees to sparkle orange, he breathed a sigh of relief. He was finally home in Jackson Hole again.

Beginning in May 2006, friends in Jackson Hole had told Mangelsen about a sow grizzly with triplets observed east of the Teton Mountains. The mama at first appeared wary. Her tribe was viewed briefly along the main highway in Grand Teton National Park that winds northward to the southern entrance of Yellowstone. Mangelsen was intrigued because he had never seen a grizzly in Grand Teton and he had been living in Jackson Hole for nearly 30 years.

In September of 2006—he doesn't remember the exact day—he caught his first glimpse of 399 when she and her youngsters, born underground the previous winter, were feasting on a moose carcass on the island shore of the Oxbow Bend of the Snake River, a world-renowned venue for wildlife watching. "I saw her briefly in the evening just before dark," he said. It was a fleeting encounter, but still extraordinary. She was there and then gone, vanishing into the black ether of night. Mangelsen felt no connection. He had no idea how profoundly this bear and her cubs would change his life and steel his commitment to conservation. "It wasn't until spring the following year that I spent any significant time getting to know her," he said.

There is no grand genealogy for the bears of Greater Yellowstone. Not yet anyway, though with sophisticated DNA-mapping techniques being developed, it will happen someday soon. 399 was born in 1996. We will never know if she had siblings or who her parents were, though it's safe to say she is probably a lot like her mother, for grizzly mothers are the teachers in the game known as survival of the fittest. More important than brawn for a grizzly matriarch is brains. Her IQ for interpreting the intentions of people has been off the charts.

According to the recorded annals of modern grizzly natural history, male bears (boars) have shorter lives than females (sows). Roaming solo, territorial boars, even the reclusive dominant ones, endure rough-and-tumble tenure punctuated by brawling with other rivals for food and mates. The National Park Service puts the average lifespan of a boar at 22 years compared to 26 years for sows. The oldest female on record in Greater Yellowstone to mother a new batch of cubs was 27. If a cub makes it to adulthood and then lives long enough to die of old age, the greatest threat to its survival isn't from another bear; it's from an encounter with a person.

"Grizzly bears reproduce very slowly—generally a female only replaces herself once in the first 10 years of her life—and consequently even a small amount of mortality can have profound effects on the population for decades to come," says Bonnie Rice, senior conservationist with the Northern Rockies office of the Sierra Club.

A young female grizzly, nicknamed Blondie, explores the mountain slopes of Grand Teton National Park.

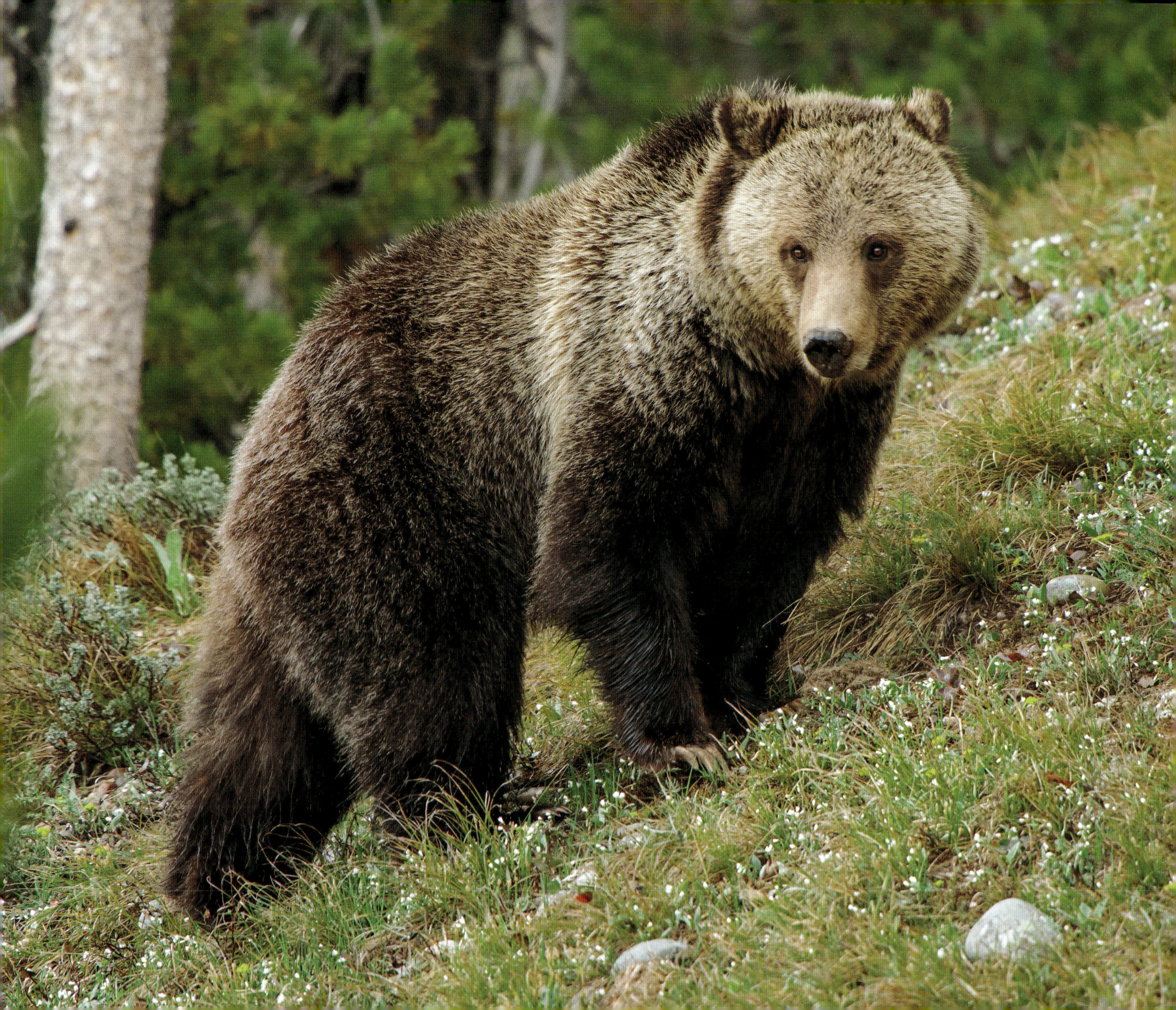

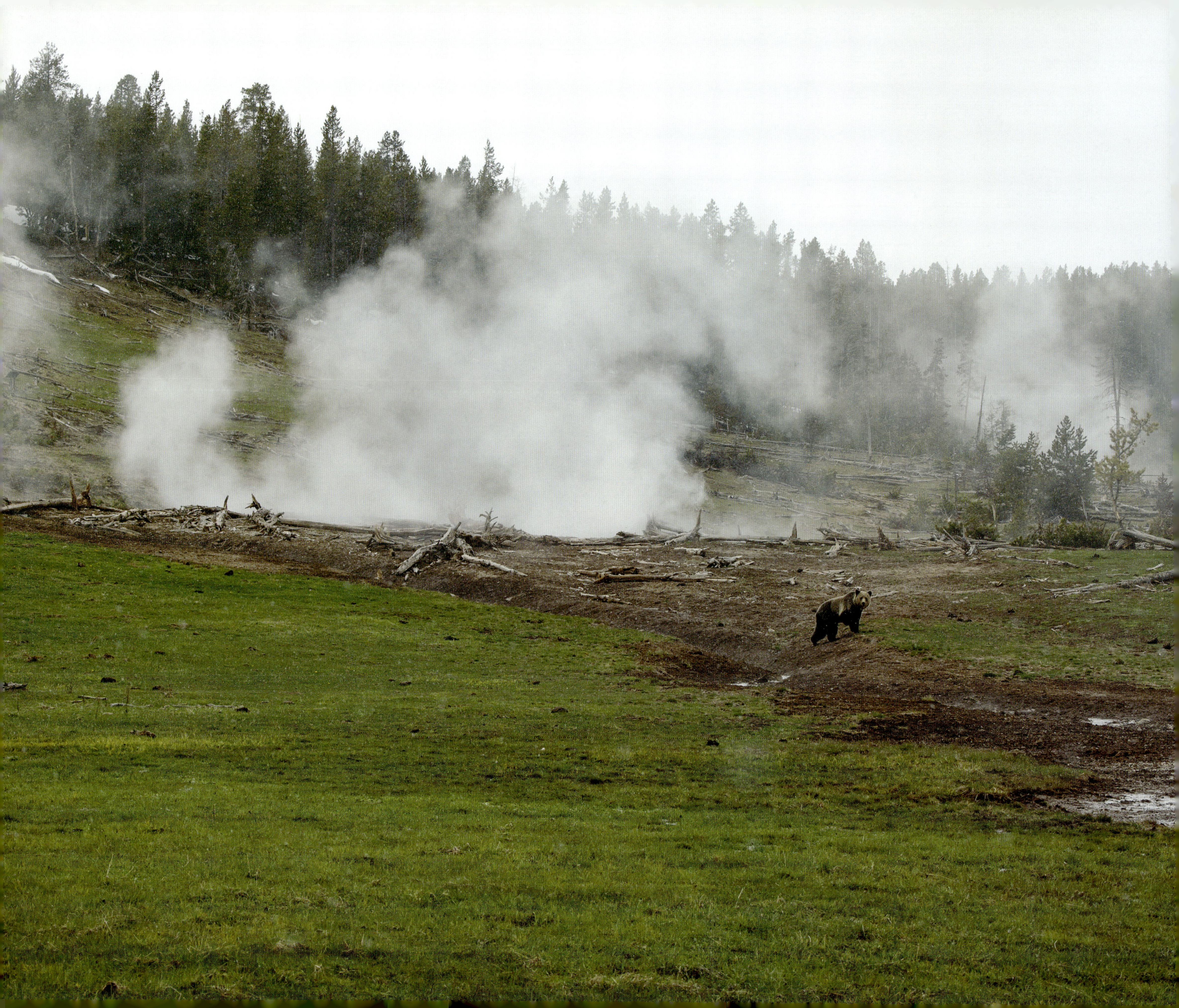

YOUNG PEOPLE DWELLING IN Greater Yellowstone have never had to ponder a landscape completely devoid of grizzlies. In his 1919 classic *The Grizzly*, Enos Mills wrote, “The grizzly is vanishing so rapidly that without protection he is likely to become extinct.” Mills cited a National Park Service estimate that as few as 50 grizzlies were left in Yellowstone, one of two strongholds for the species south of Canada. The grizzly is proudly displayed on the state flag of California, but *Ursus arctos californicus*, a subspecies of the grizzly, is today extinct. In August 1922, a grizzly was shot by a hunter in Tulare County. Reportedly, another grizzly, solo, was seen in the southern Sierras in the final months of 1924, and not one has been seen there since. Once there had been thousands of grizzlies. Seven years later, the last grizzly in New Mexico was killed by a stockman, and four years after that, the last grizzly in Arizona was felled. Sixteen years after that, in 1951, the last grizzly in Colorado was claimed.

Just two years prior, in 1949, the U.S. Fish and Wildlife Service had estimated that 750 grizzlies still inhabited the entire Lower 48, the largest cluster in Glacier National Park and the roadless Bob Marshall Wilderness in northern Montana tight along the border with Canada. Grizzlies in Yellowstone had been rendered pathetic roadside beggars by then, encouraged to feed at open-pit trash dumps in the park, behind hotels, and along the roadsides where tourists threw them junk food. These actions gave rise to Yellowstone’s lampooning image of “Jellystone” by the cartoon shop of Hanna-Barbera.

During the 1950s, as more enlightened ecological thinking convinced wildlife managers that having semitamed garbage-foraging bears was not wise, three researchers—twin brothers John and Frank Craighead and associate Joel Varney—pioneered the use of radio telemetry on Yellowstone grizzlies. It coincided with a philosophical shift taking hold in the National Park Service, emphasizing a policy of “natural regulation,” i.e., adopting a more hands-off approach to wildlife management. To get grizzlies unhooked from their addiction to human food, park managers ordered that open-pit dumps throughout the park be closed. The Craigheads favored phasing them out slowly over several years to gently wean the bears off their dependence.

Their concerns were ignored and, just as they predicted, bears that were denied their unnatural chow lines moved into campgrounds and gateway communities, behaving aggressively and dangerously toward people. During a three-year span between 1968 and 1970, the Craigheads estimated that at least 97 grizzlies were killed by humans after the dumps closed. Wyoming and Montana nonetheless continued with sport hunts based on bear population estimates that scientists said were wildly inflated.

In 1975, the Greater Yellowstone grizzly population was listed as threatened under the Endangered Species Act. Even with that shield, bears continued to die from known causes such as management actions, ranchers dispatching bears preying on their cattle and sheep, poaching incidents, and defiant antifederal resentment flaunting the cultural code of “shoot, shovel, and shut up.”

By 1981, Richard Knight, head of the recently created federal Interagency Grizzly Bear Study Team (hereafter referred to as the grizzly study team), left many shocked in revealing that there could be as few as 197 grizzlies left in Greater Yellowstone. Did the warning that America could lose an iconic species from its flagship national park make a difference? That same year, another 15 bears were believed to be killed by humans, a rate at which mortality exceeded reproduction. Knight and fellow researcher Bonnie Blanchard delivered a report noting that, in 1982, 97 grizzlies were radio collared. Of those 97 bears, 37 were eventually known or suspected of being dead. At least 84 percent of these deaths were caused by humans. Mark Haroldson, a bear research trapper who started working under bear guru Charles Jonkel and then Knight, said extreme measures were adopted to keep every female bear alive. Losing just a few sows of reproductive age could have hugely negative ripple effects that would play out for years in the form of lost replacement of animals.

Stephen Herrero, professor emeritus of environmental science and biology at the University of Calgary and author of *Bear Attacks: Their Causes and Avoidance*, expressed his concerns about hunting in 1985. He remembered how Wyoming, which holds the bulk of grizzly habitat in Greater Yellowstone, fiercely resisted closing down its killing of bears for sport in 1974 (a matter we’ll get to later). “Since we have had difficulty in regulating the kill of grizzly bears in strictly protected areas, it should not be surprising that sustained-yield hunting outside of parks and reserves has been an even more difficult objective to achieve,” Herrero wrote in *Bear Attacks*. “The grizzly is a highly valued trophy animal, the hunting of which can generate expenditures of many thousands of dollars. Yet opportunities to hunt grizzlies are limited by the bear’s inability to survive with more than modest levels of killing. Very few grizzly populations have ever been substantially hunted

With geothermal mist billowing from the vents of the Mud Volcano area in Yellowstone, a lone grizzly forages for sustenance.

without population decline." Herrero added, "One of the greatest difficulties we face in attempting to maintain grizzly bear populations and monitor their mortality is our inability to estimate how many grizzly bears there are in a given area."

Grizzlies, representing the mindset of the times, were monetized as being worth more dead than alive. Only their protection under the Endangered Species Act spared them from paralleling the doomed path of their cousins in California.

AFTER BOTTOMING OUT, the grizzly population recovered, slowly at first, and then rising more dramatically with the emphasis on sow survival paying huge dividends. However, recovery was accompanied by hue and cry, as strict regulations were implemented on logging, mining, and energy development on public lands. Some activities that previously would have been greenlighted were now denied. While the bear was blamed for lost economic opportunities, the fact is that any measures adopted to protect grizzly habitat, including the closing of old logging roads, benefitted every other wildlife species. "Back in 1981 and 1982, I never thought we would have the numbers and distribution of bears we have today," said Dr. Christopher Servheen, the national grizzly bear recovery coordinator for the U.S. Fish and Wildlife Service. "I thought we would be lucky to have *any* grizzly bears in the Yellowstone ecosystem."

Now, Frank Van Manen recently told me at his office in Bozeman, grizzlies are meeting or surpassing every threshold set as a metric to prove they've recovered. There may be 1,000 grizzlies in Greater Yellowstone—conservative estimates say there are at least 750. Grizzlies are more widely dispersed and turning up in places they haven't been in more than a century. There are likely more sows with litters of cubs than at any point since Yellowstone's creation in 1872, which predates Wyoming's, Montana's, and Idaho's statehoods. A unified conservation strategy has been adopted by federal and state natural resource agencies. Van Manen credits Servheen's leadership with getting government agencies to work together. Americans have spent tens of millions of dollars on grizzly recovery in Greater Yellowstone and studies have shown the public overwhelmingly supports it.

Looming ominously is how states in coming years will approach their stewardship. Already, there has been talk among states, and fierce pushback from bear advocates, about setting an allowable minimum number of grizzlies in the ecosystem at 500, thus enabling an attitude that bears are again expendable. A useful reference point is wolves. After wolves were reintroduced to Yellowstone and central Idaho in the mid-1990s, their numbers ballooned to upwards of 2,000 throughout the northern Rockies today, but states like Wyoming and Idaho have vowed to keep their wolf populations at the lowest levels possible while still high enough to prevent relisting them under the Endangered Species Act. In Wyoming, wolves are treated as trophy game animals in about 15 percent of the state—in the Greater Yellowstone region of Wyoming's northwestern corner—but in the rest of the state they're classified as predators and can be killed any day, at any hour, for any reason, and by virtually any method. Never before has an iconic species, protected under the Endangered Species Act with millions of tax dollars spent to recover it, been treated this way.

"I think there is probably more of an acceptance for bears versus wolves on the landscape and a lot of things play into it," said Dan Thompson, the large carnivore manager for the Wyoming Game and Fish Department. "Wolves were completely eliminated and they were brought back against the will of some. Grizzlies always remained and people got used to trying to coexist. I don't think it's about hatred toward either species. Even ranchers who experience loss to wolves and grizzlies say to me, 'I think they're neat animals. And I wouldn't mind having them around if they didn't eat my cows.'"

Here's a factoid: wolf experts say that worst-case-scenario wolf populations could be beset by losses equal to a third of their total numbers and they could replace these losses with annual pup production. Bear biologists say that a sustained annual killing rate above four to five percent of a grizzly population, particularly if it involves significant numbers of females of reproductive age, can turn a growth curve into a downward trajectory. Five percent of 1,000 bears is 50. Is that the number that Wyoming, Montana, and Idaho will allow to be hunted? And what if the actual figure is lower while the number of bears being killed illicitly is higher? Another question to ask is what if projections about bear numbers are based on certain nutritional assumptions that once existed but are no longer accurate? The life and times of 399 serve as an ideal lens for pondering the big picture.

IN LATE 1988, THE YEAR OF THE HISTORIC Yellowstone forest fires, I interviewed a wildlife biologist who had just arrived in Grand Teton National Park. He was commencing a study of black bears. His name was Steve Cain and over the years, before

A grizzly, survivor of a population death spiral just a few decades ago, lumbers through the lodgepole pine in America's first national park. At one point, fewer than 200 grizzlies inhabited all of Greater Yellowstone.

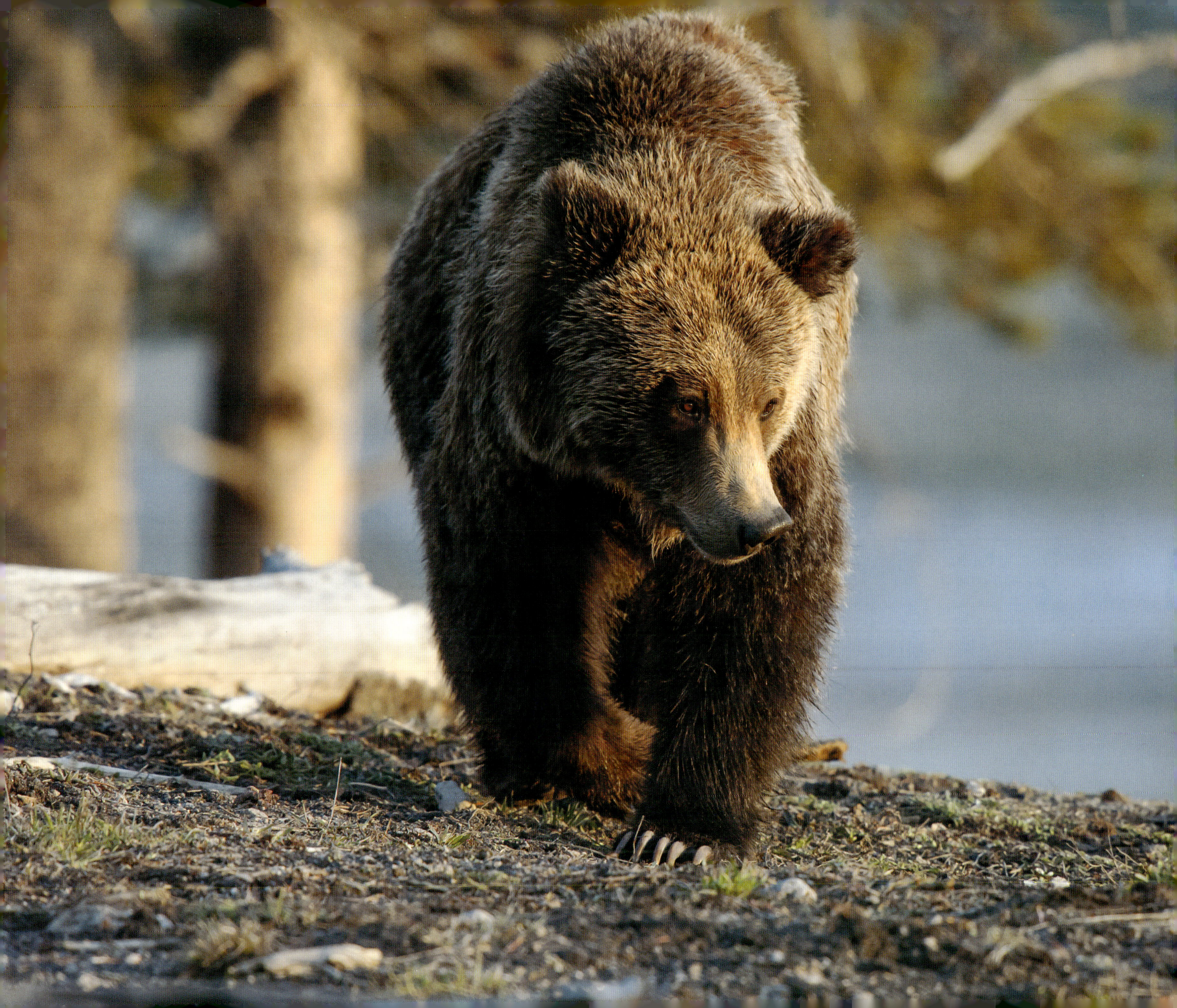

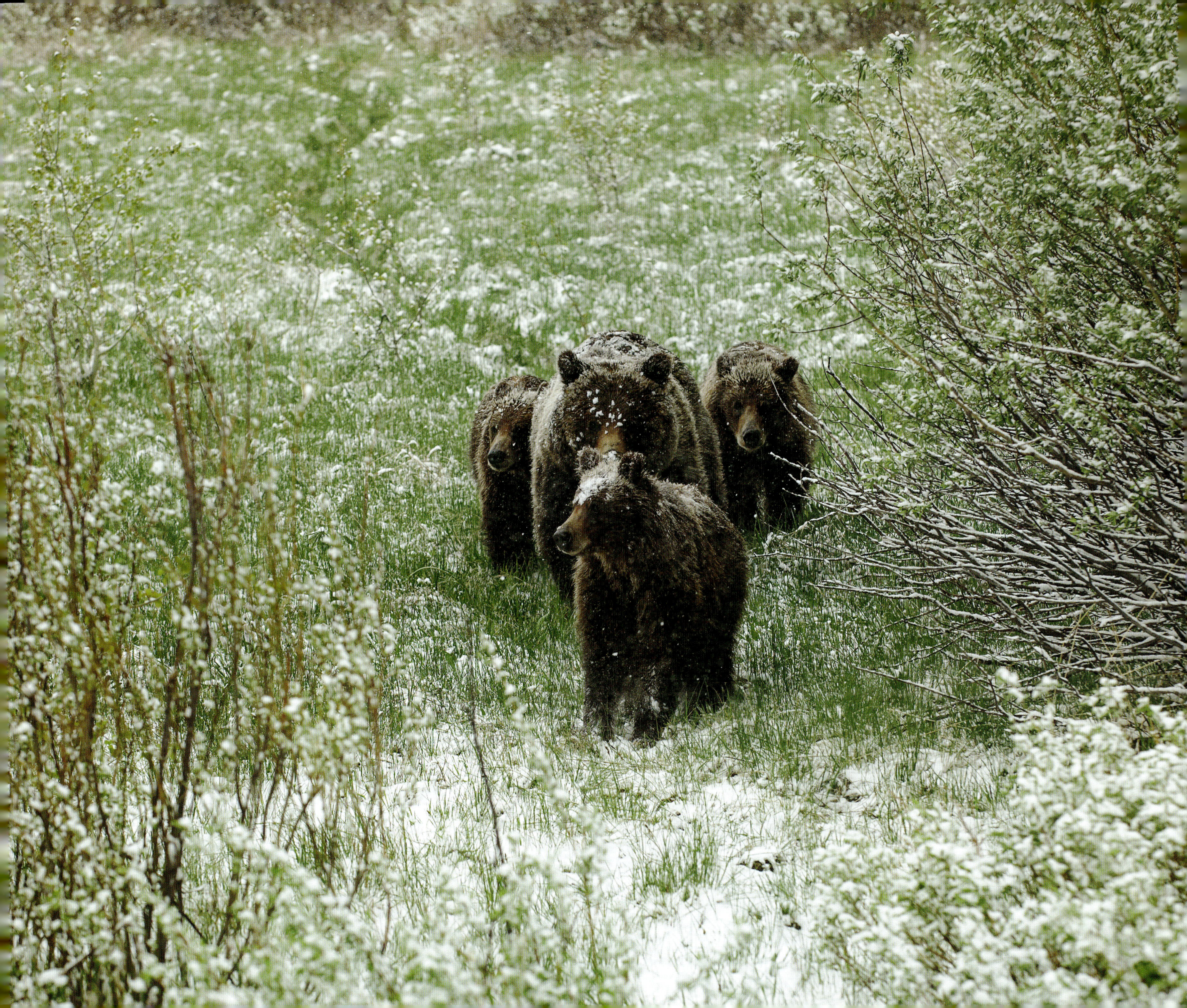

he retired in 2015, Cain became a widely recognized landscape ecologist. I will never forget this exchange we had almost 30 years ago:

"Steve, why are you studying black bears?"

"Because there are no grizzlies in Grand Teton National Park."

"Why aren't there any grizzlies?"

"Because they were basically killed off in Jackson Hole by ranchers and trophy hunters many years ago."

"Do you think they'll ever inhabit Jackson Hole again?"

"Certainly not in the foreseeable future."

Cain wanted to learn more about black bears, which were numerous in Grand Teton, to better understand their habitat needs. His findings could also be used, he thought, to gauge grizzly impacts on their smaller ursid cousins if and when they ever returned. Today, we laugh at how quickly the recolonization of Jackson Hole by grizzlies moving out of Yellowstone occurred. For the most part, it happened without significant incident. Watching it was one of the most meaningful events in Cain's distinguished career. He said he wasn't prepared for the stir caused by 399.

Until 2001, when she was five years old, 399 had no identity among humans and likely was the product of a mother who also was anonymous but part of an expanding population dispersing from the north, probably the environs of Yellowstone National Park, into Jackson Hole. Field researchers with the grizzly study team captured 399 in a baited culvert trap and fitted her with a special radio collar around her neck that emitted signals to enable tracking of her movements. Not for any profound reason, she was logged in the research rolls as 399, roughly the 399th bear given a colored ear tag.

About 800 grizzlies have been collared or trailed over the years as part of the tracking program, Mark Haroldson told me, a cornerstone of the longest and most intensive study of a large mammal population on Earth. Millions of public tax dollars have been spent examining myriad aspects of bear behavior and the research methods have also been applied to other species around the world. The work done here has influenced research on Bengal and Siberian tigers, African lions, snow leopards, polar bears, wolves, cougars, and jaguars; many of these species could be extinct in the wild by the middle of this century.

When I first began writing about 399 in 2007, members of the grizzly study team told me her origins could be traced to slopes towering over the wending path of Pilgrim Creek in Bridger-Teton National Forest's Teton Wilderness. Years later, Mangelsen would corroborate this information while sailing over the pinnacle of the 8,274-foot Pilgrim Mountain in a small aircraft, flanked by Wildcat Peak to the northwest and Bobcat Ridge to the southeast.

Little is known about 399's whereabouts in those months after her initial capture in 2001, except that she demonstrated a knack for finding a range of natural foods and had a brush with what could easily have been her doom. While reconnoitering the Gros Ventre River Valley, she fed on a couple of dead cattle in a national forest grazing allotment. The cows had perished after eating poisonous larkspur, and, fortunately, cowboys tending to the livestock didn't shoot 399 as she continued to devour the meat. The dell and undulating foothills of the Gros Ventre had been a venue for chronic conflict between predators and cattle dating back to the early twentieth century. Had this been pre-1975, she probably would be dead.

In 2002, 399 was captured by researchers twice again and released unharmed. She had a habit of casting collars. During the summer of 2004, she emerged from hibernation in spring with a single cub at her side. Later in the summer, the cub vanished, possibly from starvation or a run-in with an adult male bear or wolves. The cub was found dead in the Pilgrim Creek drainage.

Losing her first cub may help to explain why 399 in subsequent years exhibited a preference for staying close to the front country of Grand Teton National Park roads amid the bustle of people. Chuck Schwartz, who was then the leader of the grizzly study team, once hiked with me into the Pacific Creek highland to inspect a grizzly daybed site that had been plotted via GPS. I lay down on my back inside of the circle where swales of grass had been flattened, trying to imagine what the bear saw from that vantage. Schwartz noted that while living near the roadside can mean unpredictable behavior from humans, remote wilderness environs lorded over by boars can be far more hazardous. Male grizzlies have no role in raising families; their only interest in females is to breed (and, in the process, to pass along their own genes). If they think cubs are interrupting or represent an obstacle to their suiting, they won't hesitate to kill and even eat the young.

Fresh out of their Pilgrim Creek den and dusted by a spring snowstorm, 399 and clan (587, 610, and 615) wander toward Oxbow Bend along the eastern hinterlands of Jackson Lake.

Following two more instances in which 399 was trapped by researchers—in July at Pilgrim Creek and in October near Colter Bay, an inlet of Jackson Lake—2004 passed into 2005. That spring, 399 was located again moving in the hinters that rise above the eastern shores of Jackson Lake. In May, she shed her collar once more and went off the air, but that summer she was captured and collared yet again, the sixth time in five years. That's a lot of handling by humans.

Sometime before she denned in late autumn, 399 came in contact with an anonymous male bear. The outcome is now clear. When she emerged in the spring of 2006, the 10-year-old bear had three cubs the size of housecats running at her feet. Wildlife photographers in Mangelsen's tight-knit circle of friends certainly weren't complaining that she frequented the network of roads around Jackson Lake Lodge, but amid a growing commotion as summer tourist season started to kick in, they worried about her safety. (In Yellowstone National Park, as a counterpoint, grizzlies that routinely forage and navigate through developed areas often get moved.)

Among the group of Jackson residents driving up the valley to observe her was Franz Camenzind, a photographer, biologist, award-winning wildlife cinematographer, and respected conservationist. A specialist in canids (wolves and coyotes), he spends a lot of time each year watching lobos in Yellowstone's Lamar Valley. This was a scene, he said, to rival that. "I don't like to anthropomorphize, but bears like her don't come around very often," he told me at the time, concerned that she was like a celebrity. "399 represents Jackson Hole's wildlife mascot. But I'll be honest with you. I don't know how the cubs will survive this year having been so habituated to people. Their mother very skillfully made her living, having little fear of humans. People do dumb things and it's not wise to do dumb things in the company of a mother bear and cubs."

They did survive, and soon enough the cubs—two sisters and a brother—entered the research roster as 610, 615, and 587. That fall, driven by the frenzy of hyperphagia, they packed on weight. Hyperphagia, which starts to influence bears around the end of August each year and continues until they den, involves an abnormal instinct to consume as much food as possible, a product of evolutionary adaptation that has enabled grizzlies to persist in cold winter climates. Get enough food and they live healthier, longer, and with more offspring in their lineage. The opposite is true if they are undernourished at the start of winter.

We can only imagine the scene inside 399's secluded den up Pilgrim Creek that winter, with a mama and her offspring snuggled together in deep slumber, their heartbeats and breathing having slowed to conserve energy and fat supplies. They were unaware of the blizzards and subzero temperatures above ground. It is fun to wonder: What were 399 and her almost one-year-old cubs dreaming about?

Set to natural clockwork, grizzlies in Greater Yellowstone tend to begin emerging from their dens in mid- to late March every year. Boars crawl out first. Mothers with cubs tend to delay their exits until April or May. Researchers say there is probably an evolutionary reason for this. Bears coming out of the stupor of dormancy are groggy, thin, weak, and vulnerable. This delay lessens the time they are exposed to harsh elements and late blizzards that can make the search for sustenance, especially with cubs in tow, more perilous. It also reduces the likelihood of encountering hungry male bears and wolves that will eat bruin youngsters if given a chance.

MANGELSEN SPENT SEVERAL WEEKS in 2007 rising at dawn every morning and staying out past dusk watching 399 and her cubs. A triad of his photographs—*Snow Mittens*, *Grizzly Comfort*, and *Spring Bouquet along Pilgrim Creek*—speak to how visually accessible the bears were. Especially memorable was what he captured at Willow Flats.

Willow Flats has a poetic, peaceful ring to it. Next to the Oxbow Bend of the Snake River, this moist maze of willows and meadows is a hot spot for grizzlies. Why? Over many years, generations of cow elk groups have grazed in the open meadows and used the protective cover found there as a safe place to give birth to calves late in spring.

When grizzlies moved into the neighborhood, 399 and other sows began entering the labyrinth to play hide-and-go-eat with their young elk prey. The game delivered ready meals to cubs and in the process taught them how to hunt. Willow Flats happens to be located right in front of Jackson Lake Lodge. During a particularly memorable evening, 399 and her cubs were prowling through Willow Flats stalking elk as hundreds of humans watched from the hotel patio. Mangelsen and I were there together. "Except for the last few years, it has been rare in the Tetons to have grizzly bears so visually accessible," he explained that night. "To have a beautiful sow and three cubs so visible doing the things that wild grizzlies are supposed to do, and with the Tetons rising above them as a

As part of her omnivorous diet, 399 grazes on yellow biscuitroot, among a wide buffet of natural staples. **FOLLOWING SPREAD:** Teaching her cub brood how to hunt, 610 enters the meadow of Willow Flats and parts a herd of cow elk and their calves. The predator-prey drama playing out just below the veranda at Jackson Lake Lodge is as exciting as anything found on the Serengeti Plain in eastern Africa.

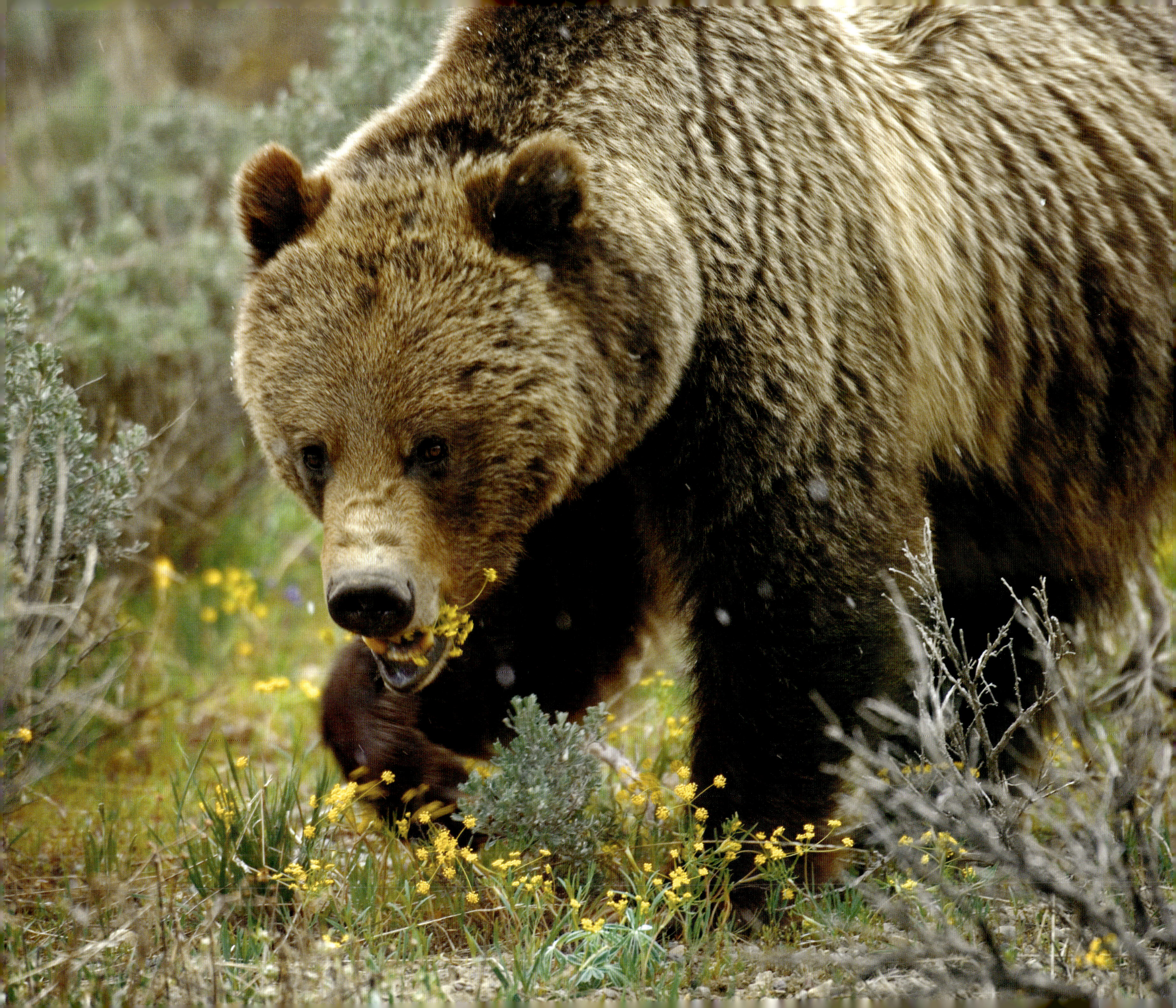

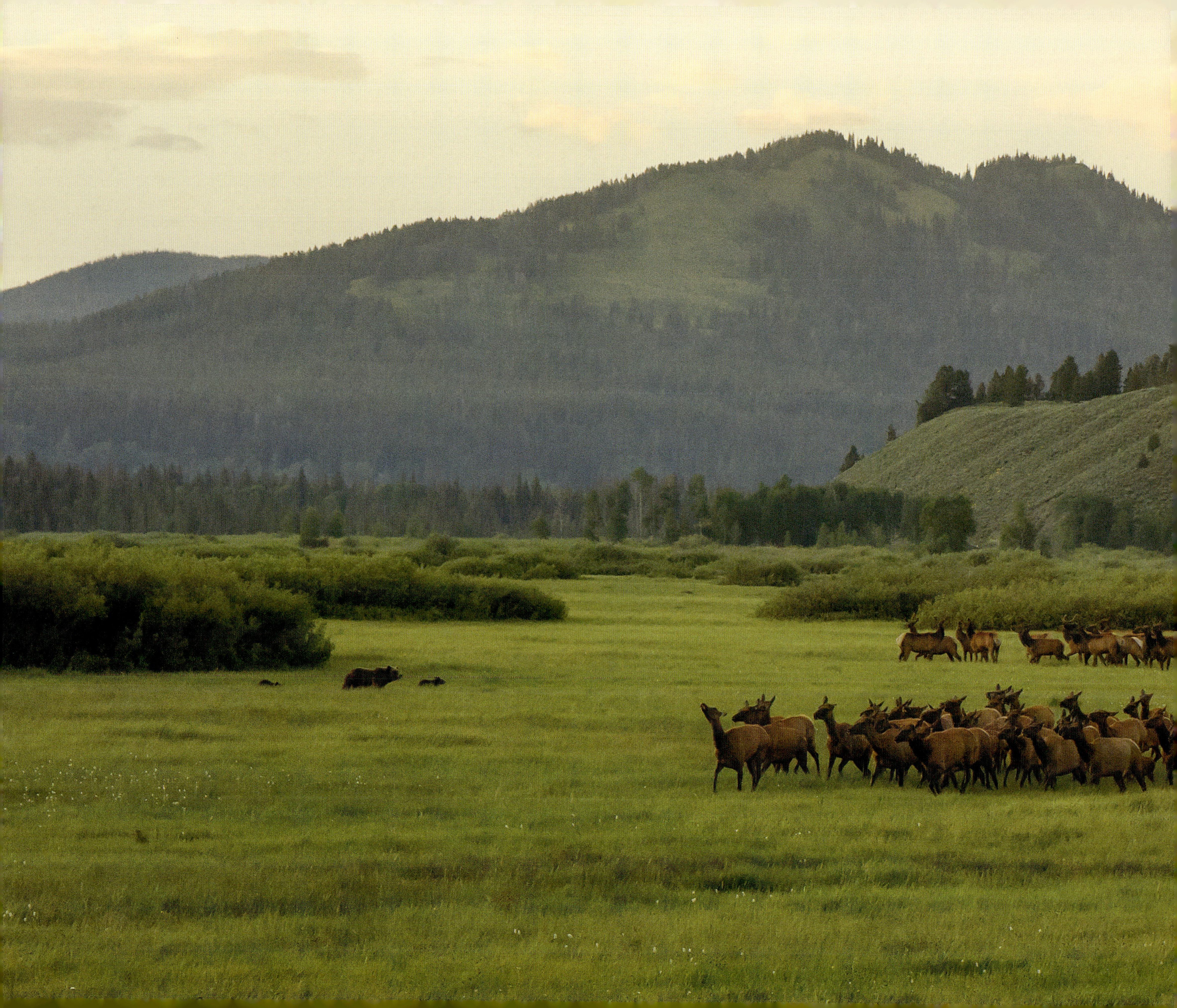

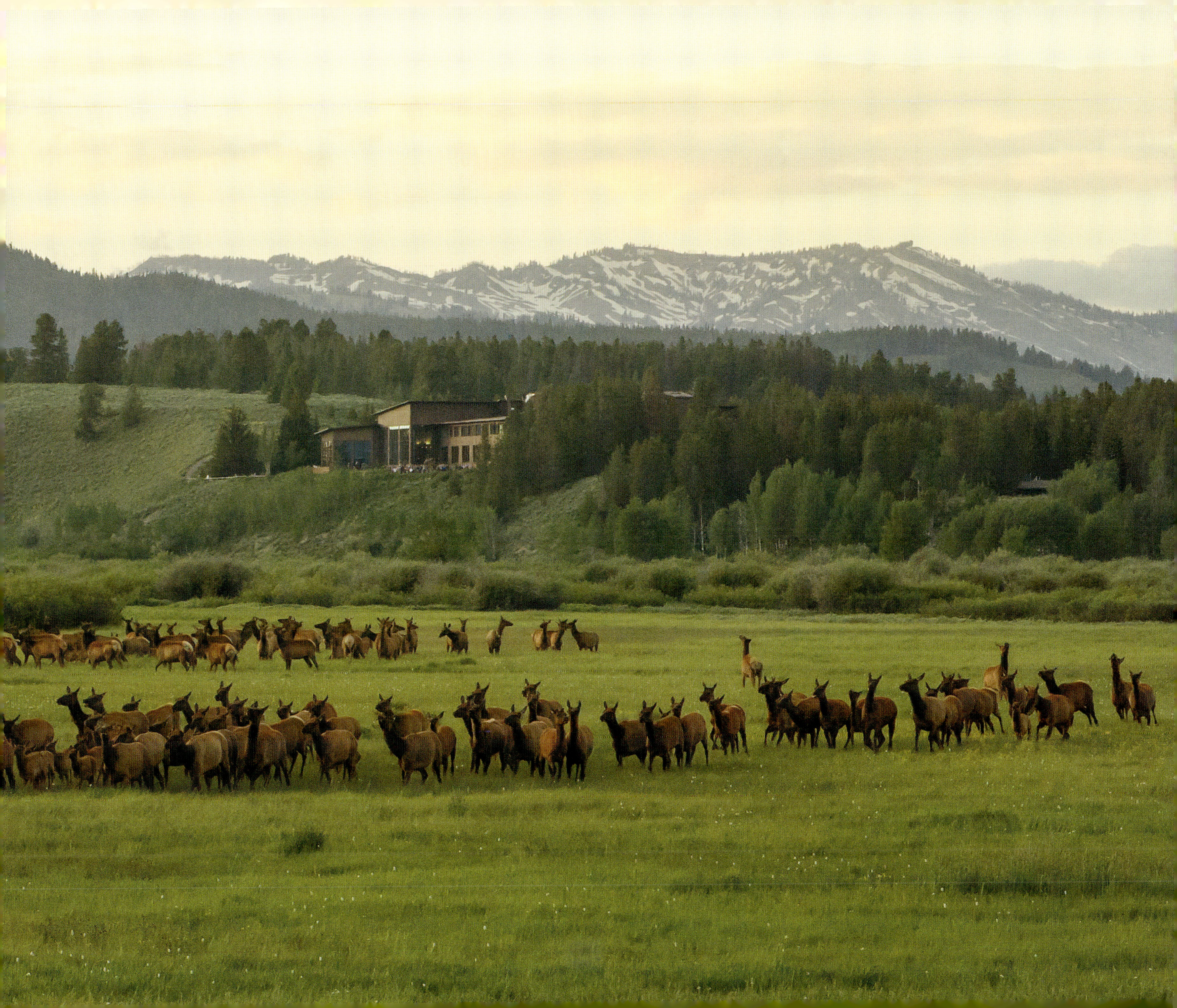

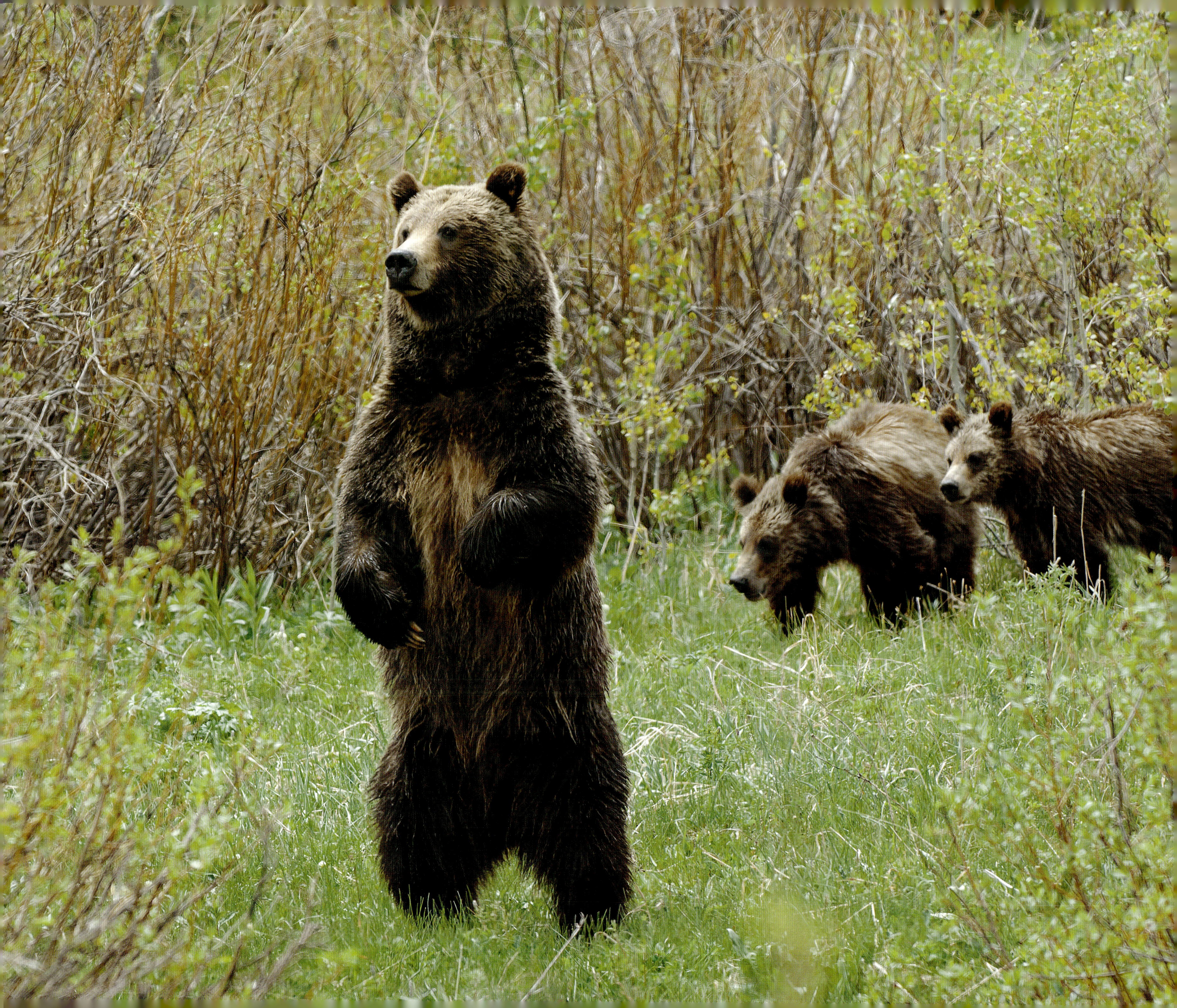

backdrop, that's as dramatic a setting as you're ever going to find." Mangelsen, better than anyone, would know.

Most days, it didn't take long for the network of photographers to figure out where 399 was, provided she hadn't ventured deep into the coniferous forest. On a Monday in June 2007, they lost track of her. Meanwhile, Dennis VanDenbos, a high-school science teacher from Lander, Wyoming, was attending an education conference at Jackson Lake Lodge.

Upon checking in at the front desk, he was warned about bear activity in the area. On the last morning of the conference, VanDenbos laced up his hiking shoes and set out for a brisk walk into a developed area down Wagon Road south of the lodge. He trekked past a set of cabins and arrived at a place where a hiking and horseback-riding trail disappeared into a thick grove of trees. VanDenbos was standing at the edge of Willow Flats. Trail closure signs posted earlier had been taken down.

Along his route, he saw a cow and calf moose trot by and halt beside him to browse. Moving on, he passed by elk cows and calves off in the distance. He didn't have a good feeling about continuing into the shadowy fringes of the Willow Flats understory. Erring on the side of common sense, he turned around and retraced his path back toward Jackson Lake Lodge.

But further along the way, he encountered a cow elk near the road behaving jittery. "It seemed kind of strange she was so close to me," he said later in an interview. What VanDenbos didn't know was that 399 and her cubs had killed the elk mother's calf. The carcass and bruins were concealed, in fact, just six feet away from where he was standing.

"I had no idea the bears were there, but I was kind of curious because of the way the cow elk was acting. I was looking around, maybe unknowingly right at the sow, and I am sure she might have mistaken my body language as being menacing to her cubs or as a competitor for the carcass," he said.

He continued to walk, unaware that he had breached the bubble of 399's comfort zone. He hadn't gone five paces before he heard a grunting guttural growl, and within a split second saw a blur of brown in full charge. VanDenbos had no bear spray—and he didn't carry a gun.

"I was thinking right away, 'This is bad. Oh, shit, this is really bad,'" he said. In a single freeze-frame moment of clarity, he saw the angry mother, her neck fur standing on end, and beside her three fuzzy teddy-bear shapes.

Backpedaling, VanDenbos yelled and waved his arms to try to make himself look bigger. 399 did not take him down but tried to stop the momentum carrying her forward. Side by side, she turned her face to stare straight into his. "I've heard people say 'Whatever you do, don't look directly at the bear. Don't make direct eye contact.'" He couldn't help himself. "I could now see all those teeth and the side of her head in silhouette and the hair on her back. To be honest, it was really intriguing, a silvertip grizzly with hair rising straight off her shoulder. I said to myself, 'Why are you thinking about *this* now?' But I have to tell you, it was, in its own way, beautiful."

Unfortunately, amid his retreat, he caught his foot on a bump in the ground and tripped, stumbling toward the ground. This movement triggered a new reaction from 399. As VanDenbos tried to stand up he was eye level with the hyperconcerned mother. "I dove straight down and pulled my arms over my head. She came and bit me in the back as I played dead. I don't know why, but I had the sense it was just a warning." The sharp nip was followed by a more powerful clamping of teeth into his right backside. "It wasn't a tearing bite into my butt, but it was forceful and I knew she wasn't messing around. I was just thinking of how I should interpret this gesture and how the encounter might possibly end," he said.

Up the hill, other people in the park heard the noise and listened as VanDenbos shouted at 399. They phoned for help from rangers. Downslope and pinned to the ground, VanDenbos felt a bear paw standing on his left calf, and then he was stung by another bite in his left rump. "This wasn't like the earlier bite," he said. "This was bite and tear. And I thought, 'Well, they are going to eat me now.'"

Husband, teacher, and human VanDenbos was in the same predicament the now-dead elk calf had been in minutes earlier. Did 399 recognize the difference between species? VanDenbos's survival, and her own, pivoted on her assessment. "It was just like a documentary you see on TV when a big predator takes something down. I was a participant and it wasn't pleasant, but I really wasn't feeling pain as such," said VanDenbos.

He had stopped screaming, accepting what would come, and anticipated another blow. Out of the corner of his eye, he saw 399 standing over him, joined by 610, 615, and 587. The heads of the cubs poked around from behind their mother, awaiting her instruction. Indeed, it proved to be a teaching moment. Braced for another bite, VanDenbos then heard

Rising on vigilant lookout and wary of big predatory male bears or people bumbling into them, 399 shepherds her cubs through the riparian jungle near Oxbow Bend.

a loud human voice. The weight of the paw that had been resting on his body suddenly was gone. Then silence. He didn't budge.

As it turned out, a woman cook and a wrangler who guides horseback rides were driving up Wagon Road and spied a huddle of bears. They had stopped and reached for their cameras only to discover the bruins were on top of a person. When they screamed, 399 and her cubs ambled slowly away.

"I didn't get a sense, based on her actions, that she wanted to hurt me, or eat me," VanDenbos said, then added that as soon as he was loaded in a truck and taken for medical care, the bears returned to what they really wanted. They went back to the carcass and dragged it across the road away from the cabins.

But a bystander inside one of the buildings heard the bears chewing. Unaware that it was a calf in their jaws, she believed it was VanDenbos. In the confusion that ensued, as VanDenbos was being rushed to a hospital in Jackson, rangers with guns in their holsters were racing to the scene. A witness was under the false impression that VanDenbos was being devoured by grizzlies. Few would have been surprised if the arriving rangers had shot and killed the bear family. Amid radio chatter, VanDenbos made it clear that he had inadvertently provoked the attack and wanted no harm to come to them. They were spared.

"No one interviewed her [399] afterward to get her take on what went down, but if she had wanted to treat him as prey and make it a lesson in hunting for her cubs, she would have," the grizzly study team's Chuck Schwartz said. "I think she probably was teaching her offspring. She showed them how a bear can respond nonlethally to people when humans represent a threat to them and their food source. Except for what the gentleman went through, I think this is a positive outcome."

The decision from bear managers was to leave her alone. 399 had been tested and proved she had no interest in becoming a man-eater. This was the second incident that could easily have resulted in her demise. Mangelsen breathed a sigh of relief that rangers didn't respond impetuously. He thanked them. The bear family continued to wander. "She and the cubs were all over the place in the days and weeks afterward," Mangelsen said. She might be at the Oxbow or eastward along Pacific or Pilgrim Creeks, walking the shores inside Colter Bay, or frequenting Willow Flats. The harrowing episode on Wagon Road only bolstered her mystique.

SOMETHING REMARKABLE HAPPENED in the aftermath of the encounter: 399 became a bona fide global wildlife icon, the most famous wild bear in the modern world. Thanks to various forms of social media—namely Facebook, Twitter, and later, in 2010, Instagram—interest in 399 went viral. Millions of people became aware of 399 and the cubs almost overnight in 2007. Americans flocked to Grand Teton National Park in droves hoping to catch sight of her. International travelers altered vacation plans just to make a special pit stop in Jackson Hole. While Yellowstone has long been known as a destination for seeing grizzlies and wolves, the phenomenon of interest surrounding 399 catapulted her into a different realm. Because of her skill—and comfort level—in navigating the developed front country, thus making her more reliably visible to wildlife watchers, Grand Teton National Park suddenly claimed its own vaunted place on the grizzly bear map.

Dr. Susan G. Clark, a professor of environmental studies at Yale University who spends summers in Jackson Hole at the Northern Rockies Conservation Cooperative, a wildlife research outfit, voiced admiration for 399: "Bears are intelligent. Here is a mother, watchful and protective, as most mothers tend to be. Around her are all these things going on—people moving in and out of her space, conveying different kinds of messages through body language. There are smells of human foods drifting in the breeze. There are vehicles moving at various speeds. There is nervous tension in the air. Given all of this, 399 is making assessments in her own mind, hundreds of them, that inform what she is going to do. I see an animal that exhibited remarkable composure in showing her cubs how to move through the chaos."

As an environmental journalist writing about the West for 30 years, I've encountered some hair-raising expressions of ecological ignorance. When they come from politicians, it says a lot about not only the elected leader but also the judgment of the citizens who put the leader in office. In the year 2000, Idaho governor Dirk Kempthorne issued an over-my-dead-body threat to fight a plan from the U.S. Fish and Wildlife Service to bring grizzlies back to the Bitterroot Mountains along the Montana-Idaho border. Reestablishing a grizzly population there is considered crucial in linking island populations of bears in Greater Yellowstone and the Crown of the Continent Ecosystem enwrapping Glacier National Park.

399's yearling cubs (587, 610, and 615) are seen on the roadside investigating the cars and looking for food with Bivouac Peak in the background. This is the first spring that 399 became visible to the general public and started drawing crowds, causing "bear jams."

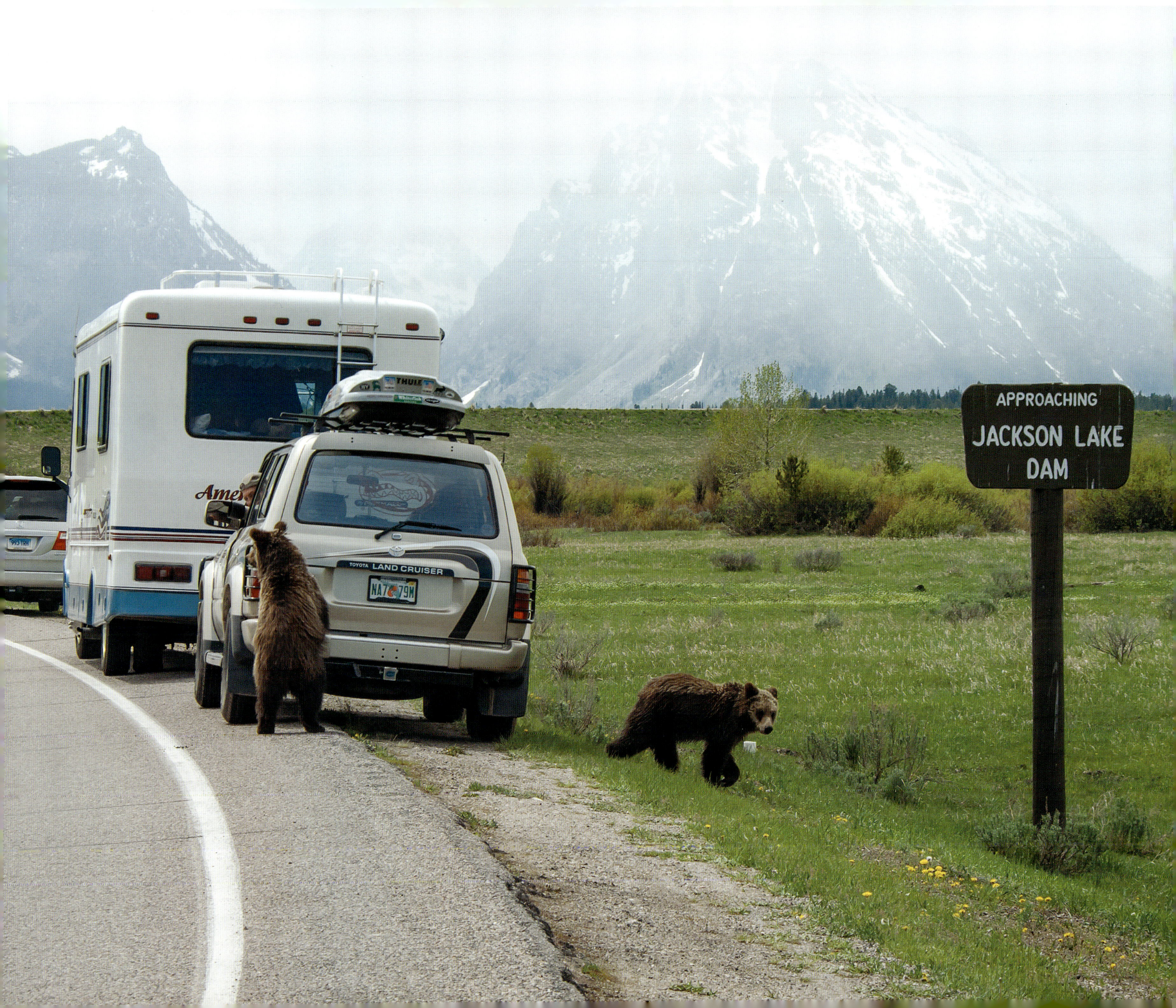
APPROACHING
JACKSON LAKE
DAM

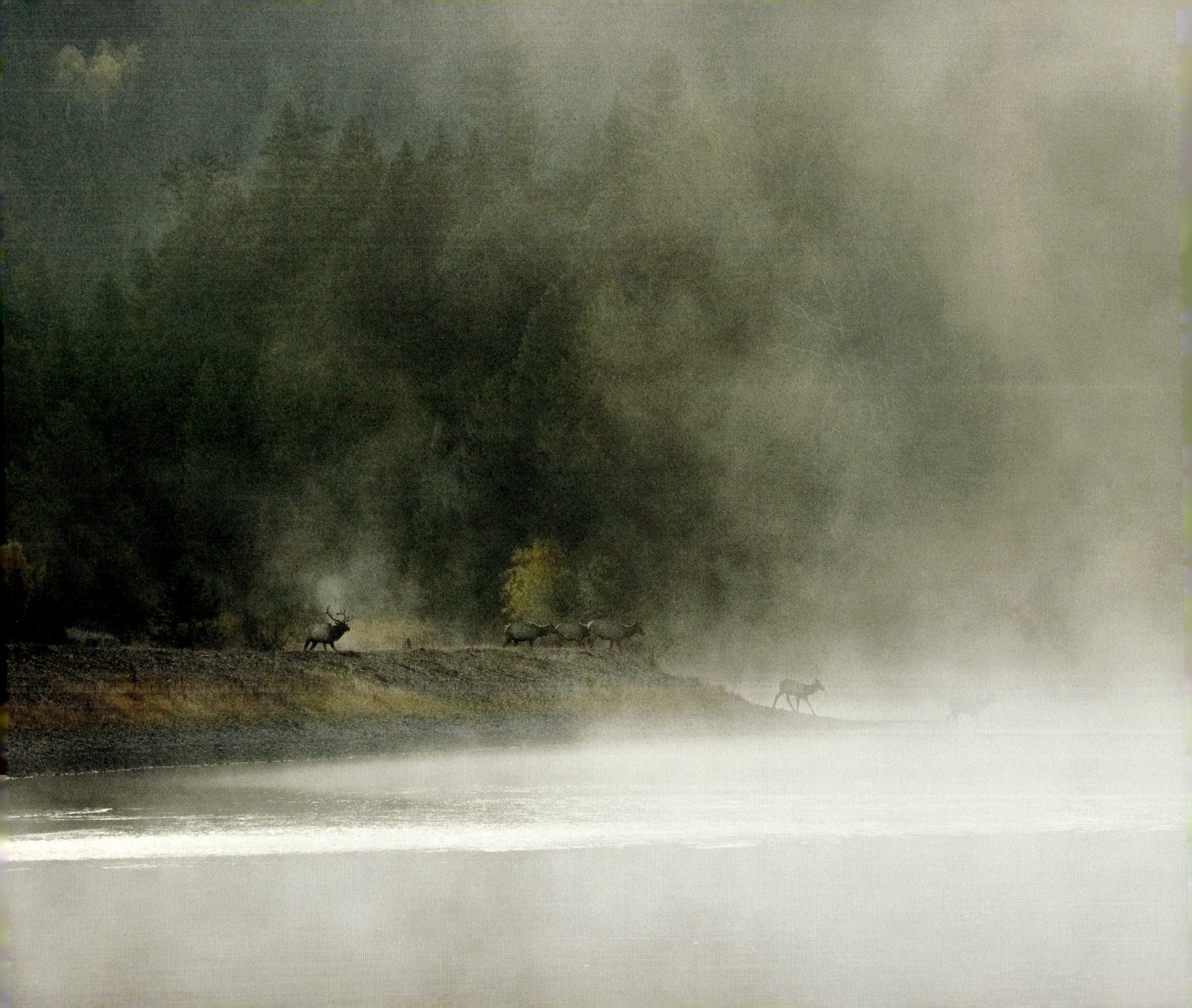

Kempthorne served up some doozies when he declared that Idaho would not accept "these flesh-eating, anti-social animals" into the state. "Whenever there's an encounter between a human and a grizzly bear, the human does not fare well," he said. Kempthorne apparently was unaware that human-grizzly encounters have ended far more badly for bears by a ratio of at least 50 to 1 for several decades. Perhaps he would have realized how dumb his fear-mongering comments were had he been able to join bear watchers in Grand Teton or Yellowstone National Park.

The downside of 399's star status, a problem not insignificant, was that huge crowds of onlookers gathered, creating eruptions of traffic gridlock called "bear jams" where cars backed up for miles whenever 399 came into sight along highways. Tourists would haphazardly hit the brakes, fling open the doors of their vehicles (sometimes leaving them ajar), and sprint with unbridled excitement to within close proximity of the sow and cubs. Not a good idea, given the reputation of grizzly mothers for being zealously protective.

The scenes, as Clark suggested, were often chaotic. Eventually park rangers would arrive to manage the spontaneous masses of humans. In many instances, before they took control, the only thing preventing people from getting too close was Mangelsen and other wildlife photographers telling crazed tourists to stay back. Mangelsen got used to the insults coming from people who didn't want to listen to him, but he remained calm and stern.

It didn't take long for the most ardent bear advocates, including a special citizen unit created by Grand Teton National Park called the Wildlife Brigade, to realize that 399 was in perpetual imminent danger. Mangelsen still gets heart palpitations recounting the close calls. "It was a miracle in those early years that she stayed alive, that she kept her cool and did not feel threatened as she and the cubs moved around cars and people, oftentimes looking both directions for traffic before leading her cubs across a busy highway," he said.

Any motorist behind the wheel who might have been distracted and not paying attention; any tourist who might have unwisely thrown the bears food scraps; any human who might have done something really doltish—such as venturing too close to 399's cubs or trying to touch them, as some did—could have doomed the bears. The worst, however, was still to come. Awaiting in the fall was the start of the annual elk-hunting season inside Grand Teton National Park.

As it turned out, Mangelsen wasn't being alarmist but astutely prescient. The idyllic Disneyesque picture of a lovable bear family ostensibly living happily ever after in a national park would soon be marred by a jarring dose of reality. In addition to human urban legends, peril was stalking 399's clan.

MANGELSEN SAVORS THE FALL; it is his favorite season because wildlife are resplendent and on the move. Bull elk engage in mating behavior known as "the rut." Dominant males armed with mighty antler racks joust, fending off rivals vying to breed with elk cows. Perhaps the most magical part of this courtship is bull elk bugling, in which they emit piercing high-pitched wails. As mist rises off the Snake River into the chilly air, the muted silhouettes of bulls and their sonic declarations are primordial.

And then hunters arrive and wildlife scatters. Yes, there is a big-game hunt held inside a national park where animals are normally protected from harm for nine months of every year. Then suddenly, with the turn of a calendar page, those same animals are chased and dropped with bullets. The current boundaries of Grand Teton National Park were established in 1950. Part of the deal legislators in Wyoming made with Congress when it expanded the size of the park was ensuring that traditional elk hunting remained to help reduce the elk herd. It is a cultural artifact as much as anything else.

Every year brings stories of another spectacle. Mangelsen recalls a specific episode in the fall of 2007 that galvanized his views about the "strange absurdity" of the park hunt and the danger it poses to bears. On the evening of October 15, a large group of people—hunters, tourists, and Jackson Hole locals—were gathered at Teton Point Overlook just east and north of Schwabachers Landing near the Snake River. A hundred yards below in the rolling sagebrush, 399 and her three cubs hunkered over a spike bull elk carcass. The young wapiti had been felled during the hunt. The cubs were now nearly the size of their mom and somewhat difficult to distinguish. They were fat from the long days of enjoying food abundance during the summer and fall. Their rich, brown, silver tipped winter fur glistened in the last rays of the evening light. "It was glorious and yet in the context of the danger all around, kind of surreal," Mangelsen said. He and Cedarholm had not seen the family this far south since early spring when they had been ranging widely in search of winter kills.

A large bull elk stands guard and bugles, challenging other bulls, as he waits for the last members of his harem to cross the Snake River to the dense forest. In the fall, rifle fire from elk hunters reverberates inside the national park, causing Mangelsen and others to worry about the safety of grizzlies and people.

They left the bruins to their feast as darkness fell over the Tetons. Returning at daybreak the next morning, they found the bear family lying near the carcass, full of meat. As darkness faded, Mangelsen and Cedarholm, much to their horror, counted 18 hunters, clad in blaze orange, mostly sitting along the bank of the Snake, facing the river and waiting for migrating elk. They were oblivious to the fact that 399 and her brood were nearby.

Several hunters were on horseback to the north and others were near their pickups in the Schwabachers Landing parking lot, all within a quarter mile of 399's clan. There wasn't a park ranger in sight. One of the hunters remarked openly to Mangelsen and Cedarholm that his hunting buddy had been the one who mistakenly—and illegally—killed the spike wapiti a day earlier. With only a tag for a cow elk, the hunter left the wapiti behind to rot in order to avoid being ticketed. The hunter then said to Mangelsen, "The problem with this park is that there are just too many damn grizzlies interrupting our hunt."

Mangelsen kept his cool. While talking, the hunter pointed to two other gun-toting buddies below the turnout, hiking unknowingly back to the parking lot. They were on a collision course with the four bears, who were hidden from their sight in the tall sagebrush. Shocked, Mangelsen told the hunter beside him of the bears' whereabouts and together they began yelling to warn the others. Still, their foolhardy presence startled 399 and her family at close range. The bears jumped up and bolted at full speed toward another row of hunters lined up along the river, who were also unaware of the grizzlies.

That day, luck was on the side of the bears. Perusing for elk in the opposite direction, none of those hunters saw the bears coming. 399 could not see them either because of the rolling terrain. "Fortunately for everyone, she miraculously chose a path between the gauntlet of hunters. Before they even had a chance to see her, use their bear spray, or fire their rifles out of fear or aggression, she and her cubs were already running past them and then across the shallows of the Snake River," said Mangelsen.

It could have been tragic, he said. "This kind of thing would never have happened in the summertime when there aren't dead animals and armed humans behaving in ways that would never be allowed for normal tourists, but Grand Teton goes from being a friendly Dr. Jekyll for wildlife and turns into a Mr. Hyde during the hunt. What happened that day was dramatic but, disturbingly, it wasn't out of the ordinary. We hoped the Park Service would close that area because the potential of somebody getting hurt or one of the bears getting shot and killed seemed pretty high to me. The park resisted. Had a tourist behaved as the hunters do, harassing wildlife, killing wildlife, or acting irresponsibly, they would have been ticketed, harangued by a ranger, and maybe banished from the park."

Mangelsen and a growing number of people who favor eliminating the park hunt know it was grandfathered when Grand Teton's boundaries were expanded in 1950, but it has outlived its original purpose. "It may have been OK in the early 1950s, but it's the product of another time. There are a lot more people around today and it's no longer appropriate," he said. "The park has changed, society has changed, values have changed, and grizzlies, one of the rarest creatures in the Lower 48, have a much-deserved home in Grand Teton, when seven decades ago, they didn't. Sending people out with high powered rifles, even 12-year-olds, to kill the same animals that are ordinarily protected in a national park is absurd. And it causes havoc for wildlife."

Here's another frightening aspect. Grizzlies are quick studies. For decades, the notion has been advanced (I have disseminated it as a journalist) in Greater Yellowstone that when bears hear gunshots they equate it with the ringing of a dinner bell and move rapidly in the direction of the noise. They allegedly do it because they sometimes get a food reward in the form of an elk, moose, or deer that has been dropped by a hunter. According to Mark Haroldson and Frank Van Manen of the grizzly study team, it sounds plausible, but it, too, is an urban legend, unsupported by rigorous research.

Cuing bears to kills is not the explosion of gunpowder but the smell that is released into the wind when a hunter puts a knife into the belly of an elk and creates a gut pile. It's a bear's nose, not its ears, that leads the way. "The olfactory senses of a bear are so incredible that it's something we as humans cannot come close to appreciating," Van Manen said. Haroldson added that the grizzly study team has conducted experiments to try and replicate an elk being shot to see if grizzlies known to inhabit an area show up. He's not saying it doesn't happen, but it cannot be asserted as a generalization. Research has shown that the likelihood of Yellowstone bears to move outside the park during the early elk-hunting season is two to four times higher, a correlation that can probably be made to the availability of elk carcasses.

However, and to repeat the assertion from above, bears are smart. Many people have witnessed bears moving in the direction of gunshots in the fall. Moreover, when

The Grand Teton National Park elk reduction program, the only national park sport hunt of major big-game animals in the Lower 48, is controversial. Here, hunters with rifles scout the bottomlands of the Snake River near Schwabachers Landing, where elk have been seen and, unbeknownst to the hunters, 399, 610, and their cubs also roam.

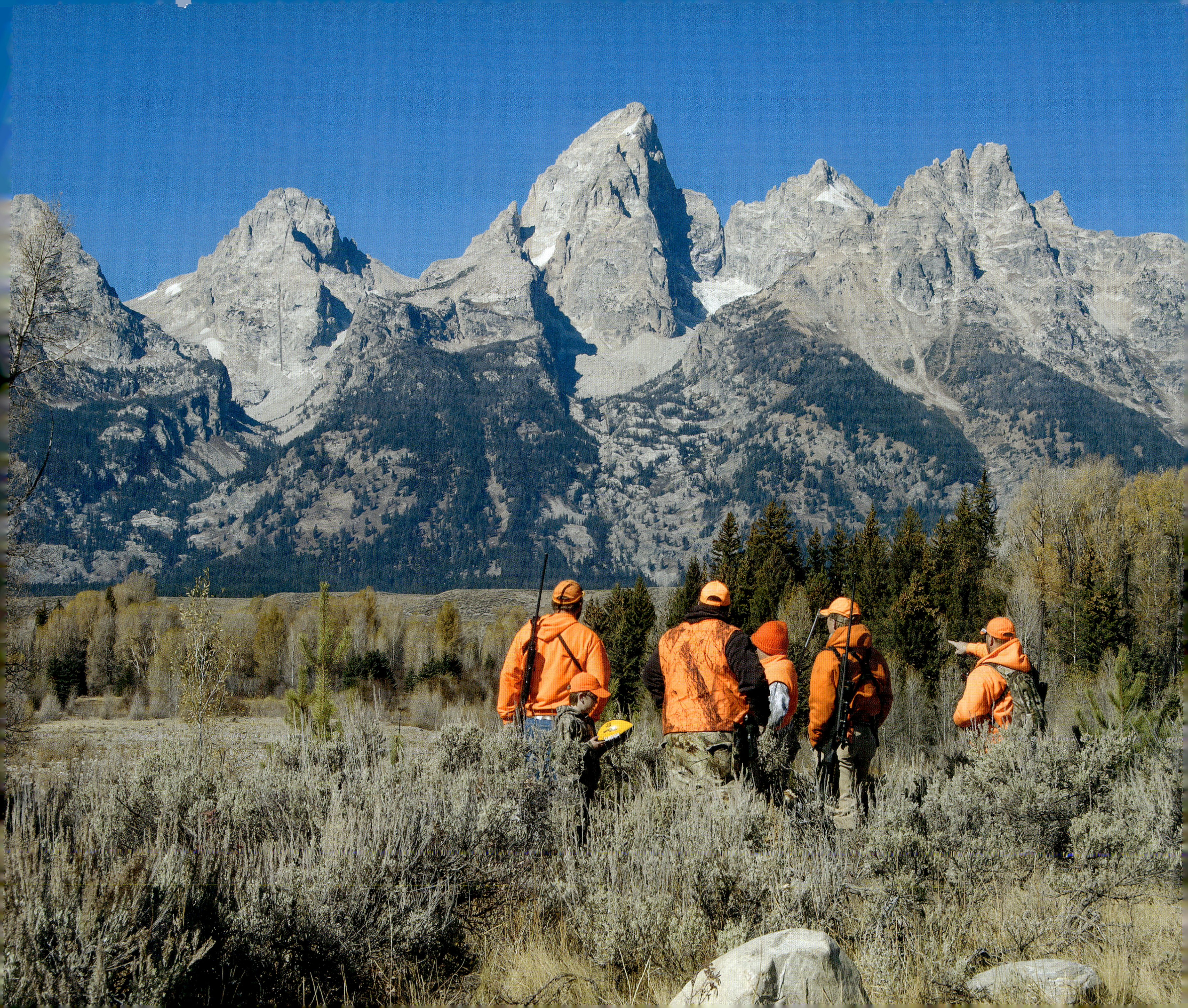

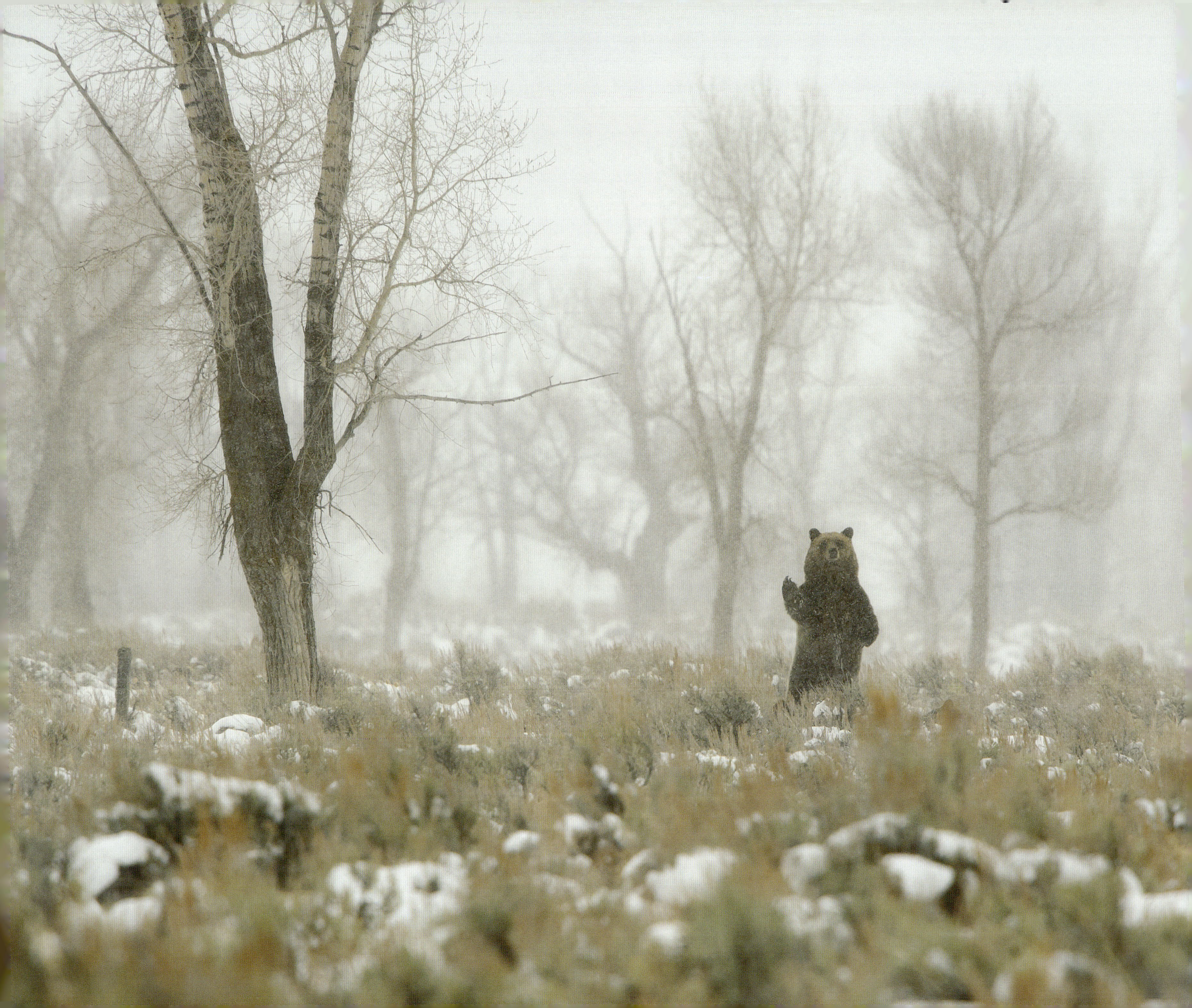

John and Frank Craighead were on fly-fishing trips to Alaska, they said big brown bears in the salmon streams knew that the zing of a reel meant dinner was being served. Anglers were told they needed to cut their lines quickly or risk having to fend off a bruin. The idea of bears being conditioned to gunshots is a concern because it puts both hunter and grizzly on a potential deadly rendezvous over a carcass, and, if grizzlies are hunted again in national forests adjacent to the national parks, a bear that goes to investigate a gunshot could ostensibly be summoned to its own demise.

SOME OUTFITTERS WHOM I'VE HAD THE PLEASURE of knowing over the years are salt of earth, proud to have grizzlies and wolves on the landscape, and do not resort to archaic ranting about predators. I contacted Jackson Hole hunting guide Forrest Stearns, owner of A+ Outfitters, who didn't want harm to come to 399, but he didn't want the park hunt shut down either. "When I first started guiding hunting trips in the mid-1970s, there were a few grizzlies around Jackson Hole, but they were the exception rather than the rule," he said. "Now we have them coming out of our ears everywhere. The recovery deal has worked very well."

Stearns requires that every hunting party undergo an orientation on how to use cans of bear spray—emphasizing that it should be used first before thinking of pulling a trigger—and how to be vigilantly aware in bear country. Studies show that using bear spray is more reliable than trying to fell a charging bear with bullets. Stearns emphasizes the regulations that demand quickly field dressing an animal, moving it away from a gut pile, and packing out the meat, and understanding that grizzlies, according to his own experience, do move toward gun blasts.

"We are doing some things now that we didn't used to. If we have to leave meat overnight because an elk was harvested late in the day, we separate the animal from the gut pile. In the 1970s, we would have left it and come back the next morning with pack horses. Ninety percent of the time if you leave an animal on the ground, bears (and sometimes wolves) will be on it. Coexistence between people and bears is a work in progress. We, in this day and age, have to be a lot more on our toes," said Stearns.

Mangelsen had devoted much of the spring, summer, and fall of 2007 to becoming acquainted with 399 and her cubs. He kept copious notes. The park hunt by itself had already delivered enough hair-raising moments. In November, he became distressed when the 399 quartet, now in full hyperphagia and aggressively seeking food, was moving between gut piles of harvested elk in the park and into the adjacent national forest where rifle season was in full swing. They were lumbering through woods swarming with more hunters, some of whom were urban out-of-staters deathly afraid of bears. From Pacific Creek, 399's foursome headed southeast again toward Schwabachers Landing on the Snake River and then eastward to Antelope Flats—all areas that resounded with rifle shots. Mangelsen again pleaded with park officials to shut down the hunts, for the safety of both bears and people. They again declined. If any of the bears ended up being slain, Mangelsen said, blood would be on the hands of the park.

So, let us address still another vexing part of the 399 saga, a double-edged sword: the very same gut piles that attract grizzlies onto the paths of hunters are an important source of nutrition that aid bears, especially pregnant females, as they stuff themselves to take in calories. Bears that have the benefit of gut piles and carcasses gain more weight and stay healthier during hibernation. 399 is a prime example.

Mangelsen described the first day of elk season in another year. It involved 399 and her cubs feeding on a roadkill elk carcass just east of the Snake River Overlook for several days. They were visible from the highway, causing a bear jam. The park was aware of her presence along the road to Lost Creek Ranch, which was filled with hunters on opening morning. She left the carcass around 10:30 a.m. "That's when I saw and counted dozens of hunters and talked to some of them," Mangelsen said. "When she went down into the woods where the hunters were shooting elk, I went to park headquarters and pleaded for them to close the area or at least post it with signs warning people there were grizzlies present. They did neither."

Mangelsen has witnessed many infractions over the years with Grand Teton's elk reduction program, in which hunters opened fire on elk and left many animals dead. He has called it a slaughter and has supported efforts to bring national scrutiny to it. If hunters are herd shooting elk, and violating hunting laws and ethics, and are often unaware of their surroundings, what evidence is there to suggest that bears would be safe? "What's going on here is so aberrant you couldn't make it up—and right in the middle of a crown jewel national park," he said.

What is she telling us? After feasting on gut piles left behind by elk hunters, 610 stands on a frosty autumn morning to assess the panorama around Antelope Flats Road.

LETTING GO, SAYING GOOD-BYE, can be traumatic for any parent and offspring. Mother bears usually part company with cubs in the months following their second birthday, but, as 399 demonstrated, it doesn't mean cutting ties completely. In the spring of 2008, Mangelsen's partner and executive assistant Sue Cedarholm, also a skilled photographer, was positioned up Pacific Creek Road and saw the triplets protesting 399's efforts to push them away. It was the last day the two-year-old cubs remained at 399's side. Cedarholm said 399 behaved as if she knew it would be best if the cubs kept their distance.

The next morning, on a hillside overlooking Two Ocean Lake Road, 399 was being pursued and courted by two males at the same time, one of average size, the other quite a bit larger. The first one mounted her to mate and then the other. The boars exerted a menacing presence around 610, 615, and 587, who were cautiously lurking in the woods at the edge of the meadow. 399 did her best to scare the trio away. A couple of days later, now fending for themselves, the young threesome was observed along different back roads.

Without warning, Grand Teton rangers approached the independent cubs with sirens blaring from their cruisers and sounded horns to haze them away from locales where they were vulnerable to being mobbed by people. The young bears scattered. Mangelsen, meanwhile, shot a series of images chronicling 399 mating with the giant male, the legendary 700-pounder whom some have nicknamed "Brutus." They swam together across the Snake River below Jackson Lake Dam.

I've been referring to the numeric identities of 399's cubs, but they were not officially bestowed until after detachment from their mother occurred. The male cub became 587 on June 28, 2008, when he was caught in a culvert trap at a residential area along the Pacific Creek drainage not far from where 399 left him to fend for himself.

After 587 was turned loose at Glade Creek inside Grand Teton National Park, he kept returning to the Pacific Creek subdivision, a residential inholding, and demonstrated no fear of humans. Like his mother, he never threatened people, and some residents who had lived there for years said they didn't mind him being around. They said you just had to be alert. And, like his mother, 587 loved to graze on the fresh green manicured lawns. A request for his removal was made by a seasonal resident who had recently bought a cabin in the subdivision and admitted to being afraid of grizzlies. He was often seen packing a pistol or carrying a shotgun on his shoulder while walking the neighborhood, which raises the question: Why move into a wild neighborhood knowing bears come with the landscape and then demand they be captured and exiled? State wildlife officials, not wanting to take any chances, complied with the request. And five years later, 587's habits, based upon what he had been taught by 399, would ironically do him in, though not before he lost a sister.

"Data shows that if we can keep young subadult bears, especially males, out of trouble for a couple of years after they leave their mother's company, their chance of survival goes back up to the level of any wild bear in the population," grizzly study team leader Chuck Schwartz said. "But if they get into trouble, like, say, getting into garbage or a tourist throwing them food, the probability of their survival is reduced 20 percent each year. A bear that wanders beyond the national park has a 60 to 70 percent survival rate compared to an animal that stays inside Yellowstone all the time." He added that Yellowstone bears don't have to contend with the possibility of being shot by hunters, a luxury not afforded to Grand Teton bears.

One of 399's female cubs ambled into a baited culvert trap on June 26, 2009, at Cottonwood Creek in Grand Teton National Park. She was sedated, had a yellow tag put in each ear and a radio collar put around her neck, and was given the identity 610. She would become, figuratively speaking, a chip off the old block. A full month later, on July 28 at the same location, 399's other young daughter was trapped, temporarily immobilized, ear and radio tagged, and logged as 615 by researchers. In 615's case, she was captured by accident. Ecologist Steve Cain said the trap had been set to catch a black bear.

TRAPPING, ANESTHETIZING, COLLARING, and ear tagging grizzly bears has yielded incredible research insights. It has resulted in bears being saved, no doubt. A topic of passionate debate, however, involves the question of how much is enough or too much when it comes to invasive research techniques with wildlife. Grizzly study team members say it's essential. Mangelsen has had several run-ins with grizzly bear trappers working under researcher Mark Haroldson. He and others have raised a litany of concerns.

For years, some of the foods put in culvert traps to bait bears were human or pet foods. Mangelsen couldn't understand why Grand Teton park rangers would do this when they so vigilantly hounded visitors—and for good reason—to keep clean campgrounds and not leave any food out that might be an attractant to a grizzly. Today, members of the

Only days after turning her cubs loose to fend for themselves, 399 is observed consorting—and mating—with a massive male more than twice her size. The pair ambled together for several days. Here, they've just taken a swim in the Snake River.

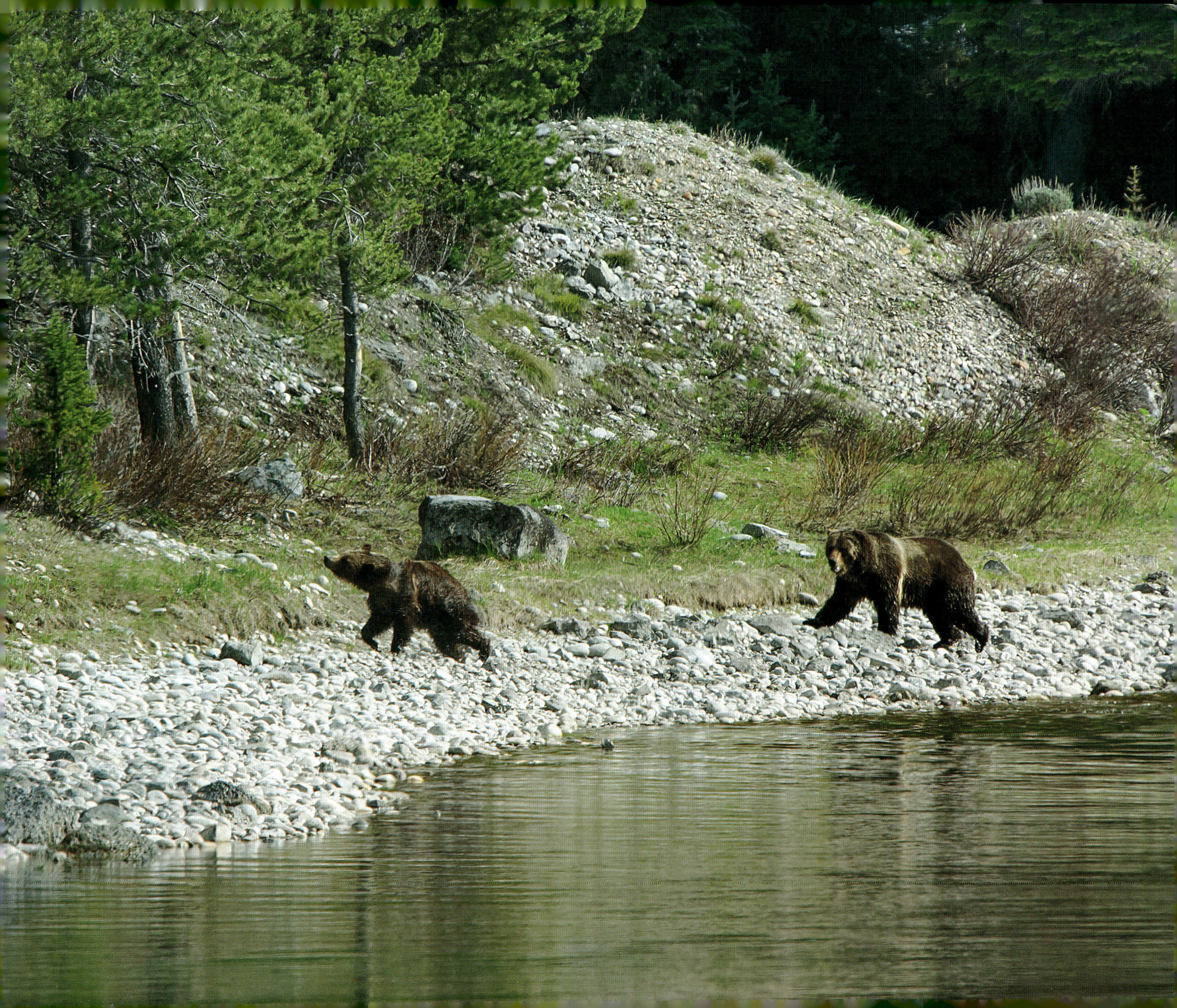

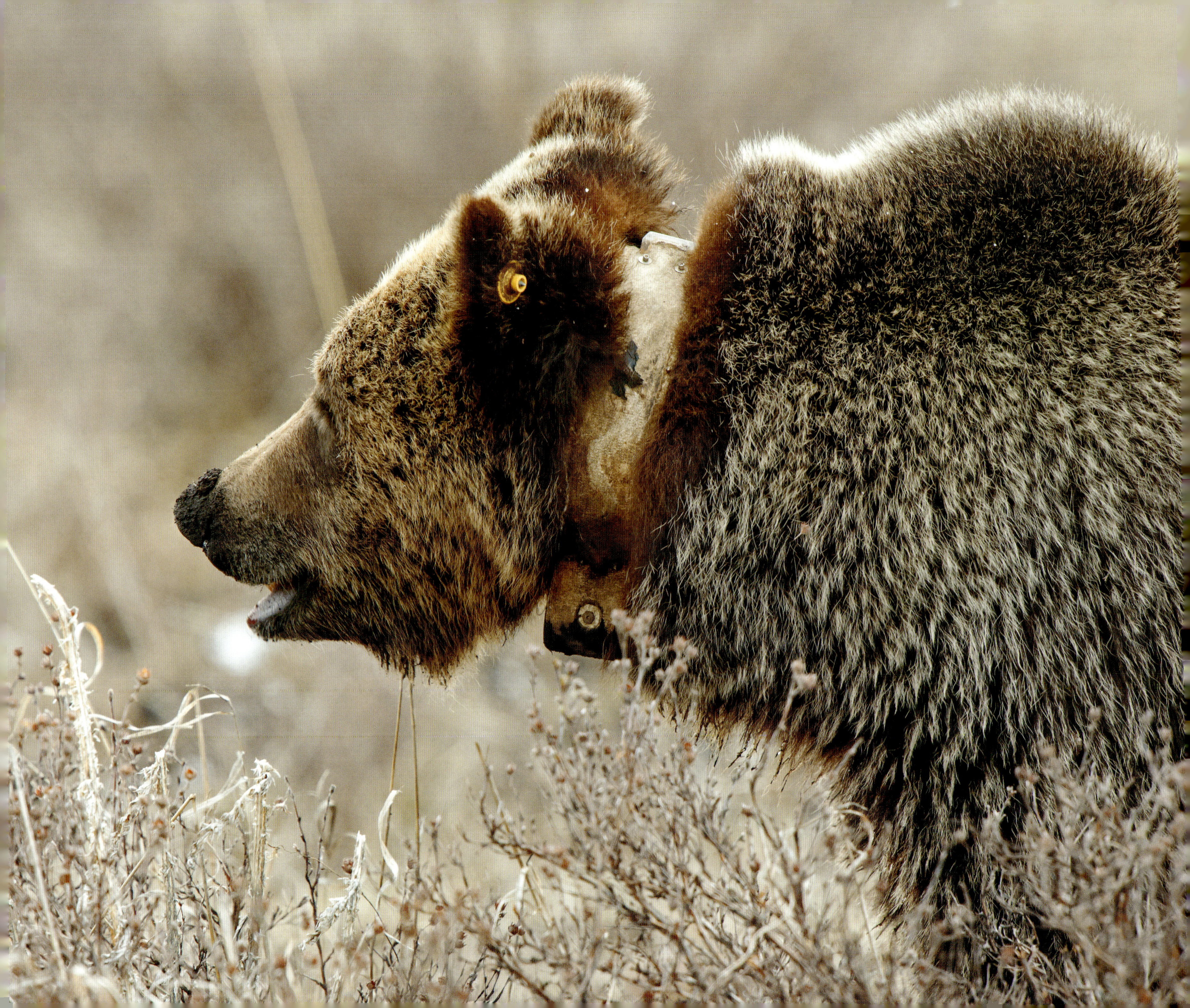

grizzly study team use wild meat as bait when setting traps in places like Pilgrim Creek. However, even wild game, whether used in trapping or left as a residual of hunting, may become a reward associated with humans because it carries their scent, Mangelsen said.

He pointedly asked the trappers why they were aggressively trapping in areas frequented by 399 and her cubs. There are examples of bears being injured, such as when the door to a trap fell, breaking the back of a bruin. Drug overdoses and dirty needle infections have killed others. Another serious concern involves the safety of cubs and people if they are in the vicinity of groggy bruins coming out of anesthesia. Cubs are especially vulnerable while their mother is caught in a trap, and there's at least one case in which cubs were preyed upon by wolves while their mother helplessly watched from her captive confines.

On one particular day, Mangelsen saw the trappers setting up their gear in Pilgrim Creek not long after 399 and her cubs had been seen. He had been there earlier in the morning and heard wolves howling in the hills above. 399 has been caught and handled more than a dozen times. At what point does this practice start to have negative or disastrous consequences on the study animals?

That day up Pilgrim Creek, Mangelsen informed the grizzly study team that some large boars had been sighted. "I asked the trappers, who are you trying to catch? Knowing where my questions were going, they said, 'Don't worry, Tom, we're not targeting 399 or 610.' But it sure appeared otherwise to me. They have enough danger to navigate, so why would you want to risk having something bad happen to these iconic bears? Taking the chance of injury is completely unnecessary. The park tells visitors to keep their distance, to not do anything that would cause wildlife to alter their behavior. But they keep messing with these bears. What more could they possibly learn from them?" he asked.

Mangelsen turned philosophical as we spoke late one afternoon up Pilgrim Creek. "I'm all for research, but I'm also for letting bears be bears, letting them have their dignity as wild animals instead of being treated as subjects in a perpetual science project. 399 has made her contributions. I think she should be allowed to live out the rest of her days without a collar around her neck. GPS is a great invention. It allows us to see the movements of these animals via satellite signals beamed back to us. They give us points on a map. But what those dry statistics don't measure or account for is a bear's sentience. And to me, that's what gives grizzlies their magic and a kind of soul. Let them be."

Grizzly study team members said they have at least 25 bears on the air at any one time. They have heard Mangelsen's complaints, but a valuable part of having front-country animals like 399 on collars is that it might be a deterrent to poachers.

Like her sister, 615 weighed around 225 pounds when she was caught in 2009. As Steve Cain told me, the advantage of having a GPS tracking device on a bear is that you know where it goes. At the time, scientists were tagging and radio collaring wildlife to gauge the effects of a proposed bicycle path through the national park.

They discovered that 610 and 615, unsurprisingly, shadowed many of the same routes used by their mother. A stroll up Pilgrim Creek was a regular part of their circuit. It was, after all, their birth drainage. According to Mangelsen, the sisters comingled, split, and comingled again. They drifted across Grand Teton and made forays over the invisible park boundary into the adjacent Bridger-Teton National Forest. One of those strolls proved fateful.

DITCH CREEK IS A SMALL STREAM that rises in the national forest and runs westward into Grand Teton National Park, where it meets the Snake River. On September 19, 2009, just outside the park boundary, a 40-year-old man named Stephen Westmoreland from Teton Village was helping to carry out meat from a mule deer harvested by a friend in the national forest flanking Ditch Creek. Westmoreland had blood on his clothing from field dressing the animal. Despite advisements from the Wyoming Game and Fish Department to carry bear spray since he was hiking in an area known to have grizzlies, Westmoreland didn't. Instead, he had a .270-caliber rifle and a .44-caliber pistol. He told authorities he had passed a dead moose as he went into the backcountry.

Making his way out of the backcountry around 11:00 a.m., Westmoreland told authorities that a grizzly, hidden in the grass and about 40 yards away, suddenly stood up and looked toward him. He didn't know that 615 was then feasting on the moose remains left behind by a hunter. On the far side of the carcass, the bear lowered to four legs, not revealing her intent.

Westmoreland, who said he was afraid and who was also in possession of a black bear tag, claimed the bear could have closed in on him in a matter of seconds. He raised his rifle and put a bullet through 615's chest. As she flailed, he fired another that struck the left

610, as her mother before her, wears a research collar and ear tags. Members of her clan have been trapped multiple times. The technique of tracking bears with radio collars was pioneered by brothers John and Frank Craighead. Around 800 bears over the years have been fitted with collars by the Interagency Grizzly Bear Study Team.

side of her abdomen. She dropped and died five feet from the moose. Westmoreland was ultimately charged with illegally killing a grizzly. Although he claimed self-defense, he was convicted by a six-person jury of peers. He could have served a year in jail and been forced to pay a $10,000 penalty, but Ninth Circuit Court Judge Tim Day fined him $500 and gave him no jail time.

Jackie Skaggs, a Grand Teton spokeswoman, told a reporter in the wake of the shooting and after genetic tests confirmed that 615 was the daughter of 399 that her lineage "doesn't make her more or less than any other bear in the region." In an interview with writer Cory Hatch, then with the *Jackson Hole News & Guide*, she added, "It's unfortunate that we've lost such a wonderful bear out of this population. [But] this particular family and this particular female are not the only [bears] in Grand Teton National Park."

But individual bears carrying good genes and savvy instincts matter. The following year 615's sister, 610, would birth her first litter of cubs. Would 615 have done the same? When a bear kills a cow, one of the assertions some ranchers make, and demand compensation for, is the value loss of breeding potential. Why is lost potential not factored in when a human kills a bear?

Mangelsen believes 615's killing is a harbinger of what will happen to grizzlies that have no federal protection and are treated as just another wildlife quarry to be taken. Individual animals that have inspired millions of people, fueled admiration for grizzlies, been the beneficiaries of millions of tax dollars spent to recover them, and functioned as living magnets for the tourism economy will be regarded as just another commodity. He bristles at the notion that the value of a grizzly could be discounted to the price of a hunting tag, and that 399, her descendants, or any grizzly could end up as a prize splayed across the floor of someone's den.

"Even federal and state wildlife managers acknowledge that 399 and those in her bloodline are remarkable bears. She has been a poster child, if you want to call it that, for her species. I think that she and those following in her tracks deserve to have more respect and accommodation from the wildlife professionals that hold her fate in their hands," Mangelsen said at the time.

As 615's death showed, it only takes one thoughtless action and 399 could be gone. "What you lose is not only a wild bear, the likes of which we haven't seen in the history of Grand Teton National Park, but a mother who is exhibiting good family values, giving her offspring the same skills and instincts she possesses," Mangelsen said. "If you eliminate her, you eliminate a line of bears that know how to walk the fine line of coexistence with people. It's bears like her that are important to the survival of the species."

That fall, the killing of 615 was the talk of Jackson Hole and it set off social-media chatter around the world. Mangelsen, Cedarholm, and friends continued to scout for the survivors. By the time the first snowstorms arrived and another hunting season had come and gone, there was a pause at this realization brought by news from the field between Thanksgiving and Christmas: 399 and two of her subadult cubs—610 and her brother 587—were still alive but separated, each observed trying to take in as much food as possible before turning into their respective dens out in the hills of Pilgrim Creek.

HOW MUCH IS THE NEGATIVE frontier reputation of grizzlies really warranted? Jim Cole was badly mauled by a grizzly—twice—in two separate incidents years apart involving two different bears and two crown jewel national parks. During the summer of 2010, we talked about both attacks on the front porch of my home in Bozeman, Montana. He signed a copy of his book, *Blindsided: Surviving a Grizzly Attack and Still Loving the Great Bear*, to my family. Little did I know that only a few weeks later he would suffer an untimely death from causes that might have been related to his injuries. Cole was a wildlife photographer, a man whom some said got too close to grizzlies. But though they nearly killed him, he never held a grudge against the bears, saying he was at fault.

The first mauling happened on September 29, 1993, in Glacier National Park. He and a friend were on a long day hike. A bear that they surprised tore open his scalp and bit into his wrist. It fled after humbling him. He walked out and recovered.

The second, far more serious mishap occurred on May 23, 2007, in Yellowstone's Hayden Valley when he was hiking alone near the swerving contours of Trout Creek. Coming over a rise, he almost literally stepped on a sow and her cubs. He was thrown to the ground and pinned. The enraged mother could have dispatched him easily by putting his head in her mouth. Instead, she raked his face with her claws, badly disfiguring him and leaving him blind in the left eye. As we chatted together, his eye patch was a reminder. Cole had been bluff charged by other grizzlies too and had repelled bruins with bear spray.

A bittersweet memory and likely the last photograph taken of this bear: 615, a daughter of 399 in perfect health, feeds on a moose carcass near the Triangle X Dude Ranch. A short while later, she was tragically shot and killed by a hunter. The man was charged with illegally killing a grizzly. His fine: $500.

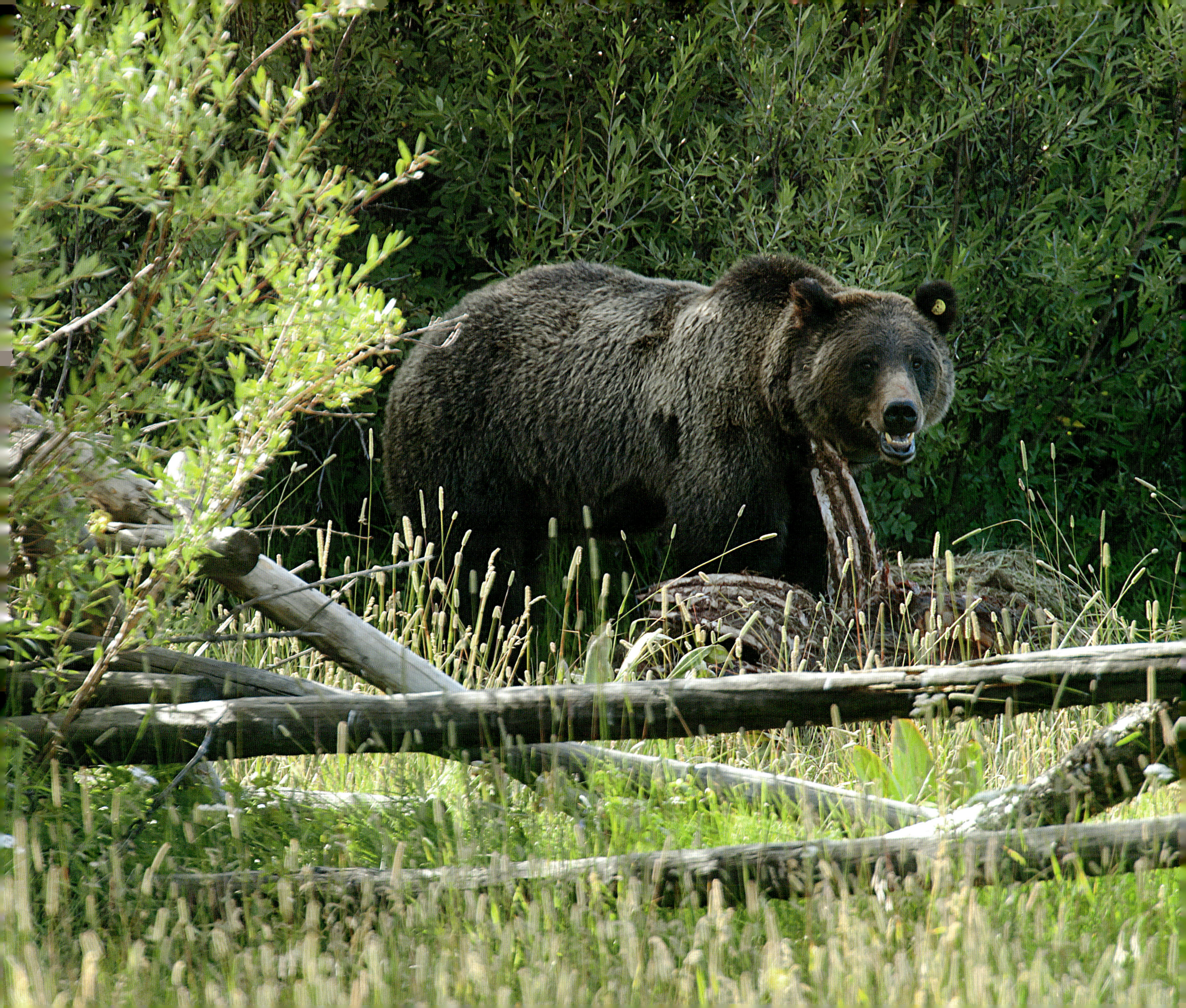

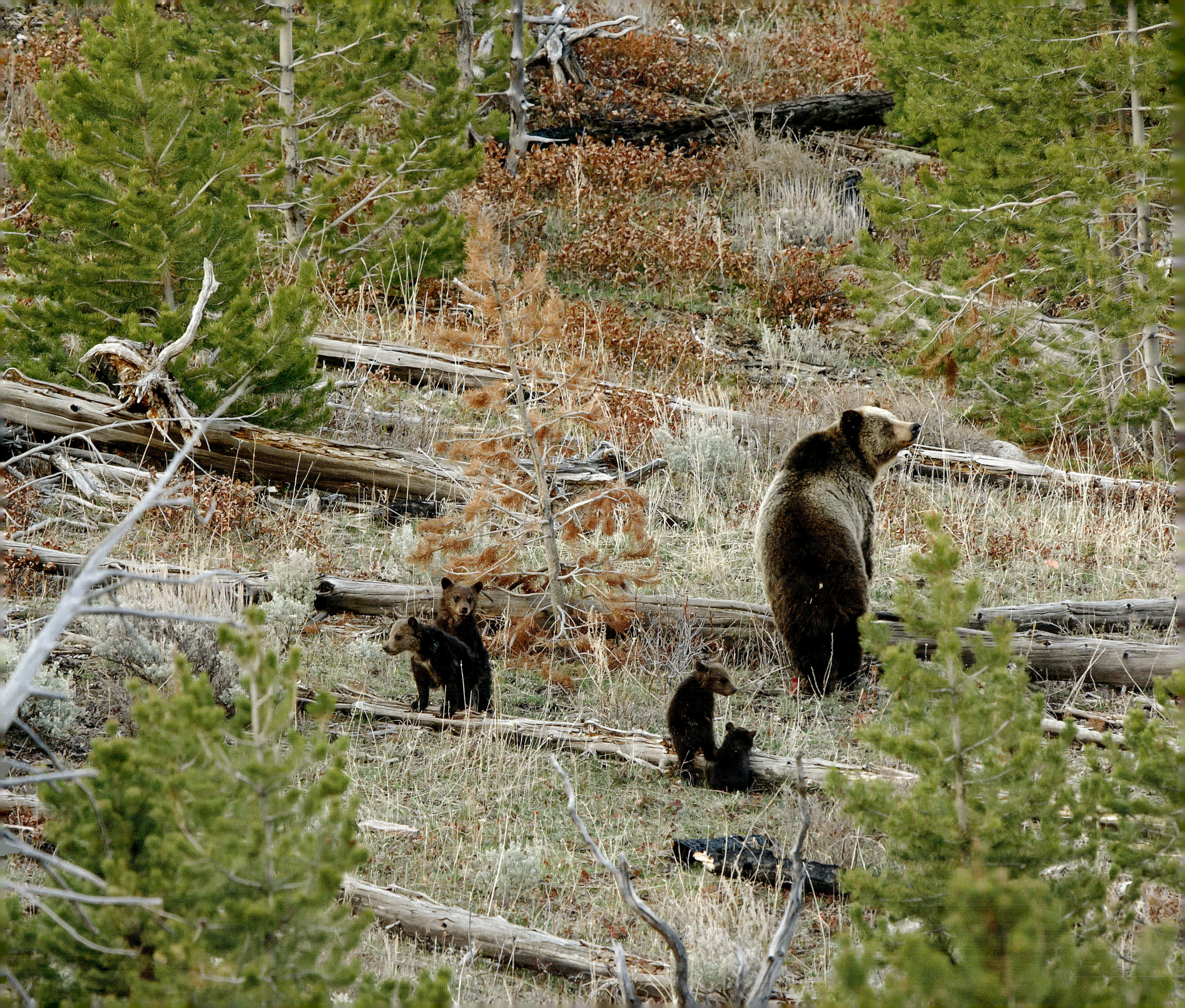

He knew he was a living anomaly. On book and lecture tours, the transplanted Chicagoan who lived in Colorado before resettling in Bozeman, always implored his audiences to respect the space of bears and lean on their elected officials to support bear conservation. "Grizzlies have few choices about where they can live," he said.

In June 2010, Erwin Evert, a 70-year-old botanist who had retired to a good life in Greater Yellowstone, set out for a walk from his home near Kitty Creek west of Cody, Wyoming, and just outside of Yellowstone. What he didn't know was that trappers with the grizzly study team had snared a grizzly, tranquilized it, took its measurements, abandoned it on the ground to awaken by itself, and then left. As the bleary-eyed bruin was coming out of its sedation, Evert was unwittingly close by. He couldn't have imagined what happened next. The bear sensed him and attacked, killing him. Attorneys representing Evert's survivors claimed there was negligence and that more care should have been taken for a bear confused and coming out of anesthesia. An investigation cleared the trappers of any wrongdoing and a federal judge dismissed the lawsuit. Evert was not carrying bear spray. His surviving relatives said he loved coming out West from Illinois to be in a place that was untamed.

In early September 2014, Adam Stewart, age 31, was conducting fieldwork on a Forest Service contract and hiking alone into Cub Creek north of Togwotee Pass in Bridger-Teton National Forest. He was found dead after he didn't return when expected and a coroner determined he perished from blunt force trauma on his head from a bear bite. A grizzly is suspected of fatally mauling Stewart, and then it and a black bear fed on him.

No bear spray was found among Stewart's belongings. Bear spray works, according to some research, more reliably than bullets. No member of the grizzly study team has died from a bear encounter. They work in close quarters with grizzlies and bear spray has always repelled aggressive bears, thus saving the lives of both people and animals.

Hiking in grizzly country, living in it, comes with the possibility that something bad could happen, but the likelihood is incredibly low. Yellowstone National Park officials often are asked just how low, and this is their explanation: Over the last 30 years, more than 90 million people visited the national park, and during that time 43 people were injured by bears. For park visitors, the chances of being injured by a bear are approximately 1 in 2.1 million. As of this writing, and going back to 1872 when Yellowstone was created, seven people have been killed by bears in the park. "More people in the park have died from drowning, burns (after falling into thermal pools), and suicide than have been killed by bears," a park synopsis of risk states. "To put it in perspective, the probability of being killed by a bear in the park is only slightly higher than the probability of being struck and killed by lightning."

Photographers, the good ones anyway, have a sixth sense for identifying the profundity of moments. Steven Fuller's camera has helped him reflect on these risks. The landscape photographer has lived in the center of Yellowstone for more than 40 years and is the winterkeeper at Canyon Village. Grizzly recovery has happened all around him. I was once visiting him at his historic old cottage above the Grand Canyon of the Yellowstone. As we were exiting his back door, an adult grizzly that we didn't know was there sauntered past us, literally at arm's length. The bear's only acknowledgment was a quiver of his ear, as if to say, "Hello, nice night, isn't it? If I wanted to, well, you know, but I have no interest in messing with you."

Fuller, while riding his horse alone in Hayden Valley not far from where Cole was mauled, once had his steed rear and roll over backward on top of him when a startled grizzly, sleeping in a daybed, jumped up and roared. As Fuller lay on the ground, beaten up but not seriously injured, the bear disappeared. Fuller has probably hoofed more miles in Hayden Valley than any civilian. He raised his now-grown daughters traipsing through the dell without any scary incidents.

"On skis in late March or April, I find bears to be a great stimulus for getting your head out of your ass and fine-tuning the awareness of your surroundings. I love having the bears here, but I don't always feel comfortable. There have been times when I just had a strange feeling and turned around and went home. Sometimes in early spring it feels like you're traversing through a massive traffic interchange of griz tracks out in the Hayden," he said.

When Fuller is on his annual trip to Namibia's Skeleton Coast and sleeping out among lions, he surrenders himself to the possibility of being prey. "Bottom line is there are so few of these marvelous terrifying beasts left on the planet. However inconvenient, uncomfortable, and possibly terminal it might be, I accept the prospect that my needs for safety are subordinate. I accept the risk in order to support them in the twilight of their

Nicknamed Quad-mom, a female grizzly forages below Dunsen Peak as her four cubs of the year play among the downed timber.

existence. In the long term, unless humankind changes its conquest of nature, all of these species are doomed." Fuller has huge admiration for Mangelsen and his work. They share the same conservation ethic, the need to want to do something to thwart the inevitability of species extinctions.

GRIZZLY 399 EMERGED FROM HER DEN the next spring, in 2010, and made an appearance at the mouth of Pilgrim Creek before wandering on and again frequenting the roadsides of Grand Teton National Park for reasons of survival. Then, abruptly, she vanished and no one knew where she was. Had she perished?

That's the challenge of trying to keep track of nomadic charismatic megafauna like grizzlies; they don't always go where you want them to or where you expect. After weeks of wondering, Mangelsen finally caught a glimpse of her again when she walked out of the eastern forest. Her daughter 610, now a full-fledged young adult and on her own, came into view too. The fall of 2010 yielded grizzly photographs for Mangelsen that ranked among the finest of any in his esteemed portfolio and they emanated from a year mellower than others preceding it. For a weeklong span in October, he was treated to a convergence of spectacular atmospheric conditions, great light, and the cooperation of some incredibly photogenic ursids.

Some 340 wapiti were killed in the Grand Teton elk reduction that year. The dispersed gut piles had bears roaming everywhere. Kate Wilmot, who oversees the park Wildlife Brigade, said there were 60 grizzlies that had primary or overlapping home ranges in Jackson Hole. 399 was gorging on gut piles and carrion in the Snake River bottoms. On October 12, Mangelsen set up well before sunrise just north of Oxbow Bend, and just as dawn was casting alpenglow on the top of the Tetons, elk came out of the forest and waded their way across the river. The result was the moody and memorable *Snake River Crossing*.

A day later, Mangelsen and Sue Cedarholm, after surveying several panoramas with binoculars and spotting scopes, dialed in on a grizzly—399—feeding on an elk carcass right next to the river around 7:00 a.m. At noon, the bruin dragged its portable larder up the bank. Stuffed and in need of a nap, the bear stretched out and slept with the carcass into the evening, yielding again some amazing and even hilarious imagery. During the night, she woke up and, deciding to be more discreet, dragged the find into the trees.

This excitement was a mere prelude. On October 14, Mangelsen was poised and ready for magic to happen. Shortly before 7:30 a.m., 399, just waking, wobbled out of the willows and chose the river fording just downriver from the elk in *Snake River Crossing*. The resulting picture, *First Light—Grizzly Bear*, had the power of a meditation and became the cover of Mangelsen's critically acclaimed 2014 book, *The Last Great Wild Places*, a collection of his best work. Light aligned, the bear moved broadside, and Mangelsen's equipment was ready. It happened in an instant. Just past 8:00 a.m., 399 hauled what remained of the elk away and buried it. The next day, Mangelsen captured the sow sprawled on the mound where the carcass was buried, protecting her prize from marauding magpies. On October 16, the carcass cleaned like barbecued ribs nibbled down to the bone, 399 walked to the shore, took a drink, and sauntered upriver. As she disappeared, a coyote, magpies, and ravens descended to pick at the scraps.

The Grand Teton elk hunt has altered bear behavior. Taking advantage of the gut piles, 399 and her daughter 610 have stayed out longer and denned later, sometimes six weeks past the time when the bears in Yellowstone, by comparison, go underground. Ecologist Steve Cain revealed to a reporter that 90 percent of bears in Jackson Hole den by the end of November. In 399's case, she typically has not retreated until mid-December.

THE EMINENT ENGLISH CHIMPANZEE researcher and ethologist Dr. Jane Goodall has said that exceptional beings thrive in wild settings. She and Mangelsen first bonded when Goodall flew into Jackson Hole to deliver a talk about the human obligation of stewarding nature. The photographer took her on an outing—the first in her life—to watch wild grizzlies in Yellowstone. Goodall was enchanted and reminded of how precious *U. arctos horribilis* is.

Mangelsen has said it's impossible to overstate Goodall's influence on him. In Africa, she has seen the aftermath of poachers and bushmeat hunters mowing down families of chimpanzees and gorillas. They too are beings that are slow to reproduce. Relegated to islands of habitat, wild chimps and gorillas are both in serious danger of winking out, the only survivors being found in zoos and rehabilitation centers. "Jane has taught me to never hold back in trying to protect the things you love, that if you act with a clear conscience, you shouldn't worry about the people you'll offend, because if you're giving voice

Few scenes better exude the majesty of both grizzly and terrain: first light strikes the summit of Mount Moran, painting the sky orange, as 399 wades a shallow bend in the Snake River, returning to her elk carcass.
FOLLOWING SPREAD: While feeding on an elk, which she buried for safekeeping, 399 swats away scavengers, in this case a swarm of magpies.

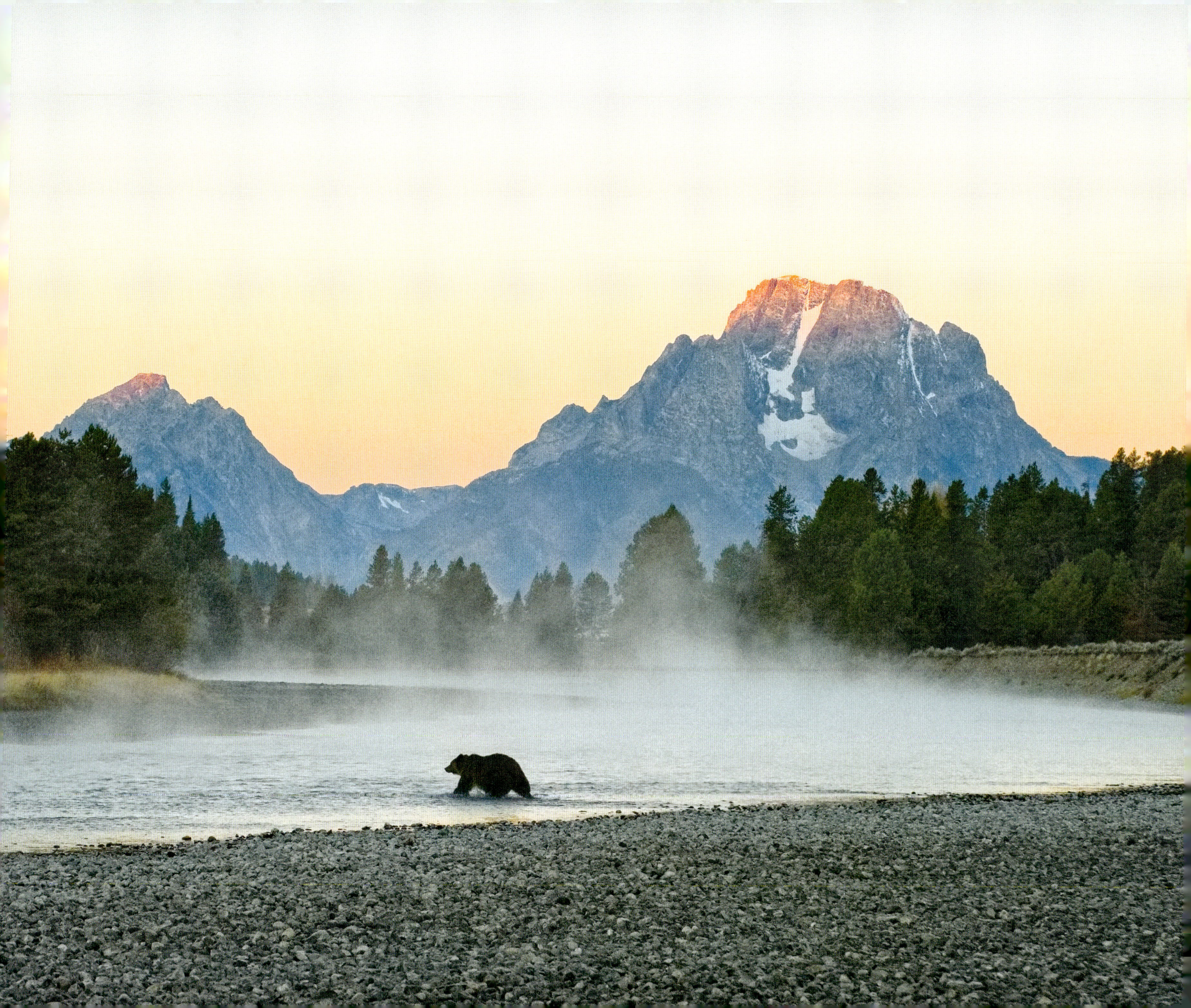

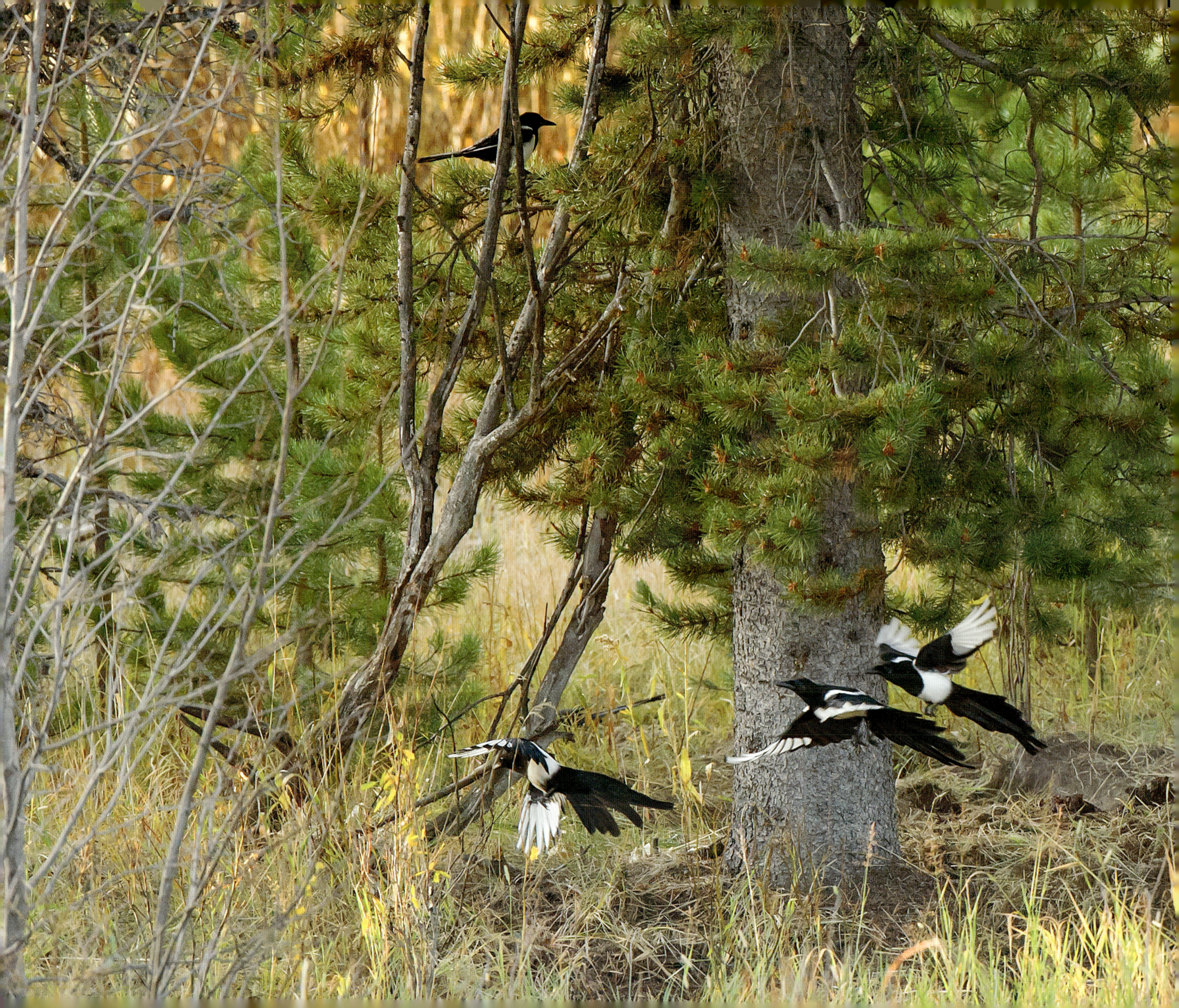

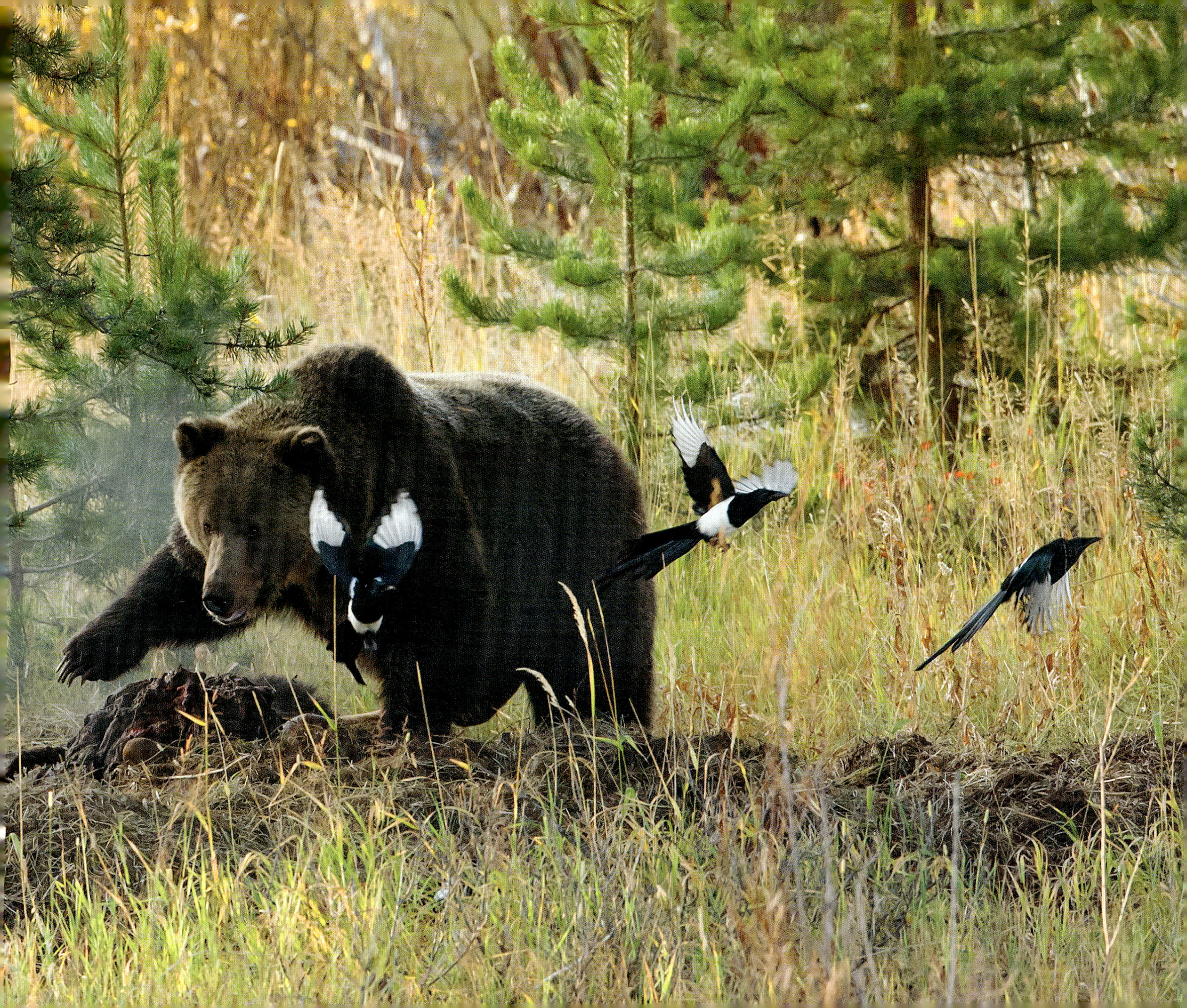

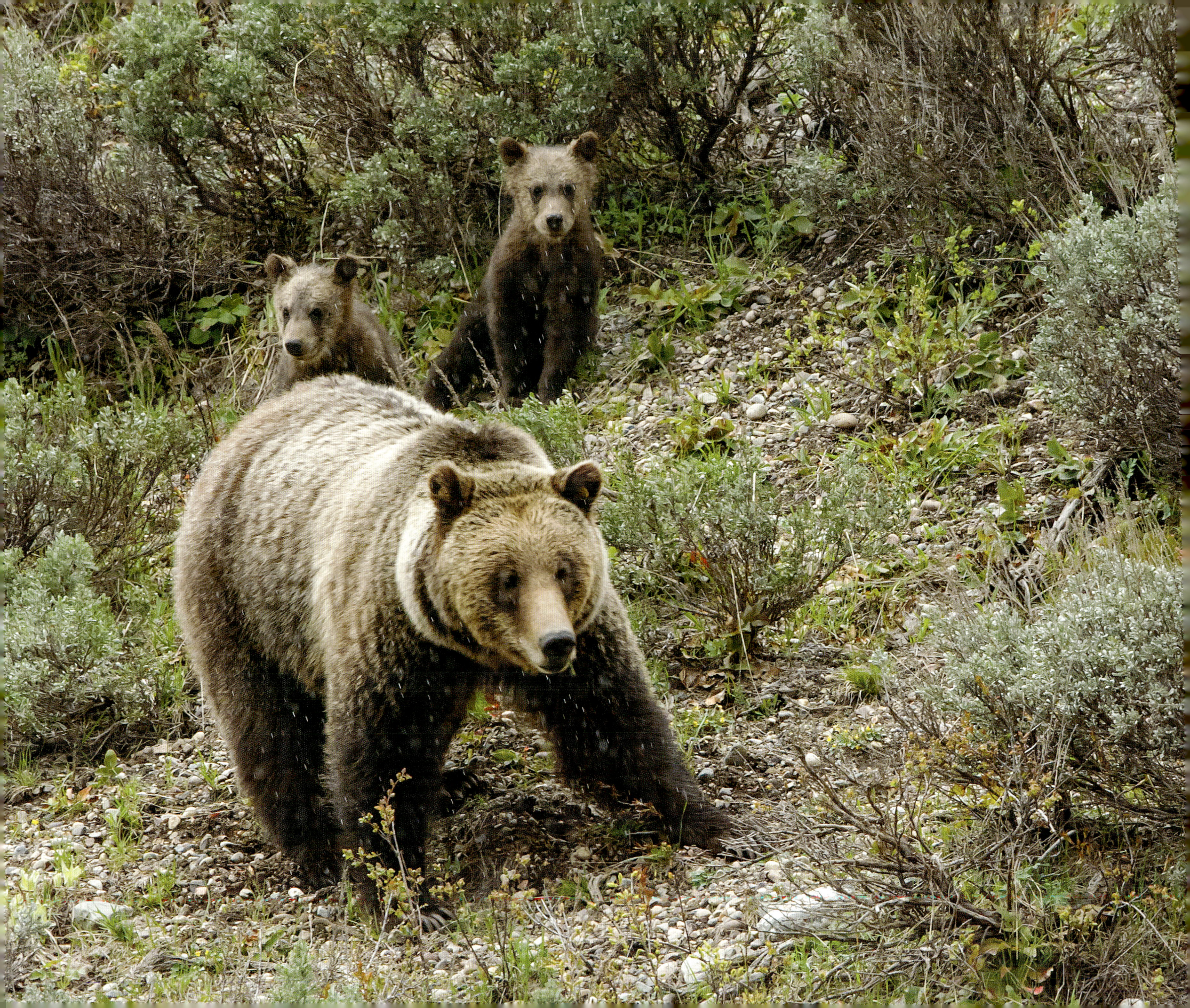

to creatures that can't advocate for themselves, your priority should be defending them and not trying to please those who just don't get it," Mangelsen said.

American essayist, Kentucky farmer, and poet Wendell Berry pondered this thought in a letter to our mutual Montana friend, Robert Staffanson. "Now that I have you on my mind," he wrote, "I have a question for you. For quite a while now I have been thinking that our use of the word 'wild' is most often used wrongly. From my observation of the so-called 'wild' creatures, I conclude that they are all going about lives for which the only accurate term is 'domestic.' That is to say that they are making homes, raising young, carrying on a kind of economic life, and, far more than most people suspect, enjoying themselves. And so I ask you: Do the Indians of your acquaintance think of the native creatures of a country as 'wild'?"

It's a rumination that should be applied to grizzlies. Most indigenous people I know regard bears and every other species of wildlife as brothers and sisters. Marc Bekoff, an ethologist and ecologist emeritus at the University of Colorado Boulder, concurs that wildness suggests an otherness, a dividing line between the proverbial "us" and "them." Humans in our self-declared exceptionalism are part of the animal kingdom. We, like most higher organisms, have outwardly defining physical traits and hidden inner landscapes, the emotional kind. Sentience is what connects our mutual awareness.

Bears know joy, grief, humor, anger, jealousy, bewilderment, tenderness, fun—lots of "feelings"—Bekoff said, noting that it is much harder to malign another creature that possesses qualities in which you can find affinity, and acknowledging it is not resorting to anthropomorphism. Harming another animal, if done in the name of purely seeking pleasure, then becomes a matter of personal conscience. Often we are willing to destroy what we don't really understand. In the clan of 399, Mangelsen has witnessed expression of this emotional range identified by Bekoff.

IF YOU THINK THE 399 FAMILY tale could not be more heartrending, well, it became a soap opera in 2011. At 7:30 a.m. on May 25, Penny Maldonado, conservationist and postmaster in Moose, was on her way to work from her home in Moran. Passing by Signal Mountain near Catholic Bay she saw the outline of a bear with two tiny cubs along the edge of the pine forest, an area often frequented by black bears with cubs. She took a quick snapshot of the family with her point-and-shoot camera and then called her friend Mangelsen to tell him that she saw bears with brand-new cubs.

Mangelsen and Sue Cedarholm immediately drove to Signal Mountain and there on the road were parked several cars but no sign of the bears. They swung into the lodging area and circled through the cabins then went back to the highway. There, 50 feet high in a pine, were two wee cubs but no mother. A few minutes later, deeper into the forest, they spotted a sow and realized she wasn't a black bear but rather a grizzly with two yellow ear tags. They wondered who the bear was but soon realized it had to be 610 with *her* very first batch of cubs!

Wouldn't it be something, Mangelsen asked Cedarholm, if 399, who hadn't had any cubs the past two seasons, also emerged with her own maternity triumph? "I thought it would be frosting on the cake," he said.

For the first time in many months, five days after 610 emerged, 399 suddenly materialized near Colter Bay around Leeks Marina. Hounding shyly around her feet was *her* second brood of triplets. The enthusiasm of bear watchers was heightened at the idea of 399 and her young daughter, now a mother, foraging in their traditional and interwoven haunts. However, it appeared that 399 spent more time in the northern part of their home range and 610 in the southern. The presence of both mothers with their cubs drew bigger mobs of human onlookers than ever before. The group of seven made national and international news. Jane Goodall e-mailed Mangelsen to say she had even seen the story and photos in the *Daily Mail* in London.

Now ask yourself: What's a new mother's worst fear? In late evening on the first day Mangelsen and Cedarholm saw 610, she was casually eating grass and digging tubers and ground squirrels. As darkness fell, she was somewhere between the Potholes and Signal Mountain. She turned around to spy her cubs and they weren't there. "She loped frantically, retracing her steps through the sage and calling out. When she couldn't find them, she bellowed and cried out like any mom losing her kids would behave," Mangelsen said. "Not only was she upset, but all of us watching her were."

The following day, after a cold and rainy night, the cubs were nowhere to be found. All the chatter was about whether the cubs had been seen and a lot of theories were floated. Did the new mother 610 force her cubs up a spruce because of some perceived danger and

610, now five years old, guides her first cubs through the old haunts of Signal Mountain.

FOLLOWING SPREAD: 399 and her triplets, including Brownie and Ash, traverse a sagebrush meadow among wildflowers near Pilgrim Creek.

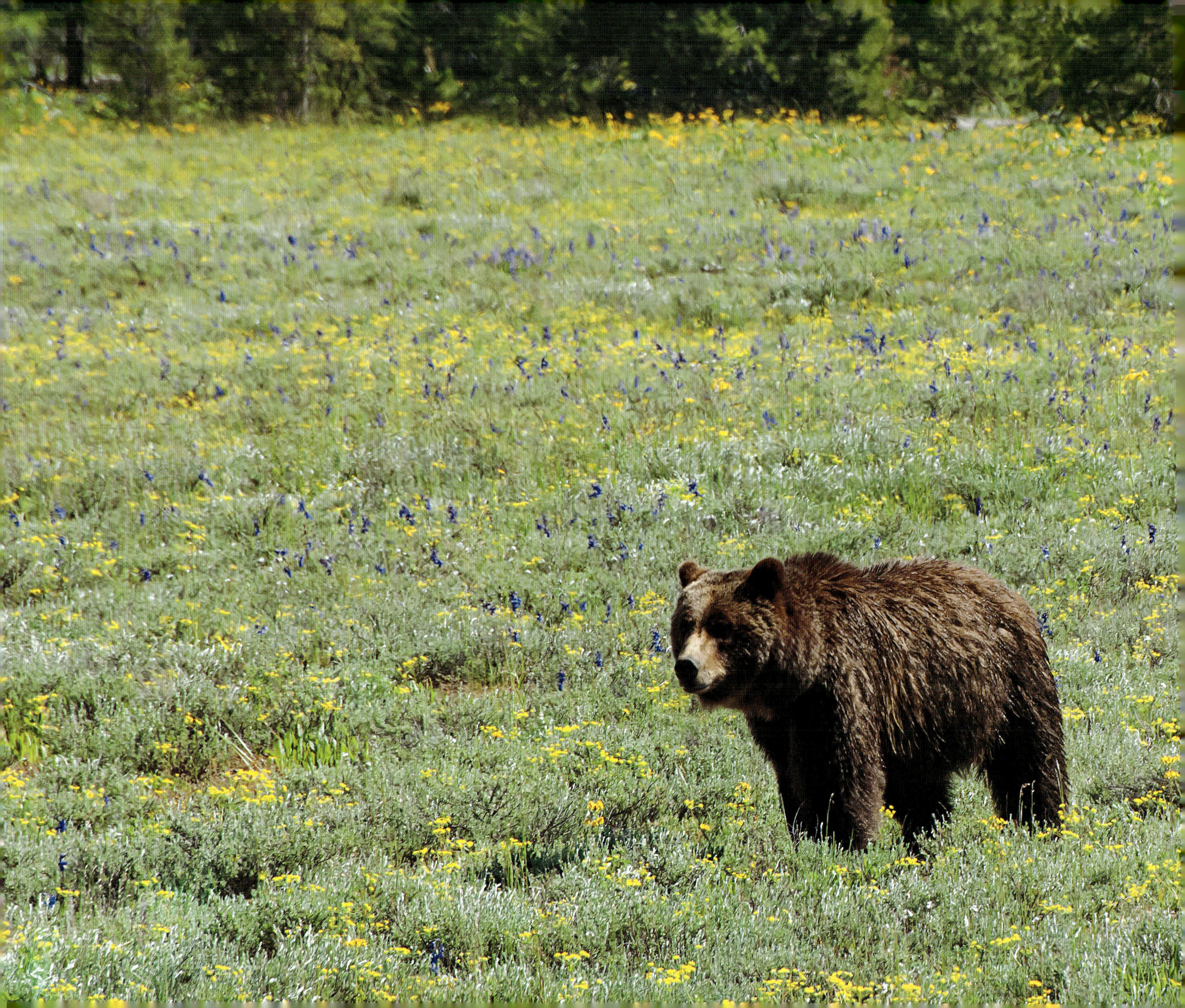

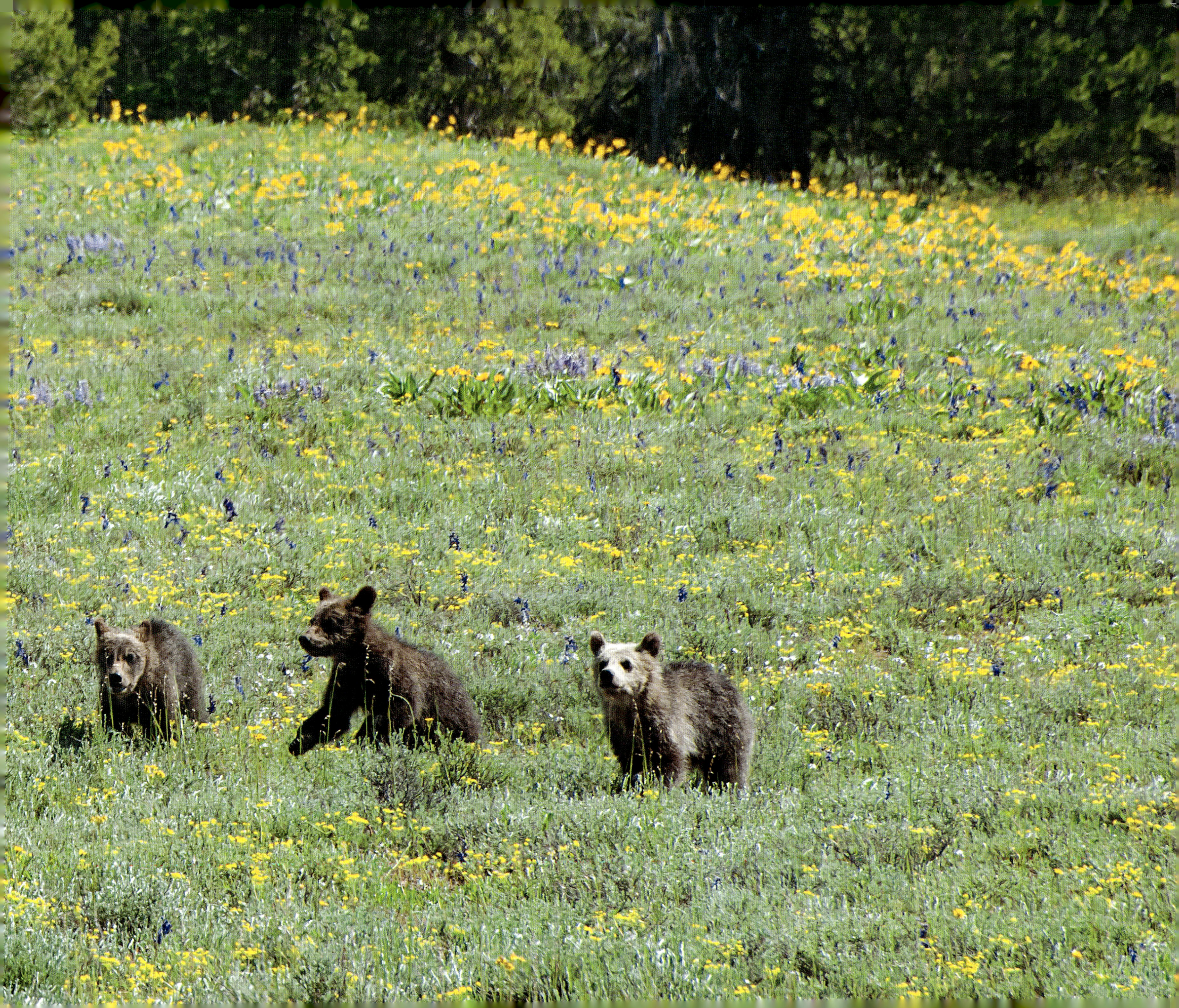

merely forget where she had stashed them? Or maybe they'd been killed by wolves in the area or injured by a boar or drowned in the lake or hit by a car and left somewhere.

After 72 hours, Mangelsen, along with friends Bob Smith and Jim Laybourn, made a last attempt to find 610's cubs. Most people had given up hope that the cubs would survive the freezing temperatures as rain turned to snow. Preparing to have lunch near the last known sighting, luck would have it that Mangelsen noticed a couple of cars stopped along the highway not far away. After learning that they had witnessed a bear crossing the road with two cubs, he and his friends grabbed their binoculars and spotting scopes and set off toward the island of spruce where the tourists had last seen them headed.

Below a ridge in the distance, among the jumbled timber through the spotting scope, Laybourn saw 610 nursing her cubs. "It was pure joy to see them and relief swept over us all," Mangelsen said, noting that he immediately called Chris Flaherty, district park ranger, who had known 399 longer than anyone and was one of her greatest advocates, to let him know that 610 and her cubs were alive and well. "She nursed them for a long time as we watched from the top of the hill then went toward the fresh green grasses along the highway where she and her cubs fed until dark."

Throughout June of 2011, Mangelsen had a bonanza with both mothers and cubs. The bear jams were epic. It was the pandemonium that originally surrounded 399 times seven. The two families took turns stalking elk calves in Willow Flats. "399 was good but 610 had surpassed her mother in her hunting skills. Both mothers taught their young the nuances of elk hunting," Mangelsen said. "At first, the elk paid little attention to the cubs bounding around while their mother laid in the grass. Sometimes the elk would even move closer to investigate the bear family, possibly to keep the bears within their sight. The bears knew what they were doing and before long the elk mothers understood the gravity of it."

The hunting was good. And then still another extraordinary event in bear interaction played out before the eyes of wildlife watchers. One of 399's three cubs was adopted by 610, giving her a trio. Mangelsen was among the first to document it. Bear researchers were intrigued. The catalyst for it might have been an altercation with a male bear that scrambled up the families. While rare, cub adoption by a different mother happens in the bear world.

In response to the massive crowds that had formed along the roads, Grand Teton National Park officials changed the viewing rules to try and ensure a buffer of space

In late evening light, 610 nurses her cubs.

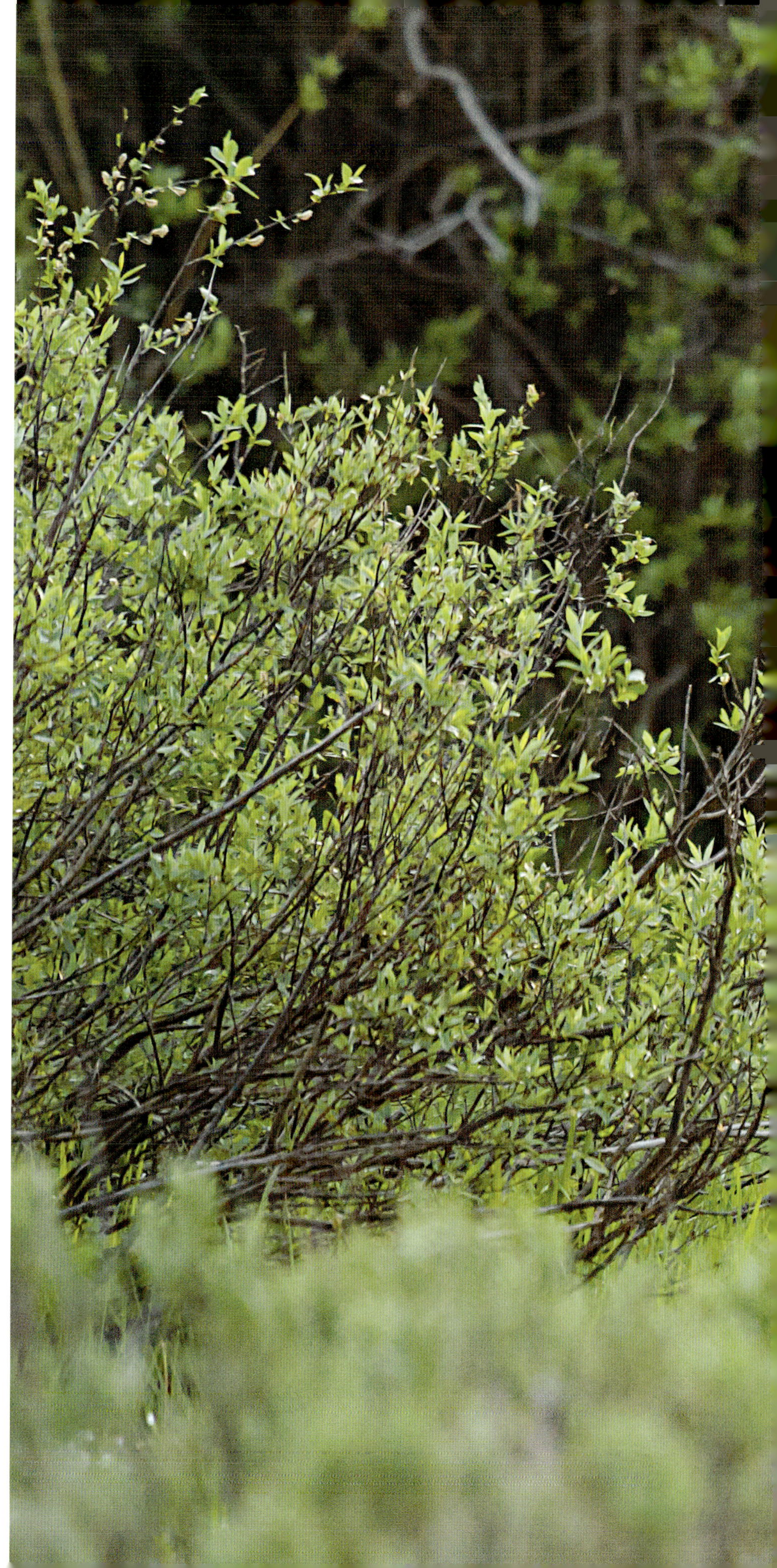

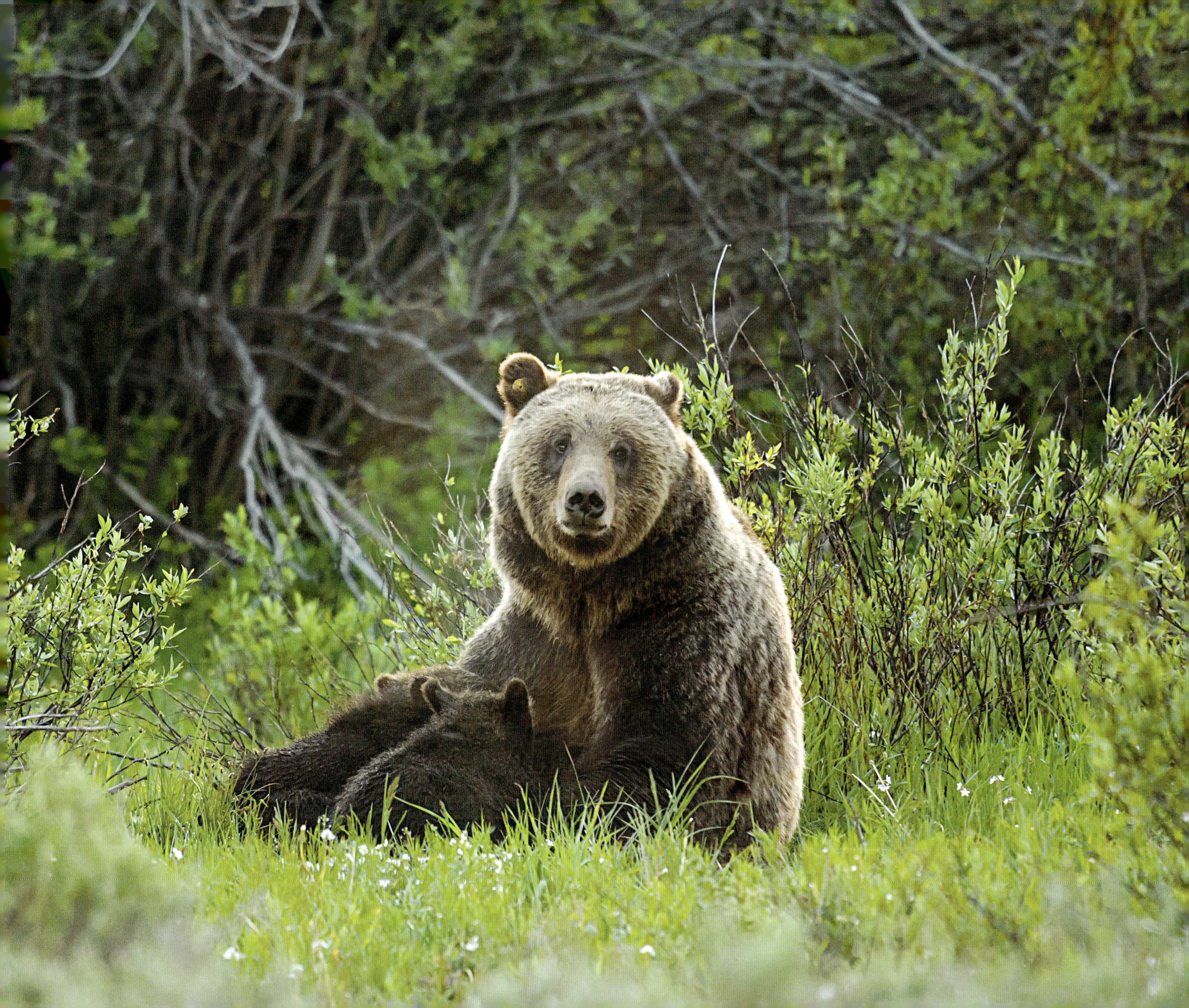

remained between bruins and onlookers. By the end of summer, Mangelsen noticed that 610 was starting to look haggard. Being a new mother can be stressful, especially with an extra cub. She was looking extraordinarily lean. "The energy that she was expending looking after her cubs and feeding them wasn't being offset by enough calories. We hoped for a good berry crop and that she'd somehow make up the deficit before she denned," said Mangelsen.

Grand Teton National Park and the state of Wyoming announced that the park would offer 710 elk permits, meaning the woods would be crawling again with hunters come October. Mama 610 drifted southwest across the park toward a corridor flanking Moose-Wilson Road, pressed up tight along the slopes of the Tetons. It was a setting rich with chokecherries and hawthorns and, best of all, the bears were on the west side of the river, an area closed to elk hunting.

Within weeks, both families were scrounging around gut piles that peppered the park from the Snake River east. There were a number of close calls. In late November, a Jackson Hole wild bison was shot by a hunter just outside the park boundary and the wounded animal staggered into the national park and died, prompting a fight to erupt between 399 and 610. They snarled at each other and nipped and took swipes with their claws and pushed, but nothing with real intent to injure.

"They were both hungry, they both wanted dibs and they had other mouths to fill," Mangelsen said. "Believe me, I've seen human families behave worse at dinnertime." All of the cubs interacted and 610 crept forward as if awaiting the sign that it was OK to dine. She was respectful to her mother's authority. Occasionally 399 would just get angry, lower her front frame, and do bluff charges. And 610 would do the same back. Even the cubs started scrapping.

"They weren't rivals. It was tough love," Mangelsen said. "I've never seen anything like it." Up in Yellowstone, park wolf biologist Doug Smith said that bear and wolf conflicts over carcasses are common. If a bison or elk is killed, it won't be long before both species converge, with bears bumping wolves off kills 80 percent of the time. Typically, they both feast but grizzlies eat first.

LATE IN THE YEAR, with the new year approaching, 399 and her cubs were still out, mother leapfrogging from carcass to carcass for the free meals and cubs tagging along behind. When a brutal cold snap descended, accompanied by a huge snow dump, many worried about their exposure to the elements. Mangelsen and Cedarholm noticed that 399 was hunkered down in the vicinity of a dozen elk carcasses scattered along the eastern flanks of Blacktail Butte in Grand Teton National Park. The carcasses were frozen solid and tough to penetrate.

As she stuffed her face, something else was on Mangelsen's mind. Dodging bullets intended for elk and unwittingly avoiding confrontations with hunters dragging fresh meat across the sagebrush are serious perils, but they aren't the only ones associated with the Grand Teton hunt. Bears can get filled full of lead in another way, by eating the gut piles and dead animals mortally wounded by hunters but never found and tagged. For humans, ingesting lead—whether through paint particles, dust, or other exposure—has been a public-health issue for years, especially for kids. Lead that gets in the blood can impair brain function, and if levels are high enough it can kill. Concerns about lead sickening millions of waterfowl and raptors like bald eagles and hawks have prompted the banning of lead fishing sinkers and BBs in shotgun shells across the country.

Jackson Hole's autumn killing fields, with huge numbers of elk being felled in the national park, national forest, and National Elk Refuge, attracted the attention of research outfit Craighead Beringia South, founded by Derek Craighead, son of famed bear researcher John Craighead. Craighead, his colleague Bryan Bedrosian, and researchers at the University of Montana wanted to know how lead ammunition in elk remains—being scavenged by eagles, ravens, coyotes, wolves, and, yes, grizzly bears—might be impacting the animals' health. It followed analysis of bear blood and tissues that showed elevated lead levels. Grizzlies inside Yellowstone had lower readings than bruins outside the park. Some birds that were tested had amounts considered lethal.

Greater Yellowstone-wide, researchers estimated that hunters discharged 500 tons of lead ammo annually. In gut piles examined in Jackson Hole, researchers found an average of 160 lead fragments. The study found no conclusive link between lead eaten in gut piles and elevated levels in Jackson Hole bears. Still, the absence of evidence, some said, did not guarantee an absence of impacts, especially for female bears stuffing their bellies on elk gut piles. In 2010, Grand Teton National Park hunters had been asked to voluntarily switch to nonlead bullets provided for free by Craighead Beringia South, with roughly half

Seen here in the middle of the photo, both bear families—399, 610, and their five cubs—share a bison carcass felled by hunters just outside the Grand Teton boundary. The wounded buffalo wandered into the park and died. Just above the hill, members of the Jackson Hole bison herd graze unbothered.

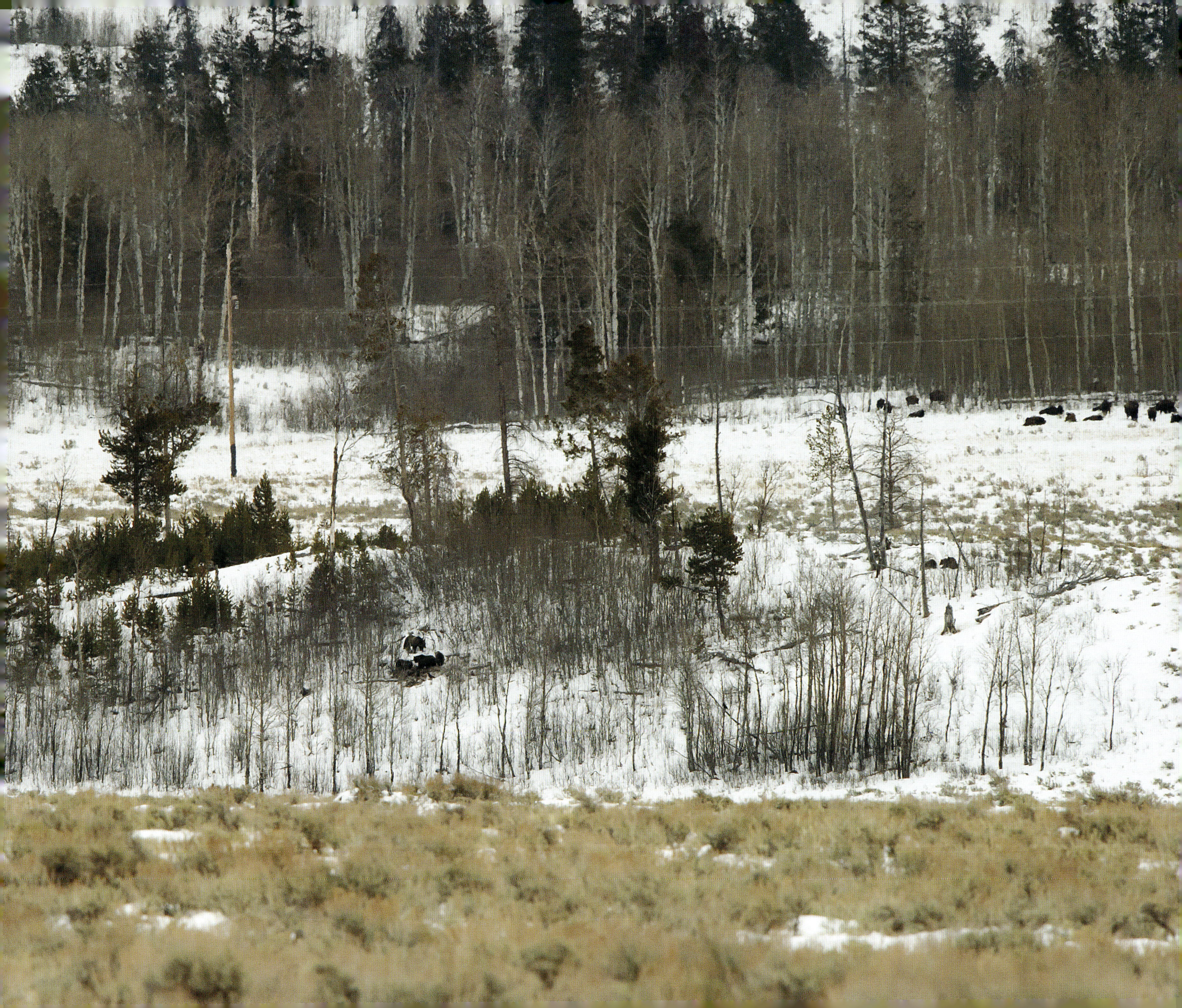

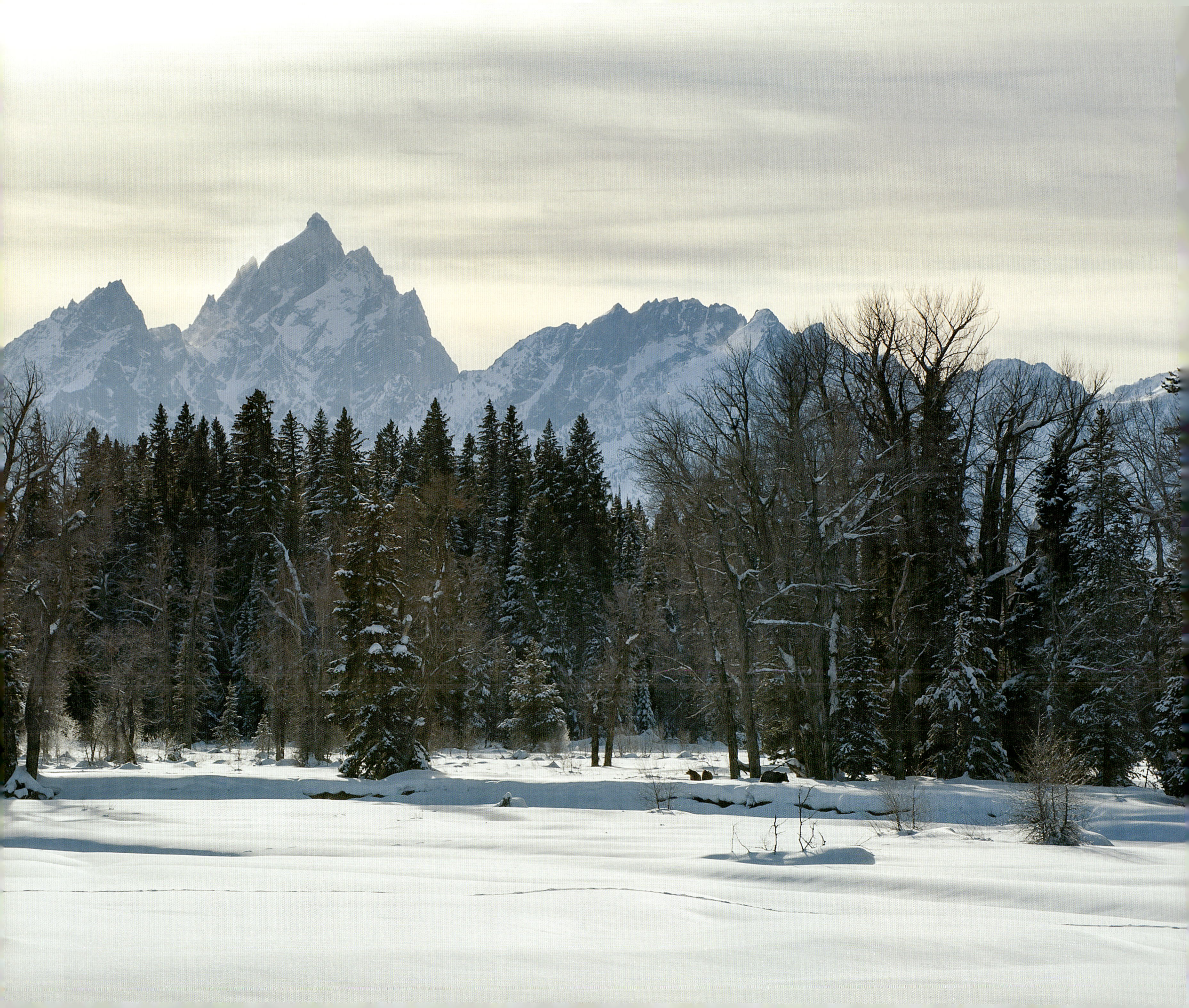

of the hunters doing so. In 2013, the park decided to ban lead bullets from use altogether during the hunt. Unfortunately, the Forest Service has not followed suit in all the national forests in the Greater Yellowstone region, even though the switch to nonlead ammo would not be much more costly to hunters.

Interestingly, it was Mangelsen's concern about this that led him to a discovery. On New Year's Eve 2011, he, Cedarholm, and a group of other photographers were out in search of 399. They captured a dramatic sequence of frames as the tribe post-holed across deep snow dunes, the cubs flopping and flailing like ocean seals on land. After the New Year dawned, the bears were nowhere to be found. Blacktail Butte was a long, long way from where they'd ordinarily be in the safe confines of the den, slumbering.

Concerned, Mangelsen enlisted a pilot. "Mostly I was curious. The inner park roads were closed, as were other seasonal roads, so where we could go was pretty limited," he said. Holding binoculars, they scanned the foot of Blacktail. They were able to decipher a furrowed line of tracks.

From hundreds of feet above the ground, they followed the tracks as they zagged away from the butte, where there was 18 inches of only windblown snow, down to the river bottom, across the Snake River Bridge at Moose to Windy Point, north to Baseline Flat over toward Schwabachers Landing but staying on the west side of the river, trending northeasterly near the vicinity of Deadman's Bar, still holding fairly tight to the river corridor, crossing the road at Jackson Lake Dam, and then through 399's old hunting area of Willow Flats. She had escorted her cubs through the closed-down complex of Jackson Lake Lodge, picking up the ice-covered surface of Pilgrim Creek, and tracing it into the mountains. "It was pretty incredible to see the trek she made in a fairly short amount of time, navigating through two and a half feet of snow," Mangelsen said.

As the tracks ventured higher, gaining half a mile in elevation, the pile of powder deepened considerably to three or four feet. Way up in the high country, the tracks turned into a stand of old spruce and didn't come out. The date was January 4, 2012, the latest date on record that a bear had gone into hibernation. "From Blacktail to that spot it had been at least 30 miles as the crow flies, and far more as the bear walks," Mangelsen said. "She blazed the trail with cubs. Quite the haul, but she covered it in a day and a half." Mangelsen breathed a sigh of relief.

THE DILEMMA OVER GUT PILES is but a microcosm of larger food issues confronting grizzlies. Climate change already is exerting its effects in the northern Rockies. Just talk to any ecologist. How warmer, shorter winters, hotter summers, shifting growing seasons, and likely much different precipitation patterns will alter the lives of bears is still only guesswork.

The consequences of availability and abundance of grizzly food staples are obviously the seminal concern. Yellowstone's chief bear biologist Kerry Gunther and a few colleagues set out recently to document the foods making up a bear's diet to establish a baseline. Grizzlies are opportunistic omnivorous generalists. That's a turgid way of saying they eat meat and plants, wherever they can find either. Based on a wide range of reference points—including the contents of 11,478 bear scats collected over 37 years and fieldwork observations in Yellowstone, Grand Teton, and six surrounding national forests—the diversity of the smorgasbord filling the dinner plate of a grizzly is head spinning. Researchers found that grizzlies in the ecosystem consume at least 266 different species—175 plants, 37 invertebrates (insects), 34 mammals, seven fungi, four fish, one amphibian, algae, and dirt.

The most frequently consumed foods are ants (31 different species), whitebark pine seeds, clover, elk, thistle, graminoids (grasses), and horsetail. Confirming Mangelsen's field notes on the foraging habits of 399, Gunther's crew also mentioned yampa roots, vaccinium and other berries, cutthroat trout, dandelions, biscuitroot, sedges, brome grass, mushrooms (primarily false truffles), and numerous other kinds of flora and fauna. Showing up in some bear droppings was also meat from domestic dogs, livestock, and poultry. The bears seem to know what's healthy for them.

"When available within their home ranges, grizzly bears directed feeding activity toward foods with high gross energy content that could be efficiently foraged, such as army cutworm moths, ungulates, cutthroat trout, clover, small mammals, spring beauty, yampa, pine seeds, biscuitroot, bistort, and horsetail. When these foods were not readily available, bears consumed a wide variety of berries, other forbs, ants, and graminoids," Gunther and his colleagues wrote. Ants, a prolific go-to food, have low energetic value but are high in protein and provide amino acids not readily available in other foods. Chuck Schwartz, former head of the grizzly study team and collaborator on the food study, suggested that the

On January 2, 2012, 399 leads her two cubs, Ash and Brownie, through deep snow toward their winter den above Pilgrim Creek. It is the latest date ever recorded for Greater Yellowstone grizzlies headed for the den, likely because the presence of elk carcasses kept them out foraging on Blacktail Butte 30 miles to the south.

biomass of ants alone may exceed the biomass of all ungulates in the ecosystem. "There are a lot of calories out on the landscape," Frank Van Manen said. "Bears just have to find them. There's way more food than grizzlies could ever take advantage of. What restricts them are the social interactions—old big male bears denying females and cubs access to feeding areas."

However, think of all the food in a grocery store. Depending upon what we load up in a cart, the food can yield very different kinds of health outcomes, determining our quality of life and longevity.

ENTER NOW DAVID J. MATTSON, a retired government ecologist who spent much of his career doing research for the U.S. Geological Survey and the grizzly study team. He worked with Mark Haroldson in the old days. When Haroldson was trapping and tagging bears, Mattson was in the field as a botanist collecting data on whitebark pine and other grizzly foods. Mattson was the first to write about the discovery that bears were eating at moth sites in the 1980s.

After Mattson left Greater Yellowstone and got his PhD, he spearheaded a study of cougars in Flagstaff, Arizona, that focused on coexistence issues involving the big wildcats and people. To express my bias here, Mattson is one of the smartest, most insightful wildlife biologists I have ever met. Because his name appears on so many research papers that form the bedrock of Greater Yellowstone grizzly bear scientific literature—a reservoir that Chuck Schwartz, Mark Haroldson, and Frank Van Manen are a part of and which is world renowned—he has standing. When Mattson, who in recent years has been an adjunct professor at Yale, expresses worries about grizzly bear persistence, people pay attention, though he has been on the receiving end of attempts to discredit.

Mangelsen attended a lecture Mattson delivered in Jackson Hole that came in the form of a PowerPoint presentation on bear foods. No one else has so insightfully distilled the discussion of bear nutrition and its relationship to bear recovery down to a language lay people can understand. Very simply, four key groups of food have, more than any other, powered the climb in bear numbers after Yellowstone bruins were weaned off garbage: the seeds from whitebark pine trees that grow in the cold alpine nethers on high ridgetops; army cutworm moths; cutthroat trout spawning in the streamside tributaries of Yellowstone Lake, the second-largest higher-altitude lake in the world; and meat (acquired by killing elk, moose, and deer, or scavenging the carcasses of dead animals brought down by wolves, human hunters, or natural causes).

Pound for pound of nutritional caloric value, in terms of protein and fat, these foods have fueled a huge percentage of bears that dwelled in the essential core of the population and upon which the entire premise of recovery has been based. In particular, Greater Yellowstone grizzlies have a higher content of meat in their diet. It may explain the relentless zeal that bears in Jackson Hole have for eating elk.

"What's really cool is that the connection grizzlies here have to those four primary foods is like a specialized dialect spoken within a biological language, but it's pronounced with regard to meat. And it helps explain how bears relate to the landscape and I believe it helps us to telegraph possible scenarios in the future," Mattson said, noting that it's something worth recognizing, respecting, and celebrating as yet another extraordinary aspect of an already extraordinary bear population.

Mattson, who is writing a much-anticipated book about the natural history of grizzlies, looked throughout the Northern Hemisphere in brown bear range to find an analogous setting involving four foods. He found only one: the region around Russia's Lake Baikal, where brown bears focus on stone pine seeds (a cousin of whitebark pine), moose, pond weed, and red deer.

For a few decades, no single food had a stronger influence on more bears than the seeds of whitebark pine trees. In years when the seed crop was robust, bears prospered and females had higher litters. After poor years, more bears died. But the whitebark pine forest of Greater Yellowstone is today more imperiled than grizzlies were when they were first put under endangered species protection. The trees have been under attack by a fungus called blister rust and mountain pine beetles. Some have also burned in forest fires. Roughly eight of every 10 whitebark trees that existed 25 years ago have been killed, leaving whitebark forests decimated. Members of the grizzly study team have noted that whitebark pine trees have been devastated in Glacier National Park, too, but bears have fared well. The grizzlies of Greater Yellowstone, Mattson noted, have a more specialized diet.

In addition to whitebark seeds, cutthroat trout in Yellowstone Lake were another vital reservoir of sustenance. Thousands upon thousands of trout spawned, and bears

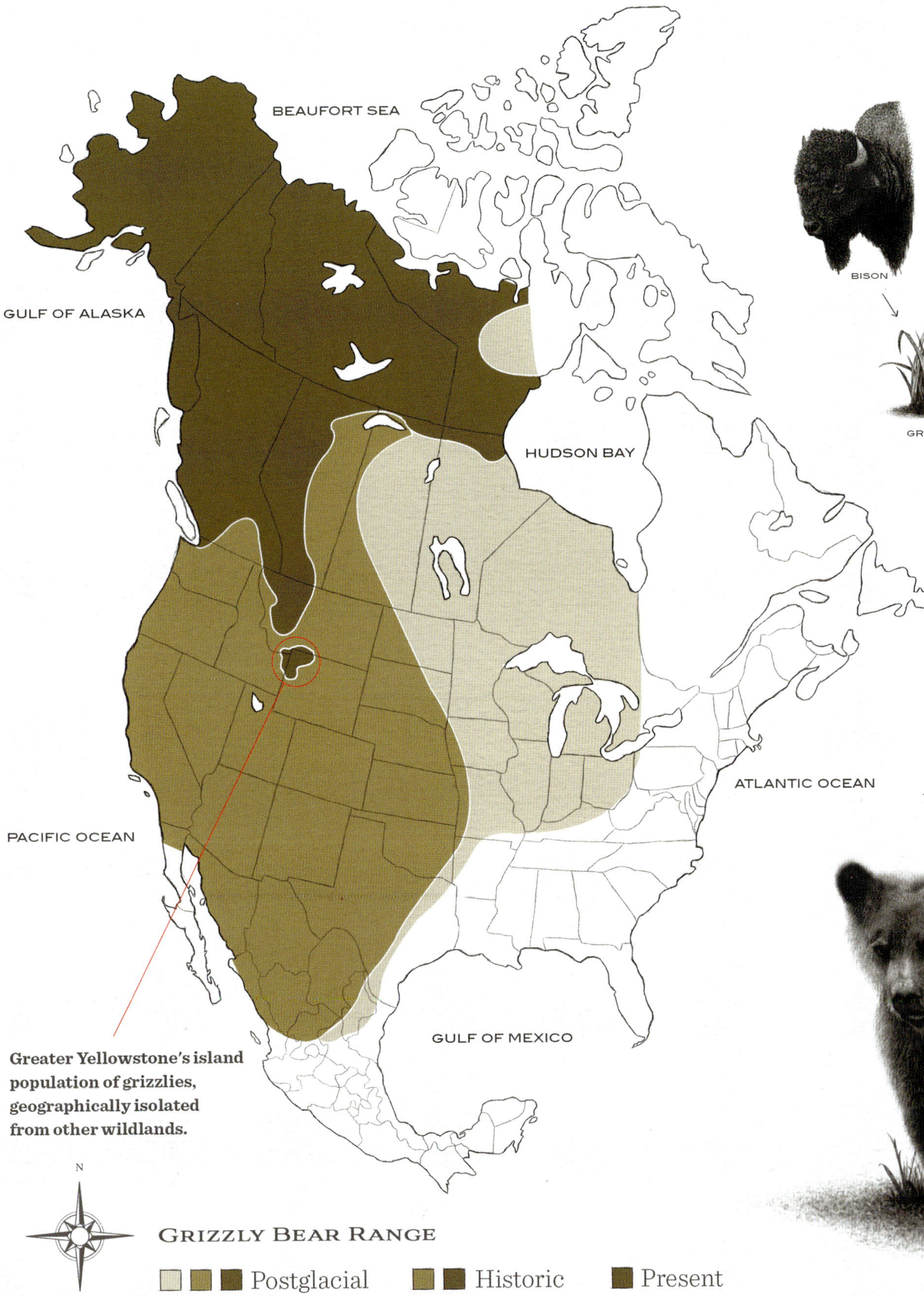
BEAUFORT SEA
GULF OF ALASKA
HUDSON BAY
ATLANTIC OCEAN
PACIFIC OCEAN
GULF OF MEXICO
Greater Yellowstone's island population of grizzlies, geographically isolated from other wildlands.
N
Grizzly Bear Range
Postglacial
Historic
Present

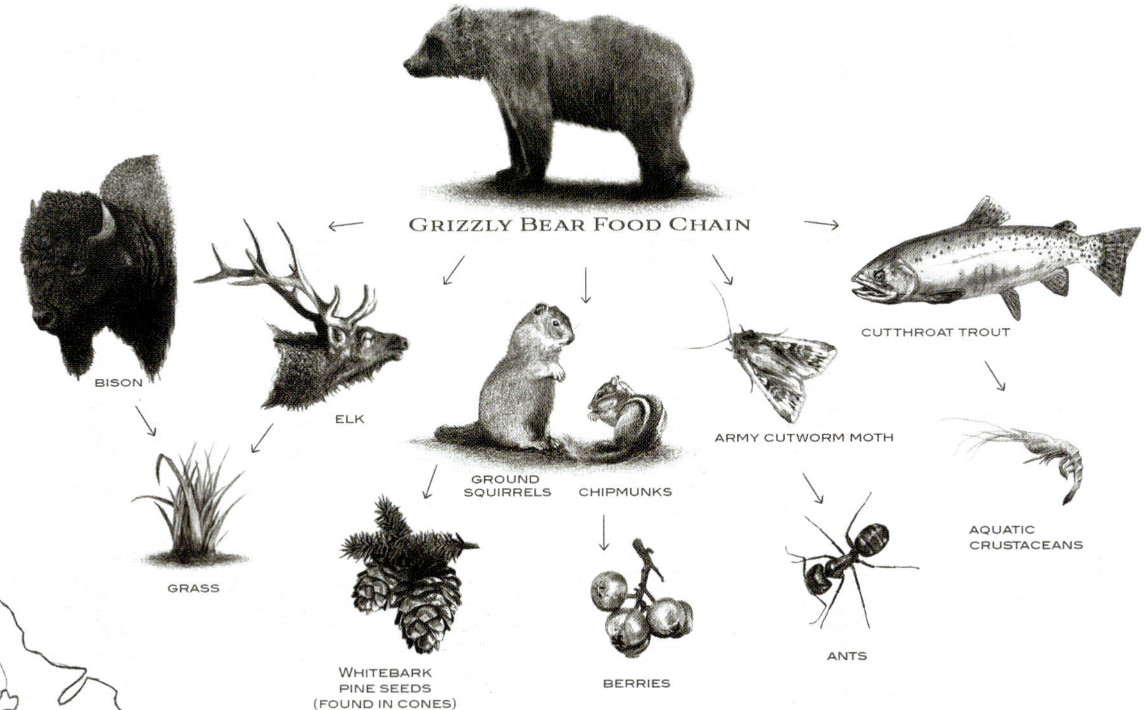
Grizzly Bear Food Chain
BISON
ELK
GROUND SQUIRRELS
CHIPMUNKS
ARMY CUTWORM MOTH
CUTTHROAT TROUT
GRASS
WHITEBARK PINE SEEDS (FOUND IN CONES)
BERRIES
ANTS
AQUATIC CRUSTACEANS

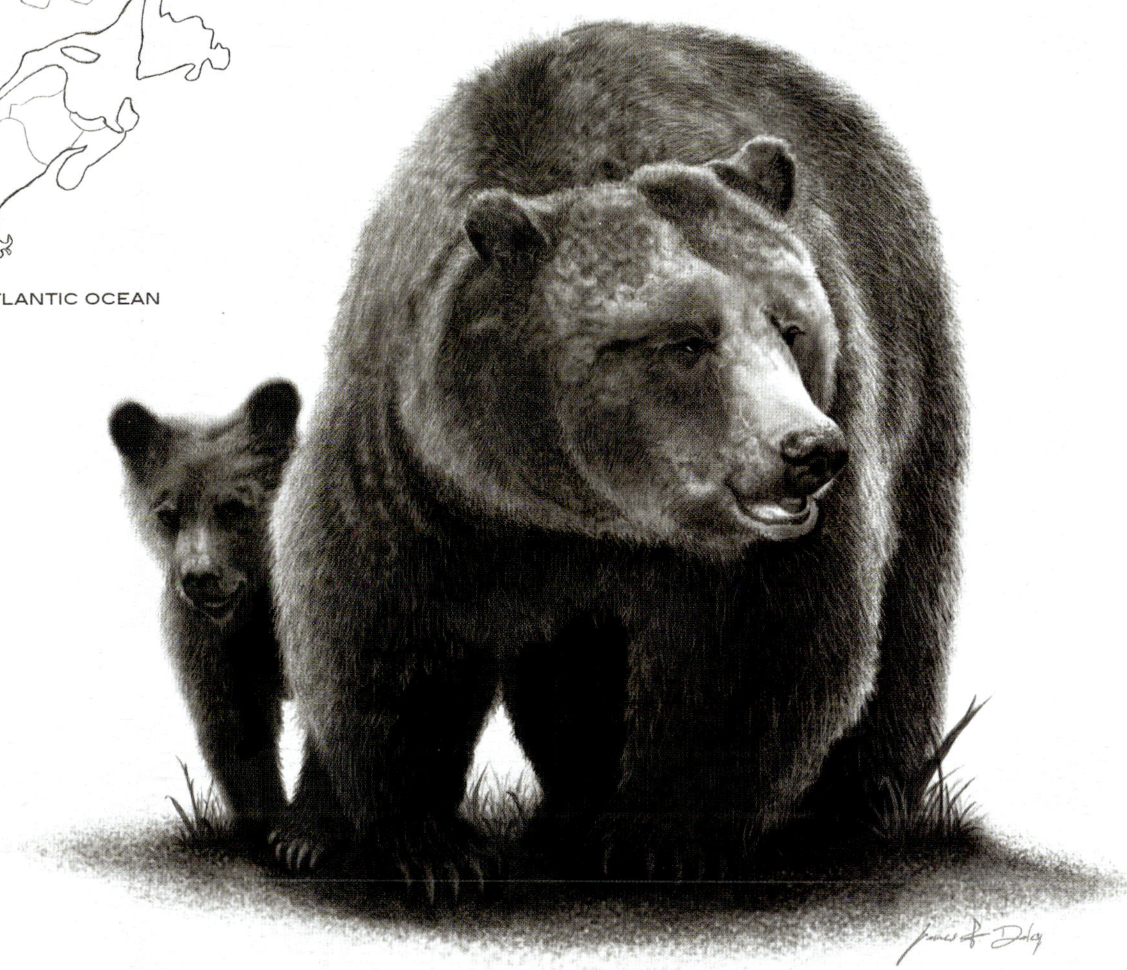

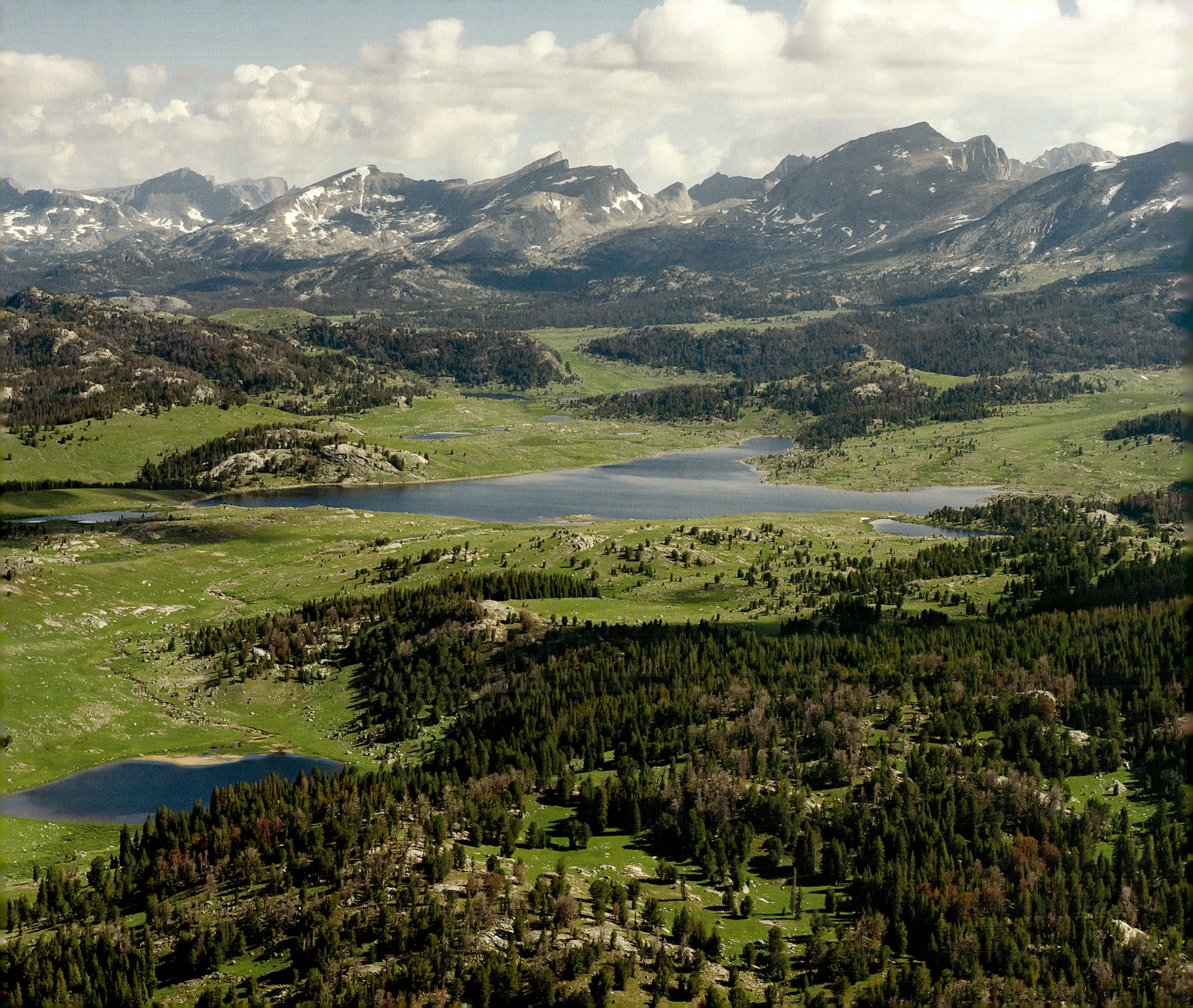

waited for them, like brown bears in Alaska do for salmon. However, the illegal introduction of nonnative lake trout into Yellowstone Lake has resulted in an alarming and well-publicized crash of native cutthroat trout. Not only was the lake a stronghold for the famous trout, but cutthroats also spawned late every spring in streams spiraling around the lake, making them accessible to bears that gathered in huge numbers. Lake trout, which are piscivorous, or fish eating, knocked back the cutthroat population by upwards of 90 percent and left the spawning process functionally extinct.

Prior to the collapse, which took hold shortly after the millennium began and lake trout numbers exploded, between 50 and 70 bears—or roughly 20 percent of the bears in the ecosystem—fished at the streams.

And then there's the phenomenon of army cutworm moths, nicknamed "miller moths." They hatch in farmlands out on the plains and migrate hundreds of miles in summer to certain high talus or scree slopes in Greater Yellowstone near alpine wildflowers. At night, they pour out of the crevices to drink nectar. Bears feast on them when they return to their bunkers. In a census undertaken between 2003 and 2013, nearly 180 different bears were observed at the moth sites. Just about every bear on the east side of Yellowstone and in national forests outside the park boundary visits moth sites. Mark Haroldson told me that the updated figure of grizzlies converging at moth sites is a whopping 220. Moths are regarded as pests in farm country, and scientific experts in lepidopterans (moths and butterflies) say they are vulnerable to insecticide spraying and habitat disturbance, as well as the effects of climate change threatening mountain foods, such as wildflowers.

According to Mattson, moths and whitebark pine are bear superfoods. Up to 72 percent of a moth's body weight is fat, making it an ideal food for bears needing to put on weight. When bears are manic with hyperphagia, male bears eat four to six times as much as females. This high-protein diet contributes to a bear's body size and makes it more competitive. It means that foraging for meat is deeply ingrained as a behavioral tradition. It means that males behave aggressively around carcasses and seek them out. It means that females with cubs that might be attracted to carcasses could be imperiled.

There are correlations, too, involving bears and whitebark pine seeds. Females that eat more seeds tend to have higher litter sizes, such as three or sometimes four cubs. Those sows that eat primarily meat have smaller litter sizes. An important consideration in calculating the value of whitebark pine is that when bears are eating more whitebark pine seeds, it lowers their contact with humans markedly, essentially producing what Mattson calls a "refuge effect." "Since we know most bears die from contact with people, it stands to reason that the less they encounter humans, the less exposed they are to risks that can kill them," he said, adding that in bad whitebark seed years, bear mortality on average rose between 1.6 and two times. That translated into an overall five percent decline in the population in a series of years following a bad seed crop.

Under the best-case scenario, whitebark pine will persist, but as a ready food source for bears, it is functionally extinct. Whitebark pine are slow-growing trees that only mature and start producing seeds after several decades, meaning that it will be a half century or longer before any seedlings planted today yield dividends for grizzlies, provided those trees are able to grow in high mountainsides becoming markedly warmer due to climate change. It means that other competitor trees are moving upslope and may outcompete whitebark.

So, in the absence of whitebark pine, what will sows do to compensate? The grizzly study team says some bears have moved over to moths. Should they shift to meat, evidence suggests it would put them on a collision course with male bears, which could mean smaller body sizes and smaller litters of cubs that, extrapolated over time, could mean a less fecund sow population than in recent history.

Haroldson and Van Manen said that hasn't happened yet. The bad news is that whitebark pine is gone, removed from being a featured menu item. The upshot is that bears seem to have adapted and persevered through it relatively well. In other words, the toll has already been exacted and bears are making do. "The core of the recovery area [Yellowstone Park] is a factory for bears and a pump," Haroldson said. "It has been an active pump, but it's not pumping them out toward the periphery of the ecosystem as fast as it used to be."

The best data shows the number of females with cubs is still growing but at a slower rate. "We all know that populations cannot grow in perpetuity," Van Manen told me. Consistent with trends observed in other wildlife, the data indicates that grizzlies are at the upper reaches of their numerical carrying capacity in the central core of the

Four major foods have served as pillars for grizzly bear abundance: whitebark pine seeds, army cutworm moths, spawning cutthroat trout, and carrion (meat), from elk primarily. This aerial view of Wyoming's Wind River Mountains shows how the die off of whitebark pine trees continues, eliminating an important bear food source.

ecosystem and that slight declines would be expected. "We predict that, over time, the population will flux dynamically, with numbers rising in some years and dipping in others but no sharp swings expected in one direction or the other," Van Manen said. They don't dismiss Mattson's contentions; they just haven't seen evidence on the ground that what he says is happening.

Finally, Mattson riffed more about meat. A grizzly study team report states: "On average, approximately 79 percent of the diet of adult male and 45 percent of the diet of adult female grizzly bears in the Greater Yellowstone area is meat. In contrast, in Glacier National Park, over 95 percent of the diets of both adult male and female grizzly bears is vegetation."

Going back to 1975 and then moving forward into the mid-1990s, that was a good period for growth in elk and bison numbers. In fact, the size of Yellowstone's northern herd was 19,000, but beginning late in the 1990s, increasing numbers of wolves descended from lobos that were reintroduced to Yellowstone a few years earlier started reducing wapiti numbers. That—combined with years of consecutive drought, the effects of an aggressive sport hunt on elk outside the park designed to knock down the herd size, and predation of grizzlies on calves—declined the elk population significantly. Today, total elk numbers on Yellowstone's Northern Range are roughly a third of what they were. Some evidence suggests that grizzlies that used to feed on cutthroat trout at Yellowstone Lake have transitioned into killing elk calves, and indeed ecologists say bears are a more formidable predator of young elk than wolves.

Taken together, Mattson's assessment is that two of the four major foods (whitebark pine seeds and trout) are gone and won't rebound any time soon. Army cutworm moths are, for the moment, stable, and attracting huge numbers of bears, but given their life history (possible exposure to pesticides) and correlation to the nectar of subalpine plants in an age of climate change, nothing about their future is certain.

Ungulate meat is available in elk but at far fewer numbers than in the past, on a landscape with more competition (wolves, cougars, people), climate change that is changing the high meadows where elk forage, and the loom of disease such as chronic wasting disease. Meanwhile, Yellowstone is under enormous political pressure to reduce the size of its bison herd and moose are in severe decline due to climate drying up their habitat.

THE RECENT SURVEY OF GRIZZLY BEAR FOODS by Kerry Gunther and his colleagues is impressive, but it builds on earlier research. Between 1979 and 1993, 4,000 different field sites were investigated to expand the understanding of what bears were eating—and David Mattson was involved in collecting and crunching the data related to biomass, nutrients, and expenditure of energy.

What is the crux over the future of grizzlies? Besides necessary efforts to ensure bears don't die in large numbers when having encounters with people, this is it: Mattson describes an echo effect. He contends that the number of bears being counted today is a reflection of good food years in the past; that it is disingenuous to extrapolate a favorable prognosis ahead based on conditions for bears that have changed. While he agrees the metrics employed to gauge recovery support removing bears from federal protection, those metrics aren't the whole picture. As mentioned earlier, stability of the bear population is based upon having a solid foundation of healthy grizzly bear mamas. Everyone is united in noting how vitally important it is to keep reproducing mothers like 399 and 610 alive.

The kinds of wild foods available to them not only affect fecundity, Mattson says, but also influence where the largest densities of bears will be found on a landscape. The benefits of whitebark pine, army cutworm moths, and spawning cutthroat trout have been that they are found in areas removed from large developments, where there is often lower tolerance for grizzlies and a constant risk of them getting into human trash. Backyard birdfeeders, beehives, fruit orchards, dog and cat food put out on the porch, trash cans, the tasty odors pluming out of restaurant exhaust vents, and barbecue grills all can lure in a grizzly.

Mattson's provocative contention is that as the major bear foods disappear, a hollowing out of the ecosystem's ability to sustain larger bear numbers will occur. Today's perception is that Greater Yellowstone is bursting at the seams with grizzlies, and bears are showing up in places they haven't been because population pressure in the core is causing subadult males to strike out in search of new territories to colonize. But Mattson worries that in the years to come an insidious counterscenario could take hold with bears under state management.

States cite this narrative of bears "bursting at the seams" to justify a trophy sport hunt of grizzlies targeting dispersers and likely focusing on big grizzlies because they're

Lessons in survival are passed down from mothers to cubs. Here, one of 610's juvenile cubs digs for roots and other tasty morsels near Willow Flats.

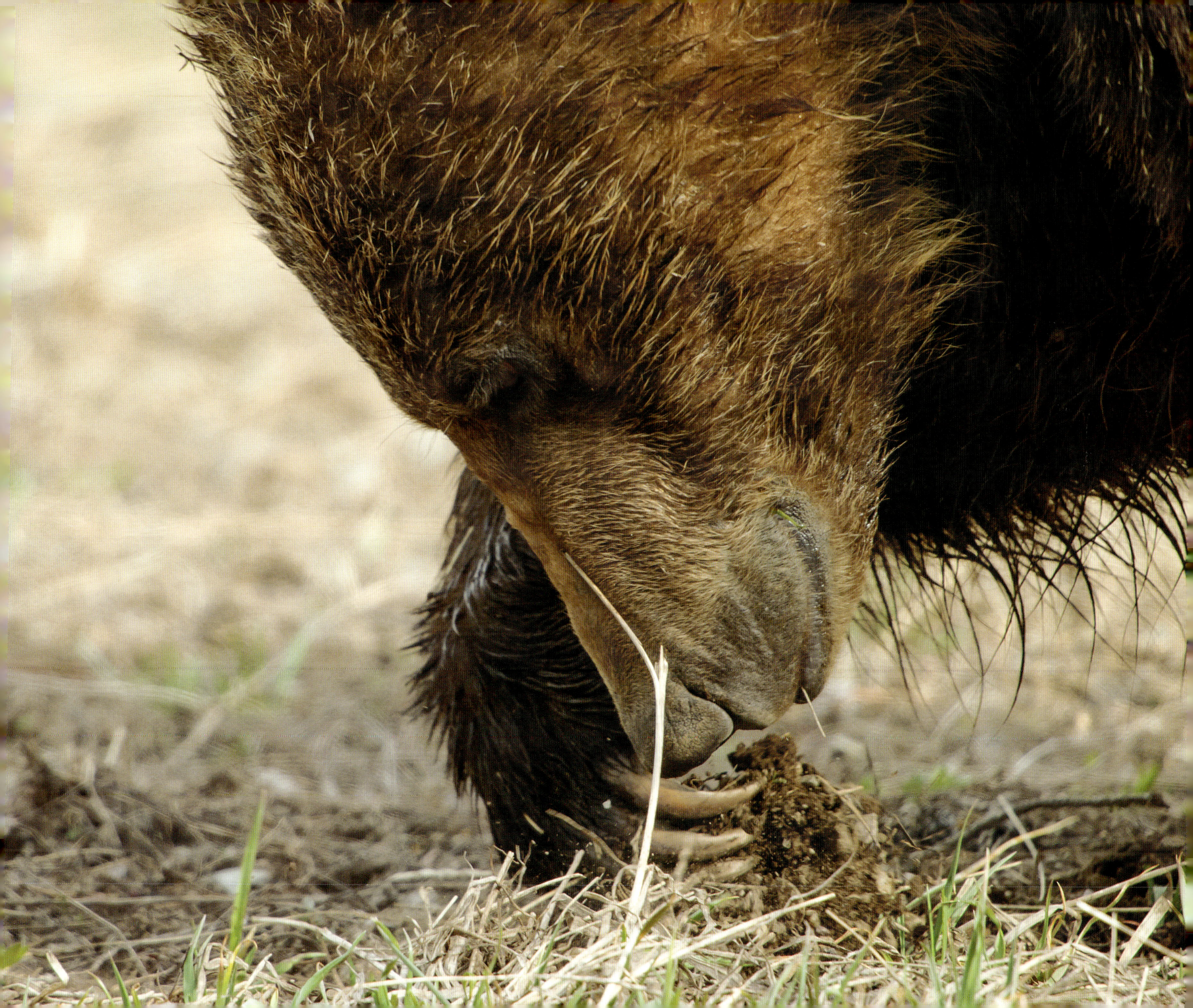

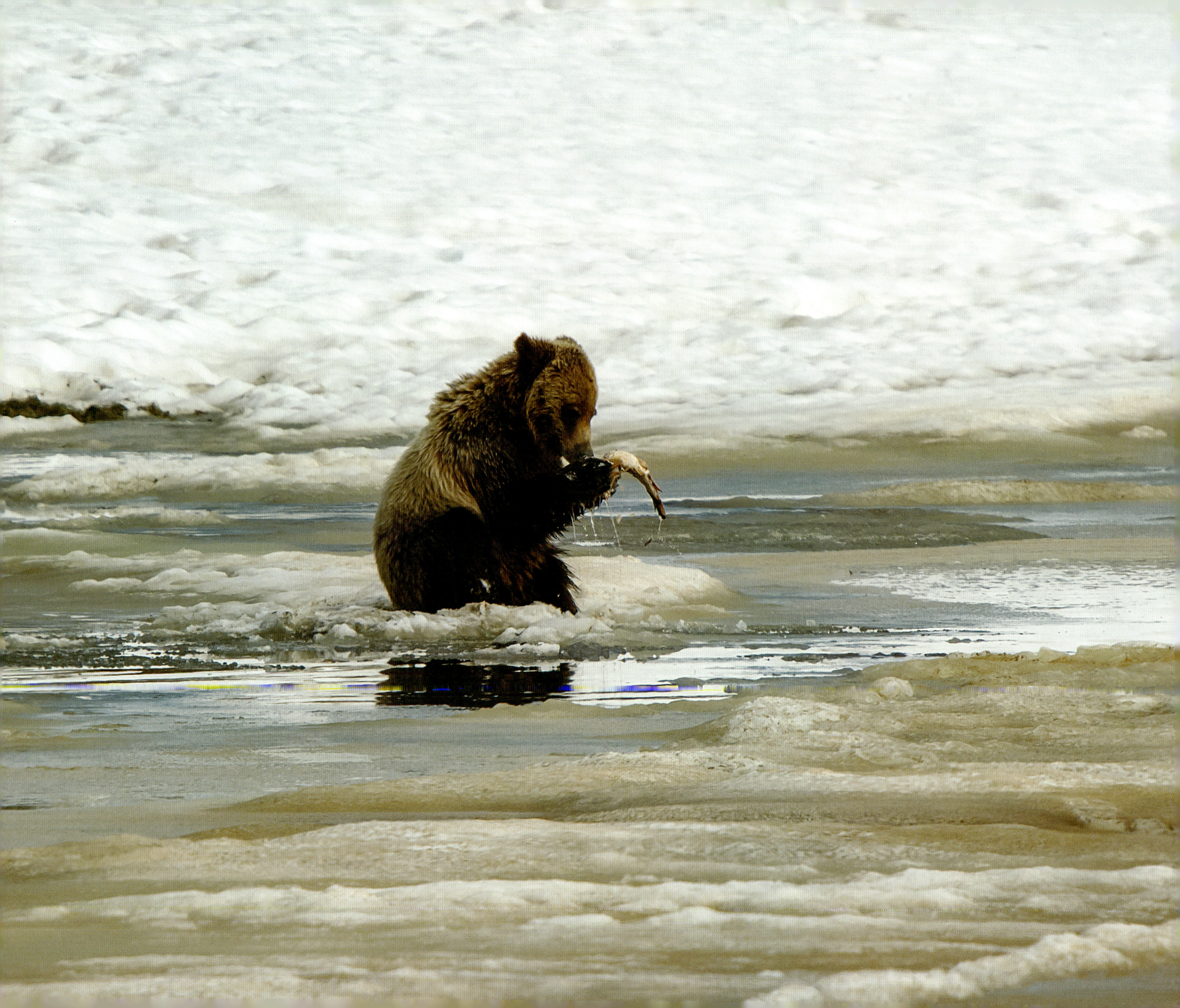

most valuable and prestigious to put on the wall. Part of this story line involves the suggestion that more grizzlies will need to be taken out of the population because they will become "conflict bears."

But what if the reason that bears are showing up in greater numbers at the periphery of the ecosystem is not because they're getting bounced out of Yellowstone, but because they're going there to try and find food that no longer exists in the core? Meat-dependent bears, if they don't find ready access to wildlife, may turn to cattle, which means they'll likely be destroyed. Female bears that turn to meat will be competing with unsociable boars, or also turning to livestock. Either way, they're likely to be giving birth to fewer cubs.

The net effect, Mattson asserts, is that the glowing picture of bear population health could be just the opposite. "Grizzlies are notoriously slow to reproduce. And we also know from history that the difference between a rising or steady population and one that tumbles into decline can be higher mortality of female bears, by just a few percentage points, over a few years. I am concerned there will be a delay, a lag time, before it fully registers with bear managers. And then it will be met with resistance from states like Wyoming that are ambivalent about having grizzlies to begin with and still view them as liabilities and threats to their cowboy culture," he says.

"I hate using clichés but it could be a multifaceted perfect storm: you have more bears being taken out in management actions, you have bears being killed by trophy hunters, you have sows producing smaller litter sizes as a response to less abundance of the foods that fueled their health. On top of it, you have a wild card like climate change. I think everyone agrees that warmer winters will mean a longer growing season for plants. But it likely will also mean bears will hibernate less. What implications does that have for female reproduction? We don't know and anyone who claims they do know has a crystal ball that I haven't seen."

The safest course, Mattson suggests, is for managers to not act impetuously, thinking harder about what state custodial care of grizzlies means and holding back on implementing a sport hunt. Conservationists have implied, and the states have refuted, that there has been an urgency to remove grizzlies from federal protection. And while scientific data is being used to prove that delisting is justified, the real breeze filling the sails of the delisting movement has been political.

Mark Haroldson says there's no indication that female bears have less body mass as a result of imperiled nutrition. Frank Van Manen, for his part, worries less about climate change, believing bears are both resilient and adaptable. More problematic would be if inward human migration results in larger numbers of people moving into bear country and being unwilling to share space. Genius astrophysicist Stephen Hawking was once asked by a young man what about *Homo sapiens* was most urgently in need of change. "The human failing I would most like to correct is aggression," he replied. "It may have had survival advantage in caveman days, to get more food, territory, or a partner with whom to reproduce, but now it threatens to destroy us all."

FOR MANGELSEN, THE YEAR 2012 was filled with both inspiration and melancholy. After 610 took on mothering duties for one of 399's cubs, the other two demonstrated an independent will and strong personalities. Now yearlings, the brother and sister (who were not collared, ear tagged, or tracked) were nicknamed Brownie and Ash.

Brownie, the brother, had a dark chocolate coat in contrast with his sister Ash's lighter head and shoulders. Mangelsen has amassed a huge selection of images of 610 and her threesome and 399's cubs from that entire spring and summer, from the moment they emerged from the mountains.

In early May, adding to the dramatic tension of their lives, Brownie and Ash were separated from 399 for reasons unknown. Seemingly without provocation, rangers had hazed them from Colter Bay's tented camp and corral area, which 399 often frequented in the spring. The campground had not opened yet. Some believe the heavy-handed action from park staff members, some of whom have been accused of conducting themselves less than sensitively around bears, left Brownie and Ash traumatized. It will never be known what exactly happened, but 399 and her cubs were never documented together again after this day. Mangelsen hypothesized that the family retreated into the backcountry and later had a confrontation with a boar wanting to mate with 399.

Missing in action for more than a week, 399 was first seen at the Pilgrim Creek Bridge traveling alone by Diana Stratton, a local photographer and avid bear watcher. Brownie and Ash turned up a few days later, navigating separately. Both had bloody claw marks on their hindquarters, suggesting that they had a possible encounter with a boar.

Grizzlies are incredibly intelligent and opportunistic. One of 399's juvenile cubs feeds on a sucker that he has just fished out of the Snake River

For several days, Brownie and Ash revisited places that they knew, from Pilgrim Creek to Moose-Wilson Road and finally Windy Point. They spent several days together foraging in the sage. 399 disappeared from view, prompting Mangelsen and Sue Cedarholm to keep a closer look on the brother and sister, not yet two years old and probably not ready for independence. The cubs hung together in areas flanking busy Highway 89, which runs parallel to the Snake River in Grand Teton. "Well, we were concerned," Mangelsen said. "They were the equivalent of a prepubescent being expected to survive." Perhaps because of the early separation from their mother their bond was stronger than that of normal yearling cubs; Ash wouldn't let Brownie out of her sight. Photos taken by Mangelsen chronicle the lead-up to yet another brewing tragedy.

On June 21, 2012, Brownie was viewed wandering the roadside between Moose and the popular Snake River Overlook, attracting attention from passing motorists. In an incredible accident of the bears simply being in the wrong place at the wrong time, Brownie was run over by a car.

A 29-year-old tourist from Pennsylvania was motoring north on Highway 89 and another driver who was headed south swerved into his lane to avoid hitting Brownie, who was on the shoulder. The northbound driver careened off the west side of the highway through the sagebrush and plowed over Brownie 100 feet off the road. Brownie died of internal injuries. Mangelsen helped investigators make identification of him as 399's son. At 130 pounds, he was in reasonable health. Ash's whereabouts were unknown and she was never documented again.

Meanwhile, in August, 610 directed her triad of cubs to the berry patches along Moose-Wilson Road. The protective mama bluff charged a tourist on foot and a bicyclist. When a ranger arrived, the bear reportedly bluff charged the park official, too. That fall, Grand Teton National Park and the state issued 725 elk hunting tags and the park again turned into a land mine. In late October, a wapiti was shot in the river bottoms of the Snake and when the hunter set out to field dress it, 610 and her cubs were already on it.

Mangelsen and his friends pleaded that the area be temporarily shut down to hunters, emphasizing the fact that there had already been a previous attack by an unknown bear over a carcass in nearly the same spot the year before. Both bear and hunter escaped without serious injury. Although the park had denied the request in earlier years, this time a five-day closure of a quarter square mile north of Schwabachers Landing was implemented. Mangelsen was characterized by some in the park as a hell-raiser and too emotionally invested in the bears, but a month later he was vindicated.

On Thanksgiving morning, he got word that a grizzly had been shot and killed near Schwabachers Landing not far from where 610 and her cubs had claimed a carcass earlier. He and Cedarholm rushed to the scene. They learned that it was a large adult male, the first grizzly killed in the park hunt. It happened while a father and his two sons were elk hunting. They surprised a bear near an elk carcass in the woods and claimed that they were charged by the bruin. The father said he pulled out bear spray and emptied it into the grizzly, but the bear kept coming. The sons then opened fire and killed it with five shots, including two to the head. The bear was covered with a tarp and dragged by horses across the sage flats to the parking lot at Schwabachers Landing, while 610 and her three cubs could be seen fleeing the river bottom to the west. The incident was classified as justified self-defense, though it was discovered that the bear spray can had expired nine years earlier. In the following days, the tracks of a grizzly with three cubs, likely 610, were seen following the bloody trail of an elk dragged back to the parking lot. Riparian areas—river corridors—are the richest pieces of wildlife habitat on any landscape. They are magnets for wildlife. Even after the death of the boar and the visible sign of continued grizzly activity, the park still did not close the area to hunting. It wasn't until 2014 that the park decided to close the Snake River bottom to hunting, hopefully, Mangelsen said, for good.

The fall was filled with more close calls, and by incredible luck, 610 escorted her cubs to the den alive. 610's mother, 399, had remained out of sight, but Mangelsen was used to her surprises.

IN 2013, MANGELSEN AND JIM LAYBOURN solved the looming mystery. Turning to the GPS location that Mangelsen had jotted down when he followed 399's tracks two years earlier, they went up into the air on May 3 and, despite a bumpy ride, flew over her den and visually documented a spectacular sight.

Now 17 years old, 399 reinforced her legacy of being a fertile bruin when she was spotted outside of her Pilgrim Creek den with her third set of triplets since 2006. In all, 399 had birthed 10 cubs since 2004. "Look at the impact she's had," Mangelsen said.

Ash and Brownie, 399's cubs, race in a game of tag as they frolic in a meadow near their natal homeland, Pilgrim Creek.

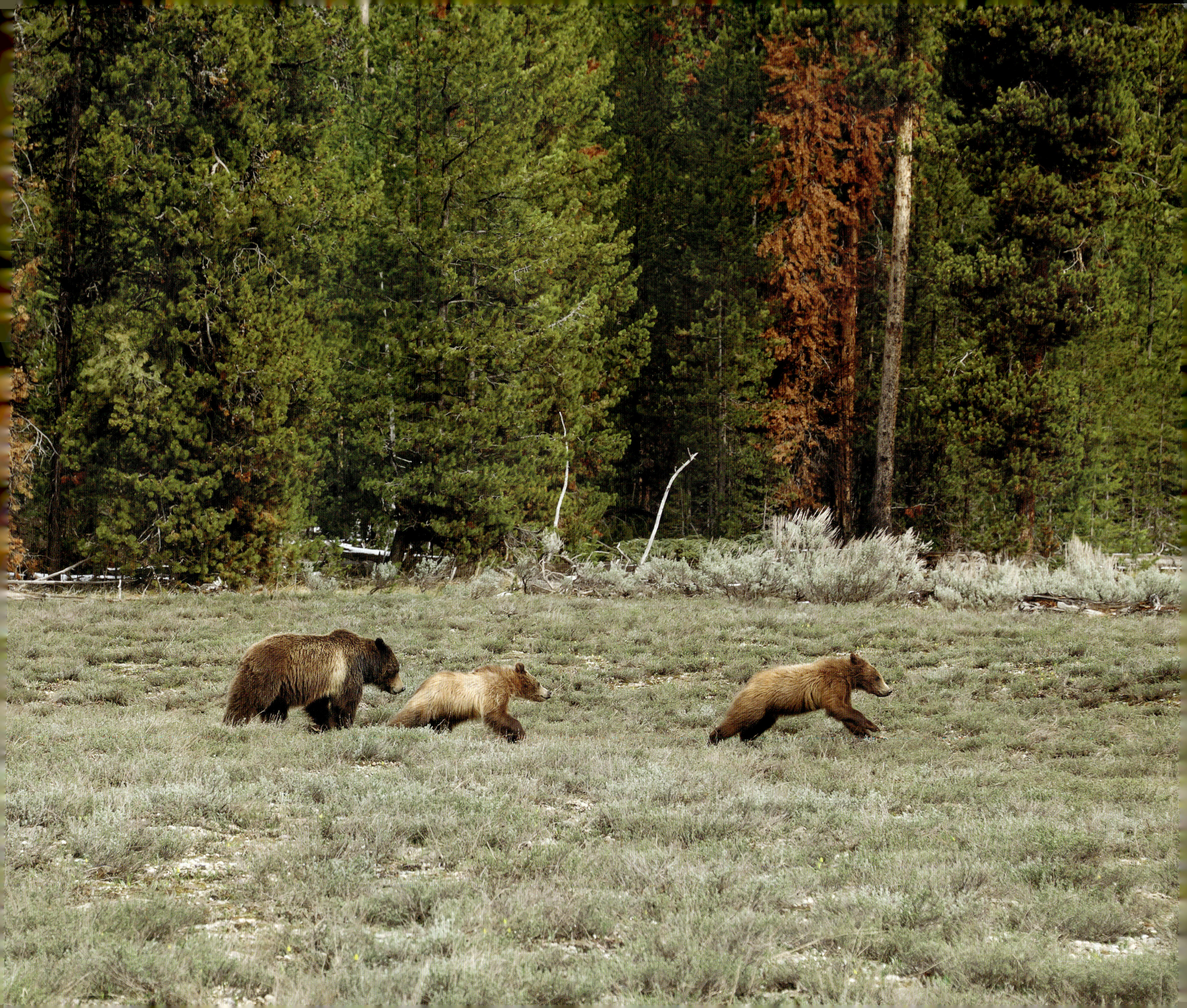

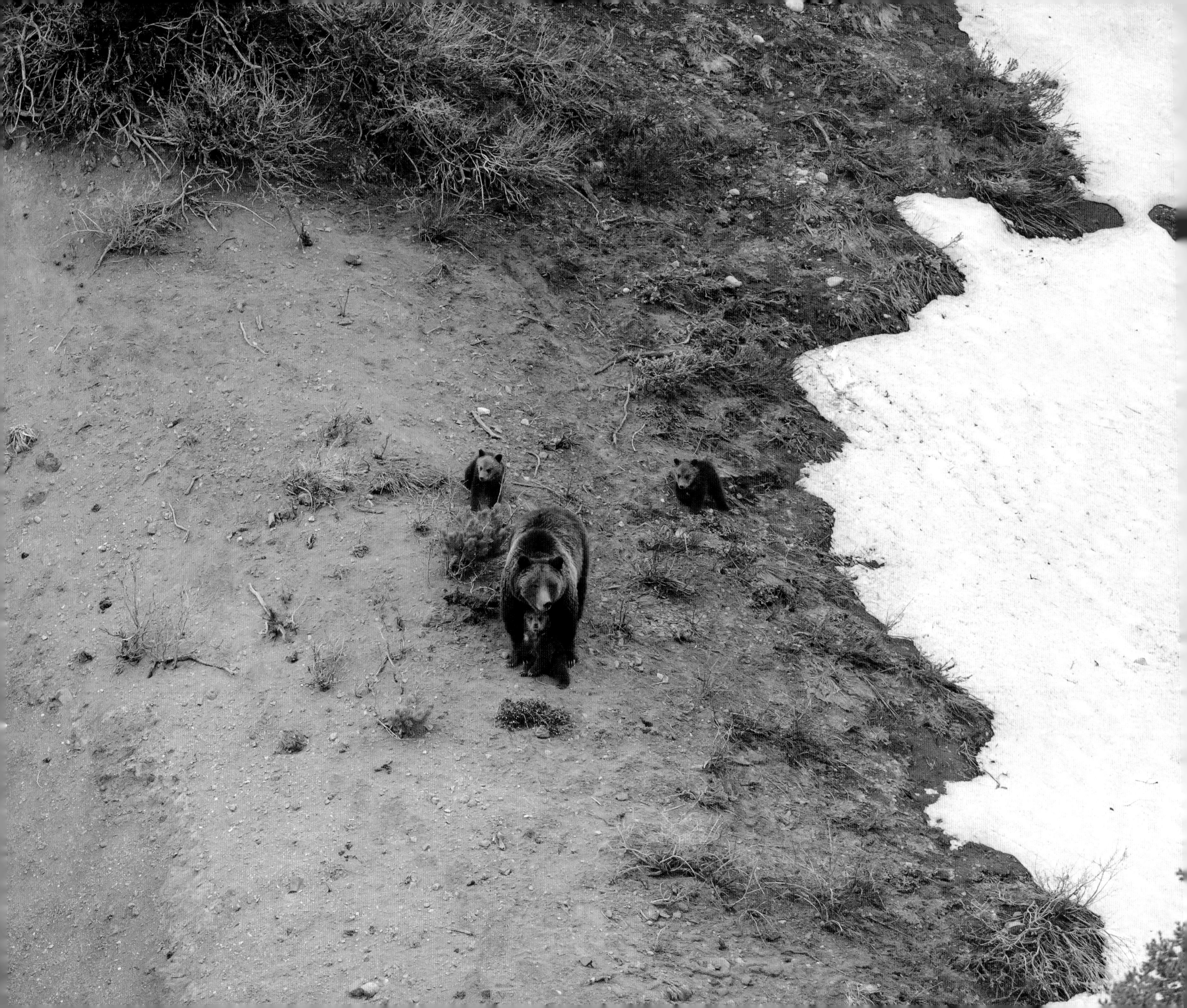

At times it was difficult to keep track of the 399 and 610 clans and differentiate who was who. Plus, amid the visual and written diary you've experienced so far, it needs to be stated that all the while numerous other grizzlies were coming and going. Another bear being tracked by researchers was 179, who might be described, maternity wise, as a sort of doppelgänger to 399. State and federal wildlife managers say that 179, now in her 20s, has delivered close to a dozen cubs but has existed largely out of the limelight. She has a very different resume.

The mother of 179, given the identity of 79, was a Yellowstone grizzly and a serial raider of apple orchards, mentoring bad habits into twin cubs. To stop her raiding, researchers trapped and airlifted her into the Thorofare, one of the remotest corners of the Lower 48 that happens to be located in the southeast quadrant of Yellowstone. The cubs—179 and a sister, 182—were transplanted into Glade Creek, which overlaps with the home range of 399. While 79 hauled out of the Thorofare and returned to the outskirts of Gardiner, Montana, at Yellowstone's northern gate, one of the young sisters—179—stayed put. Meanwhile, her sibling reached Yellowstone Lake, was trapped and released, and then turned up at Heart Lake, dropped her collar, and melted into the backdrop of anonymity again. As for 179, she has completed several circumnavigations of Jackson Hole and the country eastward toward Togwotee Pass, venturing into cattle herds but not doing anything to warrant destruction. The information gleaned from the GPS coordinates her tracking device delivers has helped the grizzly study team understand how the loss of whitebark pine is affecting the population.

LOOKING BACK, the peregrinations of 587 (399's son in her first group of triplets) were astounding. He covered hundreds of square miles, made it to the other side of the Tetons and back, and crossed the Gros Ventre Mountains, the Continental Divide, busy highways, rivers, snowfields, pastures, and people's backyards with discretion.

During the first week of July 2013, 587 was more than 60 miles away from his playground east of Jackson Lake, lumbering through broken forest and meadows grazed by cattle in the Upper Green River drainage along the flanks of the Wind River Range north of Pinedale, Wyoming.

Bear managers try to give wayward grizzlies three strikes. Three years earlier, 587 earned strike number one when he killed cattle in Tosi Creek, part of the Upper Green River system. His second offense came a short time later when he killed sheep. That led to him being trapped and transported back to Pilgrim Creek, his ancestral home. But he didn't stay put.

The following year, in 2011, he killed cattle again, strike three. Zach Turnbull, a carnivore biologist with the Wyoming Game and Fish Department, didn't give up on him. Ecologist Steve Cain noted that he was good at killing prey. He had been taught well, after all, by 399, when together they shadowed elk calves at Willow Flats.

Cain had a hunch. 587 was a victim of timing and bad luck. In early summer, the Upper Green had plenty of elk calves to feast upon, but when they grew bigger and were harder to catch, the bear likely shifted over to smaller, more docile fare—beef calves on the Upper Green's public grazing allotments. On July 7, 2013, he was euthanized for chronic livestock depredation. He joined two other female grizzlies that were also destroyed for targeting cattle.

Defenders of the bears say it's ironic that grazing nonnative cattle, being fed grass to fatten up and ultimately bound for the slaughterhouse, would take precedence over the native grizzlies that are behaving as the predators they are. The state of Wyoming spent around $300,000 in 2014 resolving bear conflicts with wildlife and compensating ranchers for livestock losses. Since 2003, Wyoming has spent $1.6 million in reimbursing ranchers, mostly cattle operations, for animals killed by bears. Dan Thompson of the Wyoming Game and Fish Department says incidents involving bear predations on livestock have been increasing. No rancher has been or is expected to be run out of business by grizzlies, but predation lessens social tolerance. What rubs many the wrong way is that exotic cattle hold primacy on public land that would otherwise be exceptional wildlife habitat. Ranchers graze their cattle at below-market rates and they receive protection from native predators.

Both Tom France and Hank Fischer of the National Wildlife Federation recognized that cattle grazing represented a land mine for grizzlies and wolves in Jackson Hole. Quietly, they worked to broker solutions: buying ranchers out of the public grazing allotments in Grand Teton National Park and Bridger-Teton National Forest, freeing up more safe terrain. It's a move that many say has made a difference between grizzlies meeting heightened hostility or reestablishing a foothold. "It's notable that these allotment closures have occurred without controversy," Fischer said. "I credit our acknowledgment that

Solving a mystery, Mangelsen took this photo of 399 and her third set of cubs from the air shortly after they emerged from their den in the Teton Wilderness high in the headwaters of Pilgrim Creek. Slightly bigger than housecats, the new cubs are guarded by their attentive mother

these grazing leases have economic value. By paying ranchers a fair value to give up problematic allotments, we have solved a problem for both parties."

Since the episode with 587 played out very publicly, a bias of the Wyoming Game and Fish Department was revealed. Thompson said his department didn't like the idea of nurturing roadside bear populations. He preferred bears to stay clear of humans. "Habituation toward people and the roadside bear situation, it's not something that we're supportive of," he told *Jackson Hole News & Guide* reporter Mike Koshmrl. "It obviously shows that they may be more prone to getting in trouble in the future."

That opinion was challenged by Mangelsen and experts like Cain, who have spent their entire careers studying bear behavior. There is zero provable correlation. "You have to wonder what kind of grizzlies does the Wyoming Game and Fish Department want? Grizzlies that are hardly ever seen?" Mangelsen asked. "Bears that are afraid of people and taught to flee for their lives because they're being shot at by hunters? I can think of nothing that leaves a bear more prone to getting into trouble than carrying on a big-game sport hunt of elk inside a national park and leaving gut piles all over the place that lure in bears. I think Game and Fish's idea is to manage grizzlies so that they are out of sight and mind, almost invisible, because it's convenient for them. Except for a few people, I don't think the agency even likes grizzlies. Its attitudes are squarely at odds with the reasons that millions of bear lovers come to this state and spend lots of money to do so."

In a scientific paper titled "Brown Bear Habituation to People: Safety, Risks, and Benefits" (published in the *Wildlife Society Bulletin* in spring 2005), noted bear researcher Stephen Herrero and coauthors Tom Smith, Terry D. DeBruyn, Kerry Gunther, and Colleen A. Matt observed, "It is our belief that bear viewing, where appropriate, may promote conservation of bear populations, habitats, and ecosystems as it instills respect and concern in those who participate."

In general, the researchers said, grizzlies able to successfully live along roadsides learn to regard humans that keep their distance, are visible, and engage in predictable behavior as benign. Bear behavior is triggered by human behavior; the smarter that people behave, the smarter and more predictable bears do.

Mark Haroldson had nothing but praise for 399: "She and her daughter are real fine bears in how they negotiate that country and they're good moms." The pair reminded him of another grizzly bear in Yellowstone, 264, that was habituated to people and frequented the front country between Swan Lake Flats and Norris, thrilling large crowds. But she got hit by a car and died, bringing a jarring end to the excitement. "That's the problem with celebrity bears," he said. "They don't know they're celebrities. They're just trying to make a living. People project all of their hopes and fears onto bears, sometimes turning them into things they are not. But with what's happening in Jackson Hole, I don't have an issue with it. I think it's great they're being seen."

As for 587, he was sent to a taxidermist in Dubois after being killed by the Wyoming Game and Fish Department. All that remains of him now is his tanned hide. It is the hope that one day he will be mounted and put on display as a stuffed specimen at the Jackson Hole Historical Society and Museum.

JACKSON HOLE'S MANTRA FOR DECADES was "the last and best of the Old West." But that slogan is obsolete, for the Old West expunged grizzlies and wolves. Booming communities in the New West now tout having such charismatic species in their midst as competitive advantages for drawing smart people to live in them, said Craig Benjamin, executive director of the Jackson Hole Conservation Alliance. "For Jackson Hole, we view the expansion of grizzly bears into the southern reaches of our valley as a positive development and a conservation success we should celebrate. But with this expansion comes the responsibility to ensure that we reduce conflicts with bears, especially in our wildland-urban interface," Benjamin said. "Conflicts between people and wildlife in our neighborhoods present a huge threat to wildlife, particularly our iconic species like grizzly bears, wolves, moose, and elk. Living in a wild place means we should take personal responsibility to reduce the chances of conflicts with wildlife on our private property." This brings us to another bear, the grandson of 399 and son of 610, the bruin mentioned earlier: 760. Part of his pedigree was having supreme confidence in navigating the human landscape, but he died in a bear-management action in 2014. Some regard him as the poster child for why states can't be trusted as managers.

In a report titled "The Life and Unnecessary Death of Grizzly 760," Kent Nelson, Roger Hayden, and Jim Laybourn, with the group Wyoming Wildlife Advocates, penned what amounted to a eulogy, death investigation, and a demand for answers. They lay out

Come autumn, grizzlies are manic with hyperphagia, the drive to consume as many high-value calories as possible before denning. Well fed and flaunting a thick coat, one of 399's subadult cubs pauses to scent mark a pole often visited by grizzlies near Pacific Creek.

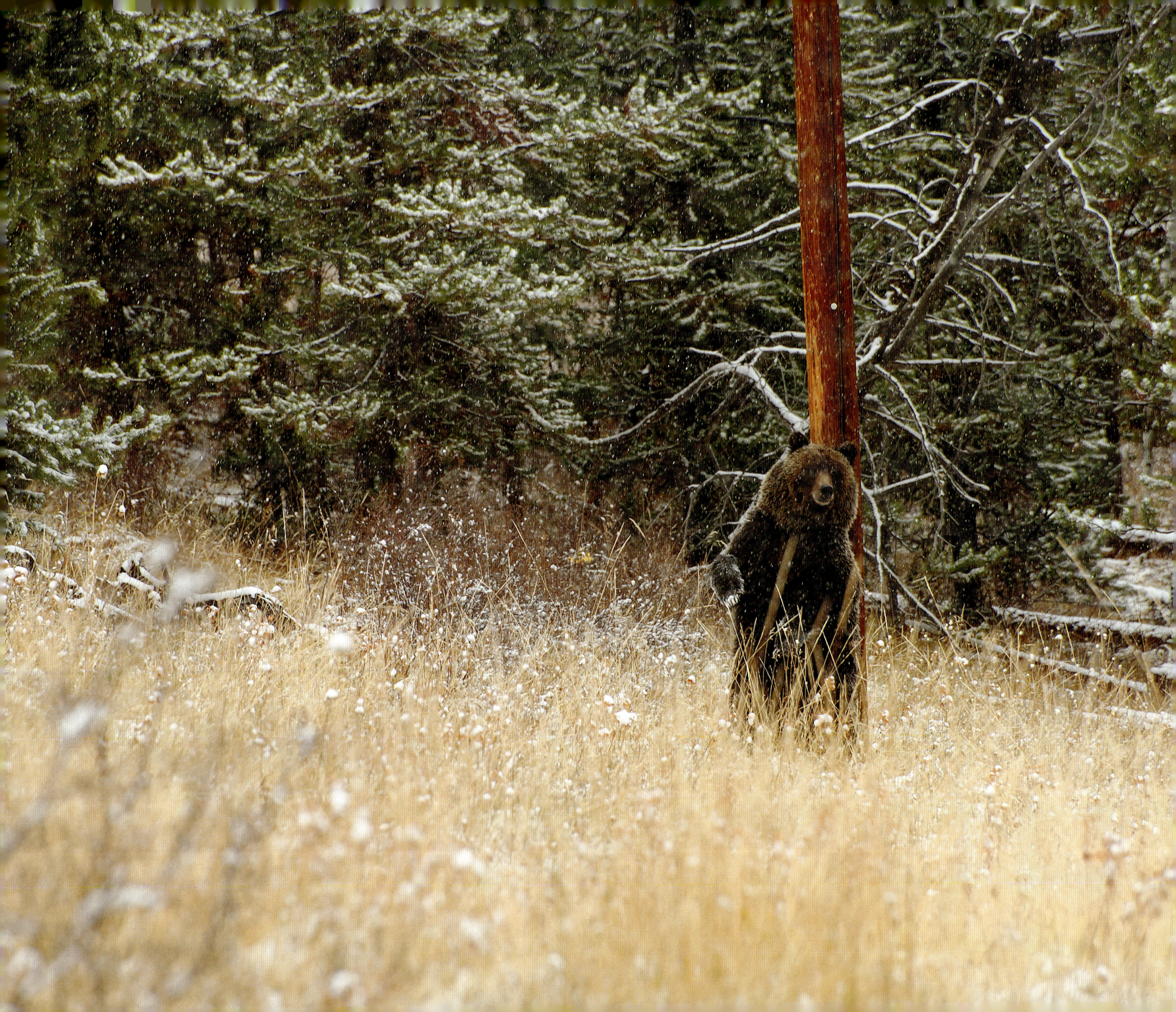

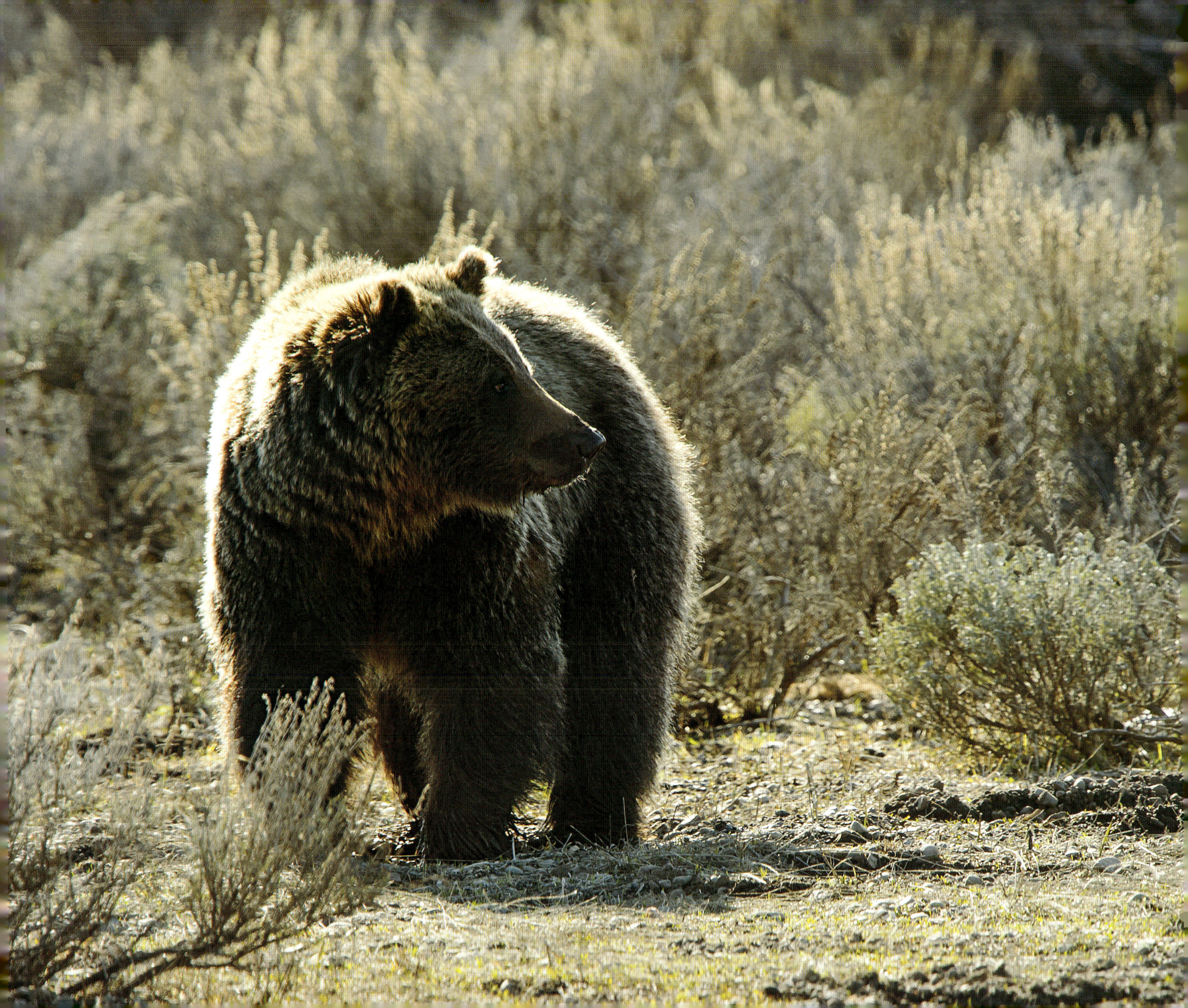

the two events that bookended the bear's final doomed path. On October 10, 2014, 760 was captured and relocated from Jackson Hole to the Mormon Creek/Five Mile Meadows area near the east entrance of Yellowstone. A little more than two weeks later, the Wyoming Game and Fish Department captured and euthanized 760 near the town of Clark, "43 miles away from where he was relocated and 100 miles from the only home he had ever known," wrote the Wyoming Wildlife Advocates.

I raised the controversy surrounding 760 to Mark Haroldson, who has either personally trapped or overseen the trapping of more grizzly bears in Greater Yellowstone than any person in history. He offered no critique, though he did say there have been many bears that were captured and relocated, resulting in them getting a new lease on life. He ticked off a half-dozen examples, including female cubs that were orphaned, moved to different locations, and ended up living long lives in which they gave birth to several litters of cubs. "If you look at the state agencies, I would say they've been good at trying to prevent conflicts and resolving them in a positive manner," added Frank Van Manen. "Almost every bear gets two to three chances."

What 760's end most challenges is not the relocation of bears but the sometimes debatable justification for it—and the kinds of grizzlies that the Wyoming Game and Fish Department profiles as being trouble makers, liabilities, and therefore expendable.

Young male grizzlies are freewheeling. 760 ventured north of Flagg Ranch as far as Heart Lake into southern Yellowstone, then turned south toward the Buffalo Valley where the Pinto Ranch raises steers and yearlings, and then crossed over the Snake River and angled to the south. In mid-September, as his mother and grandmother had done, 760 parked himself in the chokecherry and huckleberry patches near Moose-Wilson Road located in the southwest corner of Grand Teton National Park. It is berry nirvana. Human onlookers swarmed the location, causing a chaotic scene of traffic and people moving around out of control. At one point, 760 reportedly bluff charged a human gawker—another tactic that has been used in the arsenal of behavioral responses to people. Rangers arrived, and, unable to manage the frenzy, shut down the road to give 760 peace of mind.

The bear roamed out of the park drifting southward somewhere along the West Bank, the area between the Snake River and the Tetons, right through ranches, subdivisions, and open space without incident. On October 8, he was foraging in a wooded residential area known as the Aspens and crossed paths with a man walking his dog. Unfazed and acting nonaggressively, the bear continued advancing to check them out. The man told the Wyoming Game and Fish Department that he felt uncomfortable. 760's whereabouts were noted and his movements watched. Mirroring the pattern of his other kin, 760 had been trapped multiple times, first trapped and tagged and radio collared in the summer of 2013. Then he shed the collar sometime before leaving the den, and was again trapped and radio collared in the spring of 2014.

Venturing south of the berry patches along Moose-Wilson Road, 760 meandered to one side of busy Highway 390 for stretches and then over to the other side. Noteworthy is that the exurban environs are full of homes and well-landscaped grounds, trash, barking dogs, cars, and people, but, as his family members had in Grand Teton National Park, he made no menacing moves. Indeed, it is logical to ask at what point does a bear cross an invisible line into a realm that is no longer acceptable. According to Mangelsen, there is a difference between a grizzly steering through the suburbs and inhabiting the park, but he believes it was understood that given who the bear was, and it being late fall, 760 would likely soon head back north again.

In the Aspens, 760 fed on berries, a small carcass, and the remains of another. He likely did not differentiate between an animal here and one that had been killed in the park hunt. Their patience expended, Wyoming Game and Fish set a snare, tranquilized 760, and hauled him on a long journey to the North Fork of the Shoshone River, in the Shoshone National Forest east of Yellowstone. It was a drop-off site bear managers knew contained a high density of other boars. Hayden said that's important to note because it means they were dropping him into an area where he would meet hostility from his own kind, let alone be in unfamiliar surroundings.

"You can't blame him if he eventually decided to hightail it out of there," Mangelsen said. "In the buildup to denning, most bears rely on a homing instinct piqued by places, smells, temperature, and the light of day. Up there, he must have felt like he had been dumped onto Mars with scary bears that didn't provide him with a welcoming party."

He raced up Grinnell Creek into the Absaroka Mountains. Wyoming Wildlife Advocates obtained GPS coordinates from Wyoming Game and Fish showing that 760 headed north and east out of the Absarokas toward lower ground near Clark, which is

The grandson of 399, 760—whose 2014 death was controversial—hunts over pocket gopher warrens near Willow Flats.

where ranches and farms press up against public land. Speculation is he might have been fleeing male bears. He found his way toward the banks of the Clark Fork of the Yellowstone River, having no orientation to his surroundings. And, this being in the fall, he had no den site awaiting him. One day, 760 sniffed into the air and caught a whiff of ungulate meat. He followed his nose and found a hunter-harvested deer hanging from a tree.

It must have looked like a dangling invitation of a free lunch. He reached up and pulled it down, feasting on it, perhaps his first hearty meal in several days. The owner of the carcass, to his credit, didn't pull out his rifle. He made it to his pickup truck and tried to haze 760 away. The hungry bear, who had never encountered treatment like this before, didn't budge. He continued eating. He was, after all, in hyperphagia. The hunter called Wyoming Game and Fish, who set a leg snare near the carcass. 760 was caught, tranquilized, and held in a barrel trap in Cody. Following the well-established protocol, Wyoming Game and Fish phoned Dr. Christopher Servheen, the national grizzly bear recovery coordinator for the U.S. Fish and Wildlife Service. He issued permission to have 760 euthanized.

When word reached Jackson Hole, the bear-watching community went into an uproar. 760 was considered a member of royalty. "This was a bear that grew up around people in Jackson Hole and never got into trouble until it was relocated to a completely unfamiliar area, a region with a strong resident grizzly population," observed Franz Camenzind, the biologist, cinematographer, and conservationist. "Looking back, it seems the agency relocation of bear 760 created a challenge the bear could not overcome."

Hayden and his associates lay out a litany of complaints that called into question the judgment and procedures employed to deal with 760, including the apparent haste in classifying him as a nuisance bear. Mangelsen commended them for demanding answers, and, in the bigger picture, he wondered what it would mean in the years ahead when managers would have less rigid protocols to follow before destroying grizzlies. He wonders how adept Wyoming Game and Fish is at appreciating the qualitative differences between grizzly behaviors that consider more factors.

"With the incident occurring in the Jackson Hole suburbs, the risk of something bad happening for the bear and people was just too high," Wyoming Game and Fish Department large carnivore coordinator Dan Thompson told me. "We have a lot of country out there for bears and it's a good thing. We need to recognize the importance of social tolerance for bears and wolves. I heard the claims that that bear would never hurt somebody or that it was only a gentle bear, but somebody has to be held accountable for the risk. One bad event and progress could be set back. God forbid a bear hurts or kills a person. It will not be good for bears and it will bring negative connotations."

CITIZENS IN WYOMING, MONTANA, AND IDAHO have a mirror being held up before them, the same one that Stephen Herrero offered in 1986. Bonnie Rice of the Sierra Club speaks aloud the topic no one wants to discuss. "Perhaps the biggest question is should this majestic and rare animal—the ultimate symbol of our wild heritage that today occupies less than two percent of its historic range—be hunted for sport?" Rice says no. She doesn't think the grizzly should be removed from federal protection. She's not opposed to hunting big-game animals, such as elk, for meat. Just not grizzlies. Mangelsen concurs. "This is not to say grizzlies won't be 'managed,'" he says. "We know that bears are going to be removed in management control actions, especially grizzlies that chronically get into livestock herds and bears that hang out on the back doorsteps of someone's house, getting into their trash and eating the dog and cat food. But removals must be justified. Although I think all of us conservationists know they are necessary, we shouldn't be treating the grizzly as a 'consumptive resource.'"

The Wyoming Game and Fish Department has already listed prospective grizzly bear licenses on their website for approximately $650, Mangelsen says. "In the future, we definitely envision hunting as a management tool, but it will not be the end-all," Dan Thompson says. "It's not any kind of silver bullet for bear management. We're not going to do anything to threaten the population because we've done so much to get to this point. We're not going to jeopardize that by being greedy in the number of bears that might be harvested in a hunt."

One philosophical line of argument used to reject the hunting of Greater Yellowstone grizzlies is spelled out in a recent scientific paper titled "Wolf Hunting and the Ethics of Predator Control." Lead author John Vucetich is a renowned researcher who has studied wolves on Isle Royale in Lake Superior and across the Upper Peninsula of Michigan. "Because wolves—and other predators—are living creatures, the morality of killing wolves and other predators depends on being able to provide a good reason to do

A handsome bruin, 760 had learned how to navigate the interface between wild country and people. Even though he was never aggressive, his presence near human developments made some nervous. The justification for his removal is a source of intense, ongoing debate.

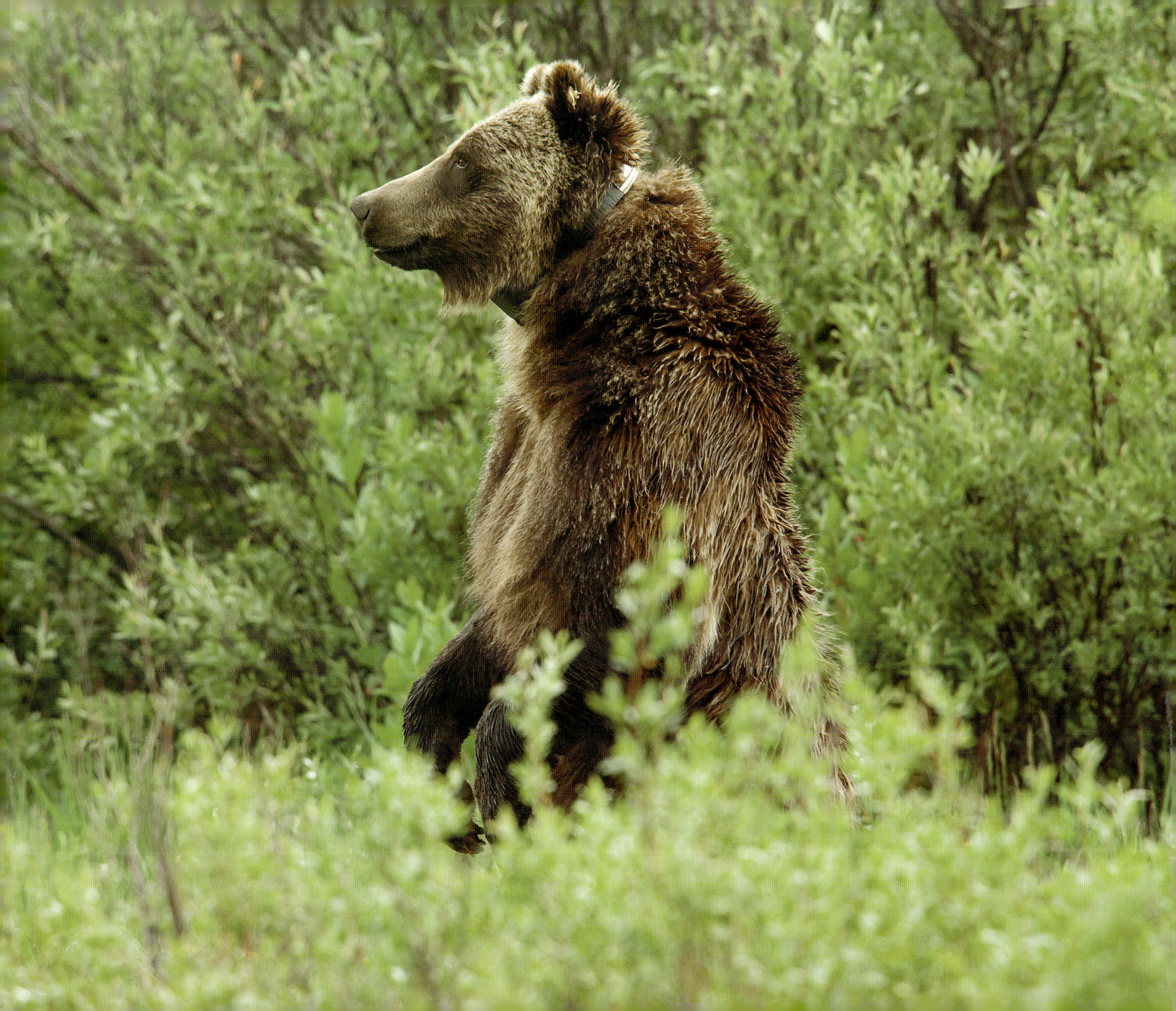

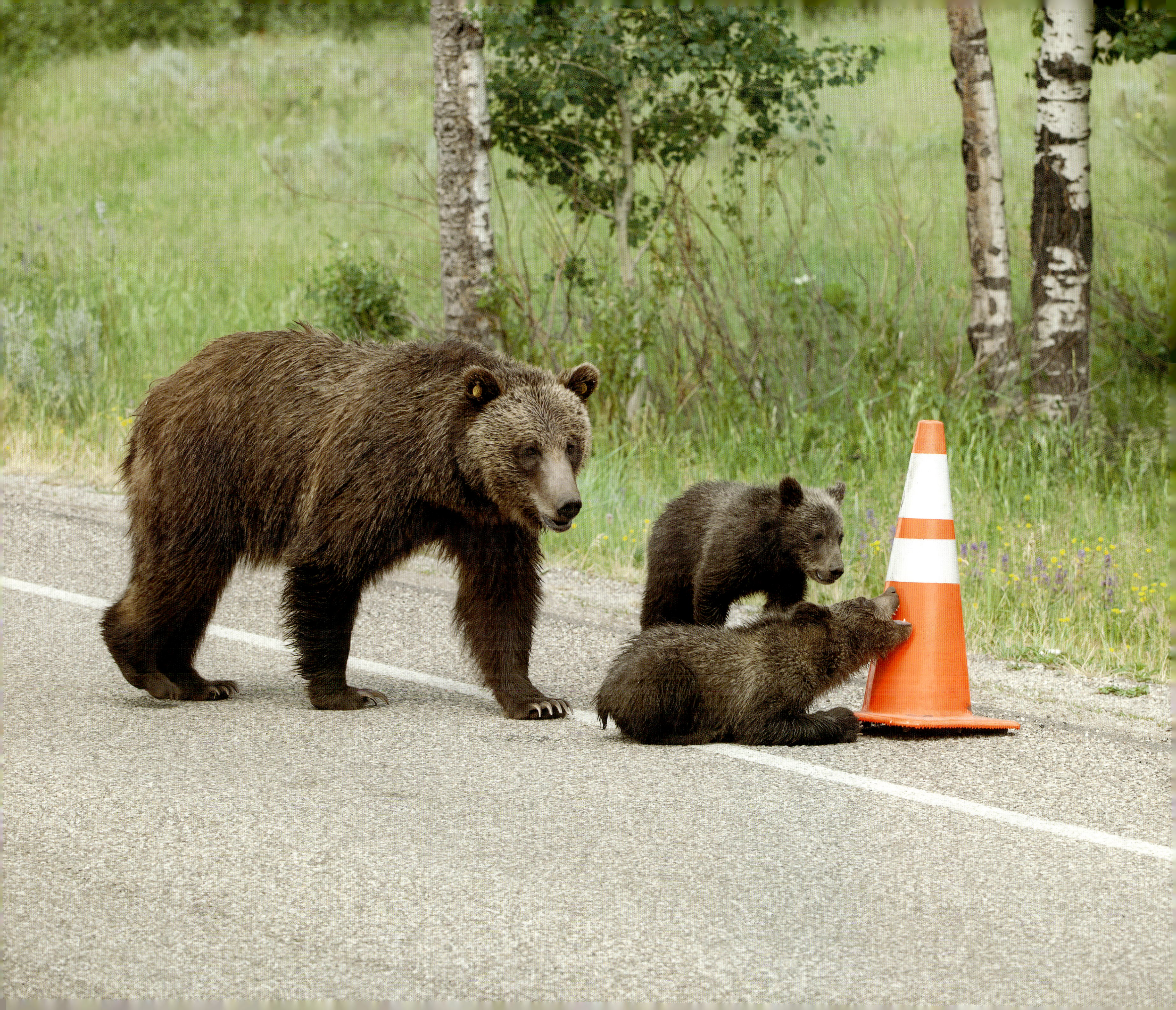

so. The analyses presented here and elsewhere suggest that good reasons have not been offered," he and coauthor Michael P. Nelson wrote. "This conclusion may raise the question, 'Who gets to judge what counts as a good reason?' That question is misplaced. In a free society, every citizen is free to judge what counts as a good reason. The critical question is not who gets to judge, but rather, 'By what rules and standards is one obligated in judging what counts as a good reason?'"

Americans have spent millions of dollars resuscitating grizzlies in Greater Yellowstone. It has been more than 40 years since they were last hunted in the West. Hunting grizzlies is not a deeply ingrained cultural tradition in contemporary society. Therefore, there is no compelling reason to hunt bears, as managers should be trained and capable of taking care of the population. Hunting ostensibly would entail random killing of bears. It likely would alter bear behavior and it could most impact the very bears, like 399, that are models for the kind you want living on the edge of civilization. Frank Van Manen isn't convinced it will happen to 399 and her descendants, but the behavior of modern European brown bears is a testament to the flight response of bruins being hunted.

"The ethics of hunting are complicated," Vucetich and Nelson wrote. "Even ardent supporters of hunting disagree among themselves, for example, over the appropriateness of hunting methods that maximize the possibility of a clean kill to minimize suffering and the appropriateness of methods that emphasize fair chase." For 399 and 610—because their response to humans is more nuanced, arguably more sophisticated and tolerant—their trusting nature of humans would, ironically, make them more vulnerable and perhaps the first to be killed.

"Is that really what Wyoming would want to do with bears that are ambassadors for the most popular wildland destination in the state and Lower 48?" Mangelsen asks. "It seems absurd that for a few cheap thrills and a $650 license, somebody could come in here and take a beloved asset, so synonymous with all of the good things Jackson Hole stands for, kill it, and cart it away like a stack of cordwood in the back of a pickup. Have we not progressed as a society? How obscene is it to think that in someone's 'man cave' bears like 399 or 610 could stand fully stuffed and manipulated by a taxidermist to imply they are alive? And then, over drinks at a cocktail party, the person who shot the bear could boast that he brought down one of the most famous grizzlies in the world."

"Hunting *is* the elephant in the room," grizzly study team researcher Mark Haroldson says. "Whatever your position, it is based on your personal values." Like it or not, grizzlies are a symbol of maligned federal presence in the West for some. Walk into the legislative chambers of any rural western state capitol and the antifederal sentiment floats in the rhetoric. Some state leaders begrudge the federal government for taking over control of grizzlies and forcing citizens to live with them (even though the problems have been minimal compared to the fear-mongering about bears that pervades any discussion of them). There are old-timers in saloons who hold court saying their grandfathers proudly erased grizzlies from the landscape and made grazing areas safe for cattle. There are politicians, whether they personally believe it or not, that know they won't get elected unless they grandstand against grizzlies and feed an onerous characterization of the Great Bear.

STEVE PRIMM, A BIOLOGIST AND WILDLIFE conflict-resolution specialist with carnivores in the Madison Valley of Montana, praises the work of government agencies in protecting grizzly habitat. "We face a paradox with the 'message' of delisting," he said. "There may be a lot of people who are expecting that delisting is going to give them the opportunity to push grizzlies back geographically and numerically. Yet the conservation strategy that states had to embrace before they could assume management, and other governing documents, commit us to a very different course than simply regarding bears as being expendable again."

History serves as a reference point for marking how far we've come as a society. Yet in the present day and age, with rapid antifederalism, antiscience, and antienvironmental rhetoric fueling crazy talk in western state legislatures, Primm's caution about provincial attitudes surging again is well founded.

In 1974, Rex Corsi, then Wyoming's chief game warden, complained openly to a national newspaper reporter about potential federal meddling in telling his state how to steward bears. With no hard numbers, no accurate scientific method of counting bruins, and only fly-by-the-seat-of-his-pants instinct, Corsi thought there might be 50 grizzlies in the state—more than enough, he insisted, to justify the issuing of 12 trophy-hunting tags.

If all of those aspiring to kill a grizzly had been successful, it would have amounted in one year to a legal harvest equaling nearly 25 percent of the population, not including

Cubs can turn any distraction into a game of fun. But when you inhabit the front country along a busy highway, it comes replete with peril. Mama 610 is an expert chaperone.

poaching, which was a rampant problem, ranchers killing bears to protect livestock, and others putting down bears in self-defense.

Up in Montana, 666 grizzly licenses were sold in 1974, even after state officials liberally estimated that just 800 grizzlies probably inhabited all of the Lower 48.

Corsi and his contemporaries in Montana had reason to worry. The year before, Congress put a new law on the books; it was called the Endangered Species Act and was initially enacted to protect bald eagles but, in only a matter of months, also provided a shield of protection for grizzlies.

As it turned out, states had been dangerously optimistic in calculating not only the status of bears throughout the northern Rockies but also the number that could be safely harvested without jeopardizing populations. Wyoming's legal allowable quota was more than five times the level that makes the difference between a population that is stable and holding its own and one that is falling into decline. Montana's issuing of so many tags—too many, based on the argument that not all hunters will slay their quarry—was hugely troublesome to bear researchers like the Craigheads and Stephen Herrero.

The year 1974 was the last that grizzlies were hunted for sport. Max Planck, the great German thinker who earned the admiration of Albert Einstein and won a Nobel Prize in physics, once remarked, "A new scientific truth does not triumph by convincing its opponents and making them see the light, but rather because its opponents eventually die, and a new generation grows up that is familiar with it."

I asked Dan Thompson—who has the difficult task of managing the big three large predators (omnivorous grizzlies and carnivorous wolves and cougars) and doing it in a way that keeps the peace with people whose lives and habitat intersect—his opinion on the subject. Thompson—who expressed open fondness with grizzlies, saying what's happening in Greater Yellowstone with conservation is something to behold and champion and celebrate—has hope. The destructive attitudes of the Old West, he claims, are gone. Thompson and the state realize keenly that the relationship Jackson Hole has with bears, especially the clan of 399 and offspring carrying her bloodline and adaptive behavioral traits, is special.

Not long ago, Roger Hayden recounted a conversation with a colleague who mixes socially with a Jackson Hole outfitter: "The outfitter allegedly said he couldn't wait until bears are delisted and the state [Wyoming] enacts a sport hunt. He can't wait to get a tag because the first bear he's going to target is 399 because he hates that bear. He hates what she represents. He hates environmentalists. And he hates the federal government telling people in Wyoming what to do." He is not alone in voicing his sentiment.

Of course, apart from that shocking declaration of intention, if true, no one has probably pointed out that the outfitter makes his living on the largesse of the American public, the vast majority of which love grizzlies. But through their generosity, the outfitter is provided with an inspiring office (federal public lands) and a product (public wildlife) for him to profit on and at what amounts to a giant taxpayer subsidy.

Such attitudes are something that Mangelsen rubs up against often, causing him heartburn, but it's offset by an ever-growing and prolific love for the grizzly. "I believe that, in the end, love trumps hate and I'm betting that the bears of Jackson Hole will endure because the old hatred will fade away as those who used it to formulate wildlife policy toward predators pass from the scene," he said. "In a way, I pity the people who can't find it in their hearts to appreciate what grizzlies are, right before our eyes. I feel sorry for them. Anyone who can't let that in must be really suffering."

On the other hand, those who still insist on persecuting predators (bears, cougars, wolves, and coyotes) are only proving the point that they cling to ignorance. They are committing offenses not just against these animals but also against science, the recognition that wild nature holds priceless value in our lives and that the world around them is changing. The more people who know the story of 399, Mangelsen said, the better. It can and should be a part of society's oral tradition, so that citizens can say they lived in the time of 399 and they rallied to support and maintain a feat that architects of the Old West said could never be done: the heroic rescue of *U. arctos horribilis*, which isn't actually such a horrible bear after all.

"Grizzlies symbolize the wild, rugged beauty and spirit of this cherished region," said Scott Christensen, conservation director of the Greater Yellowstone Coalition. "As a keystone species with a need for vast open landscapes, they tell us that if they're here this place is healthy and still functions as an intact ecosystem, one of the last on Earth."

While the amount of money that Wyoming spent on paying ranchers who lost cattle to grizzlies isn't insignificant, it must be considered within the context of the $1 billion annual value of nature tourism, featuring grizzlies and wolves as marquee attractions. A

What insight does the horizon hold? A grizzly in Yellowstone surveys a distant valley. Once there were 50,000 grizzlies. Today, the Great Bear in the Lower 48 inhabits just two percent of its historic range.

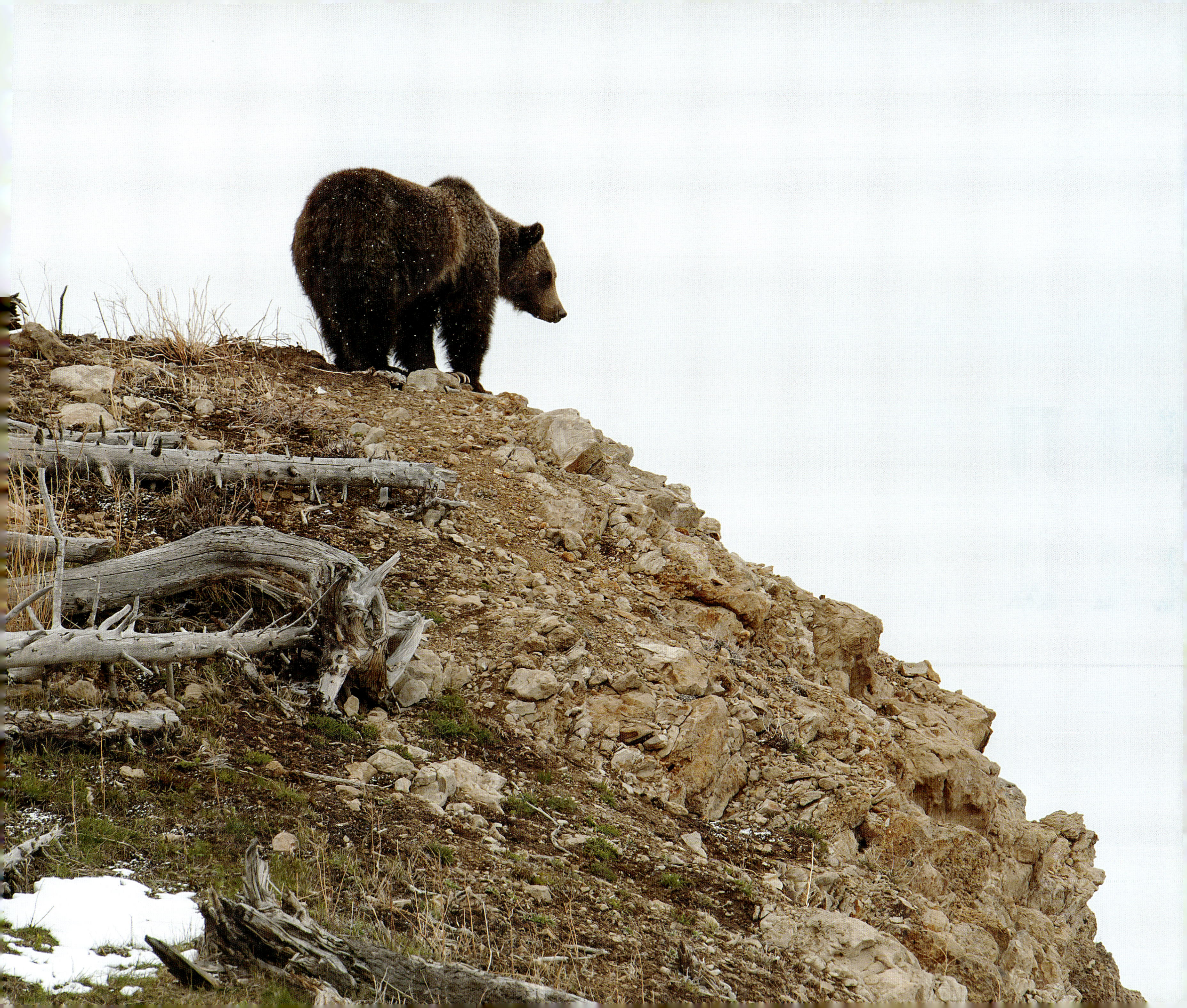

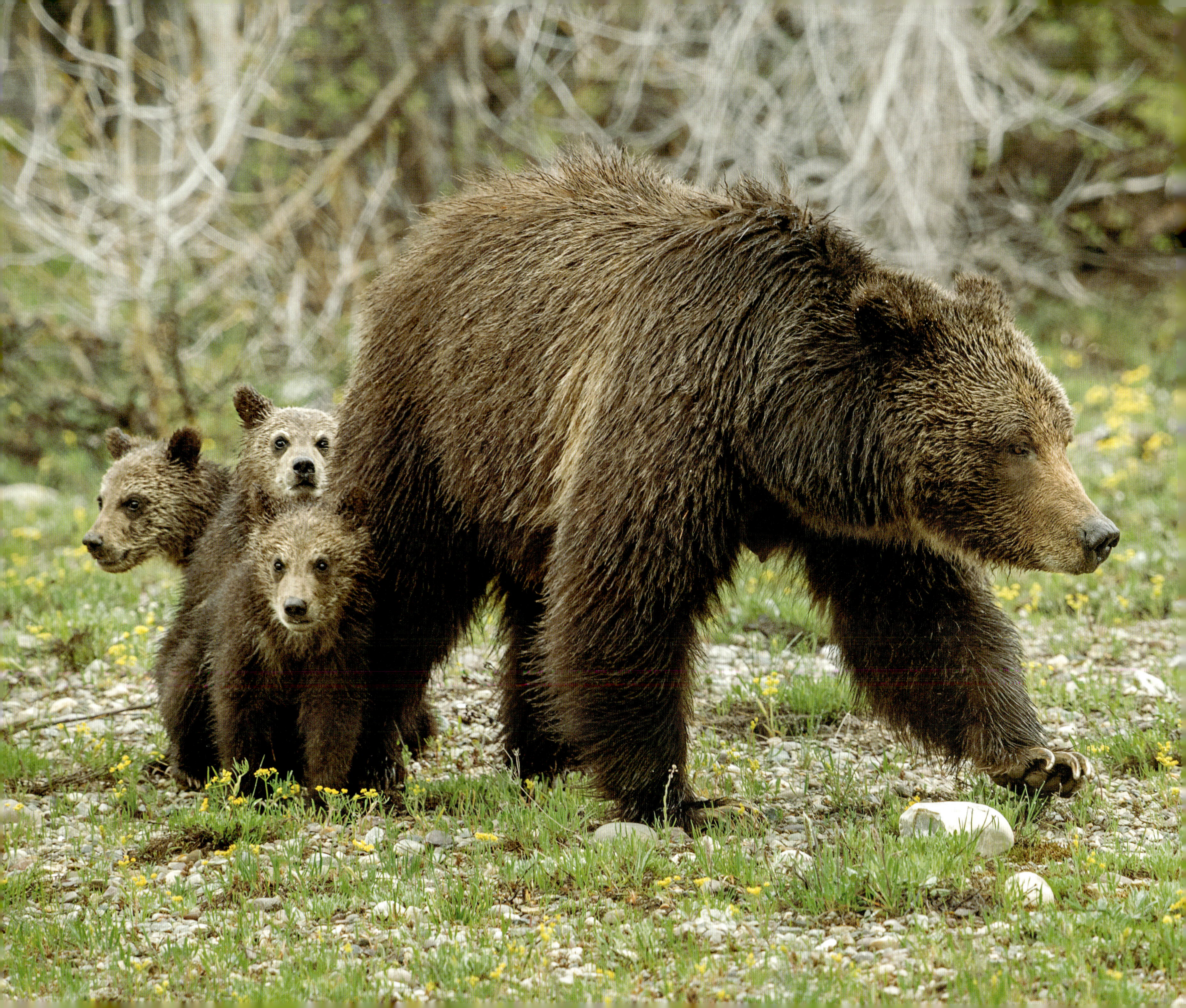

study published in the *Journal of Environmental Management* indicates that grizzlies in Yellowstone alone contribute millions of dollars in economic value. Tourists would be willing to pay $41 in addition to the park entrance fee of $25 to see bears. According to researchers, results indicate that if bears were no longer allowed to stay along roadside habitats, spending in the local economy by park visitors would decrease by about $10.1 million, resulting in a loss of 155 jobs. According to Hayden, grizzlies are worth far more alive than dead. The millennial generation now coming of age will be the first in the history of the West to regard grizzlies as being overwhelmingly positive economic assets rather than impediments to progress. Jonathan Schechter, who runs a think tank in Jackson Hole focusing on the intersection of economy and ecology, has a maxim he invokes for a number of nature-related issues in which common sense is disconnected from reality: "Economies change faster than perceptions; perceptions change faster than politics." The soaring global interest in Wyoming's bears, wolves, and protected areas should be telling Wyoming politicians what they need to know and they should be leaders in ushering forth a new kind of discussion for their constituents. Wildlife conservation yields tremendous dividends year after year.

Americans who reside in grizzly country need to step back and reflect on what they are witnessing, because the rest of the world is watching, paying attention, pointing a finger with envy. Grizzly recovery is what hope looks like for those trying to save African lions, Indian tigers, and polar bears.

"The reason it's spectacular is the fact that it's a big animal, so scary and powerful, and it brings a lot of considerations to how we think about the landscape," said David Hallac, Yellowstone's former science chief who now serves as superintendent of Cape Hatteras National Seashore, which is dealing with rising seas caused by melting ice linked to climate change. "There's not a lot of other places in the world where you could do something almost unimaginable like grizzly recovery and it's only possible because, by and large, humans are able and willing to coexist. This is a lot different from restoring a small plant or a little bird; it's an animal that, at a population level, needs so much space. It's pretty amazing to think about. The grizzly has been a tool for making us better managers, to see the need to knit the landscape together across differing jurisdictions. Grizzlies have made government more accountable to the people."

The journey of 399 and her offspring may be the most closely documented of any wild grizzly family in human history. These bears have alighted imaginations, charmed their way into our own sense of place and longing for wild nature, given us a better perspective on the value of rare species in a crowded human world, and provided endless fodder for thinking differently about grizzlies than our fear-mongering ancestors did. Yet reality forces us to look inward at ourselves. Can we make a place for grizzlies in our hearts?

Yes, 399 and her line of cubs has remarkably added 13—and counting—new bears to the total Greater Yellowstone population, but three-quarters of them, at least, have already perished from clashes of various kinds with people. Meditate on that for a moment. Is that a prescription for recovery?

THE SURVIVAL OF GRIZZLIES DEPENDS less upon what bears do; it's more about our willingness to be accommodating. And it's a challenge that applies as much to grizzlies as to lions, tigers, elephants, rhinos, and every other dwindling creature on our planet. This book is meant to be a valentine of hope. So, now that we are at the end, consider this scene:

Carried with Thomas Mangelsen in the pocket of his camera vest are oval-shaped bumper stickers that he hands out to kids and young people, bringing them into the fold of a special fan club. The stickers—which say "I Saw Bear 399" and "I Saw Bear 610"—are collector's keepsakes and thousands upon thousands are scattered around the world on cars, tacked to bulletin boards in college dormitories, and stowed away in trinket boxes. These mementos are only awarded to those who have actually experienced the delight of laying their eyes upon those special bruins. The stickers commemorate an experience they will never forget; they will become part of the oral traditions of families as we move deeper into the twenty-first century.

Today, should a family with children feel inspired to come to Jackson Hole and head out bear watching, there's a strong likelihood they might stumble upon not just the Great Bear but also a two-legged silvertip holding a vigil that doesn't end. Mangelsen may look grizzled but he's probably just entranced in concentration, being attentive to the nuances of the shoulder-humped bruins around the vicinity of Pilgrim Creek. Approach him. He won't bite.

399 has been called an extraordinary mother by bear researchers, with at least 13 offspring in her lineage. Unfortunately, a high percentage of these bears have perished from various kinds of run-ins with people. She is a muse for thinking about grizzly conservation.

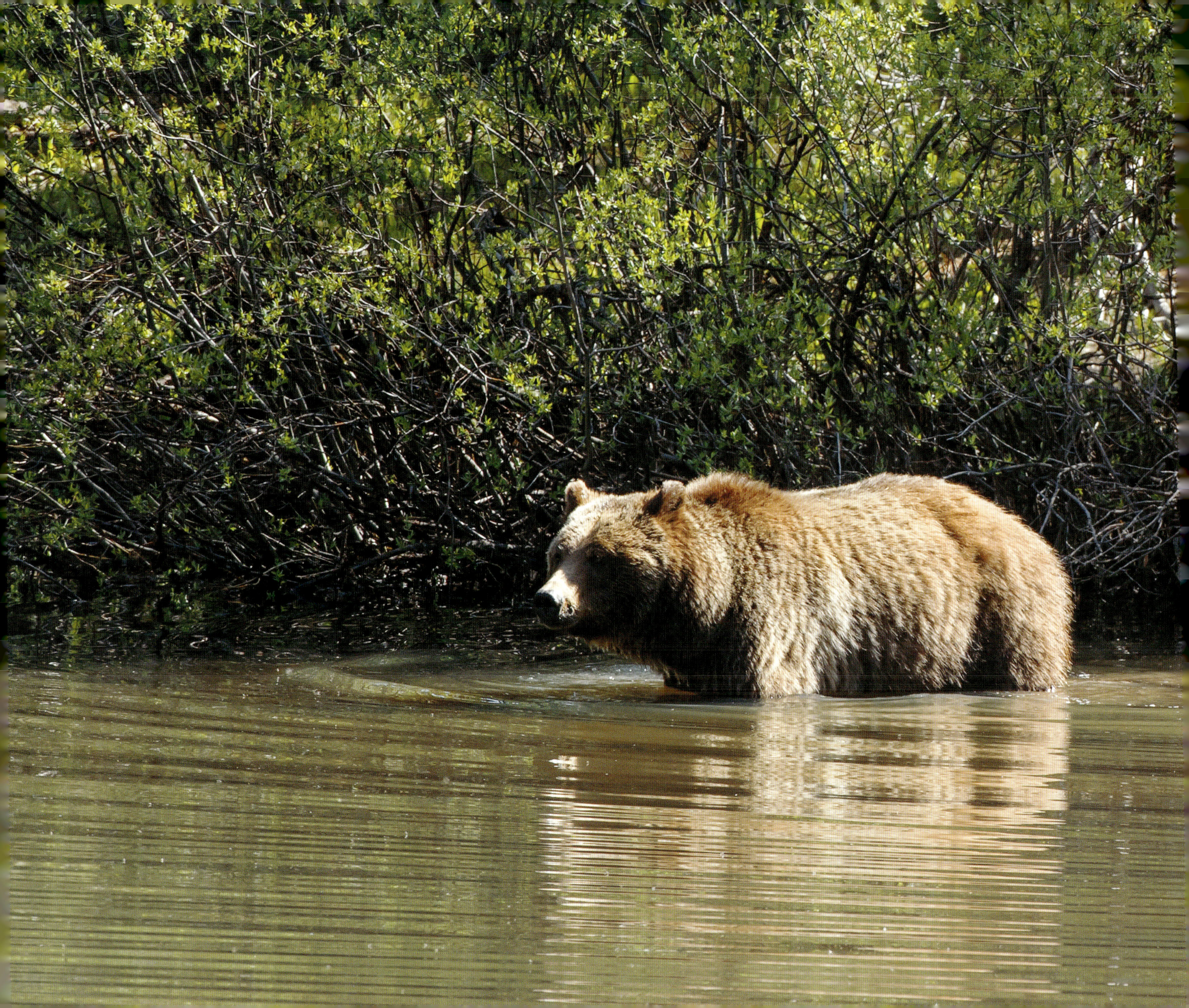

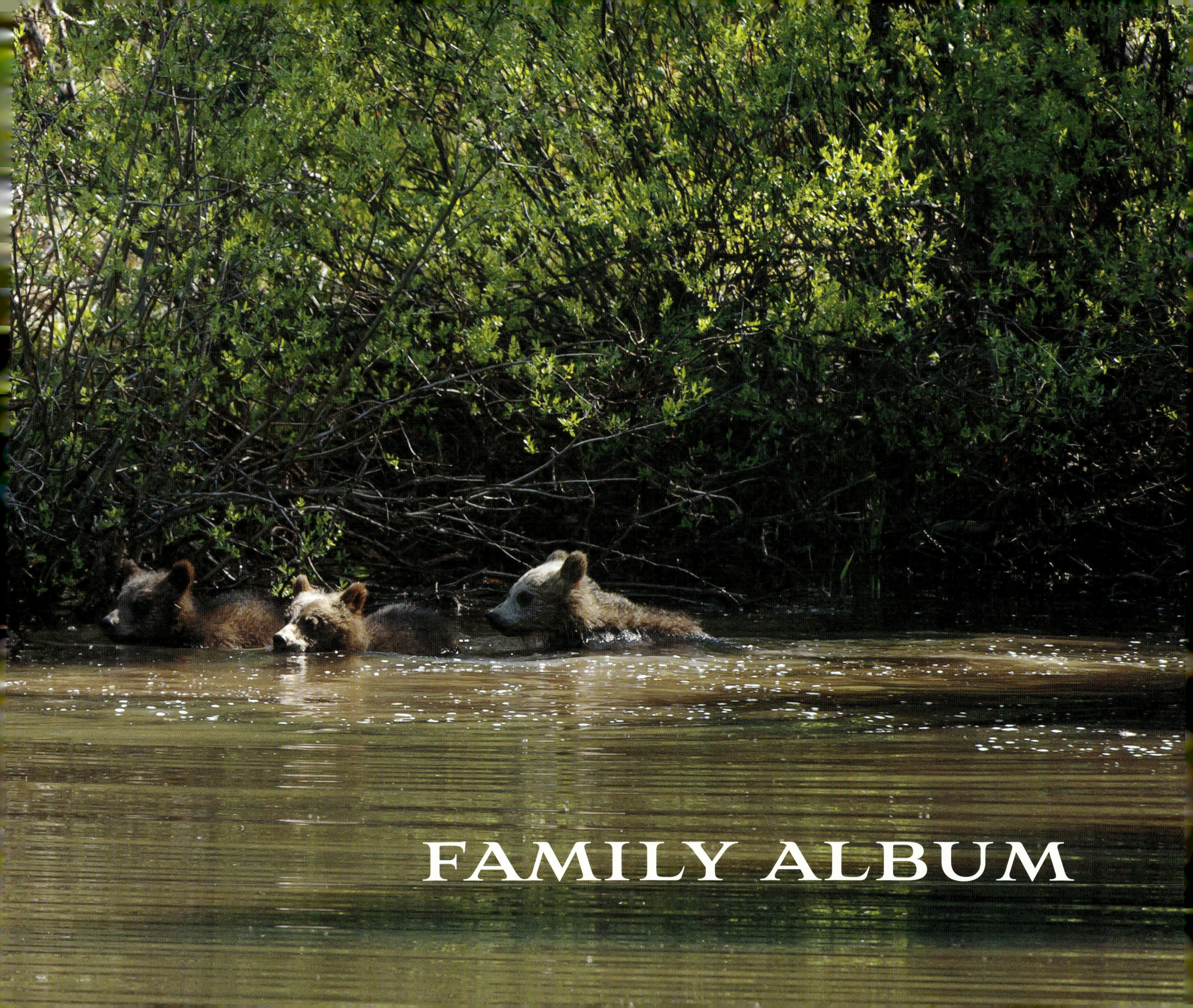

FAMILY ALBUM

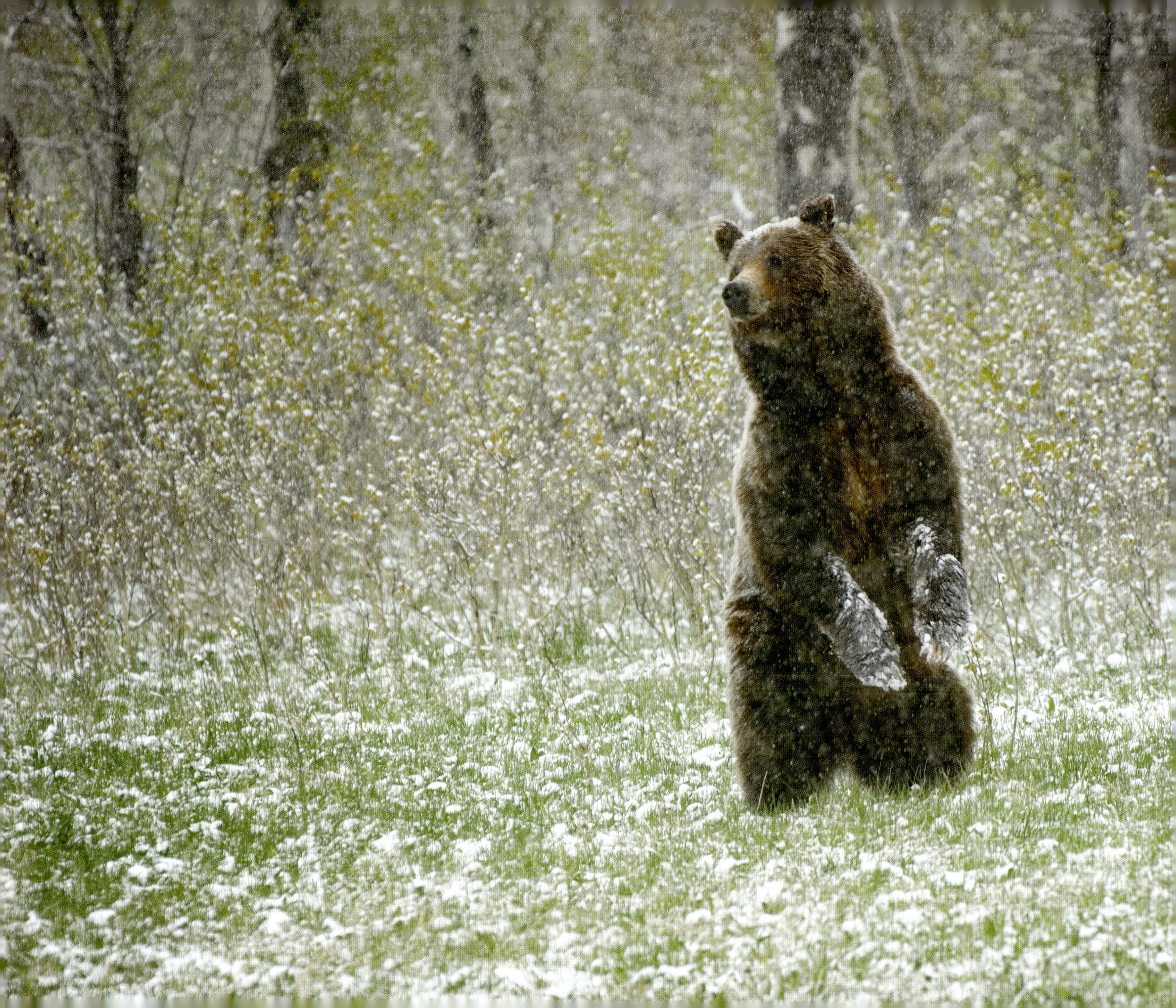

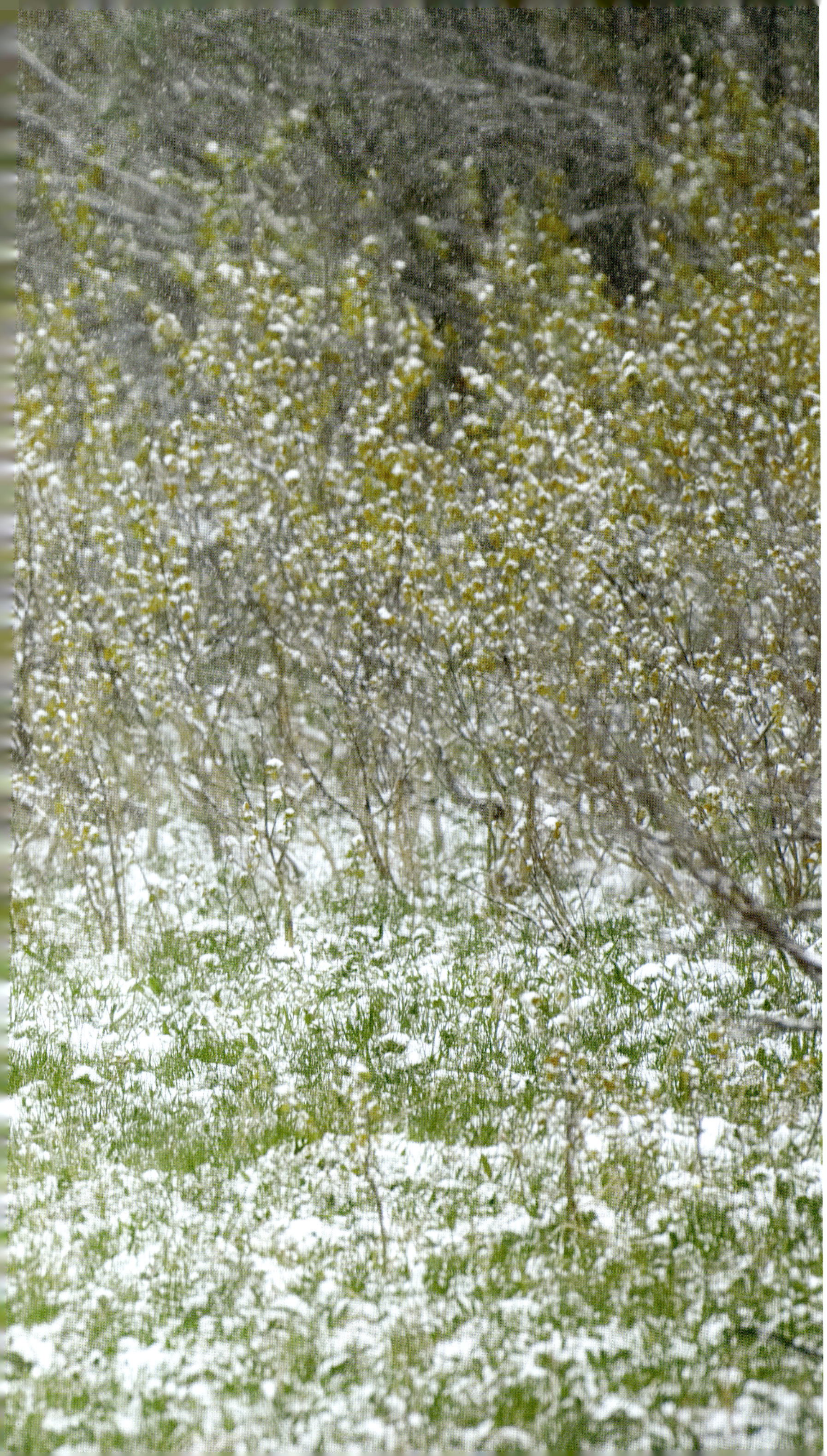

PREVIOUS SPREAD: On a warm morning in late June, 399 and her triplets cool off in a pond south of Colter Bay.

During a late spring snowstorm, 399 stands and calls to her cubs.

FOLLOWING SPREAD, LEFT: Two of 399's yearling cubs play in an aspen grove near Jackson Lake Junction.

FOLLOWING SPREAD, RIGHT: 399 lays back while nursing her cubs 610, 615, and 587 in a meadow near Oxbow Bend.

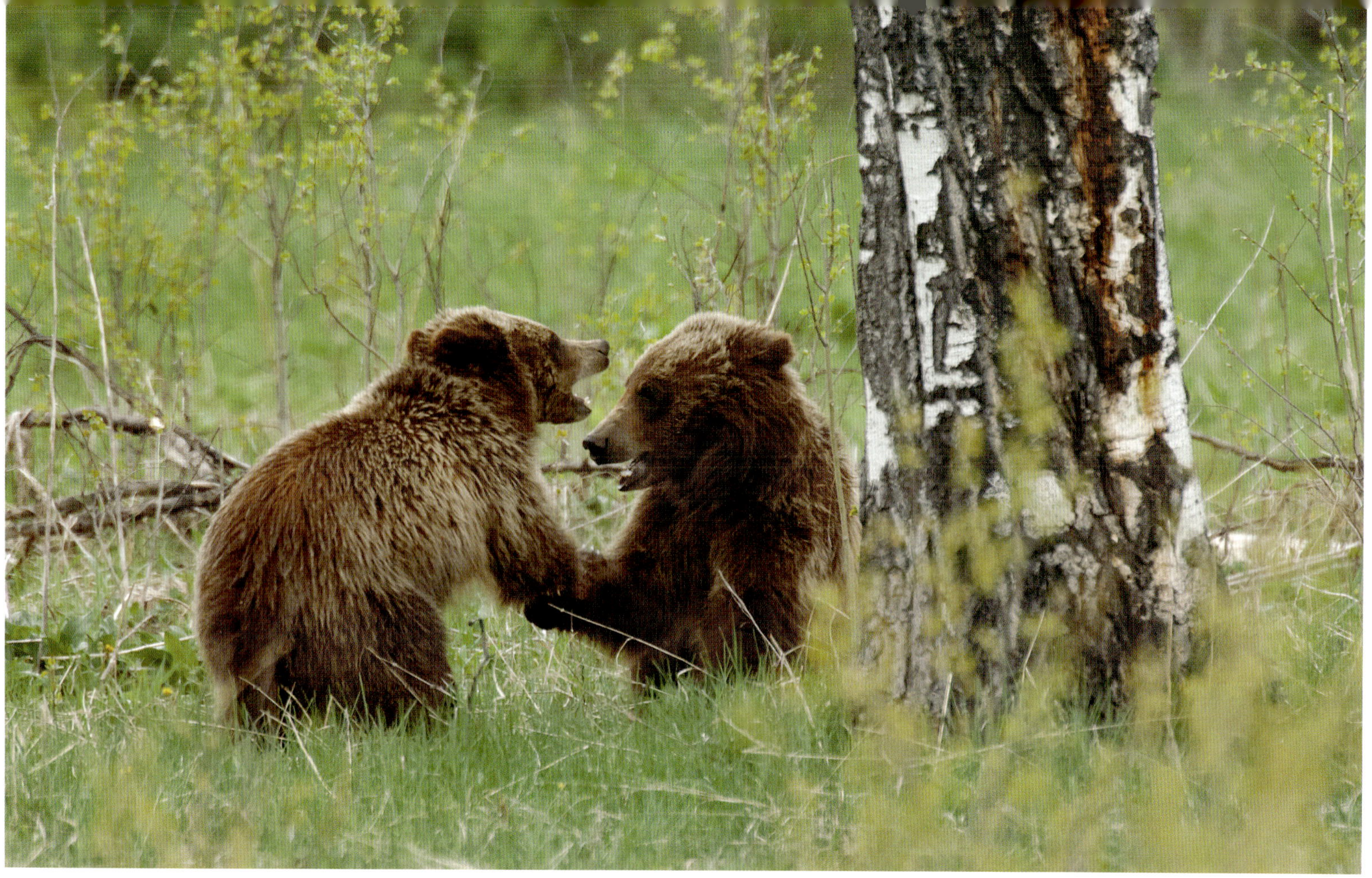
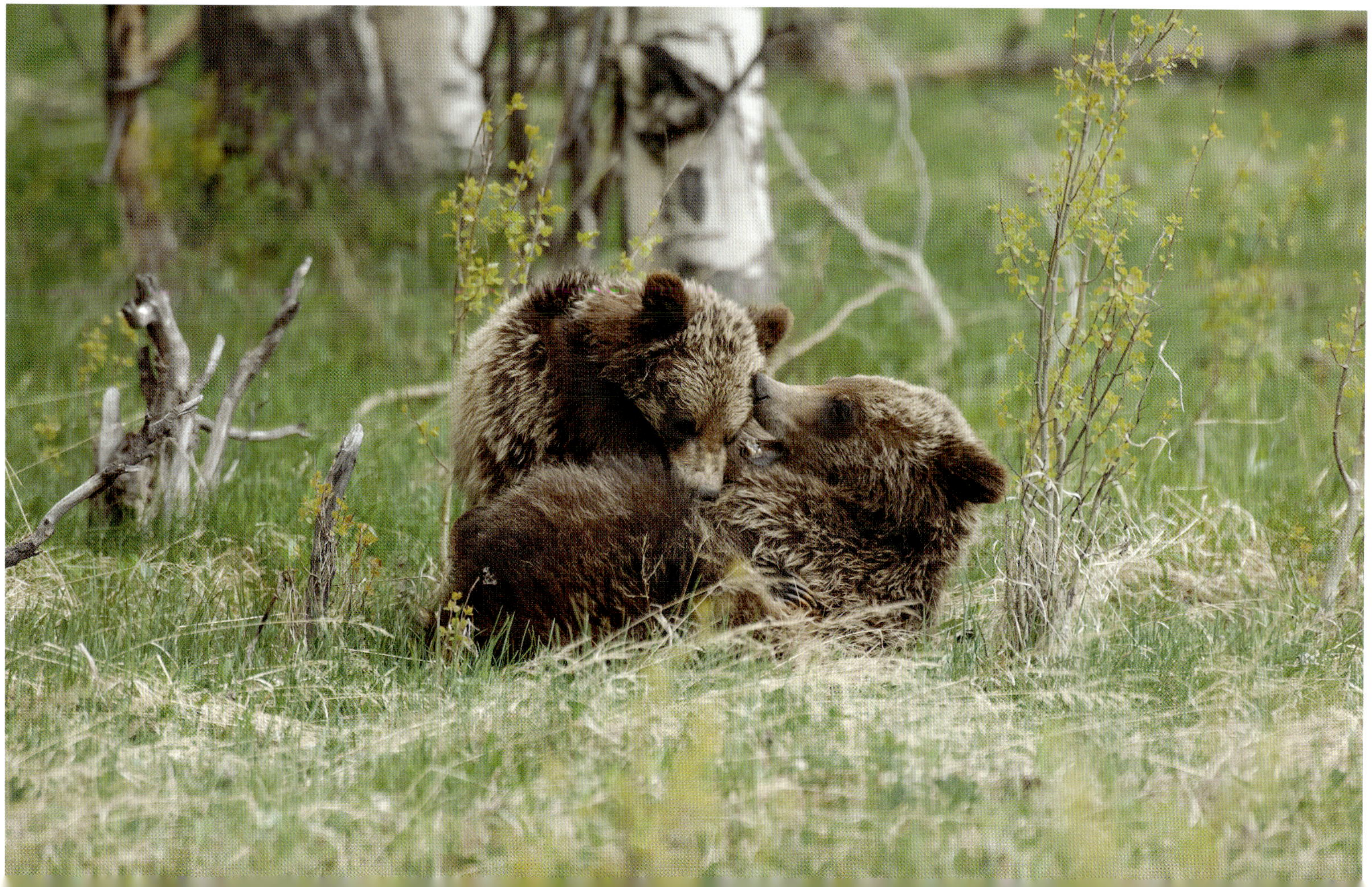
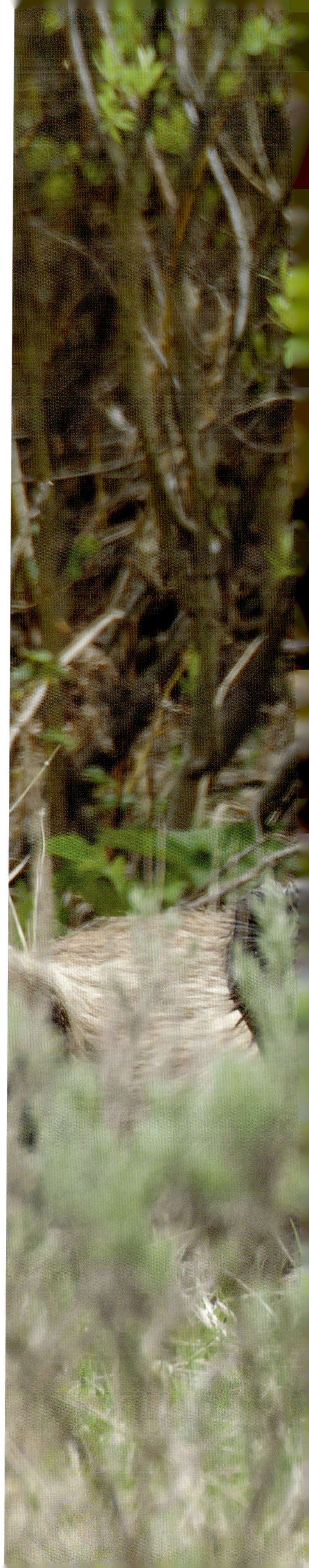

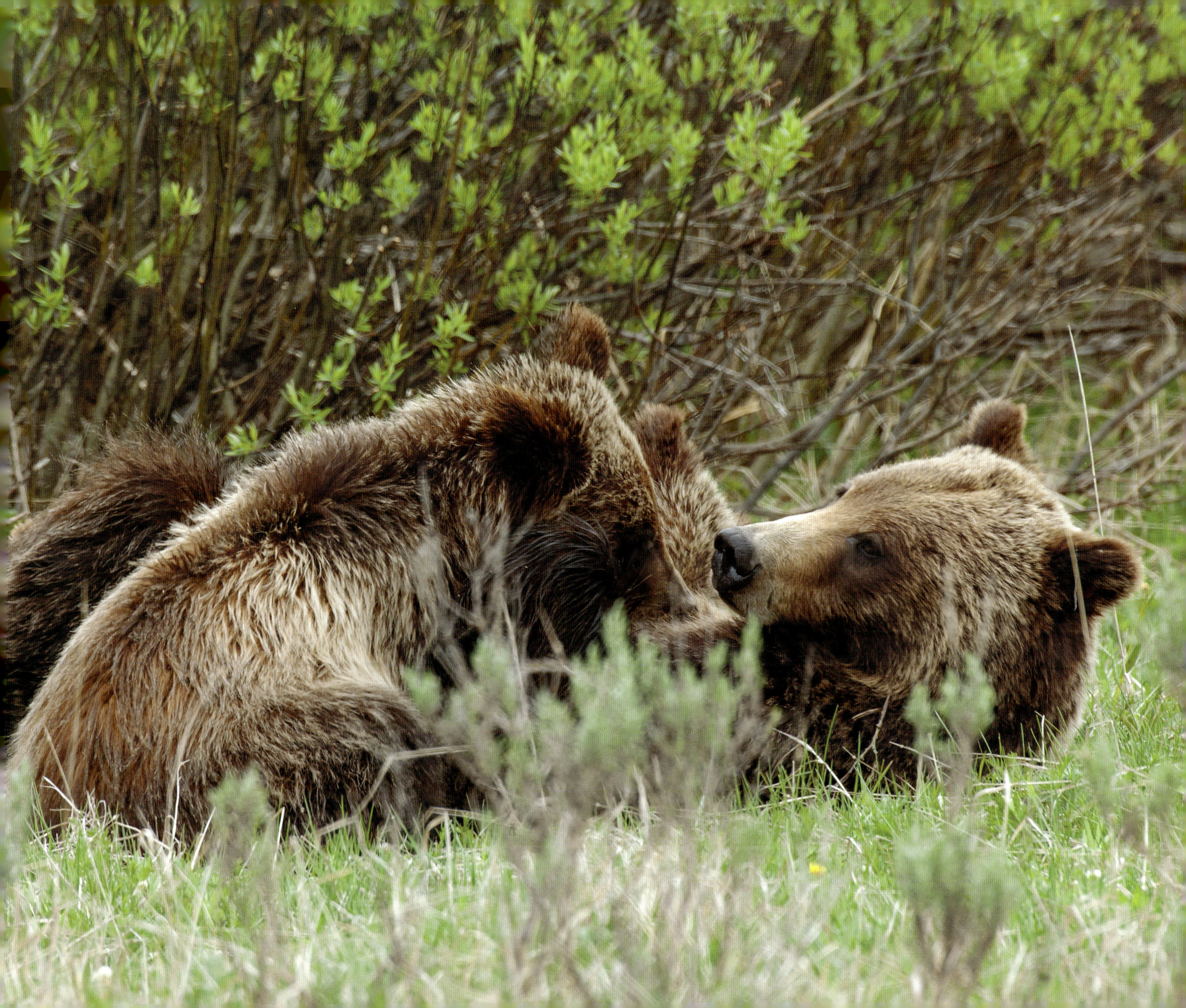

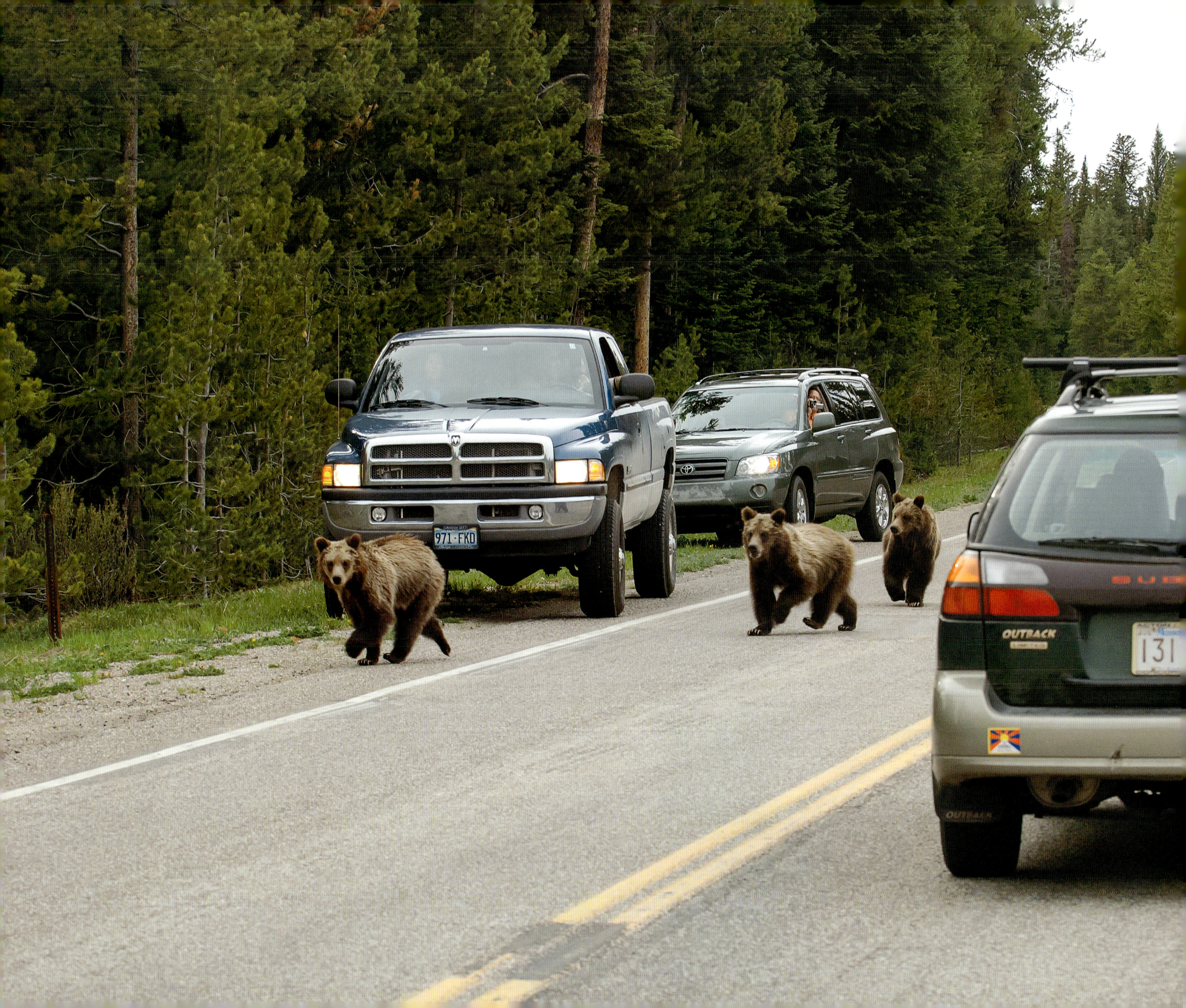
971·FKD
OUTBACK
131

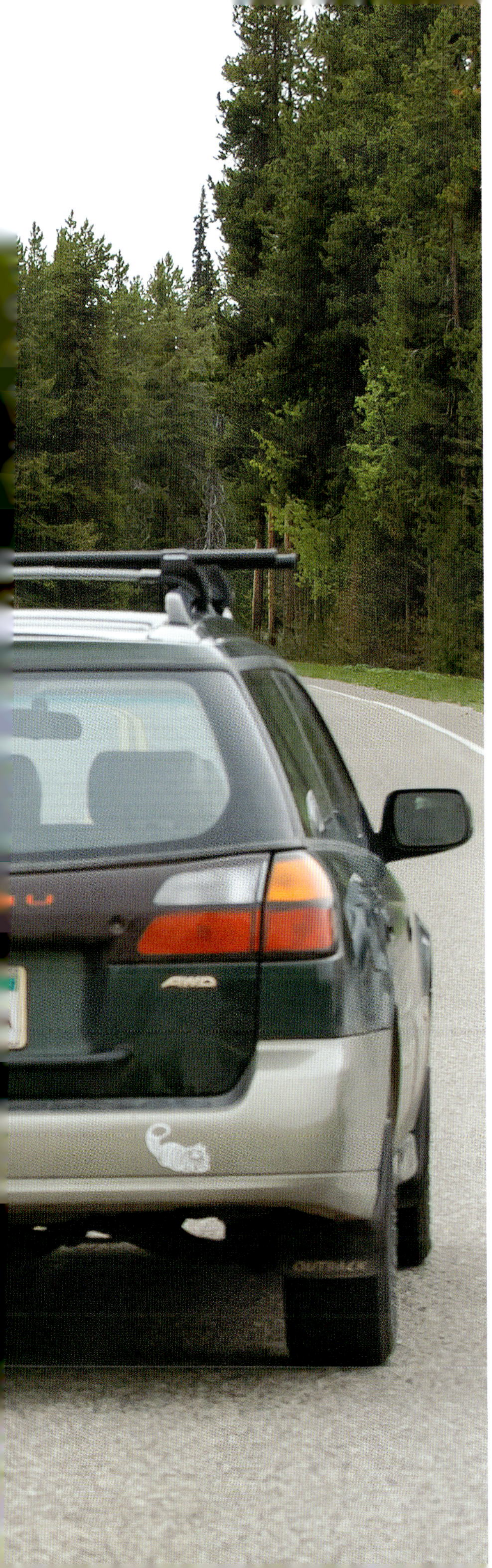

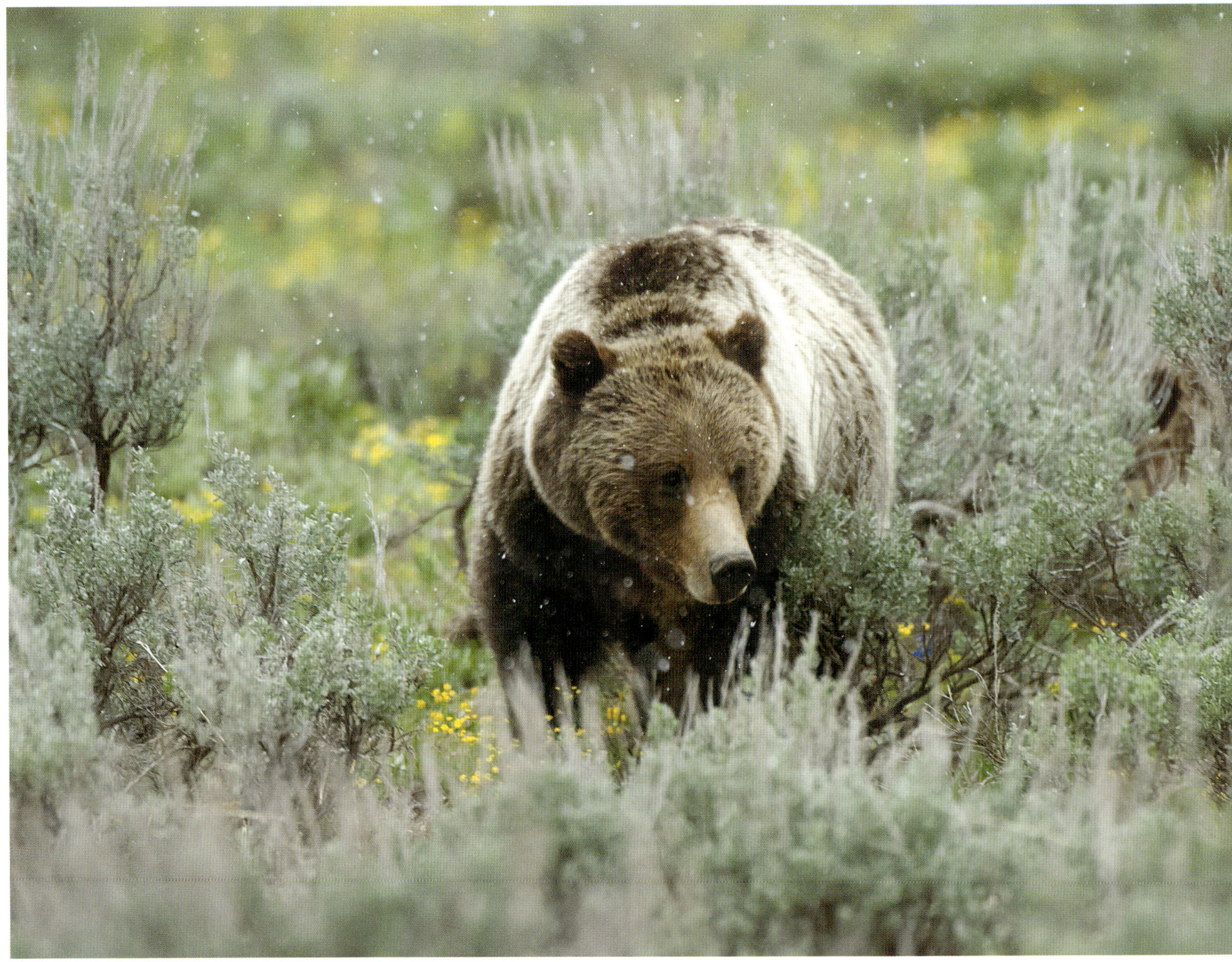

OPPOSITE: After losing sight of their mother in the tall sagebrush, 399's yearling cubs dash across the highway in search of her.

The dramatic and frantic separation lasted more than 30 minutes. Once reunited, 399 immediately nursed and comforted the cubs and then continued grazing on the biscuitroot growing among the sagebrush.

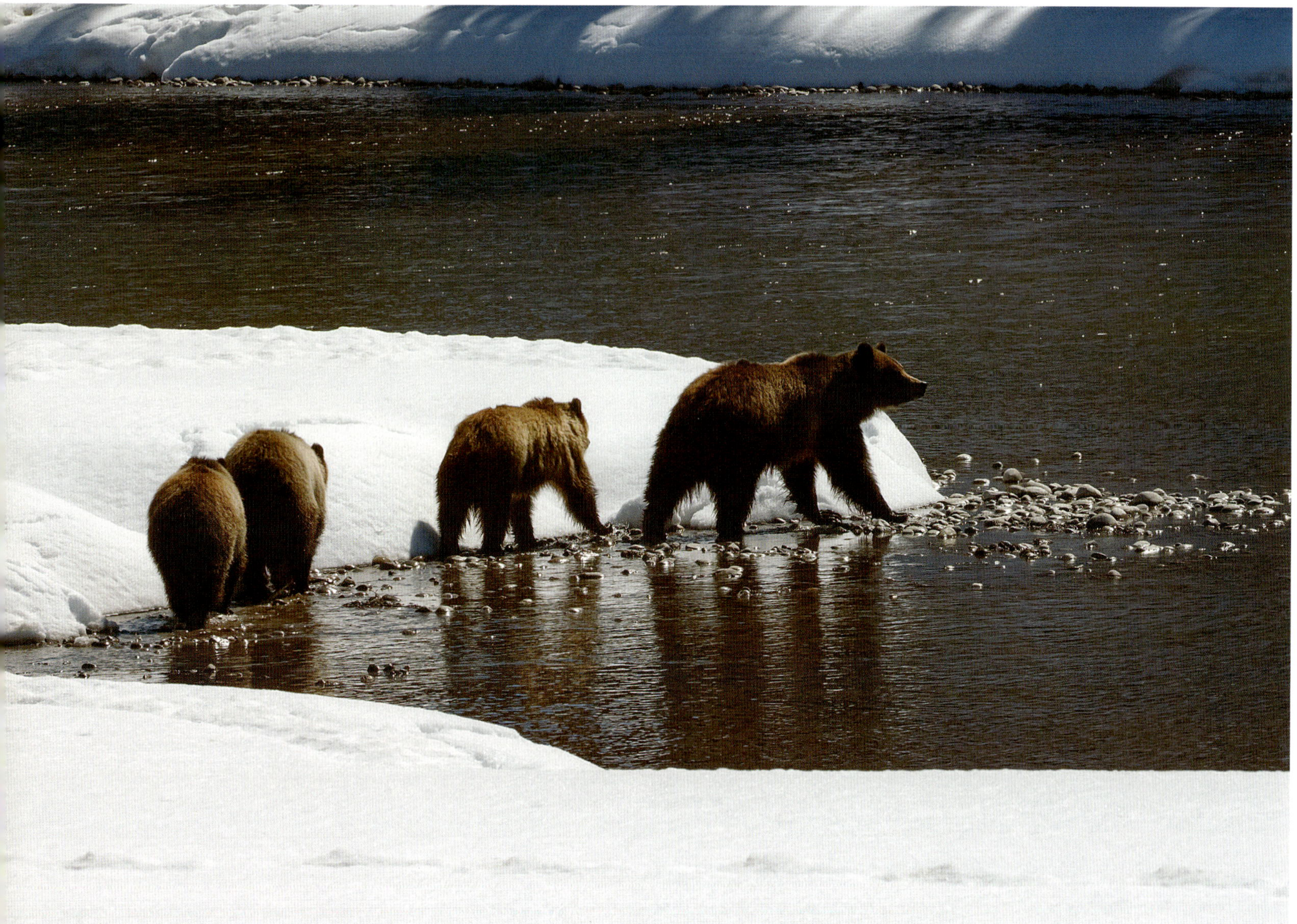

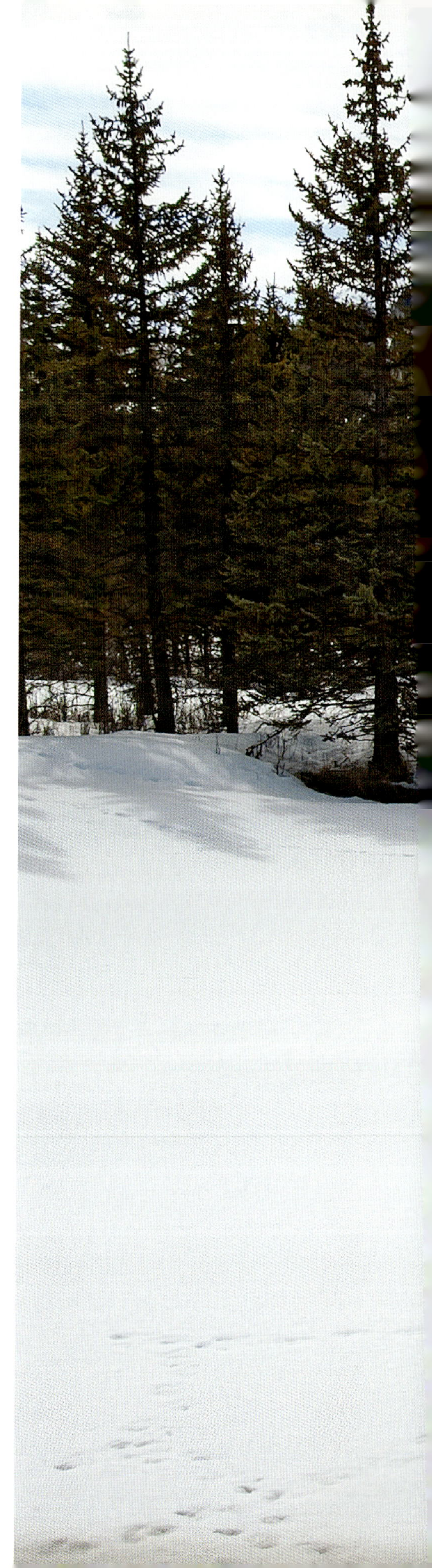

Soon after emerging from the den in mid-April, 399 escorts her juvenile cubs, 615, 610, and 587, along the shoreline of the Snake River at the junction of Pacific Creek in search of winter-killed ungulates.

OPPOSITE: As the day warms, 399 and her cubs post-hole through the soft snow across the Buffalo Fork of the Snake River.

FOLLOWING SPREAD: 399 scans the far hillside as one of her cubs takes a break from the long walk to scratch its back among the snow-covered sagebrush.

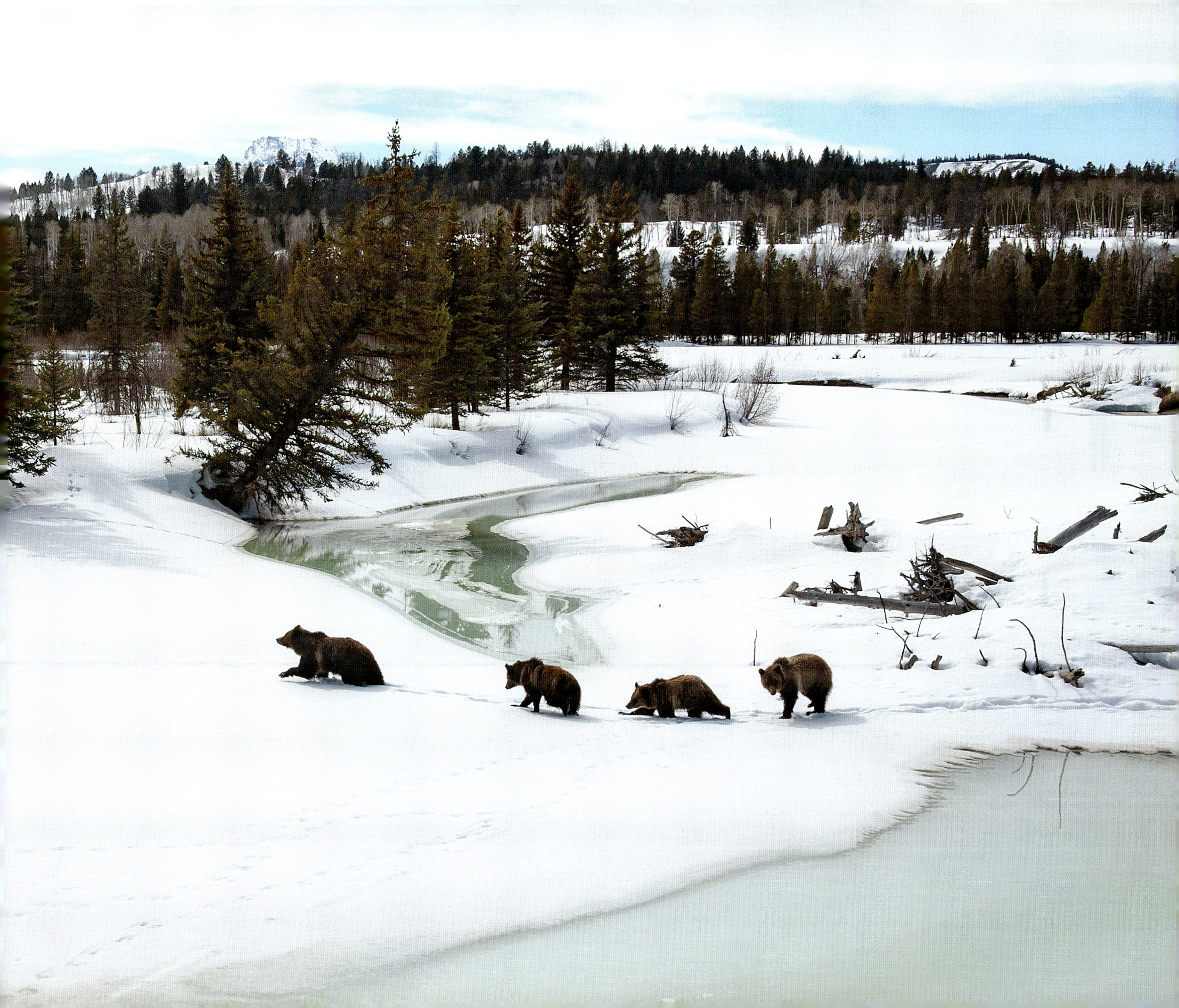

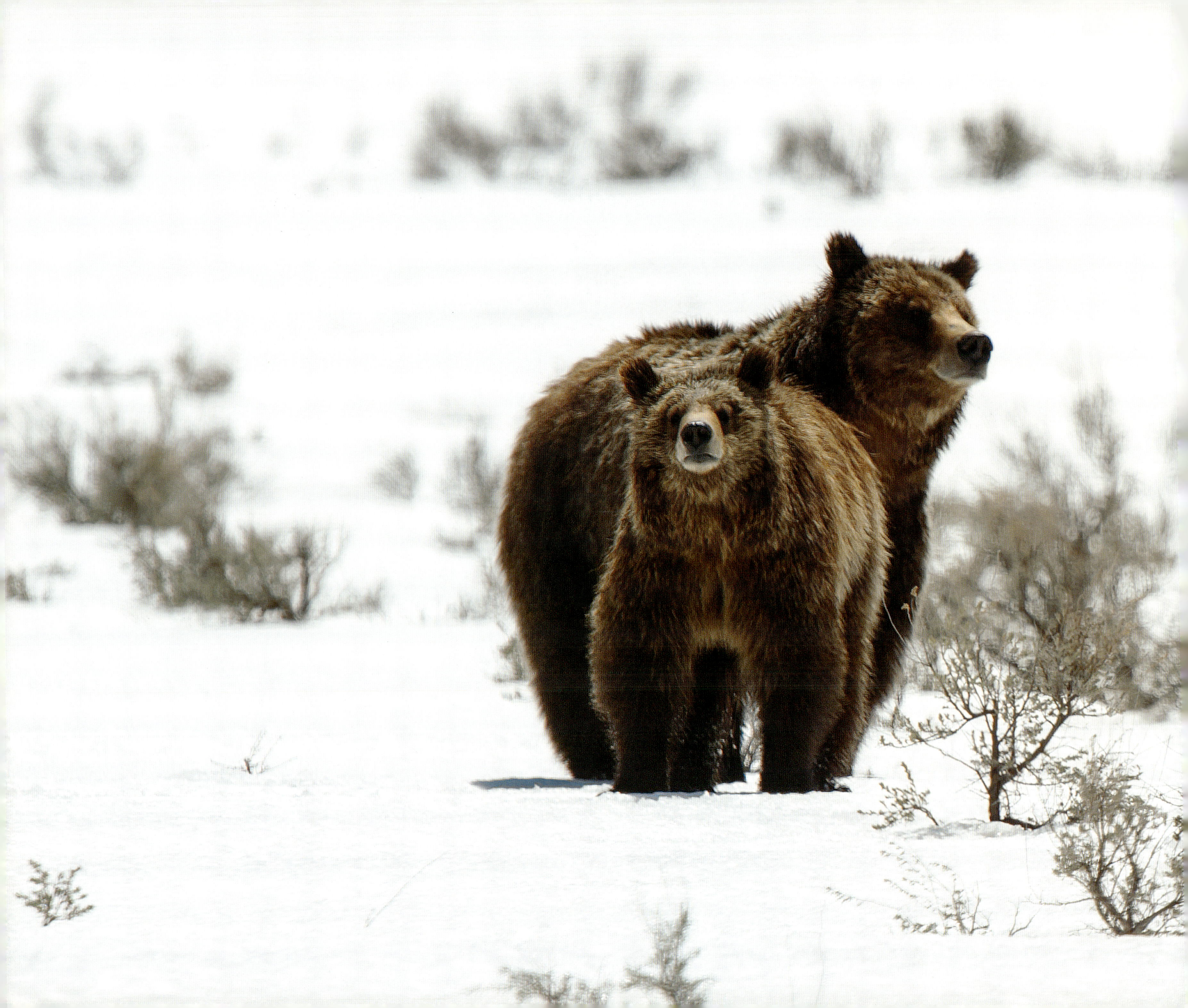

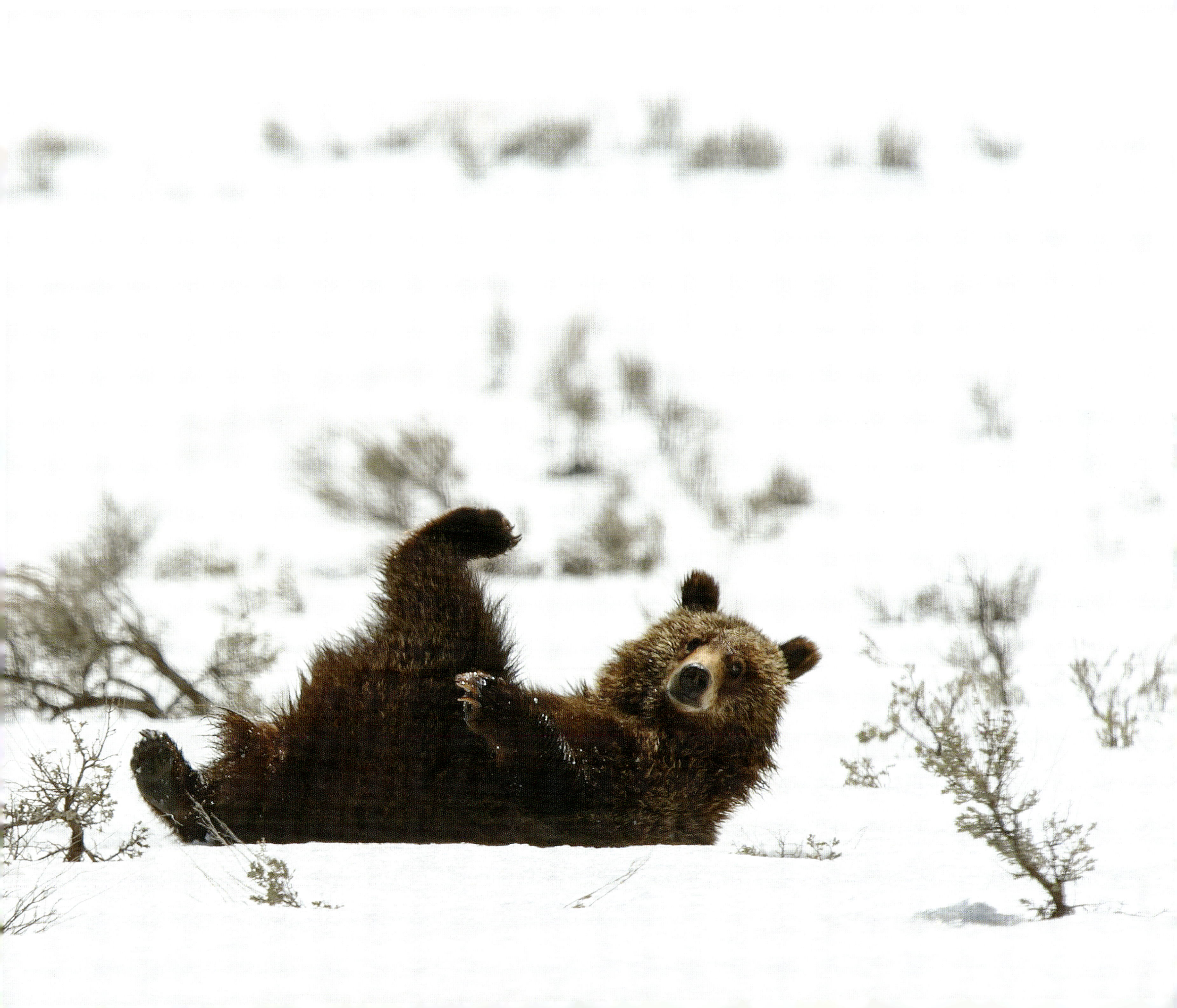

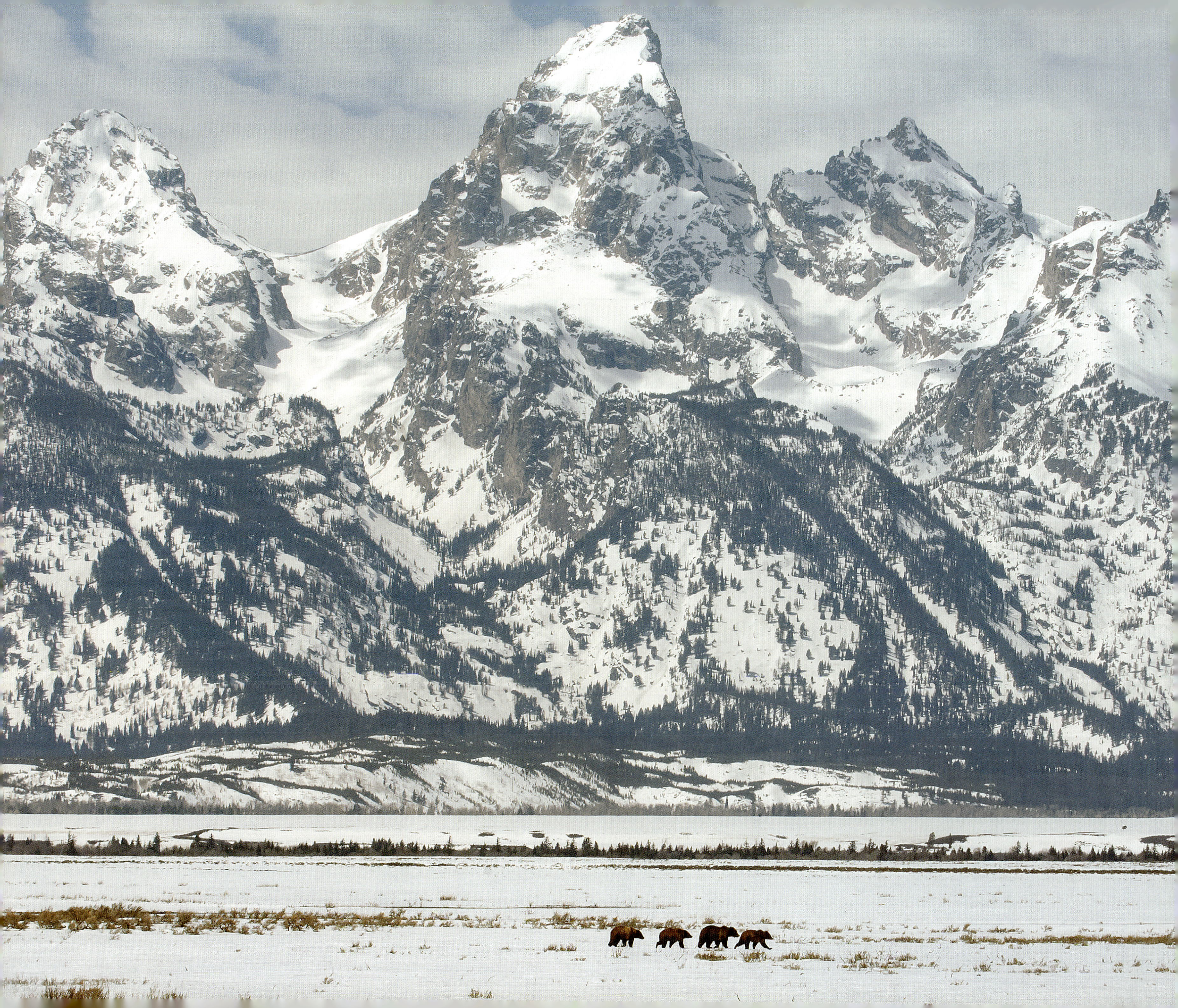

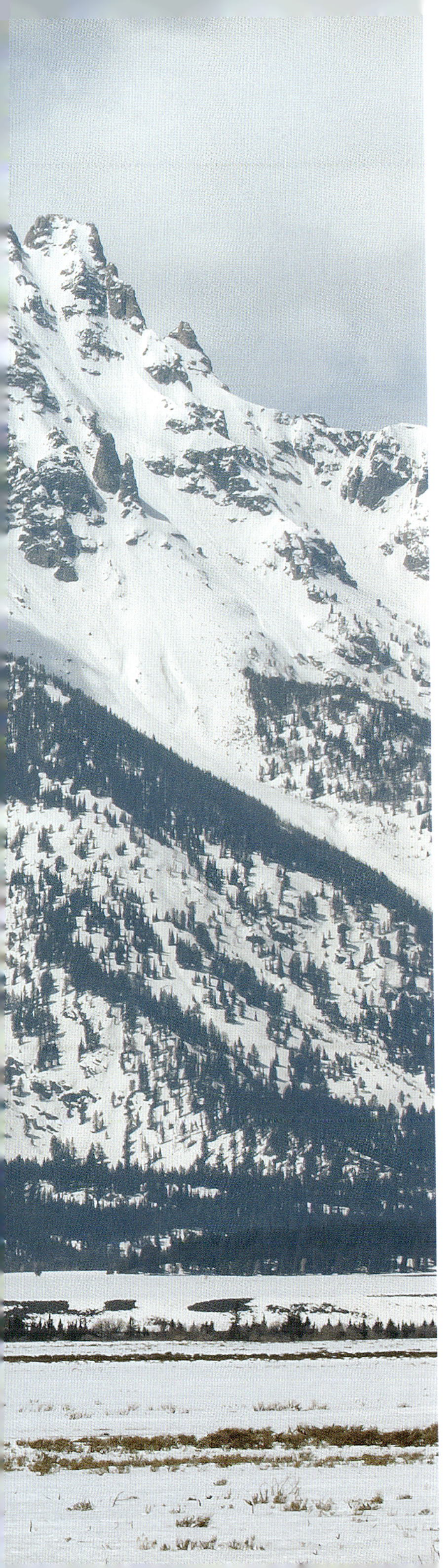

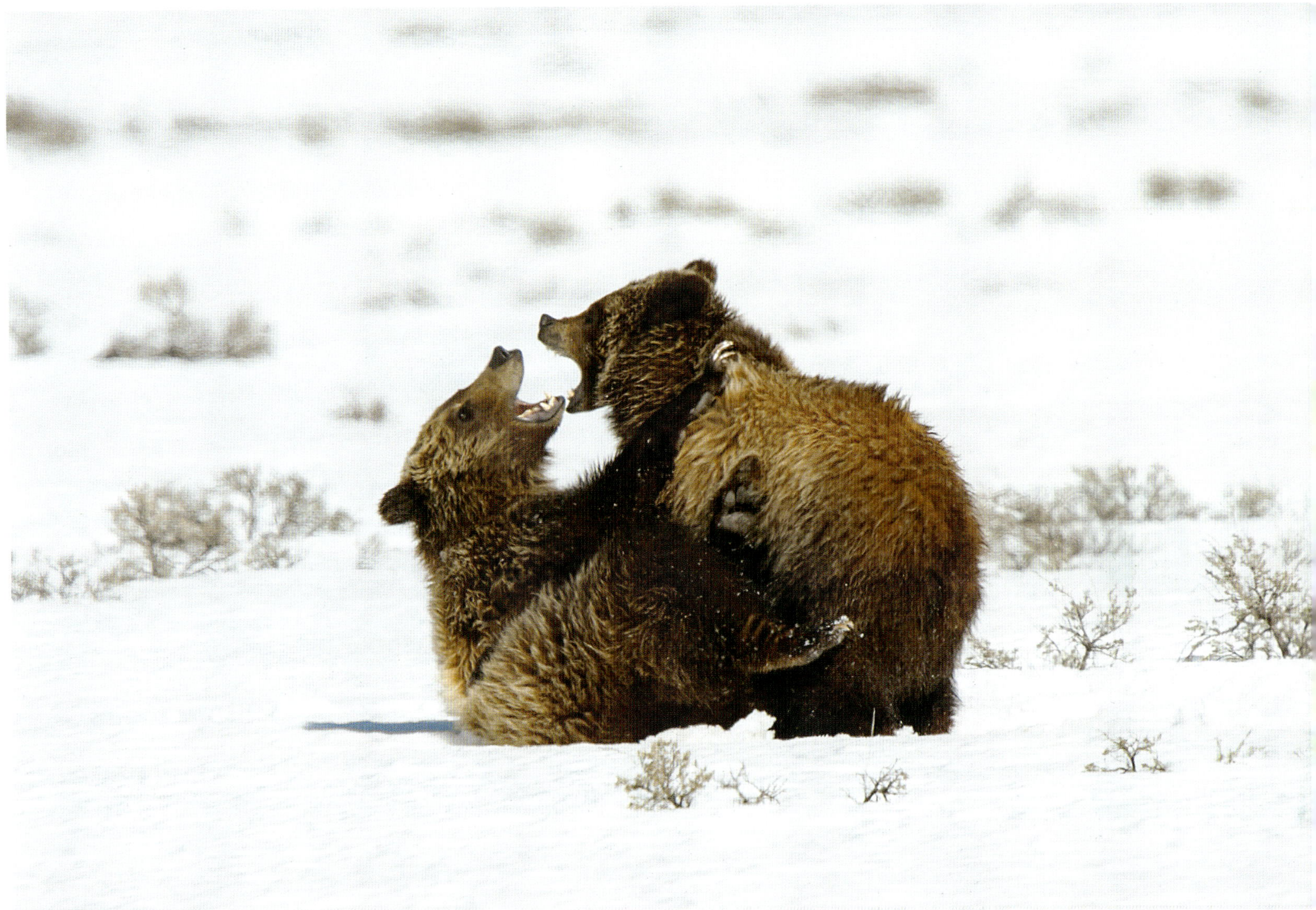

OPPOSITE: Looming large is whether a recreational sport hunt of grizzlies should be brought back in Wyoming, Montana, and Idaho. A hunt has been prohibited for more than four decades, and critics say it should never be resurrected and would put bruins like the family of 399 in peril.

Two of 399's cubs wrestle.

FOLLOWING SPREAD, LEFT: As evening falls, 399 looks out over Oxbow Bend.

FOLLOWING SPREAD, RIGHT: 399 and a cub share a tender moment, reinforcing their bond.

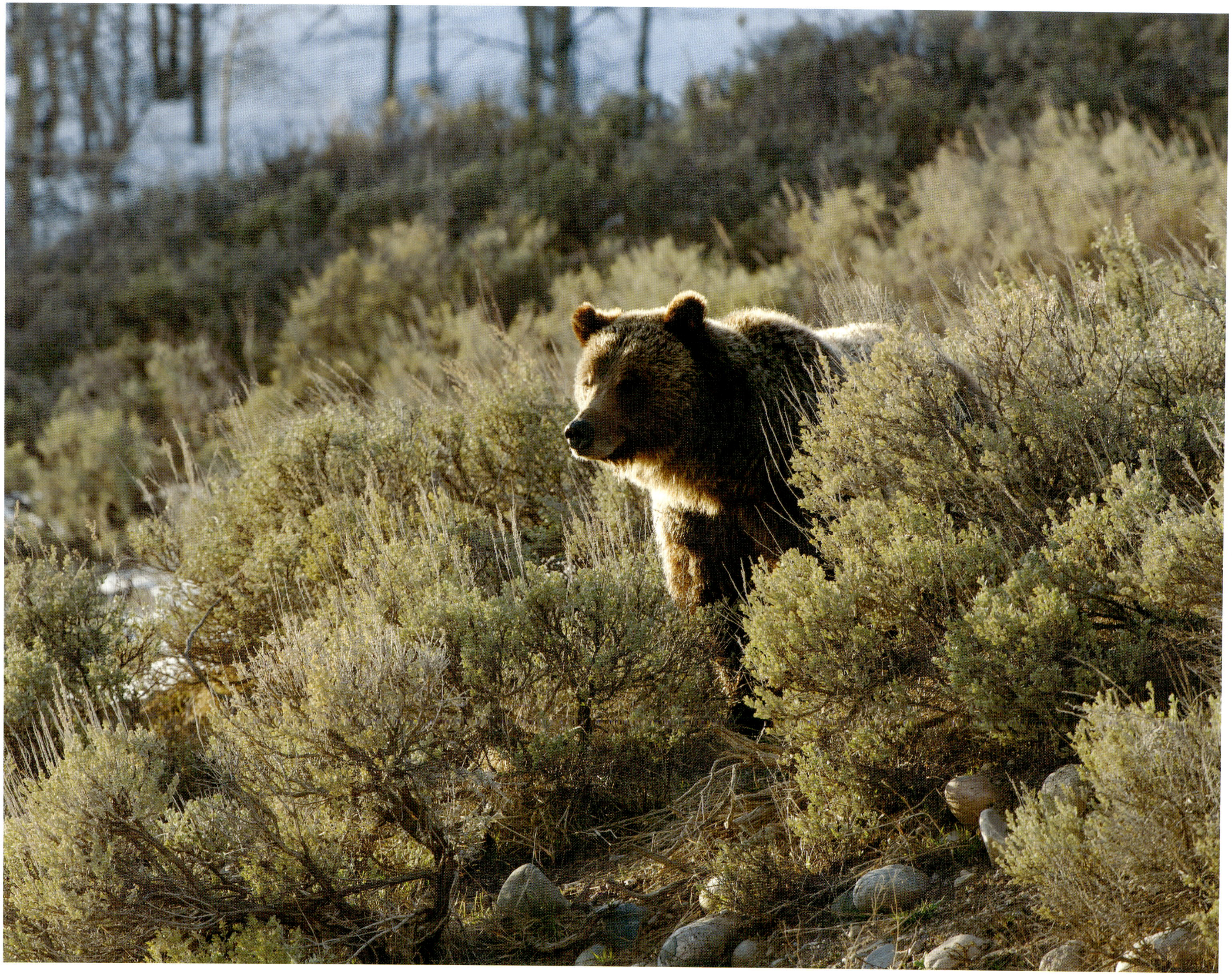

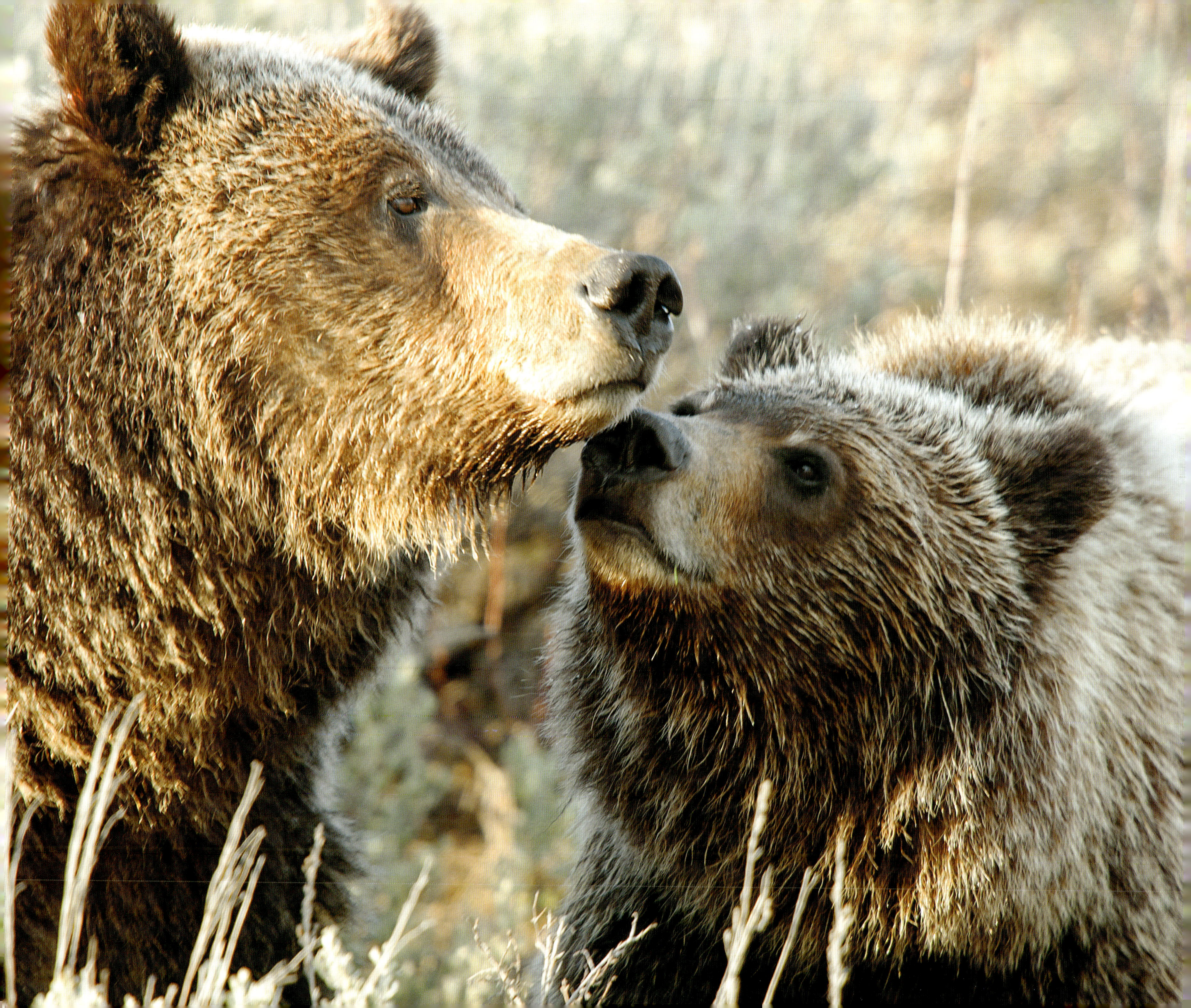

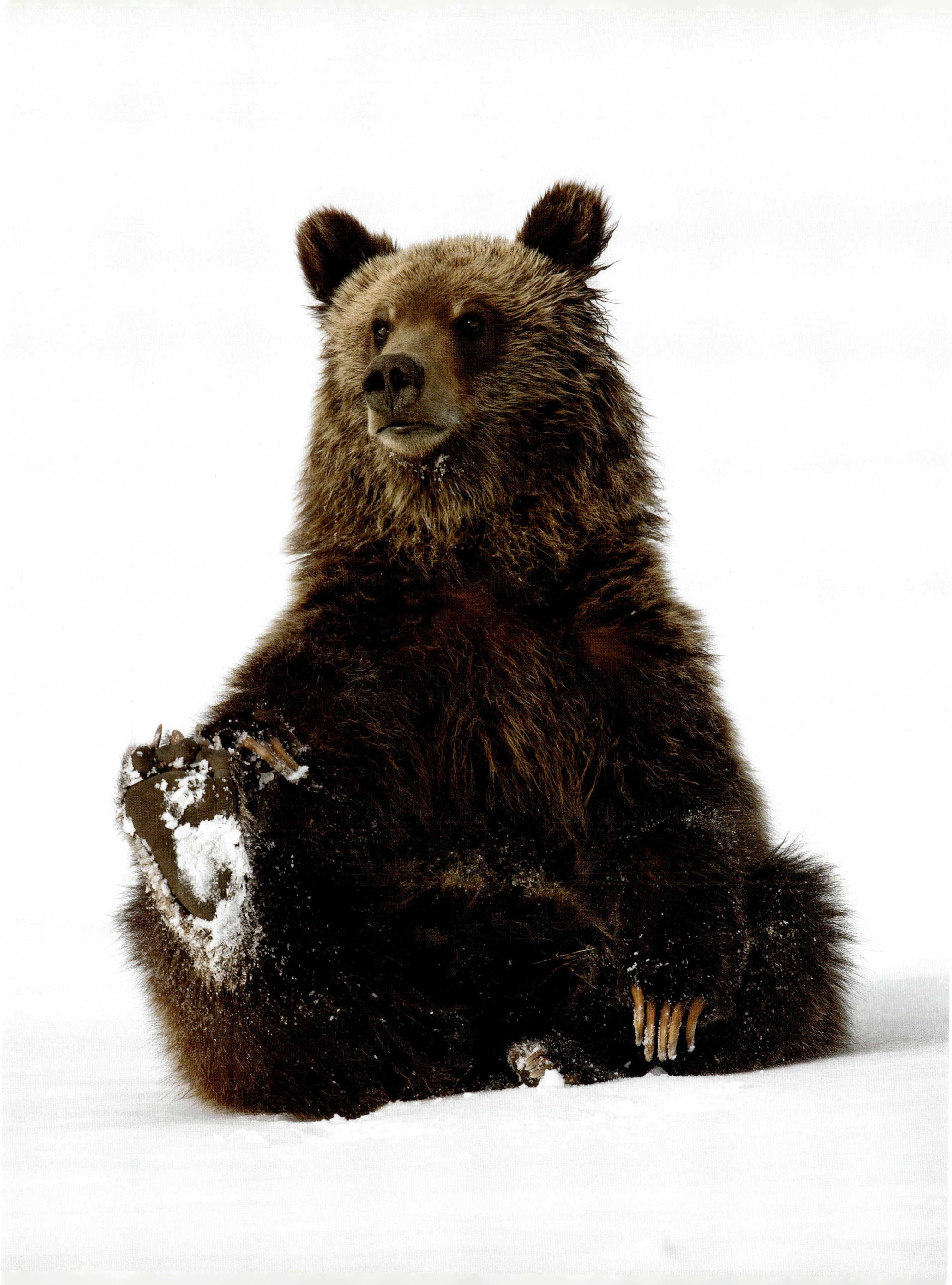
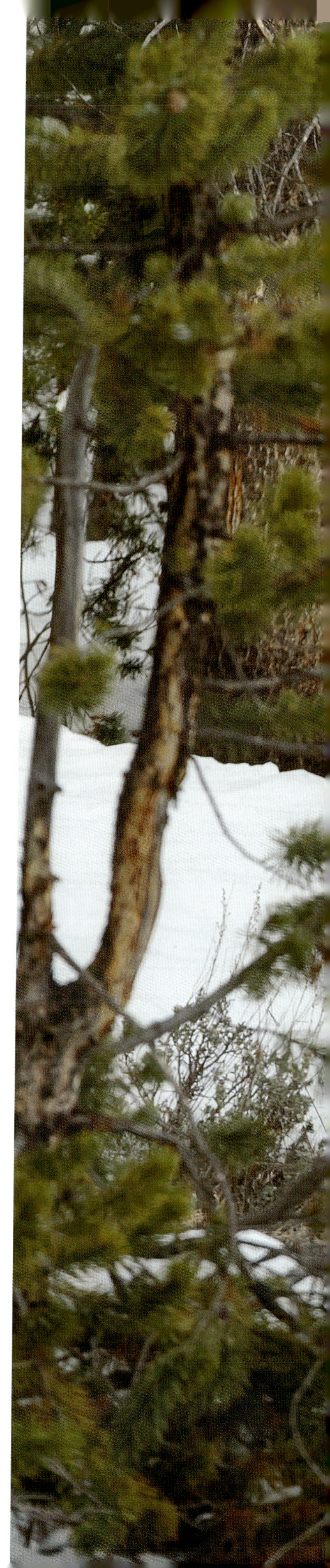

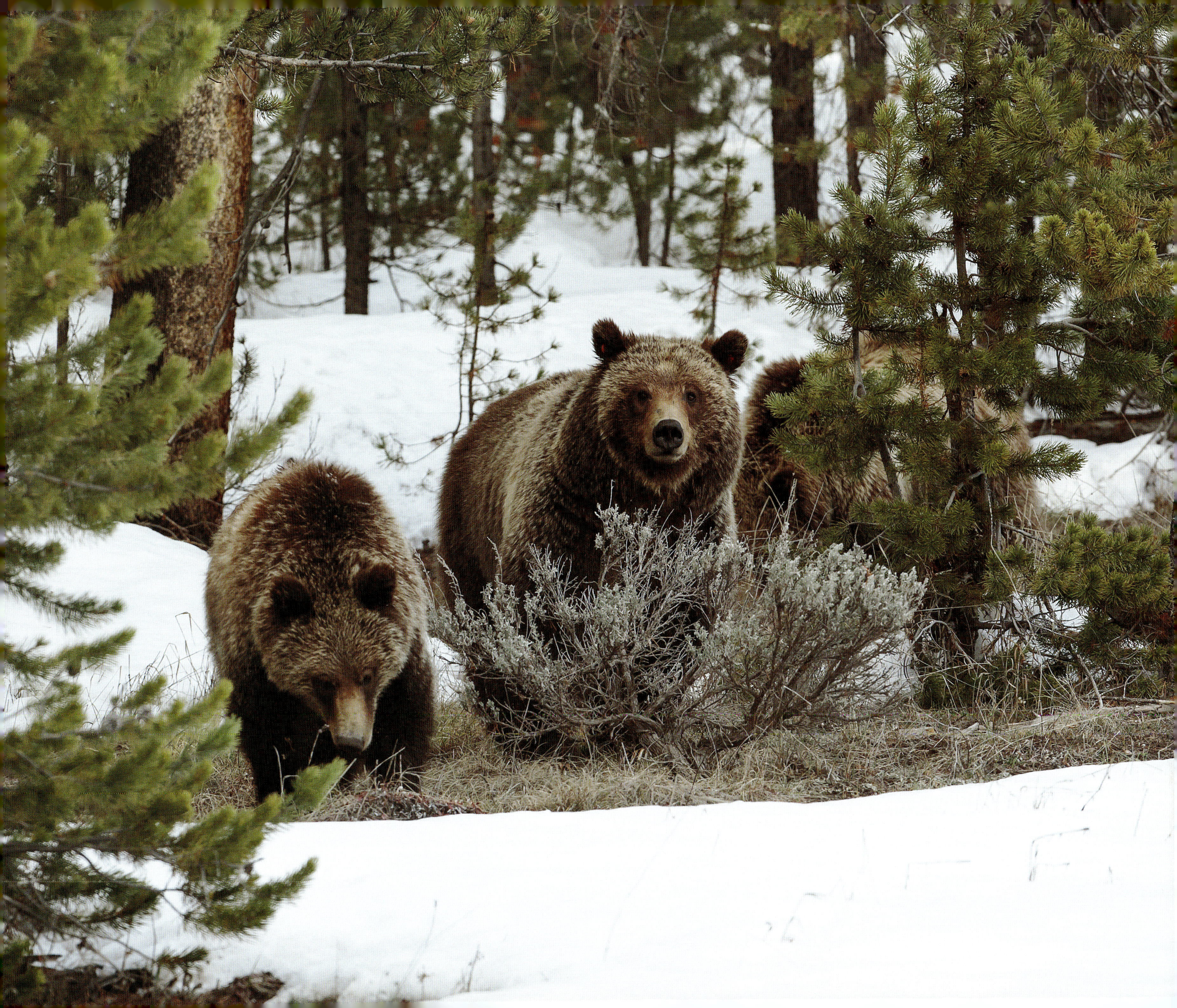

PREVIOUS SPREAD, LEFT: There's still time to play, as 399's cub sits Buddha-like, relishing his last few weeks under his mother's watchful eyes.

PREVIOUS SPREAD, RIGHT: Unaware of the long road ahead, these cubs will face a whole new series of challenges come summer, when their mother leaves her cubs so that she may breed again.

399 and her juvenile cubs find a bonanza of suckers and trout that became entrapped under the ice at Oxbow Bend when the water dropped rapidly during the winter.

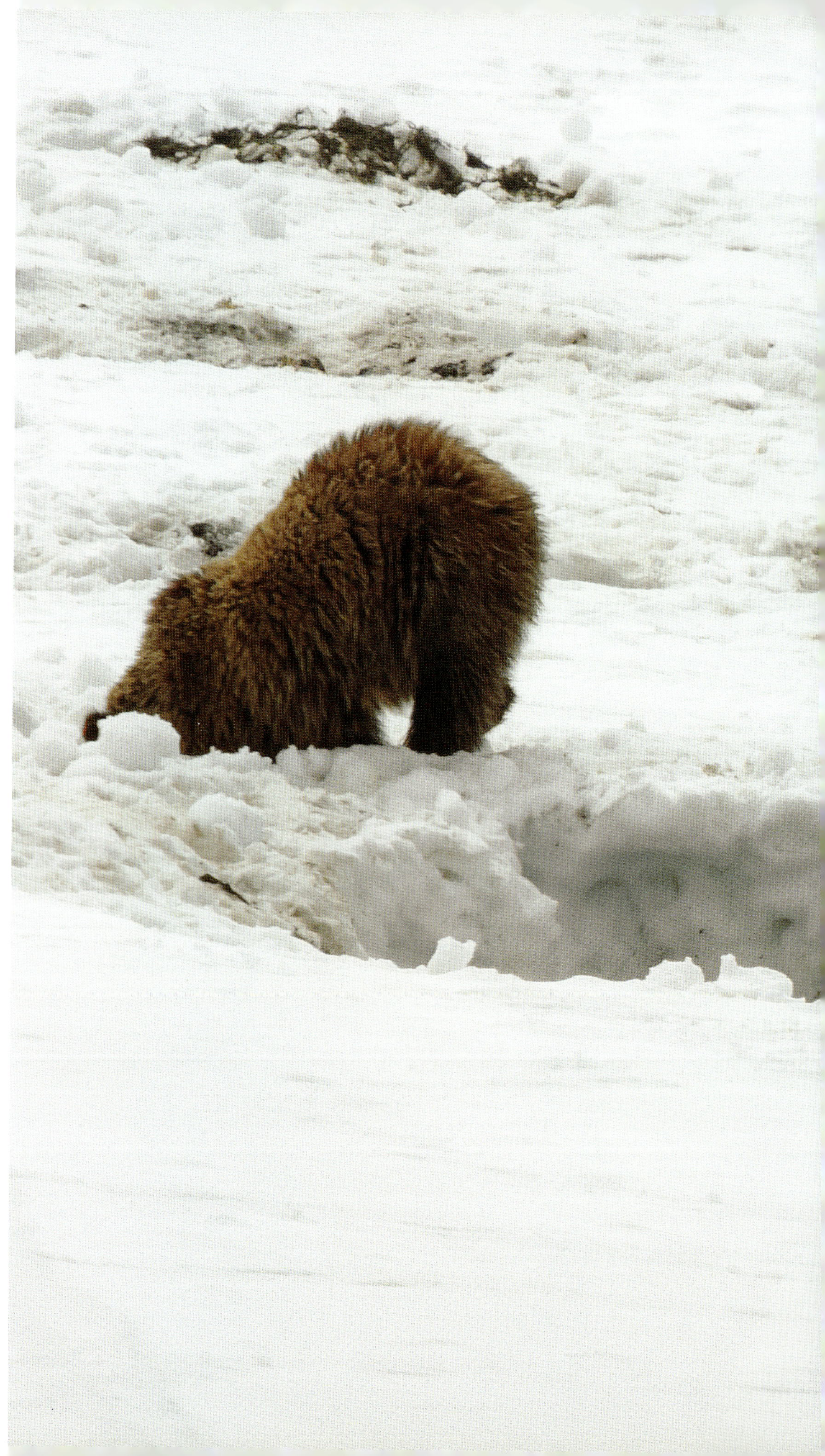

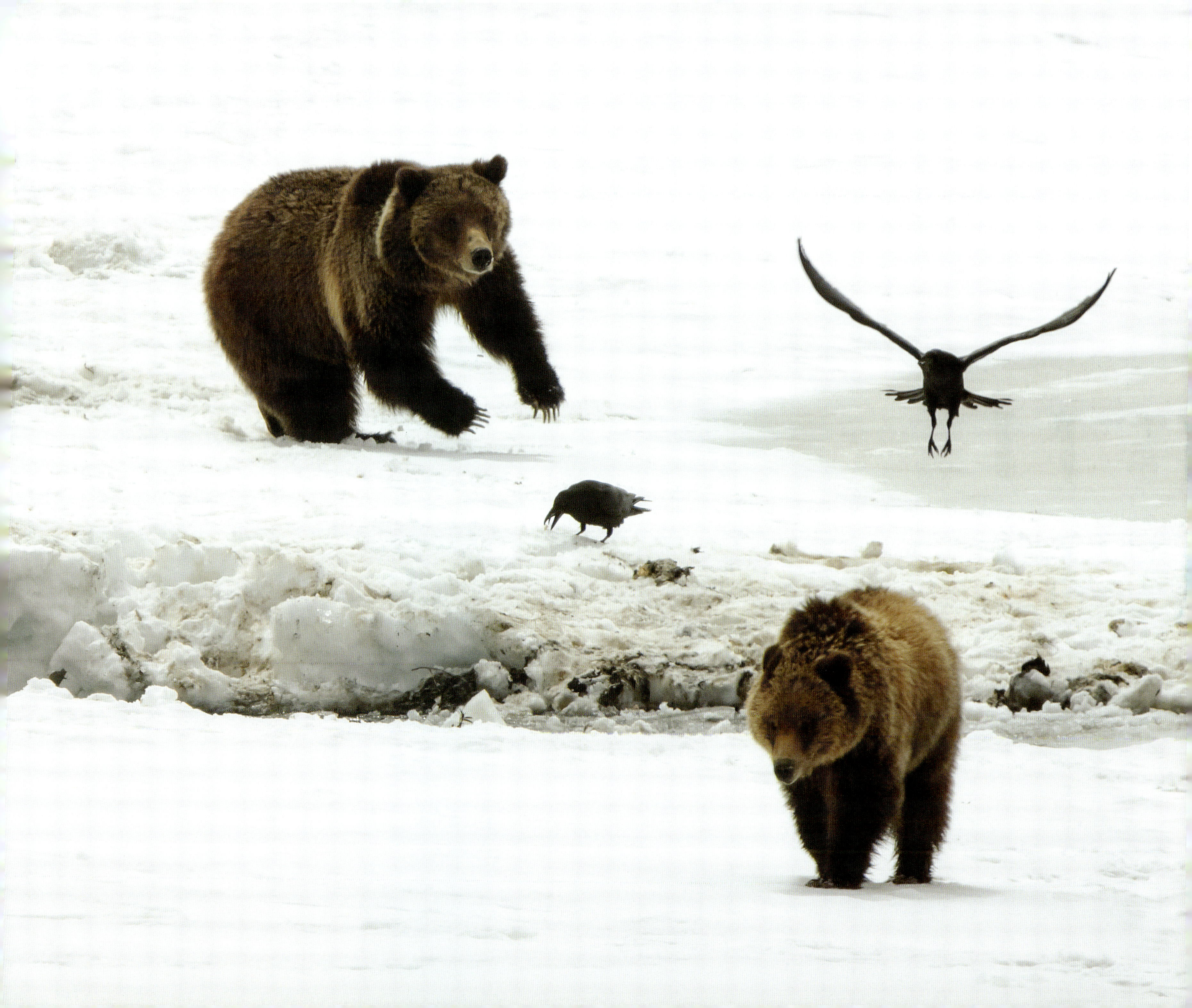

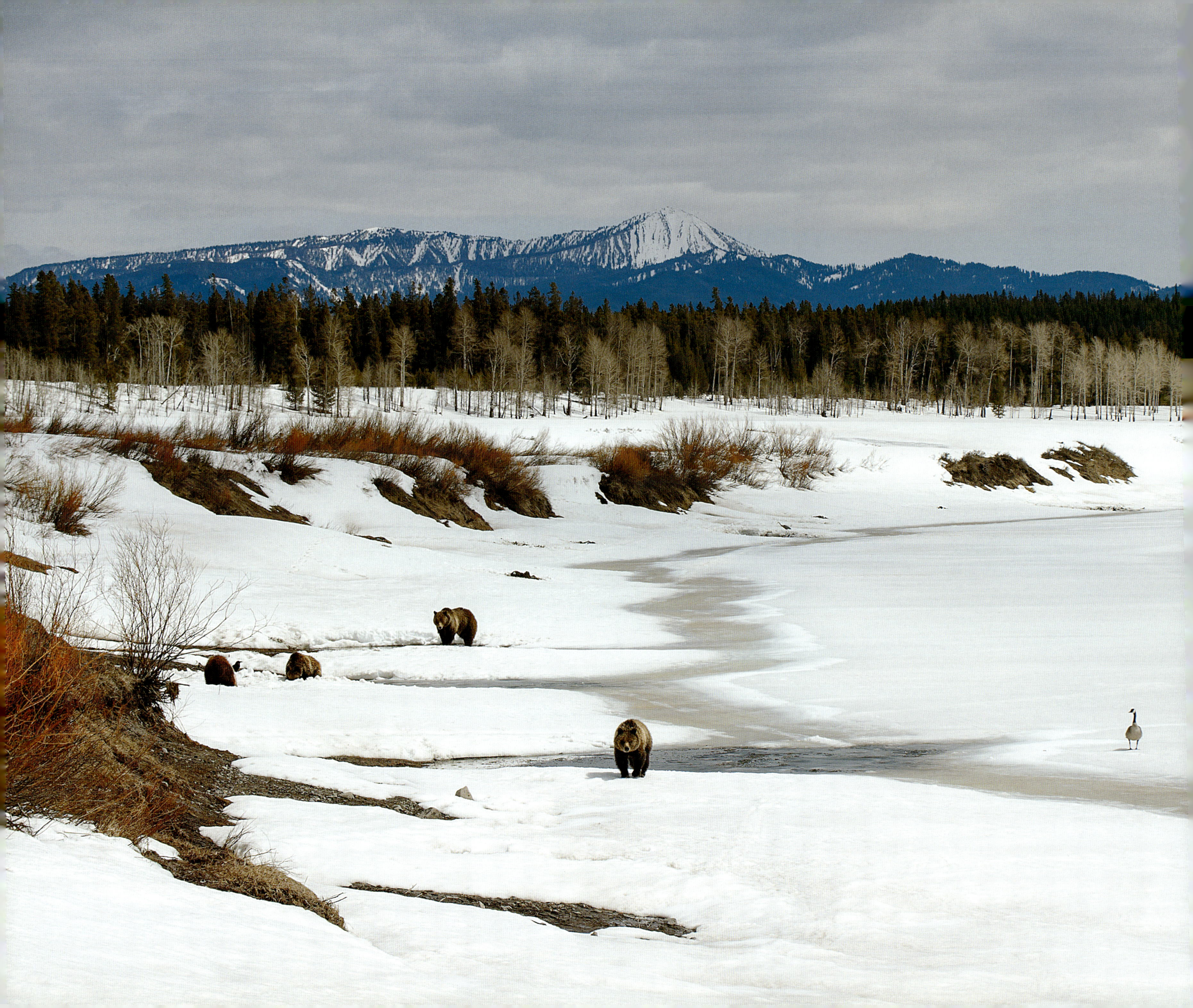

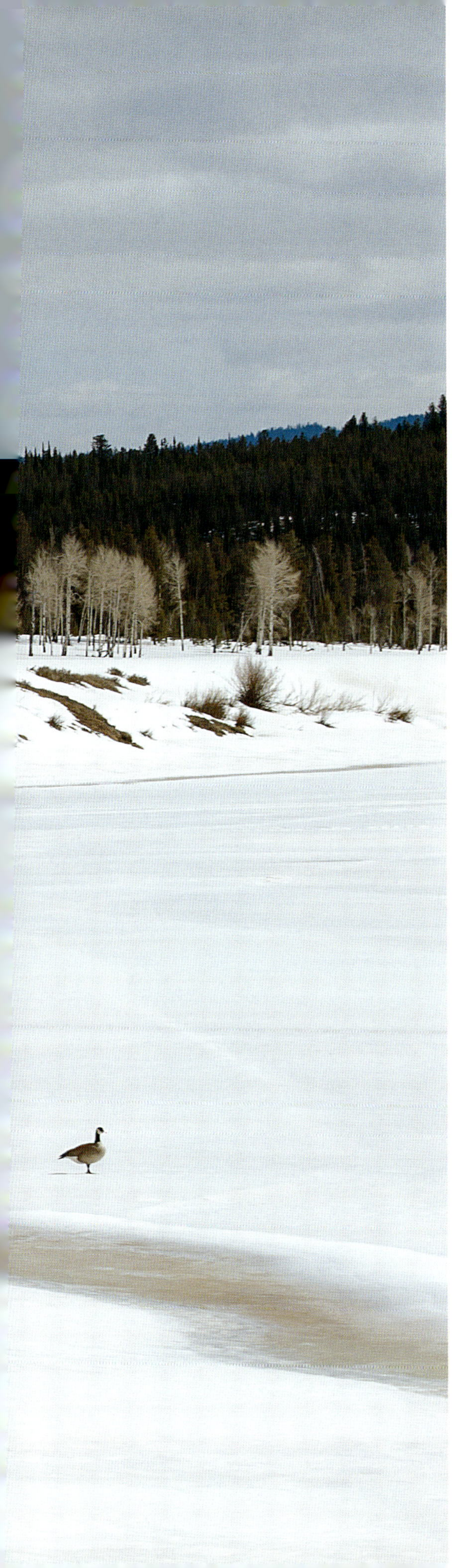

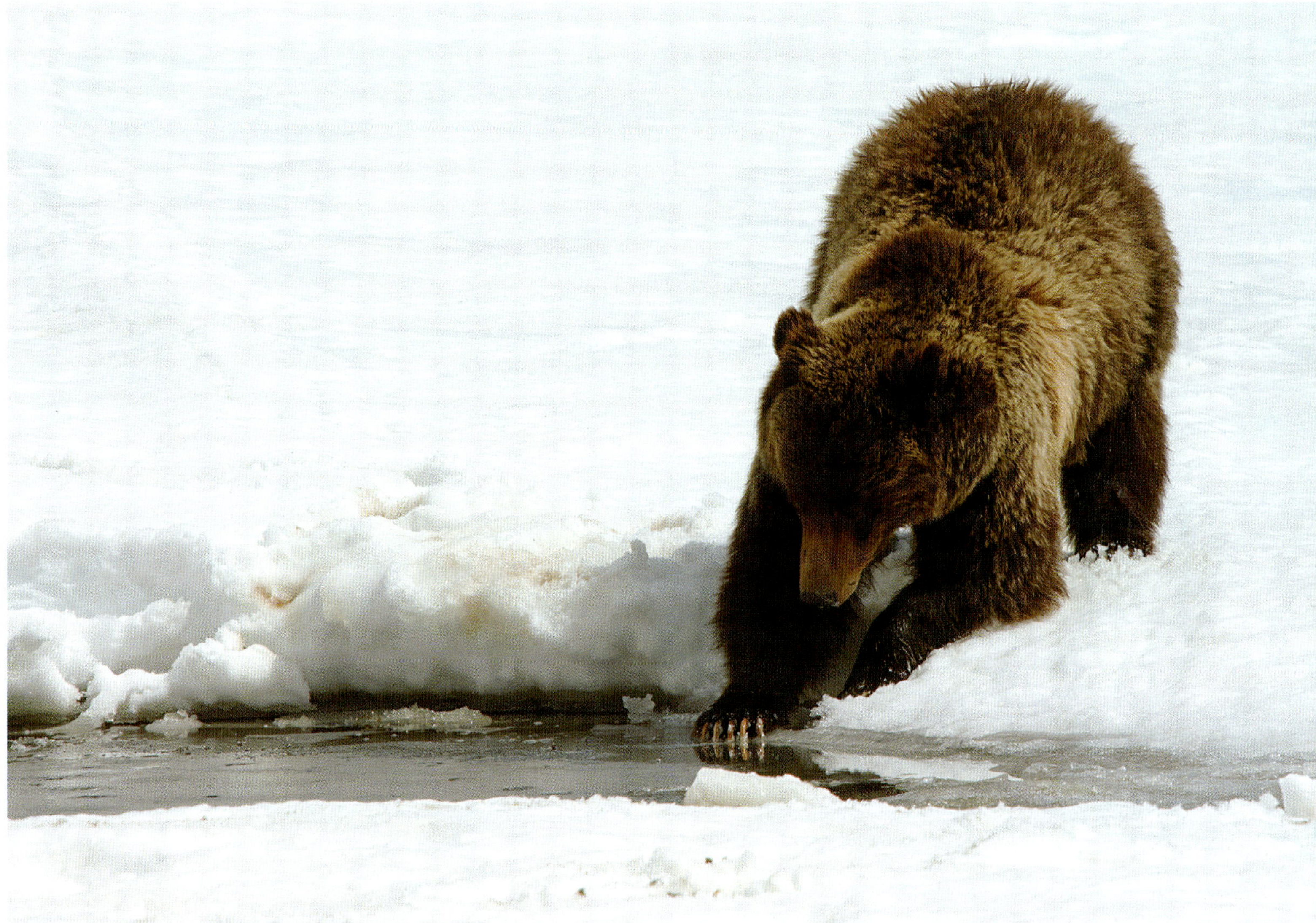

OPPOSITE: Thousands of fish died and were preserved until the ice started to thaw, creating a surprise banquet for 399's family. Digging out the fish attracted scavengers such as ravens and coyotes to the feast.

A skilled angler, 399 breaks through the ice in search of fish frozen beneath the surface.

FOLLOWING SPREAD, LEFT: After feasting on fish, 399 nurses her juvenile cubs for one of the last times. Two weeks later, she will force them out on their own and be observed mating the following day.

FOLLOWING SPREAD, RIGHT: 399 breaks up a quarrel between two squabbling cubs.

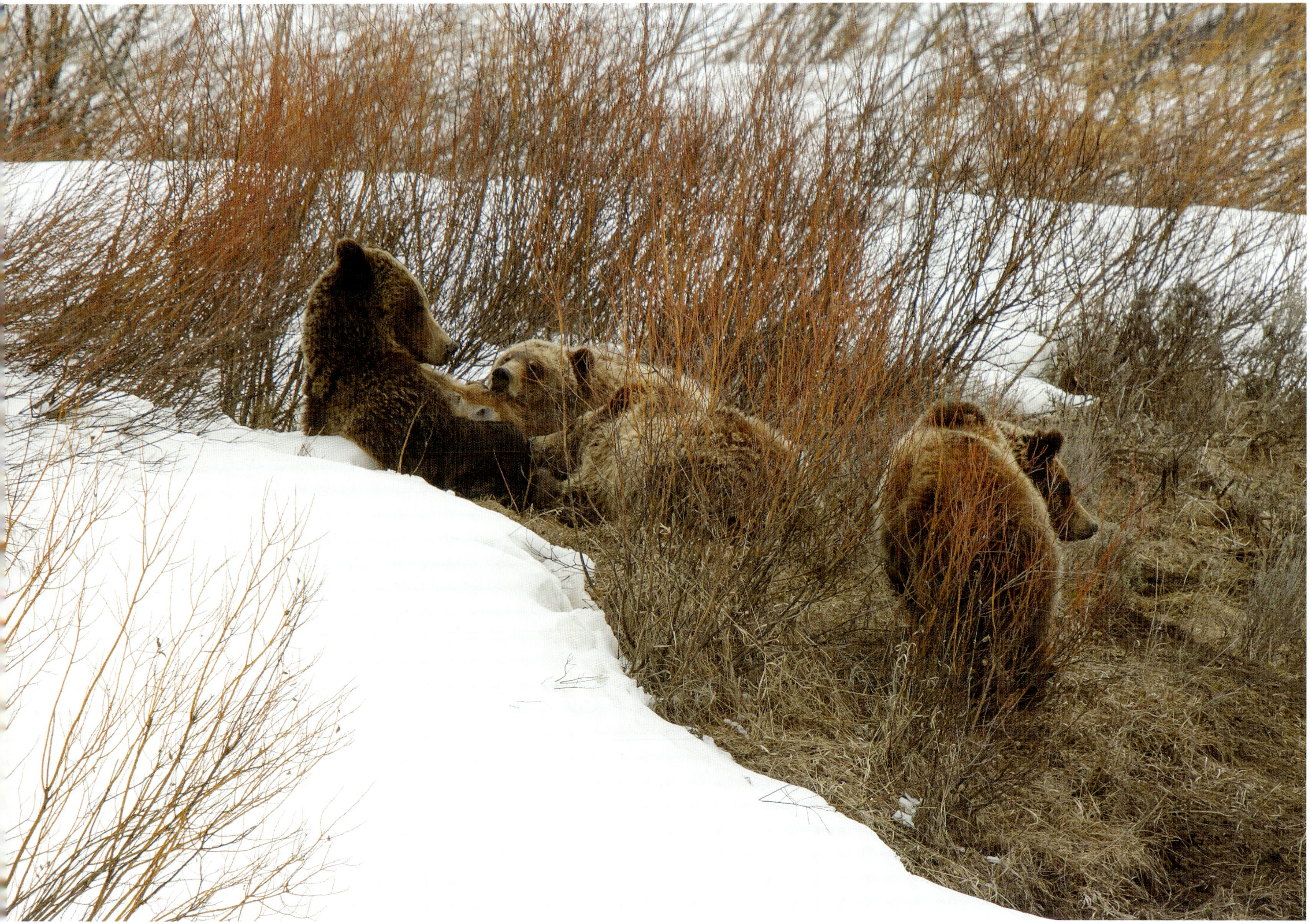

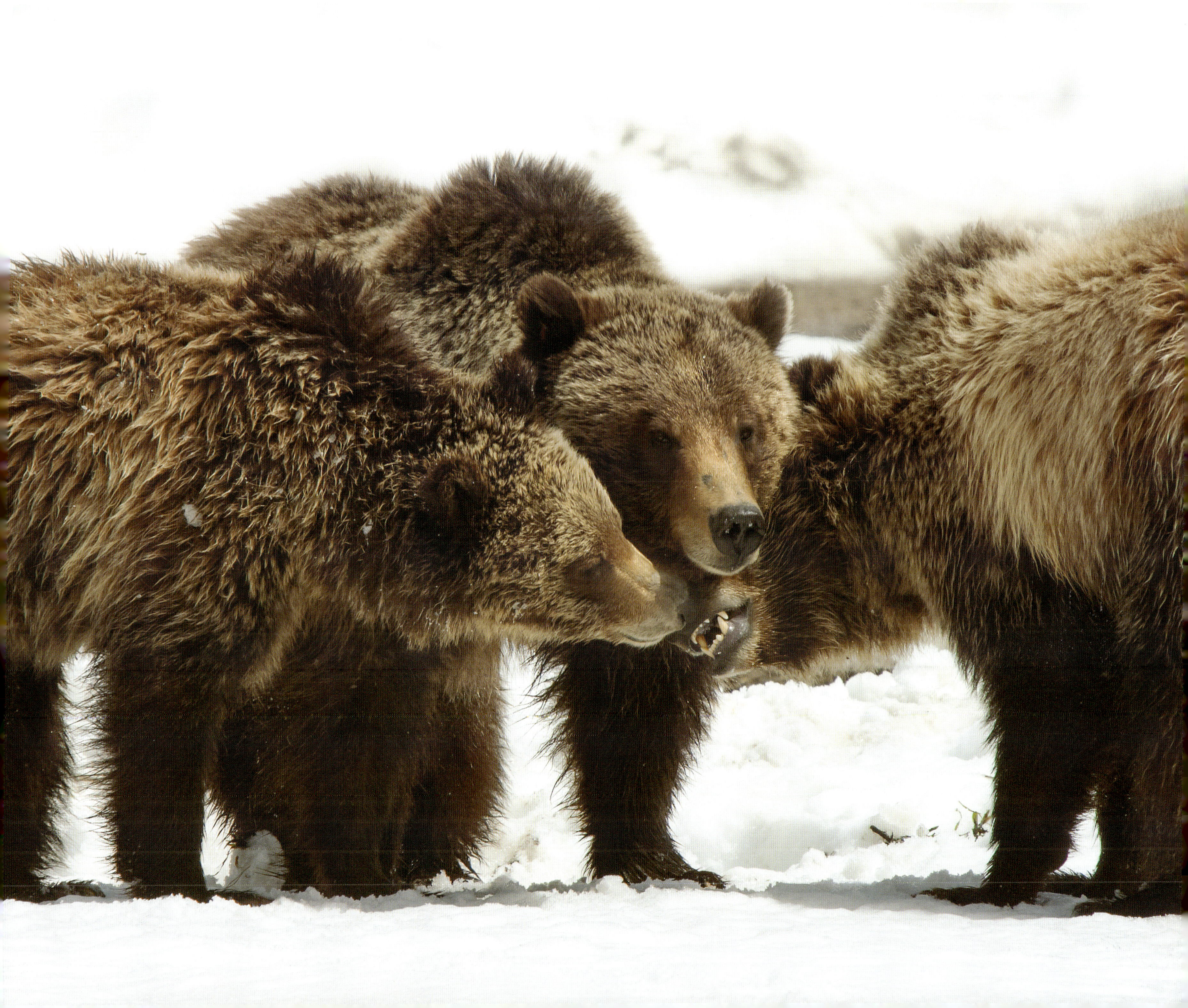

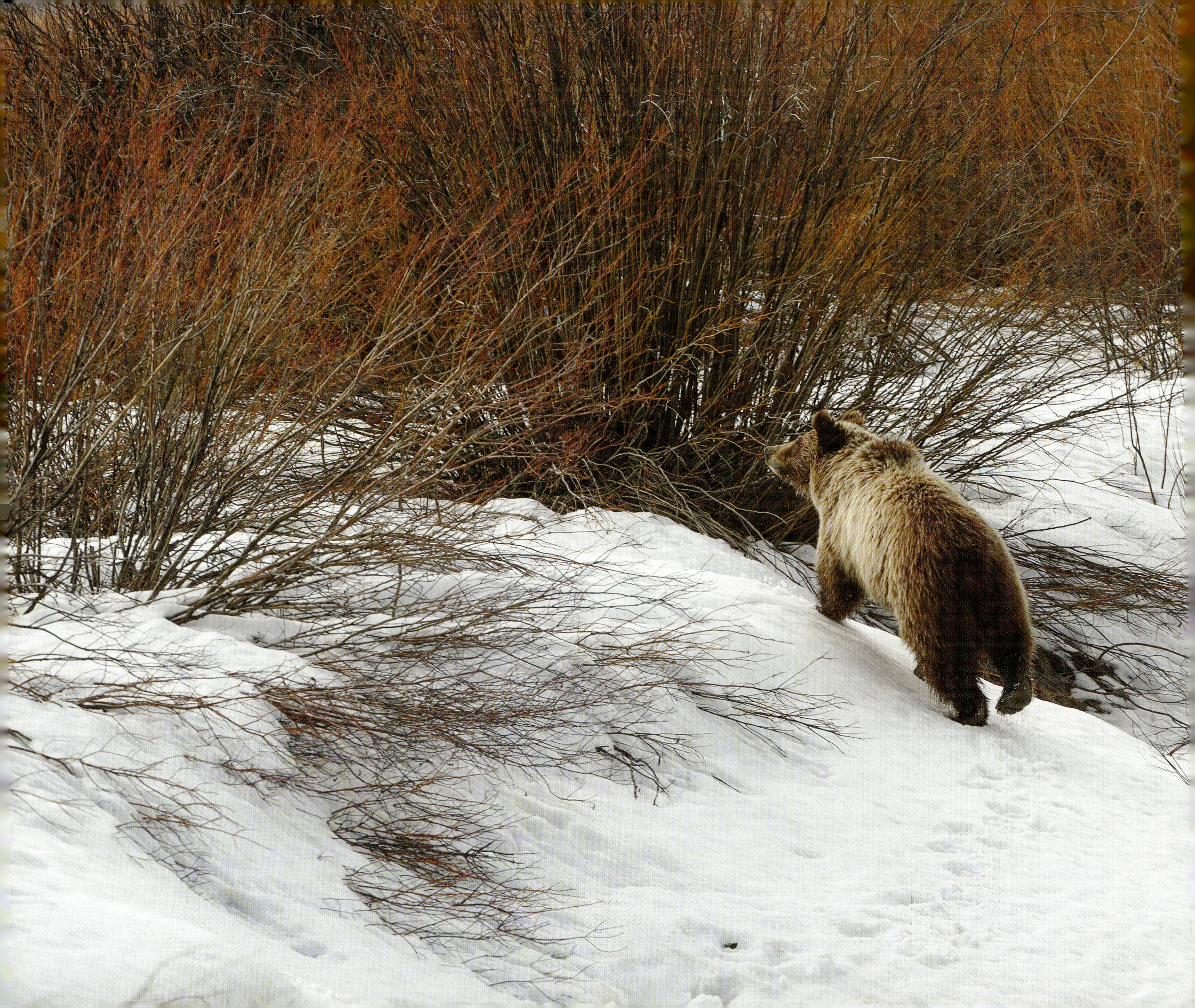

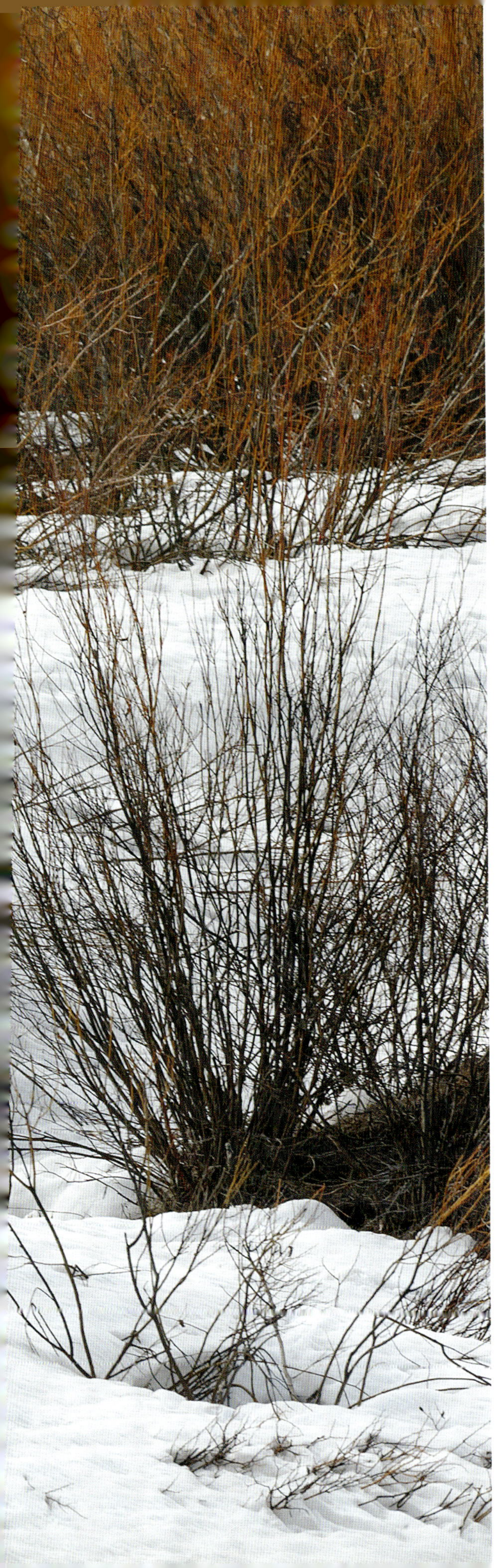
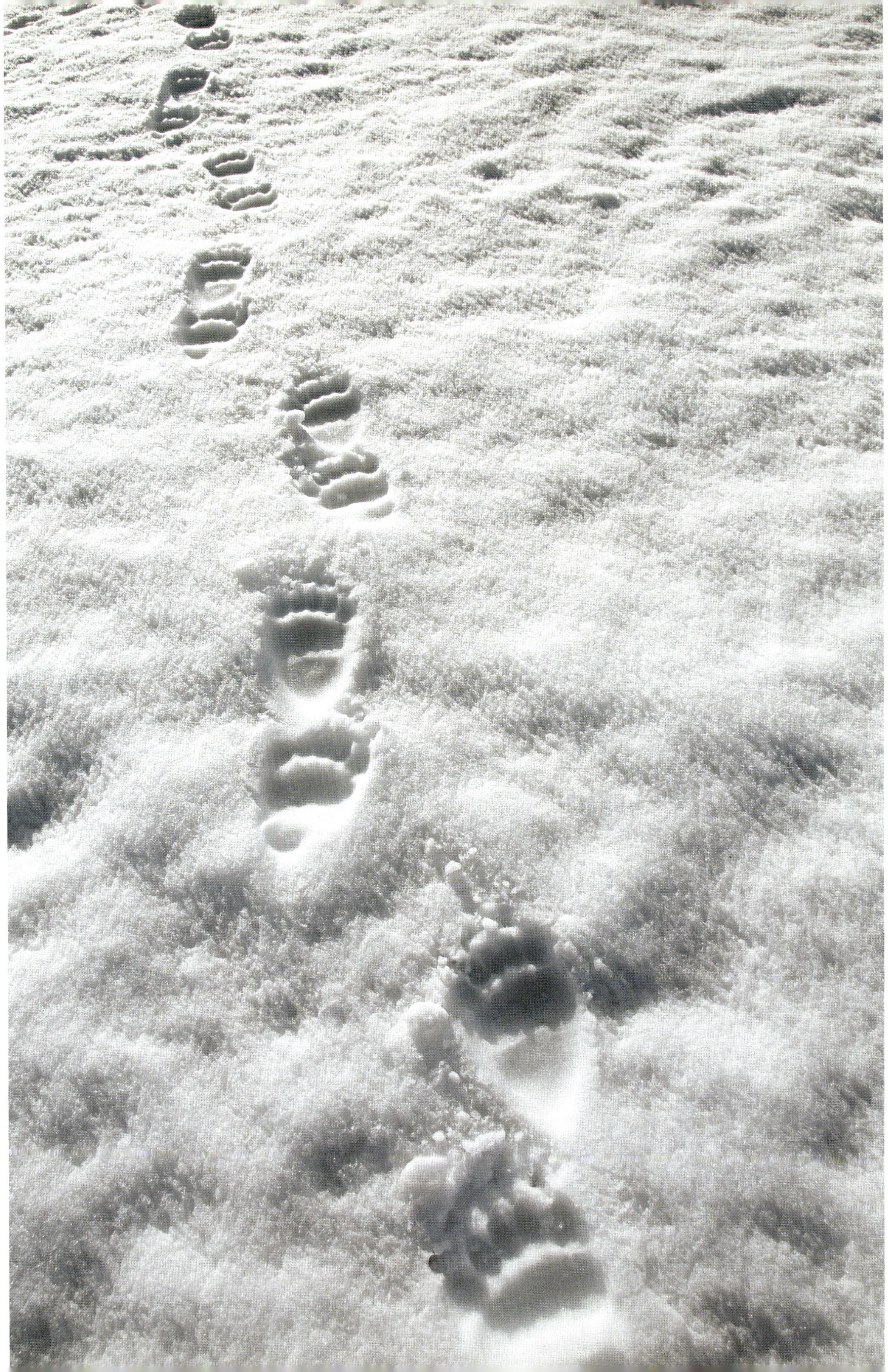

PREVIOUS SPREAD, LEFT: One of 399's cubs passes through the willows at Oxbow Bend on a late spring evening.

PREVIOUS SPREAD, RIGHT: 399's paw prints appear in the snow at Willow Flats.

Evening reflections show 399 and her cub foraging for fish at Oxbow Bend.

FOLLOWING SPREAD, LEFT: Warm days and rotting ice bring an end to the bears' fishing bonanza.

FOLLOWING SPREAD, RIGHT: What's the most prolific food source on the landscape? Answer: ants. One biologist says the biomass of ants in Greater Yellowstone probably surpasses the biomass of all the elk in the ecosystem. Here, 399's cubs dig for ants under three feet of snow.

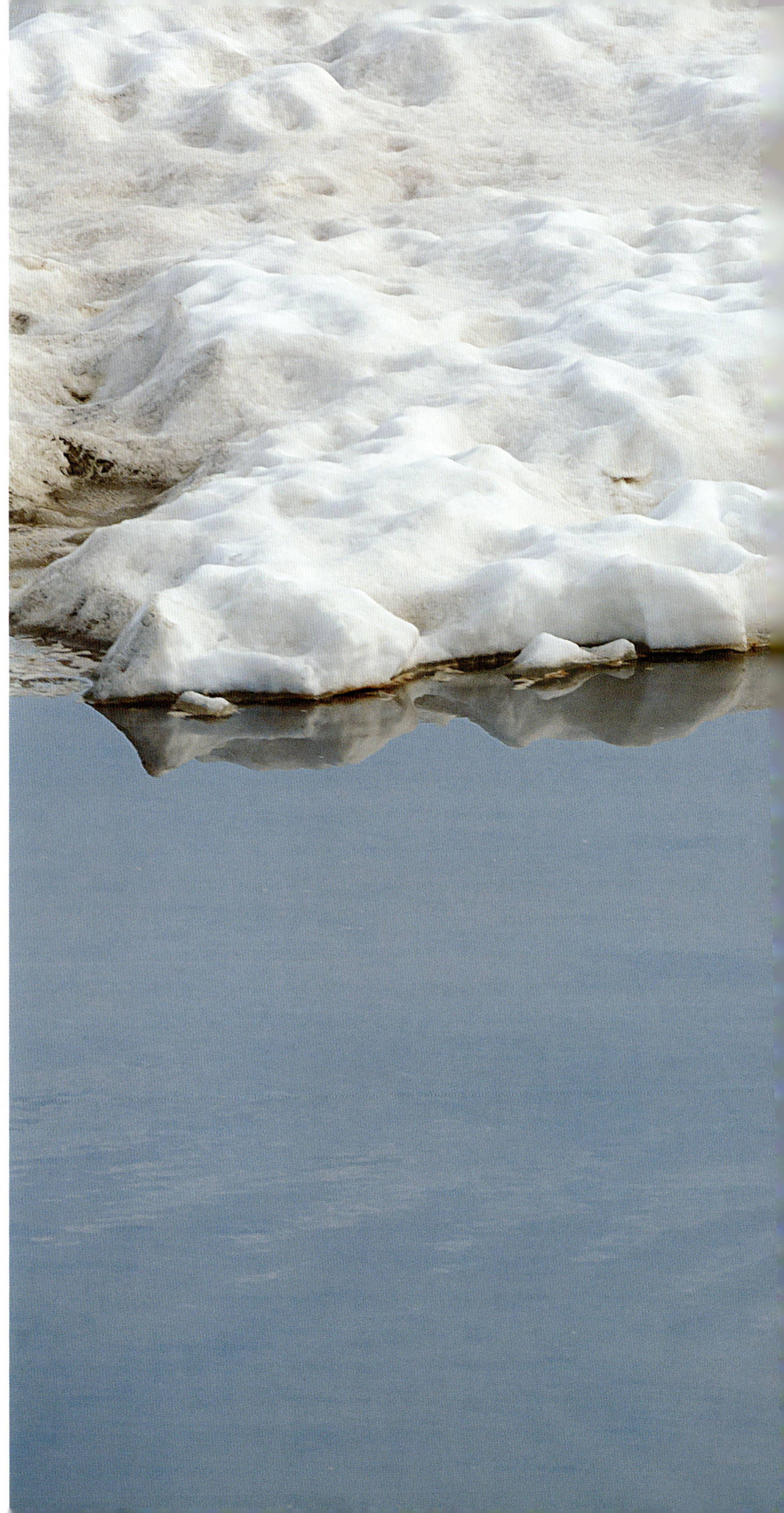

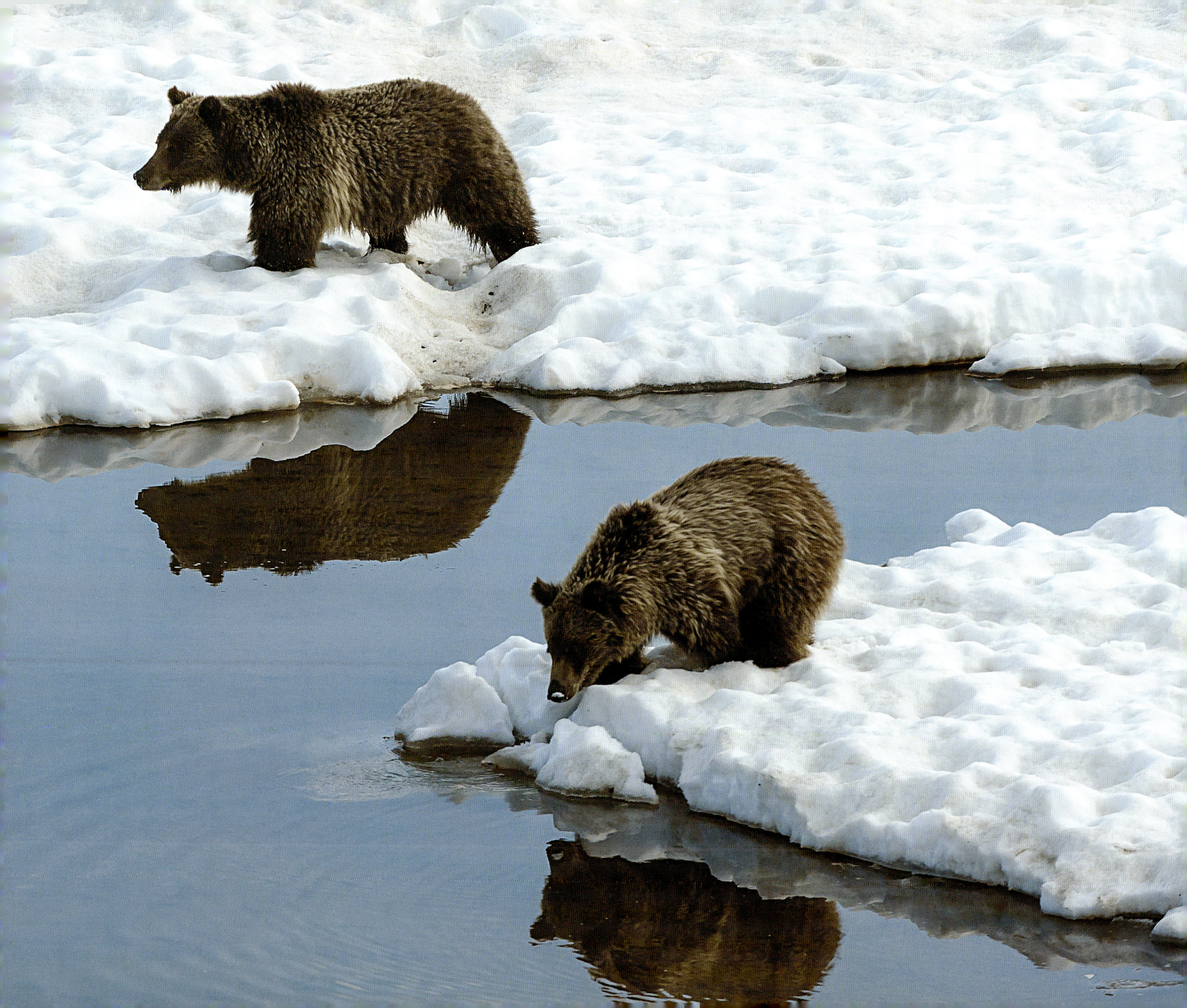

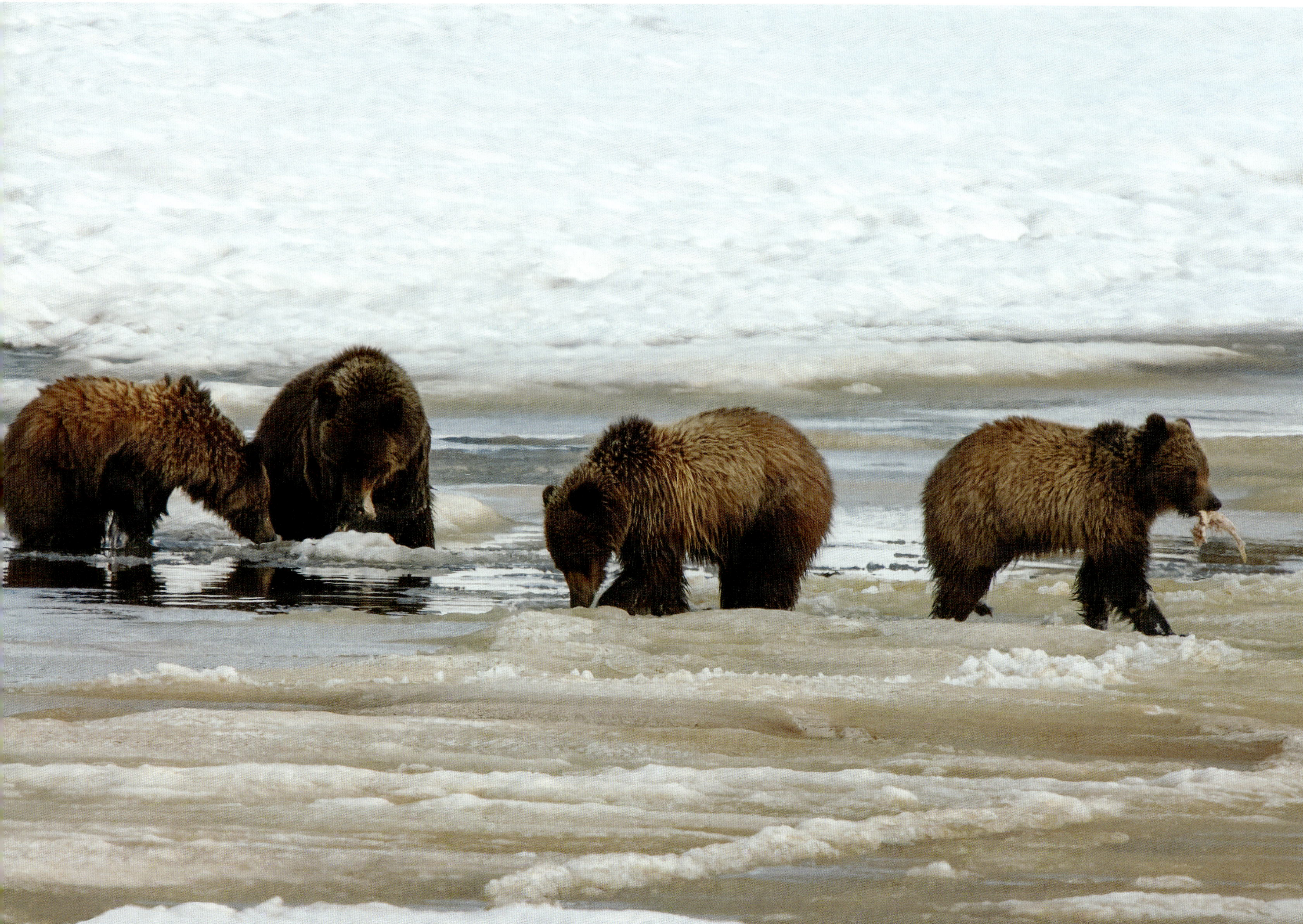

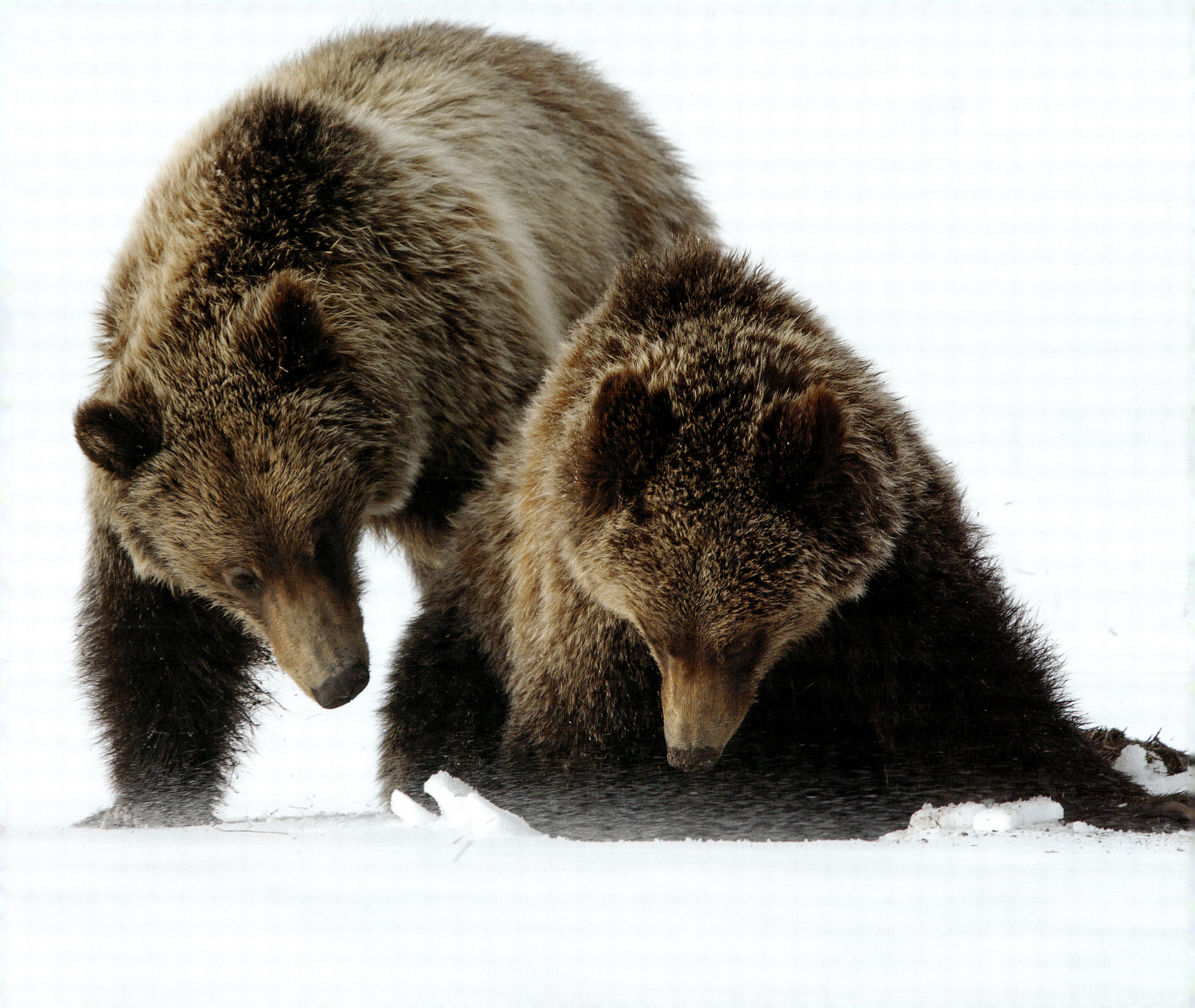

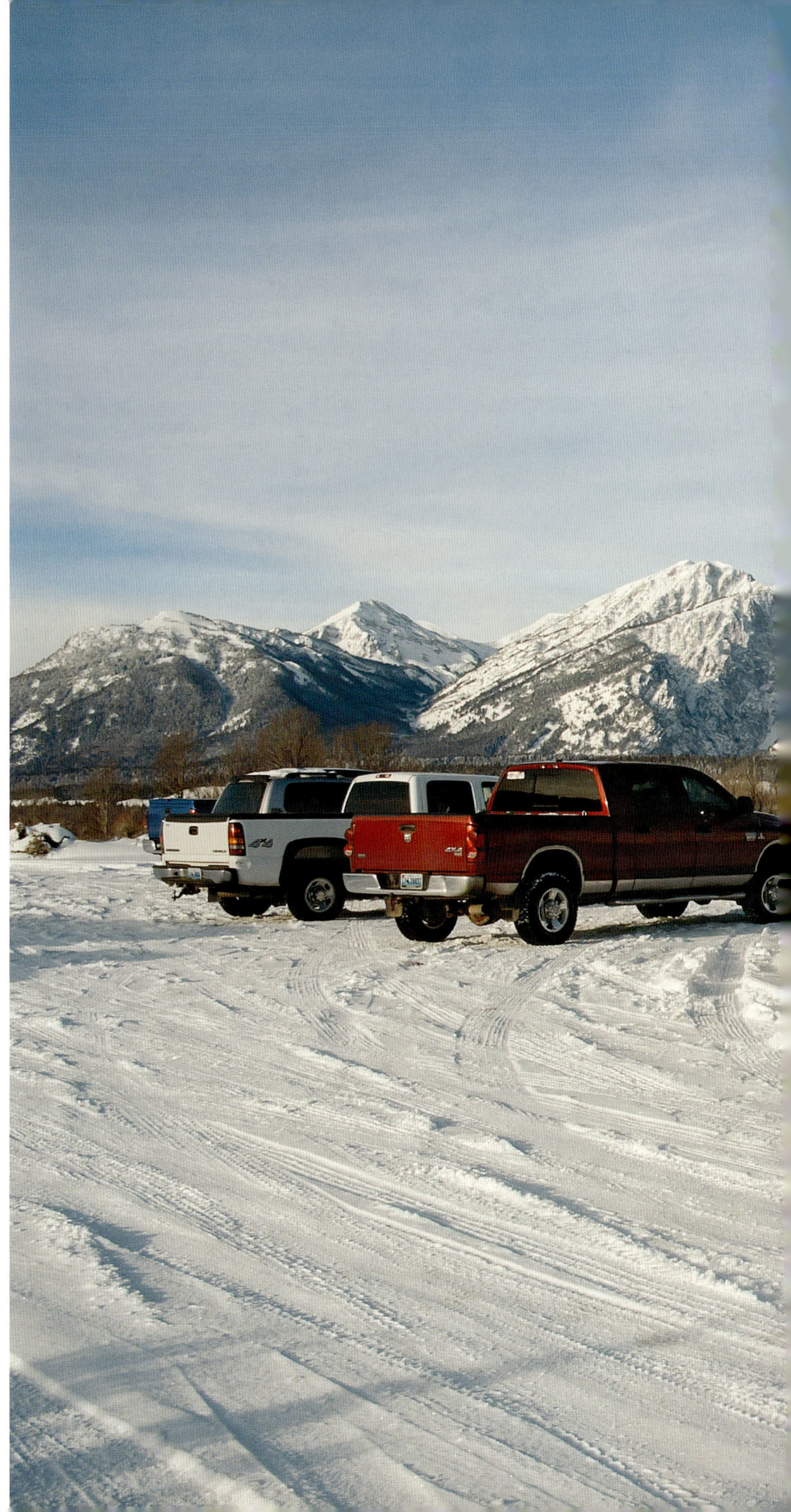

A wapiti shot by a hunter in Grand Teton National Park's elk reduction program is dragged back to a parking lot above Blacktail Ponds Overlook. In the past, grizzlies picking up the scent of dead animals have had many dangerous run-ins with hunters.

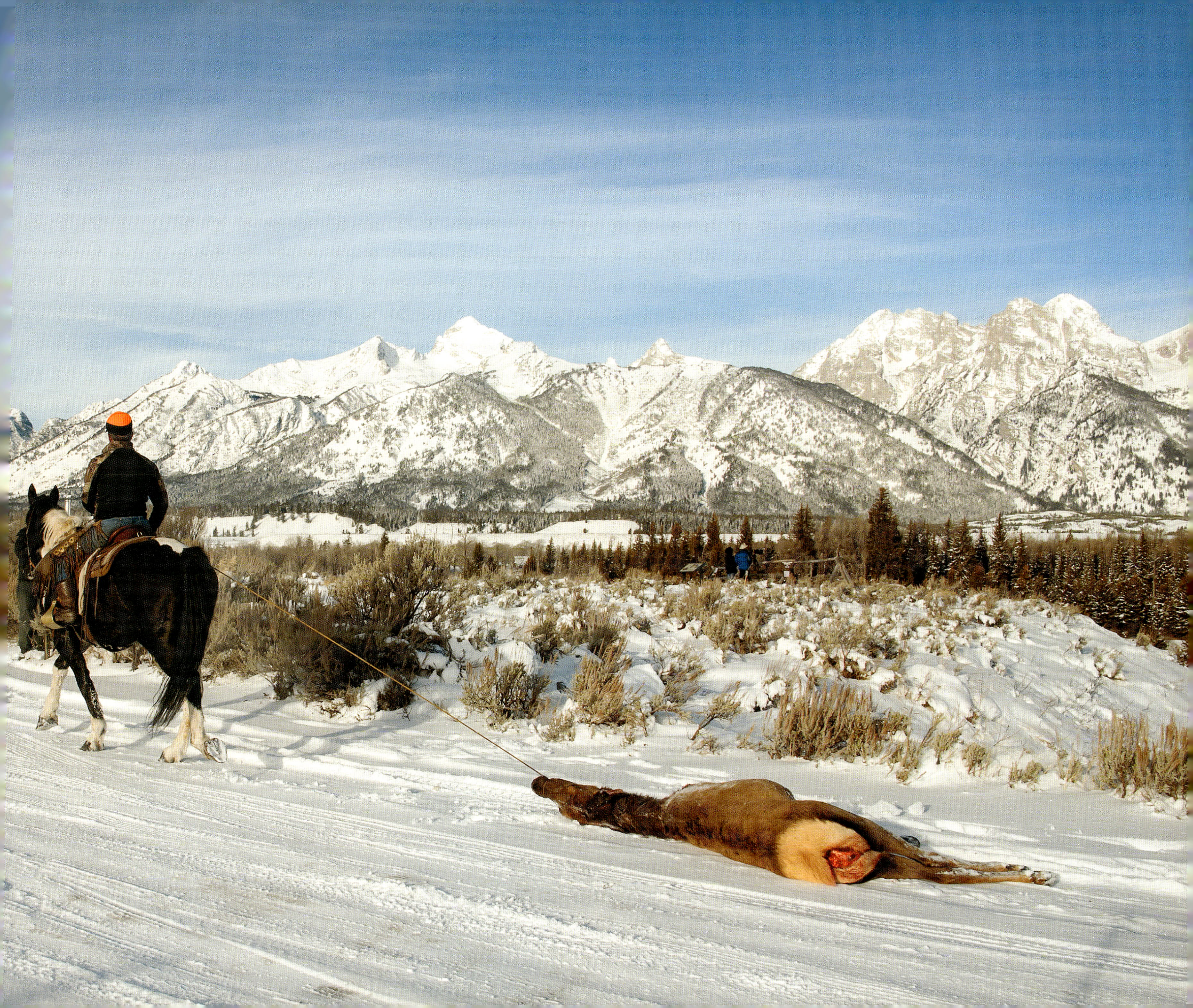

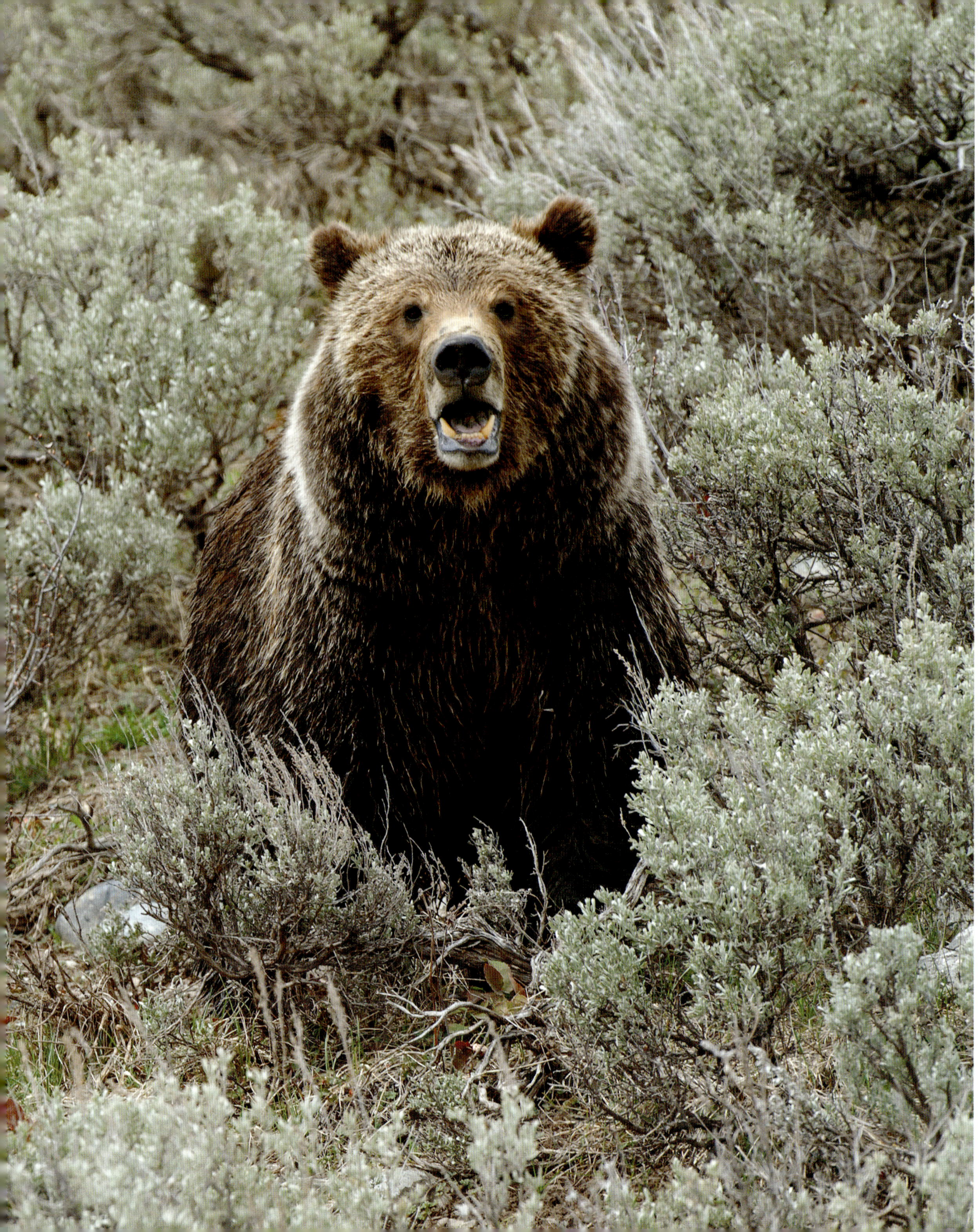

From a hillside above Oxbow Bend, 399 huffs at her cubs, reminding them to stay close.

OPPOSITE: Remember your greens: one of 399's cubs rolls over a rock, discovering fresh shoots of grass.

FOLLOWING SPREAD: In the spring of 2010, 610 walks across Willow Flats shortly after coming out of hibernation. She had no cubs that year and had been trapped, radio collared, and given yellow ear tags by Grand Teton National Park the previous fall.

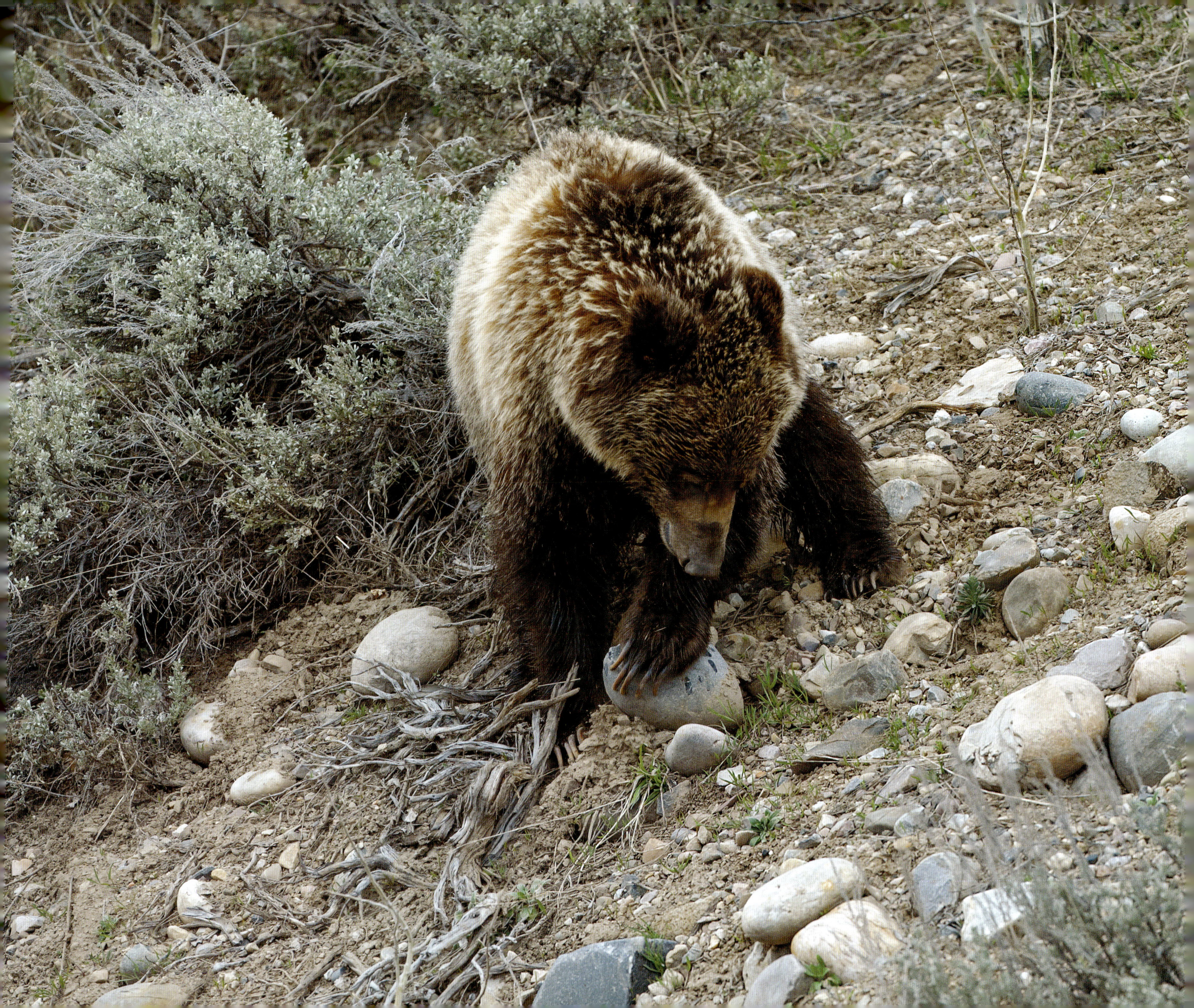

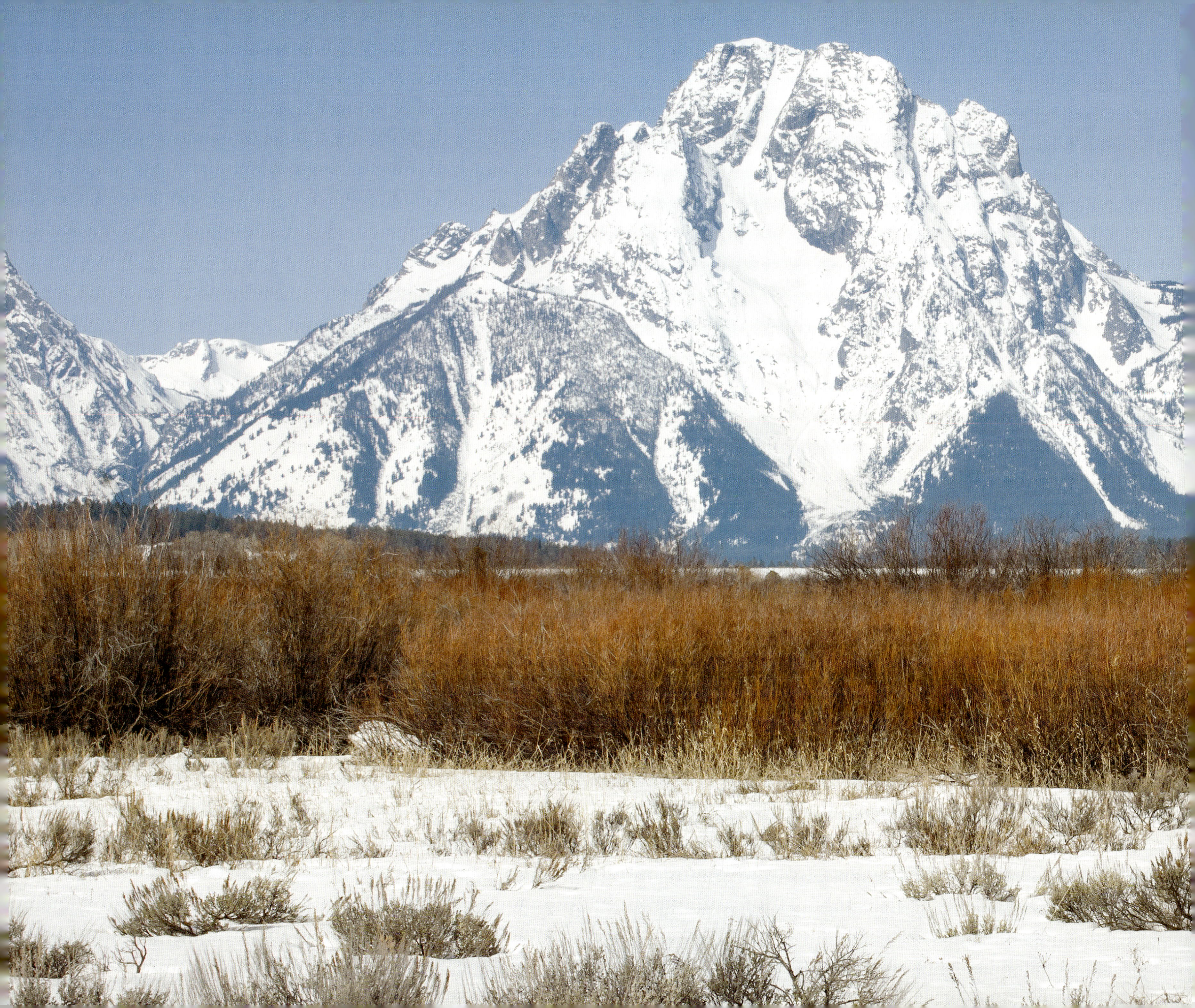

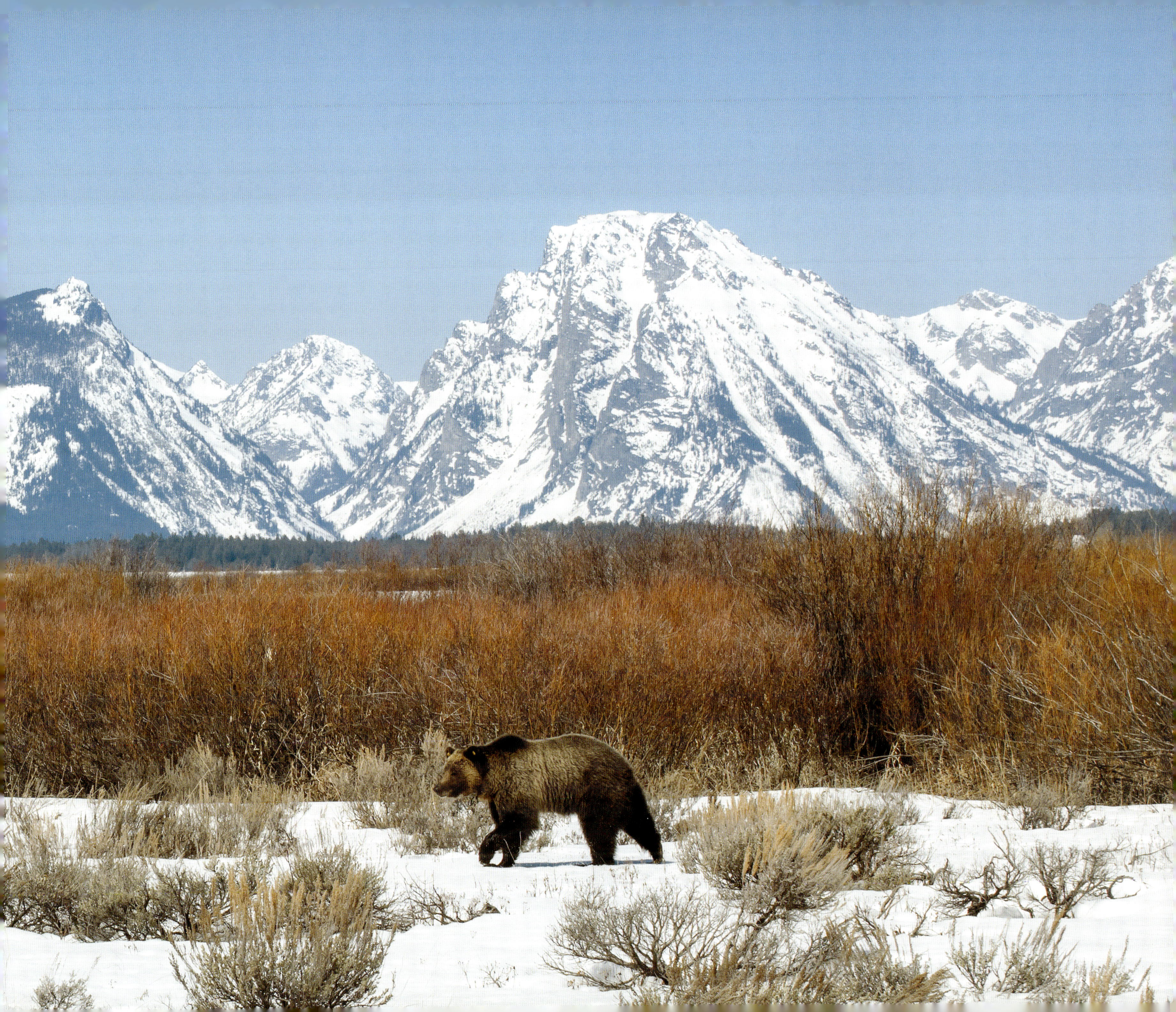

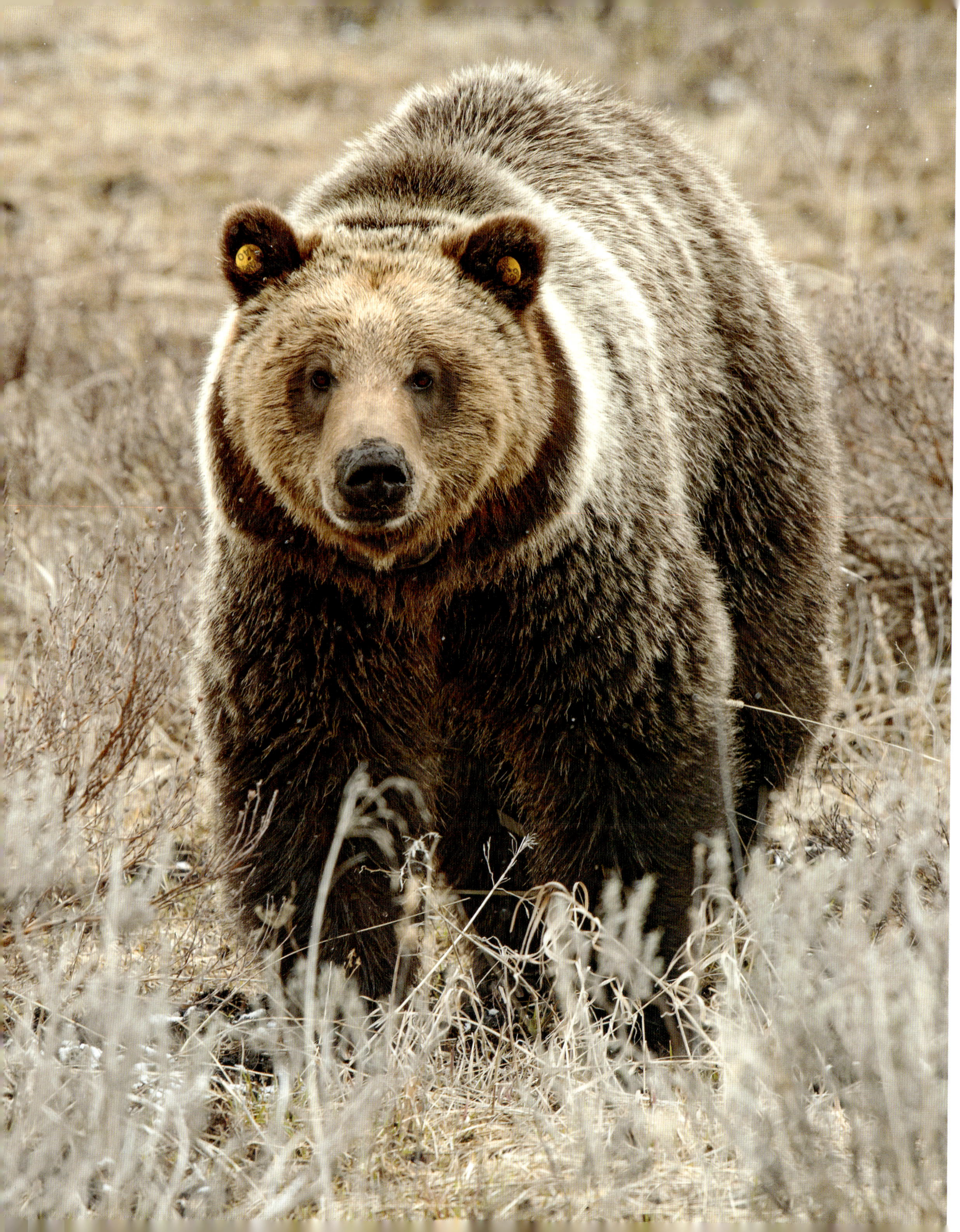
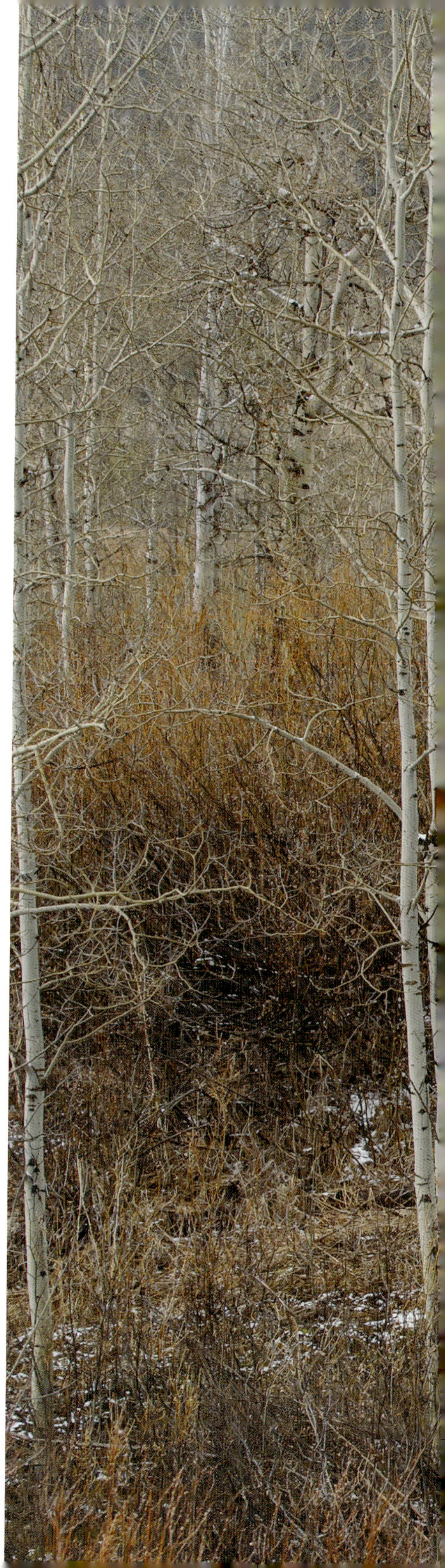

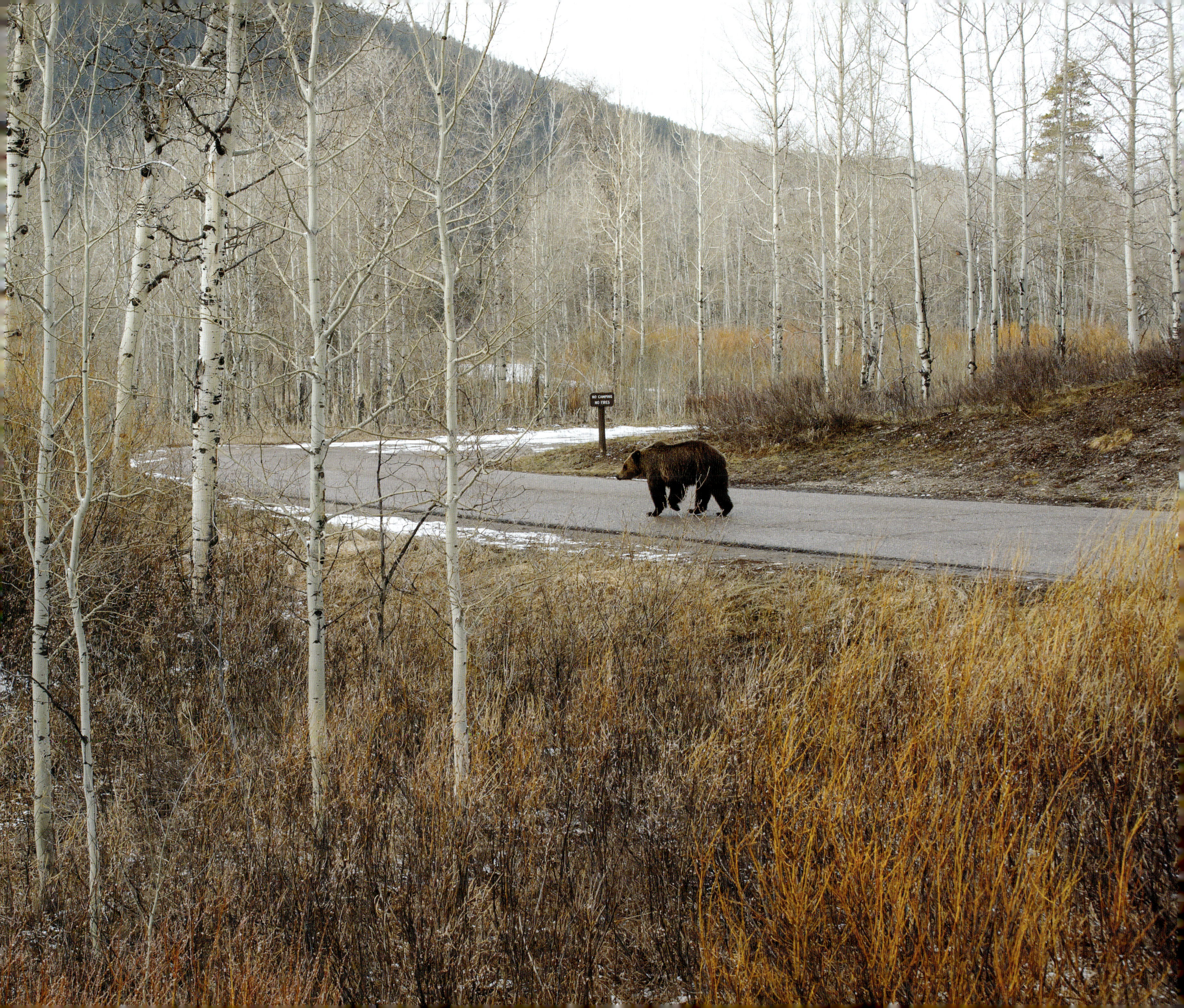
NO CAMPING
NO FIRES

PREVIOUS SPREAD, LEFT: Far from being unpredictable man-eating menaces, experts say the vast majority of grizzlies want only to be left alone by people. 399 and her daughter, 610, have demonstrated remarkable tolerance toward humans invading their space.

PREVIOUS SPREAD, RIGHT: On a rainy morning, 610 wanders alone down Cattleman's Bridge Road toward the Snake River.

399 drags a cow elk carcass to the rocky shoreline of the Snake River near one of the regularly used elk crossings. She likely ambushed one of the cows as it swam the river in the early morning darkness. Here, she stops to catch her breath and feast on the prize kill as a magpie tries to steal scraps.

FOLLOWING SPREAD: 399 stands guard over her elk carcass while a kayaker snaps photos as he drifts by.

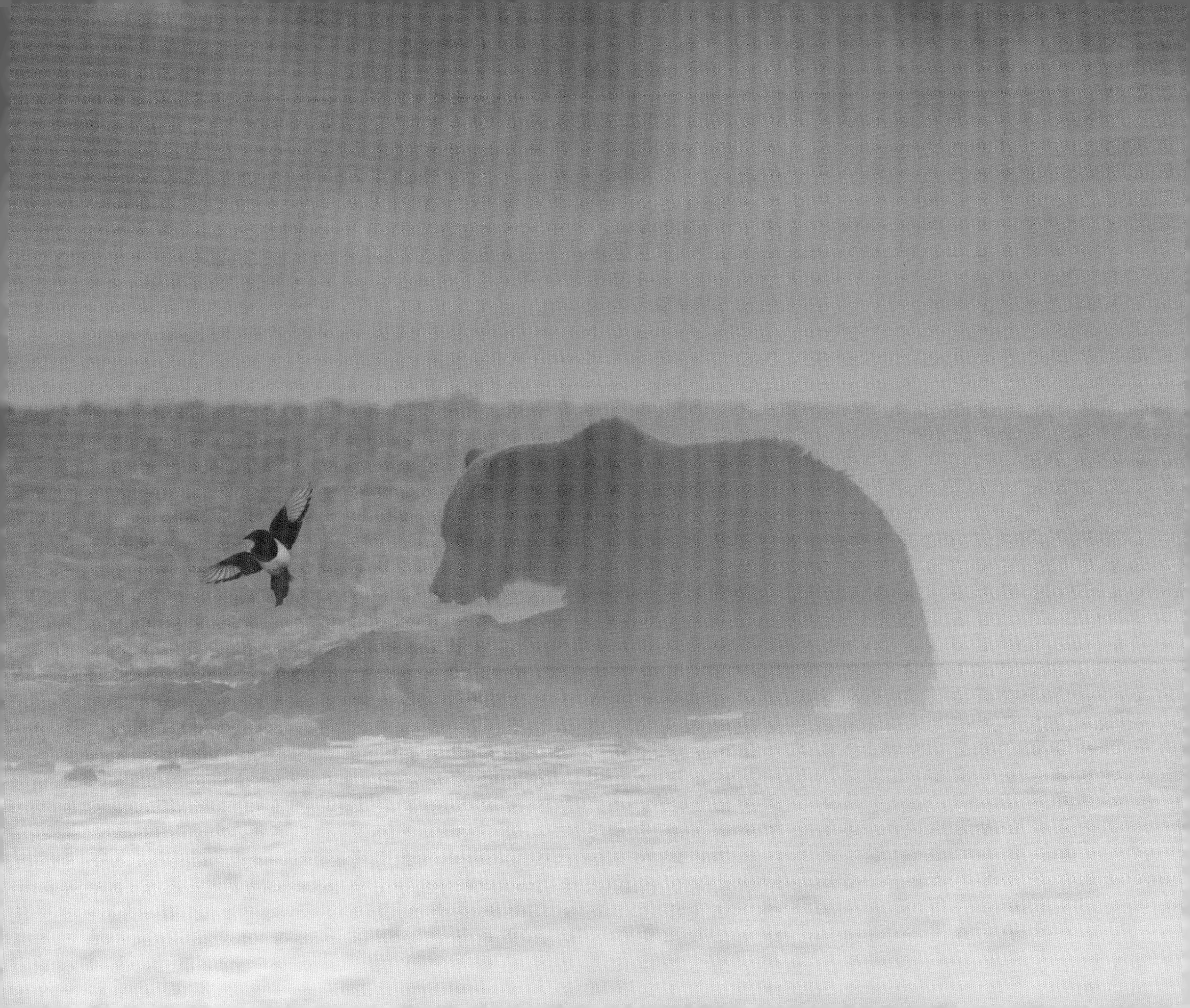

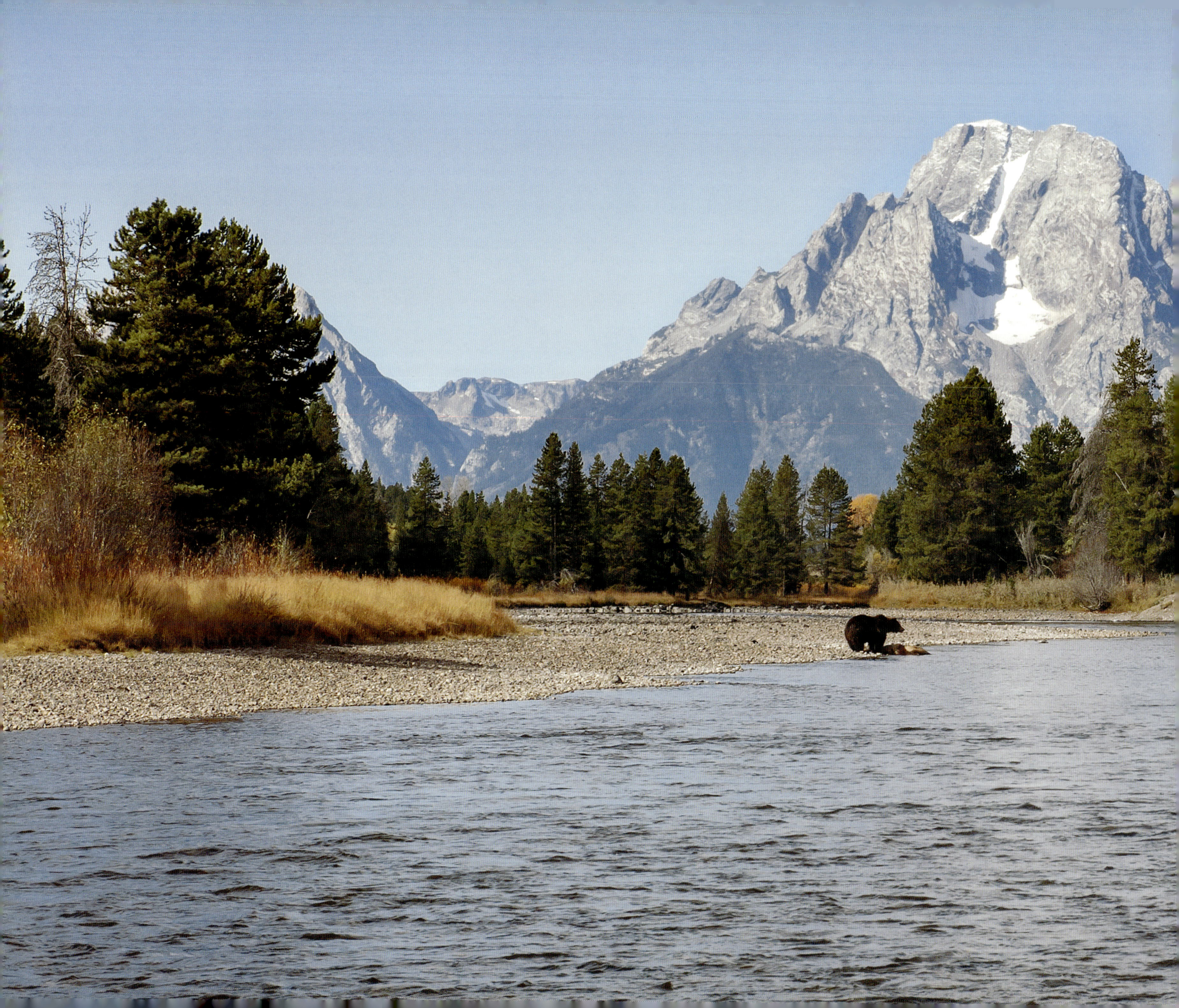

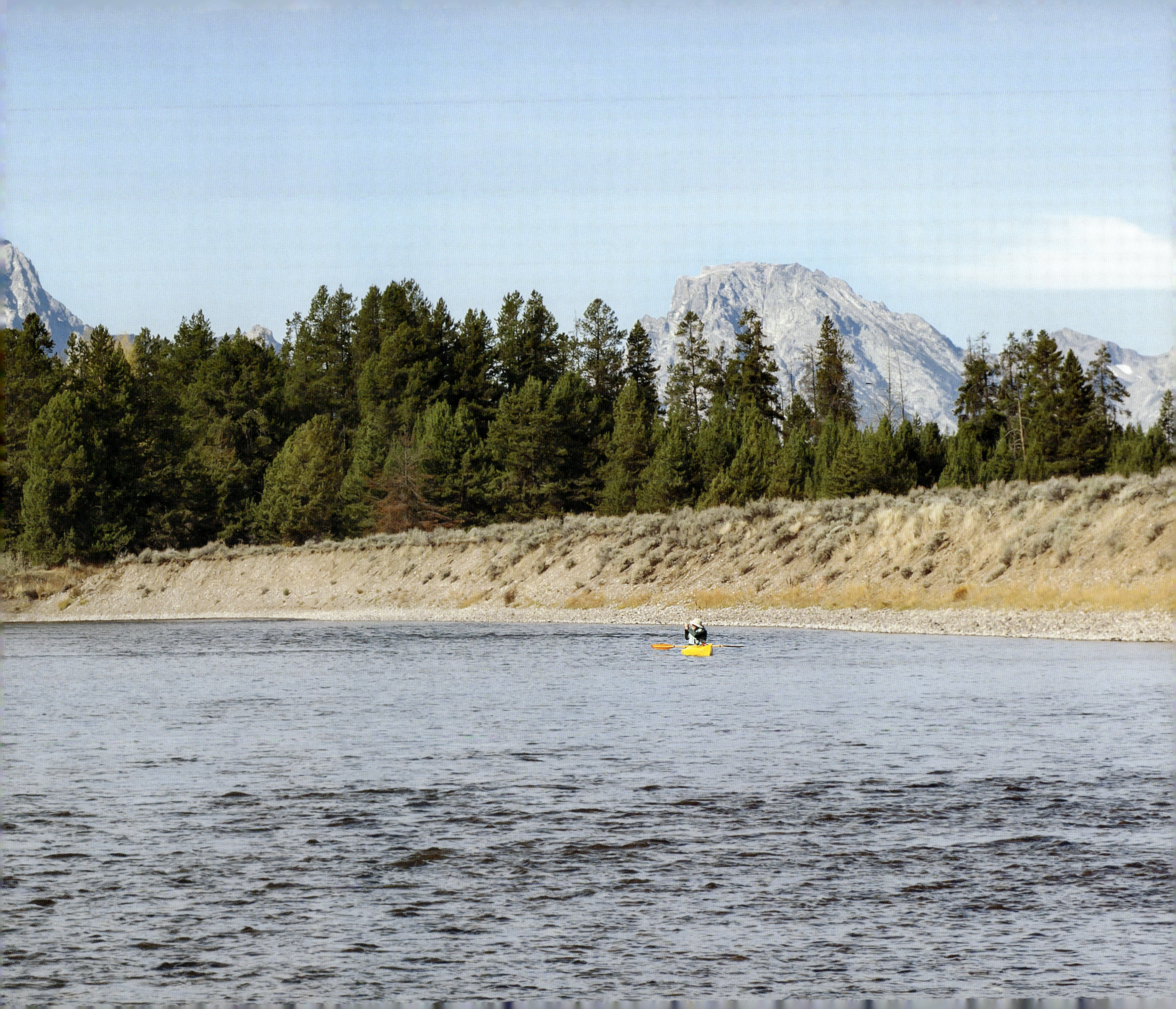

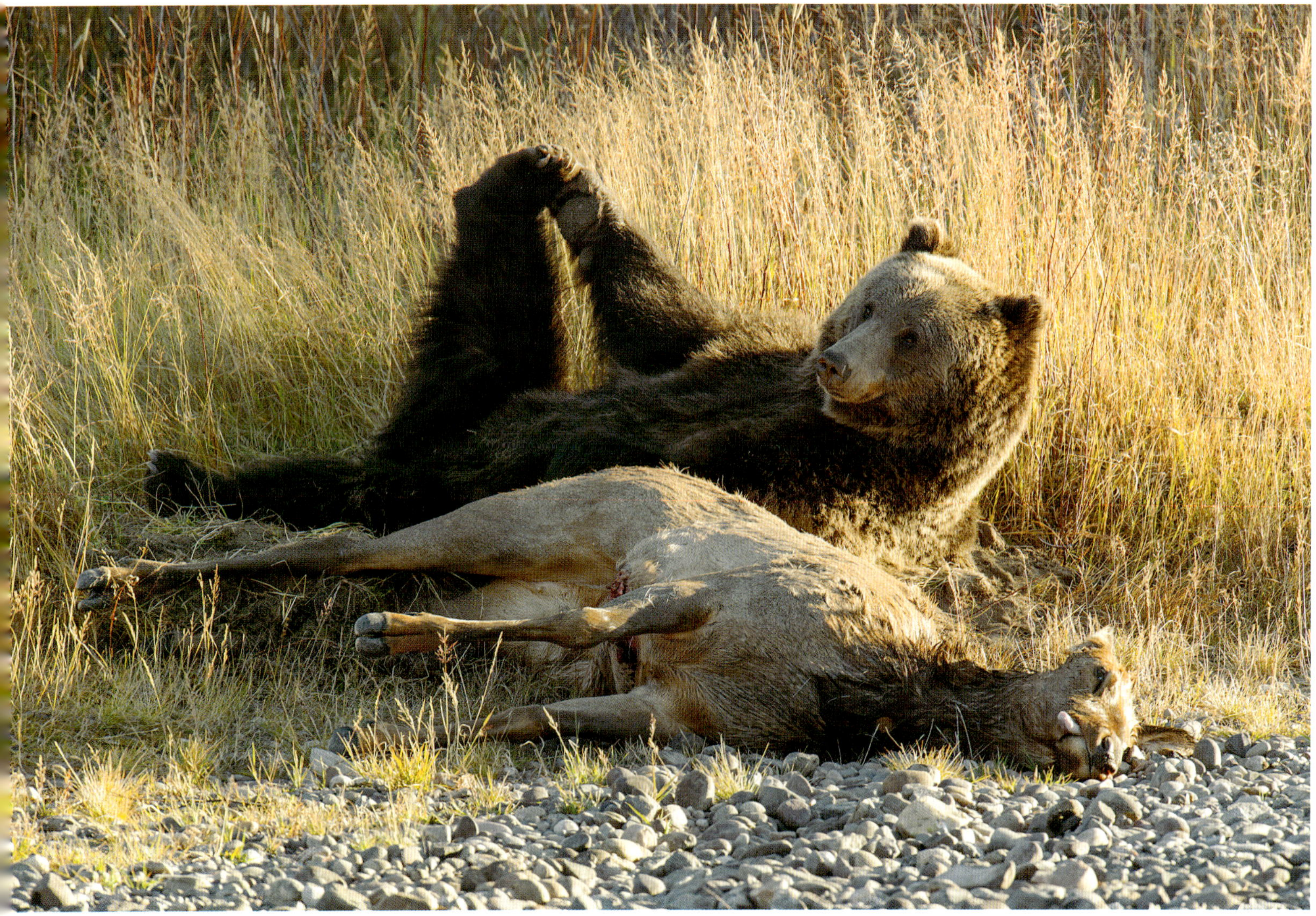

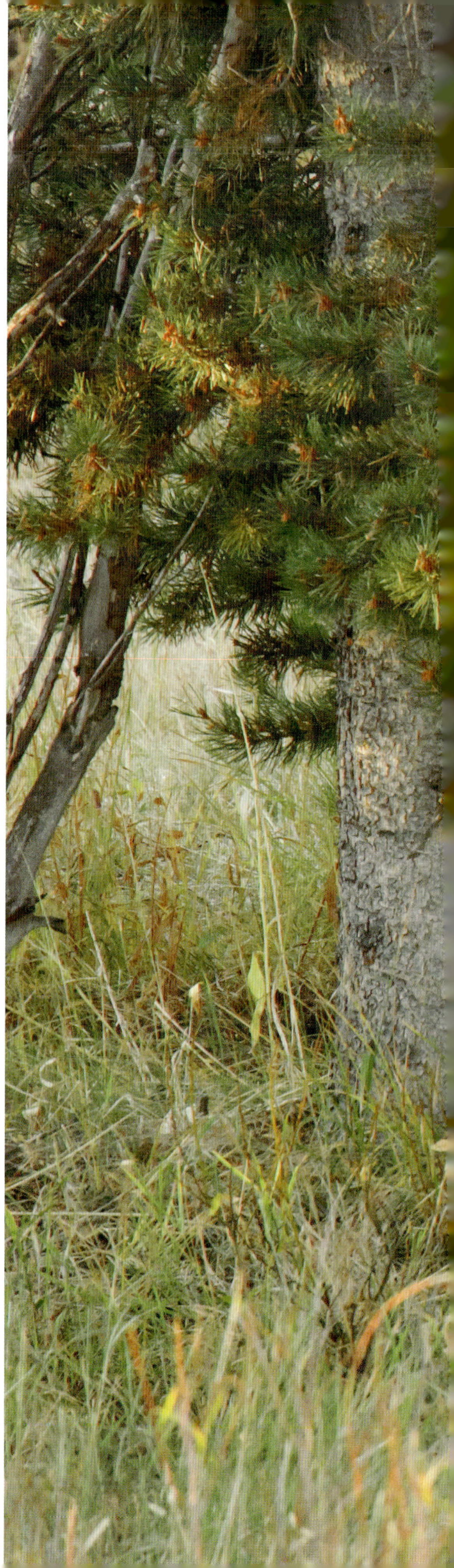

At the end of the day, feeling satiated and rested, 399 appears to do yoga while protecting her quarry.

OPPOSITE: After three days, 399 lies on top of a huge mound of grass and dirt she created to cover her elk carcass. Burying the carcass not only keeps it cool but also serves to minimize and hide the scent from would-be predators and scavengers.

FOLLOWING SPREAD, LEFT: Soon after coming out of their den, 610 leads her tiny cubs across the highway near the Potholes on a rainy, snowy May evening.

FOLLOWING SPREAD, RIGHT: Lacking the confidence of Spiderman, one of 610's adventurous cubs hangs on for dear life as he descends a large conifer, returning to the safety of his mother.

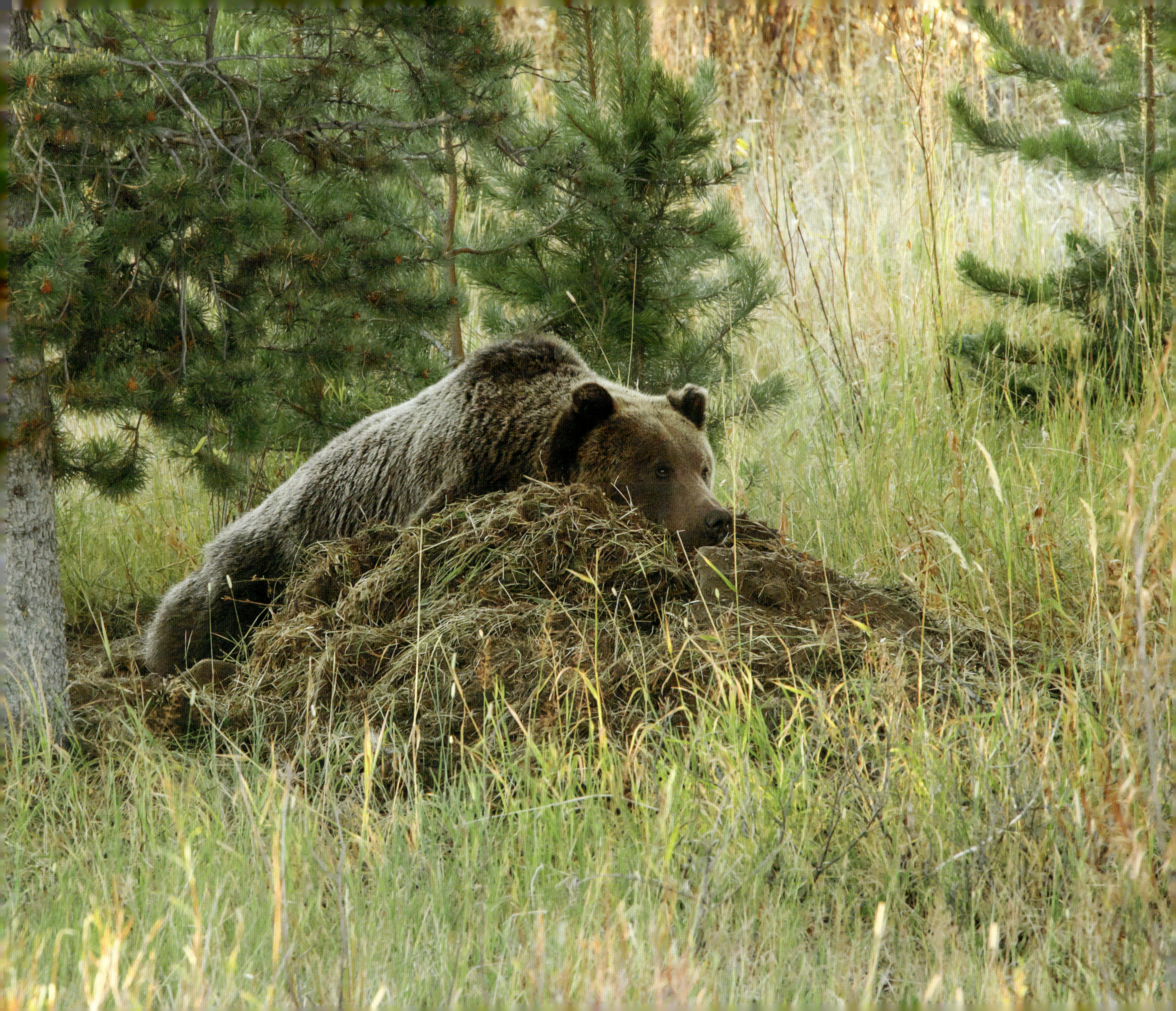

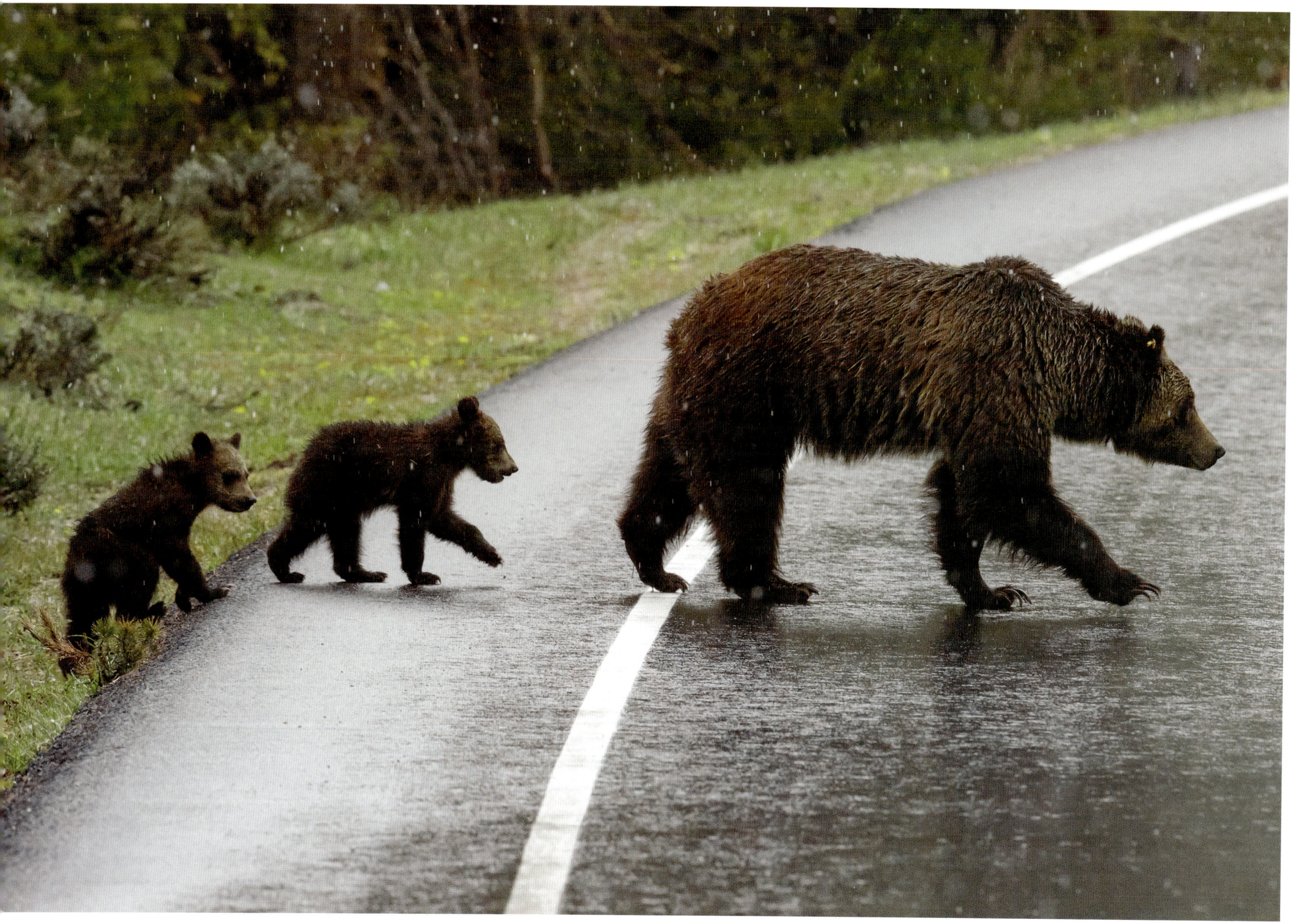

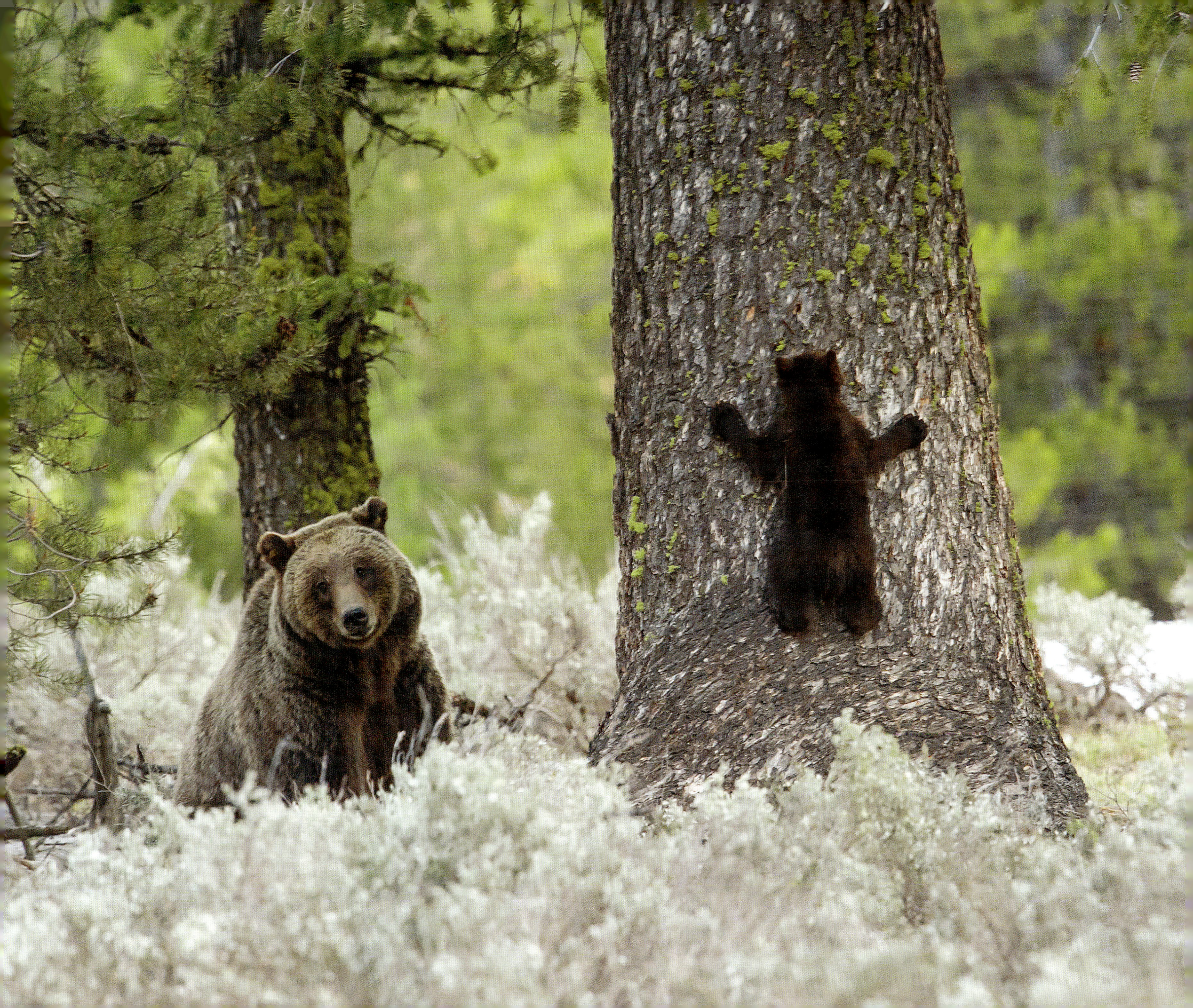

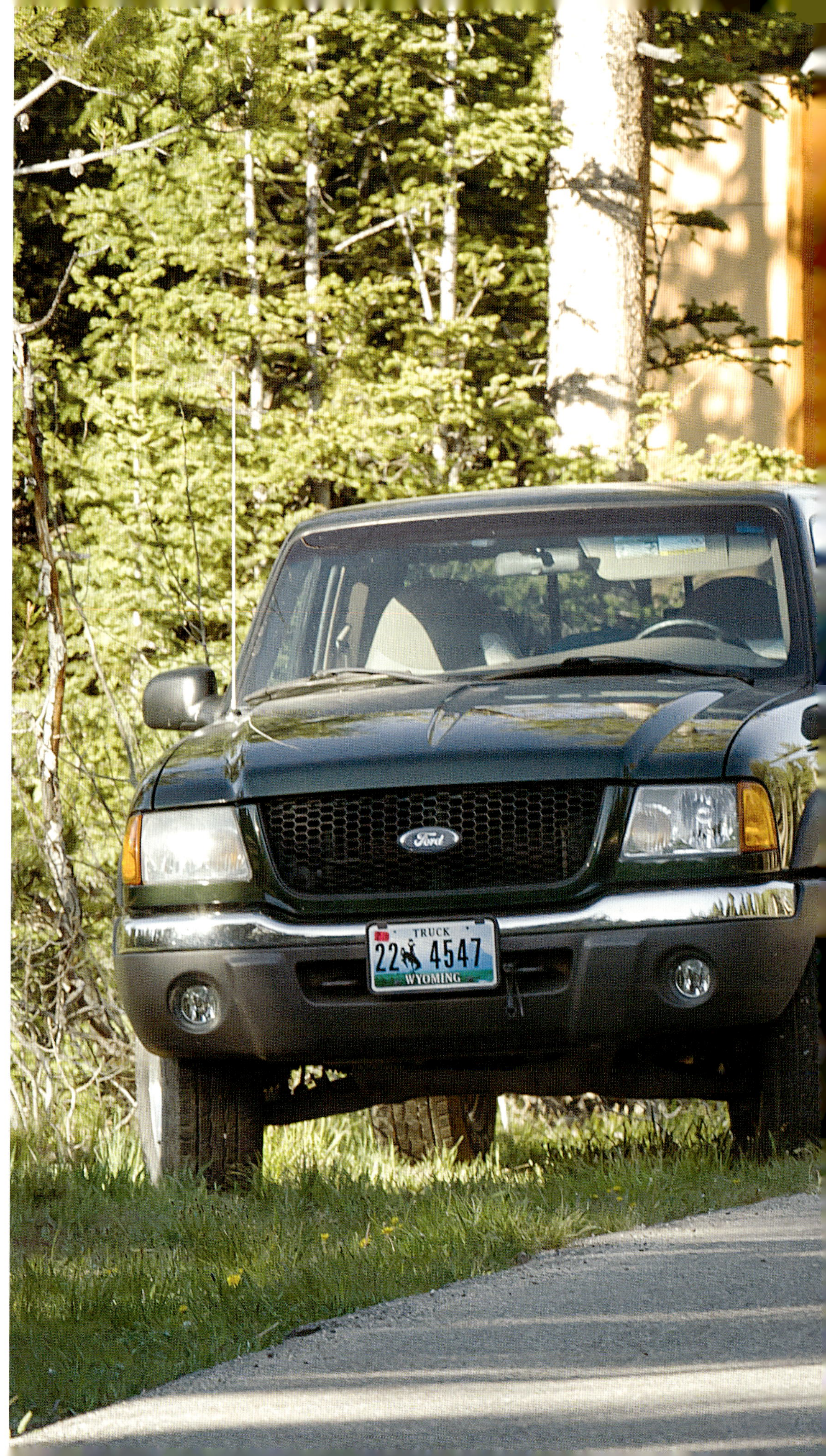

399 crosses the road with her second set of triplets on the first day she appeared, June 5, 2011. The grizzlies sharing the bloodline of 399 have attracted an international fan club, enchanting onlookers from around the world.

FOLLOWING SPREAD: 610 pauses and waits for her tiny toddlers as they struggle against the current of the Snake River.

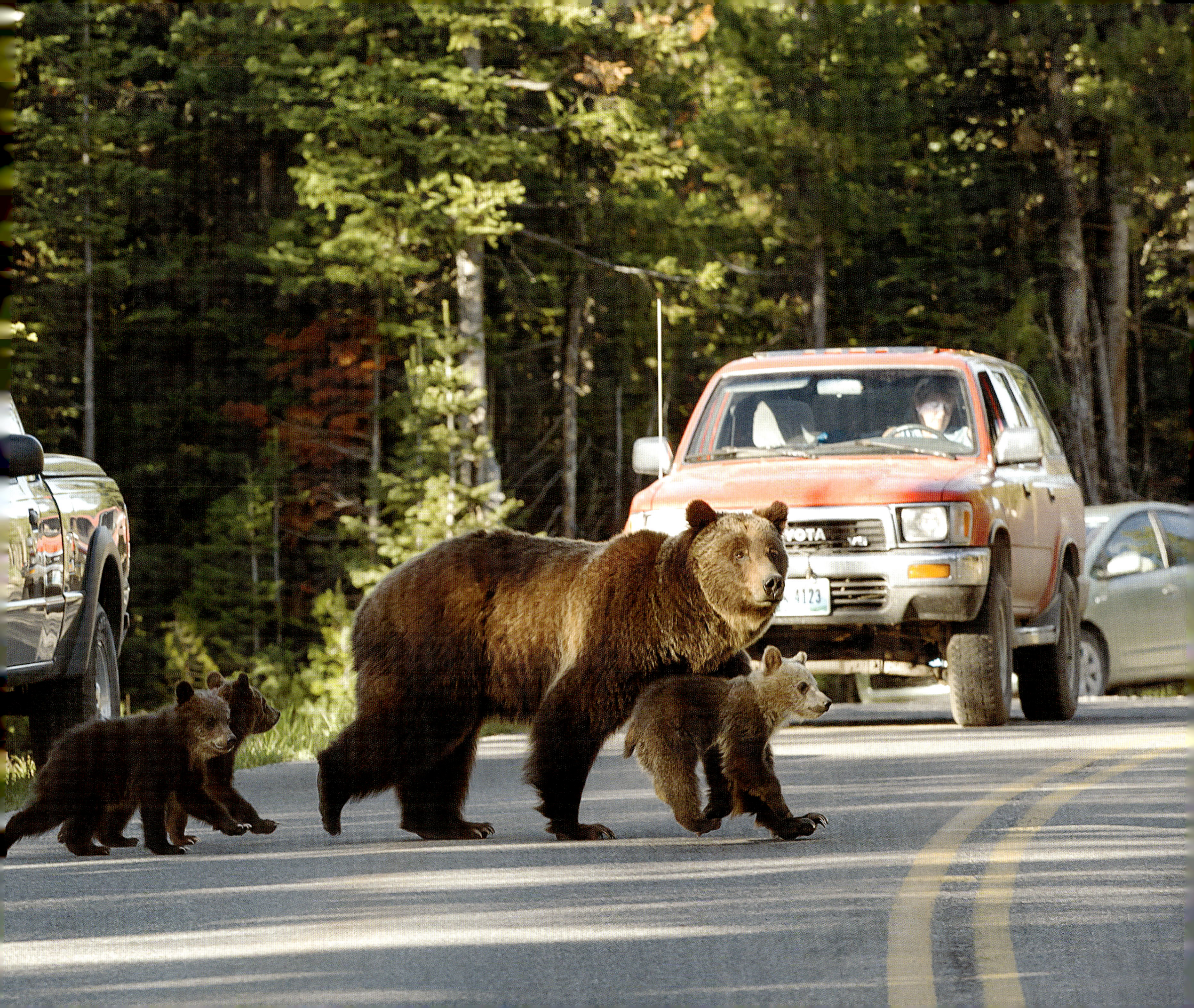

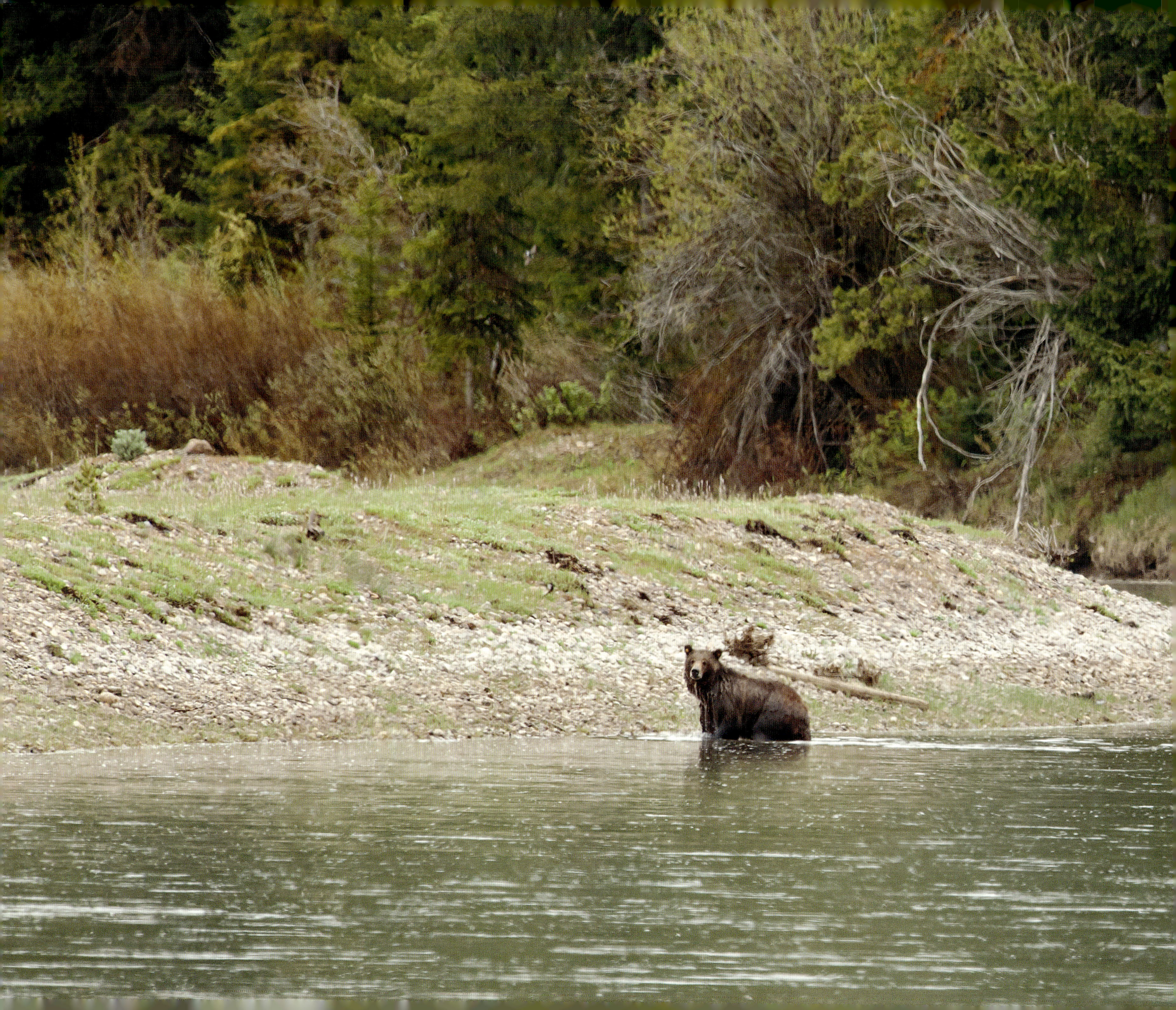

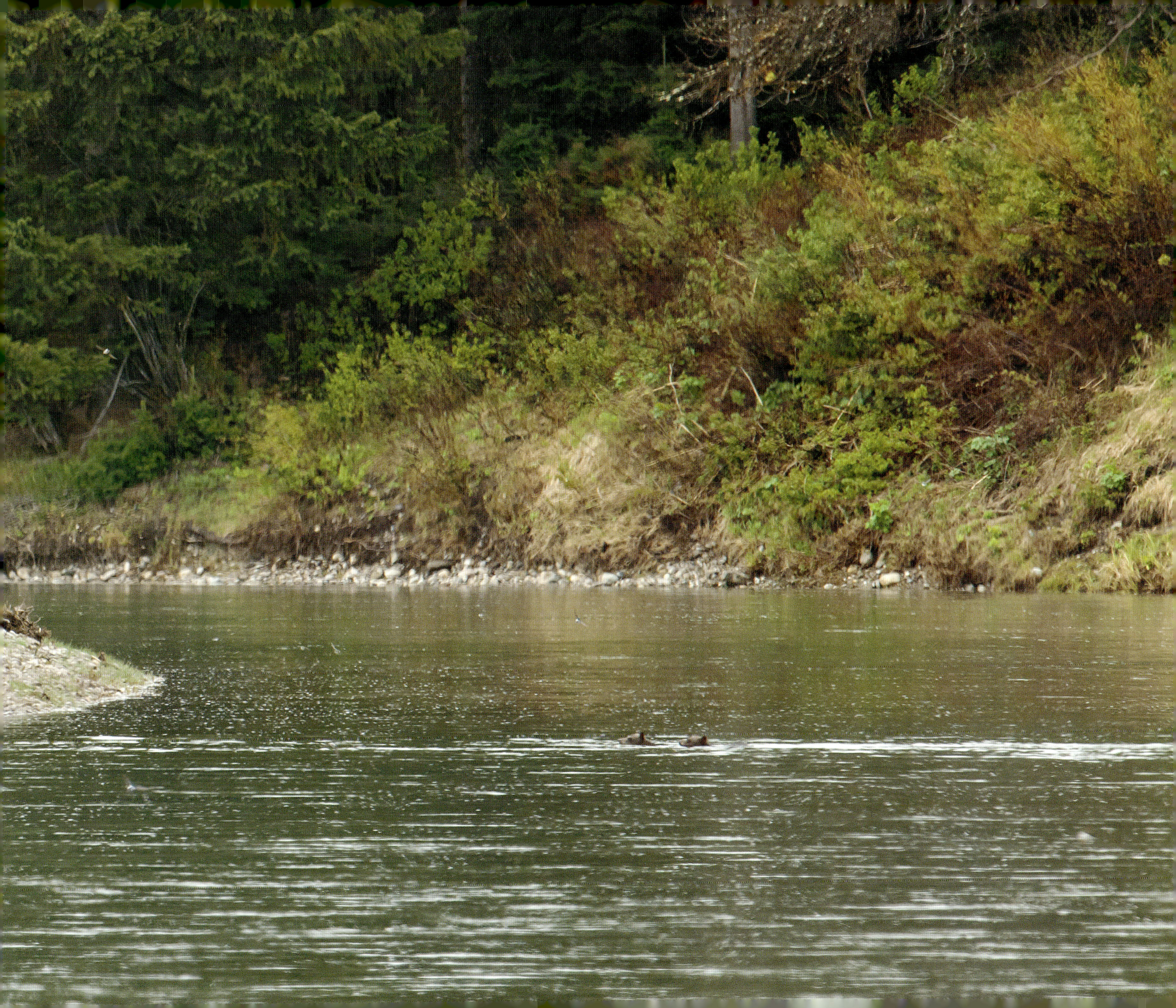

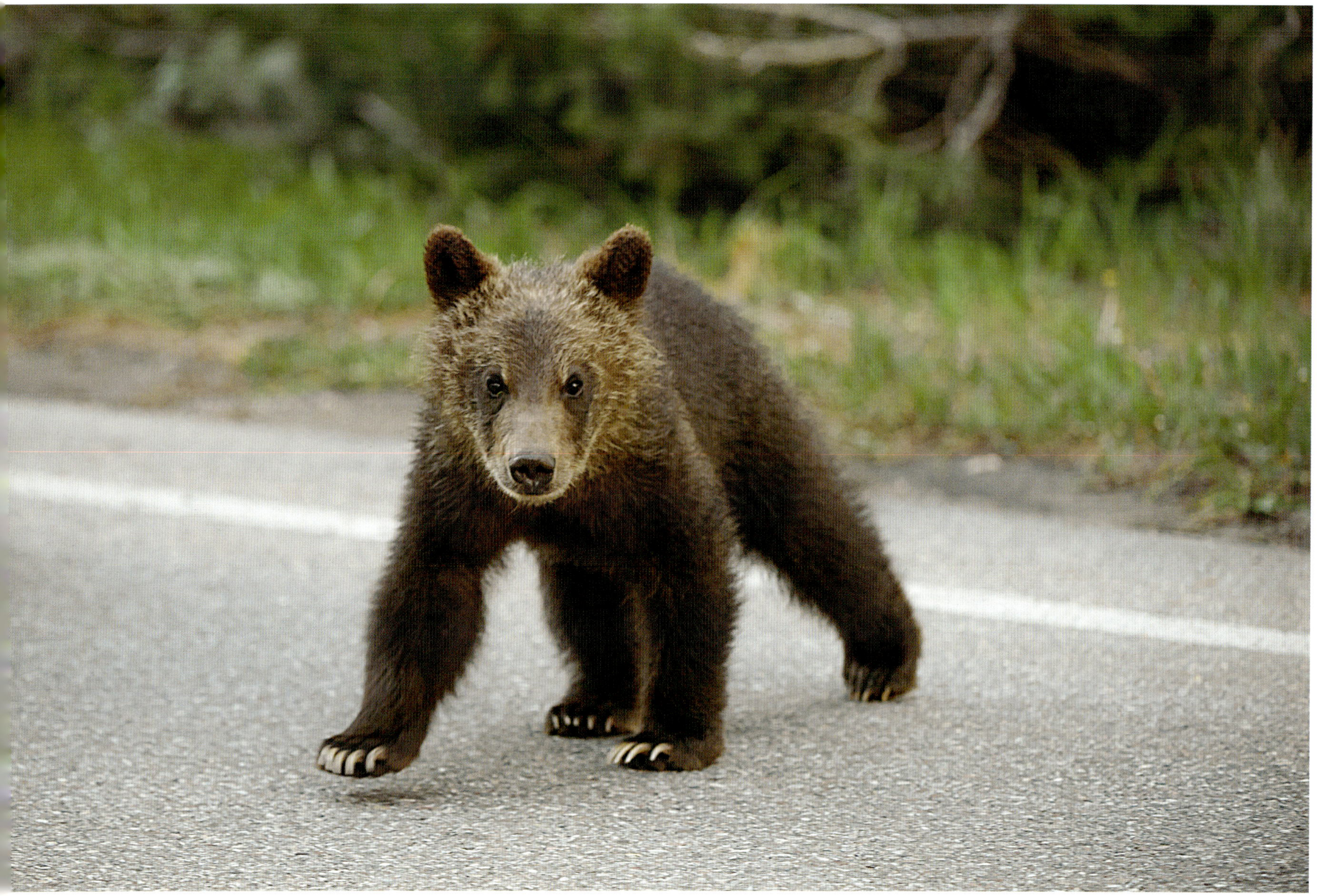

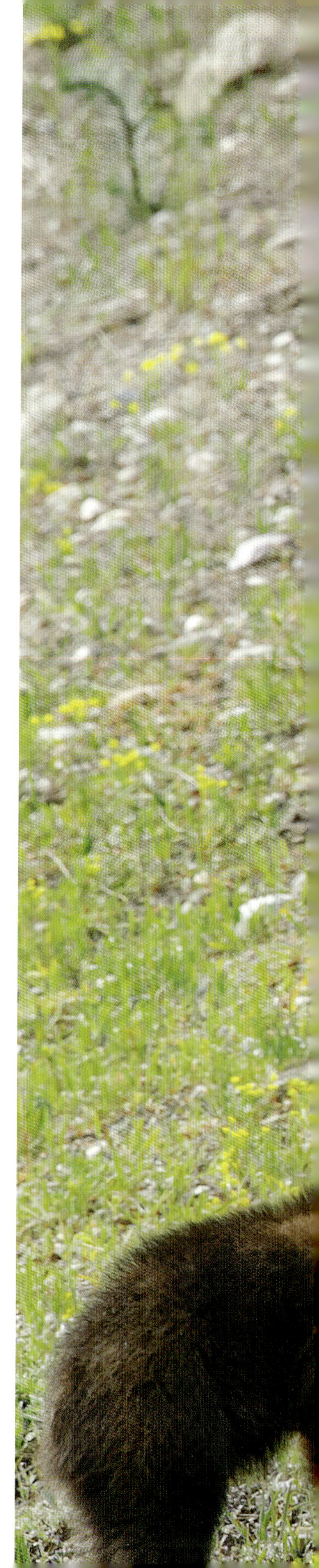

If a grizzly does not ingest sufficient amounts of calories and build its fat supplies before denning, it might not survive the five-month hibernation. Availability of high-nutrition food not only affects survival but also influences a female bear's ability to get pregnant and the number of cubs she'll have.

FOLLOWING SPREAD: With her nose pointed high, fifteen-year-old 399 sniffs the breeze as she pauses in a meadow near Pilgrim Creek; taking their mother's lead, the three cubs mimic her, learning the ancient ways of the grizzly.

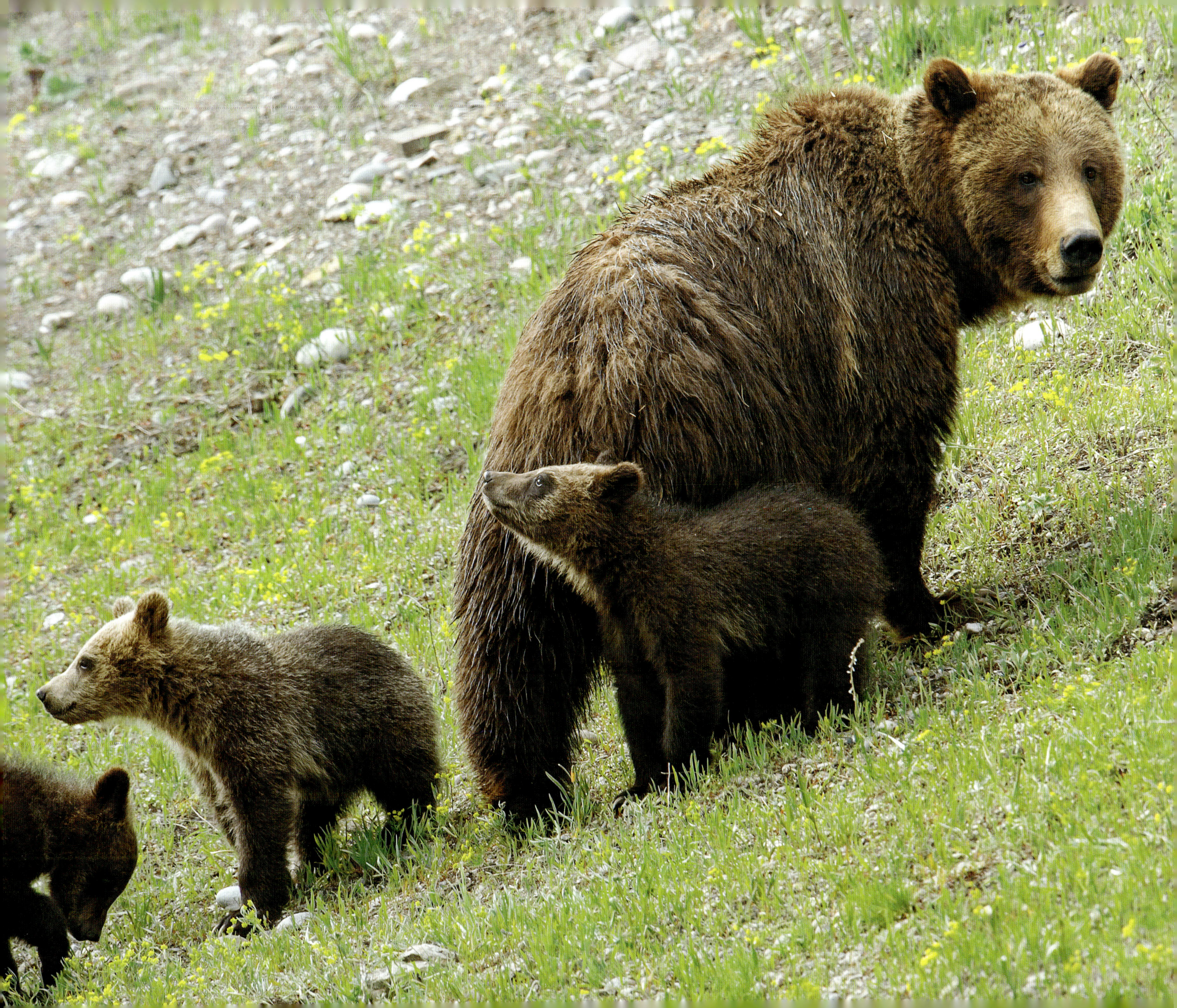

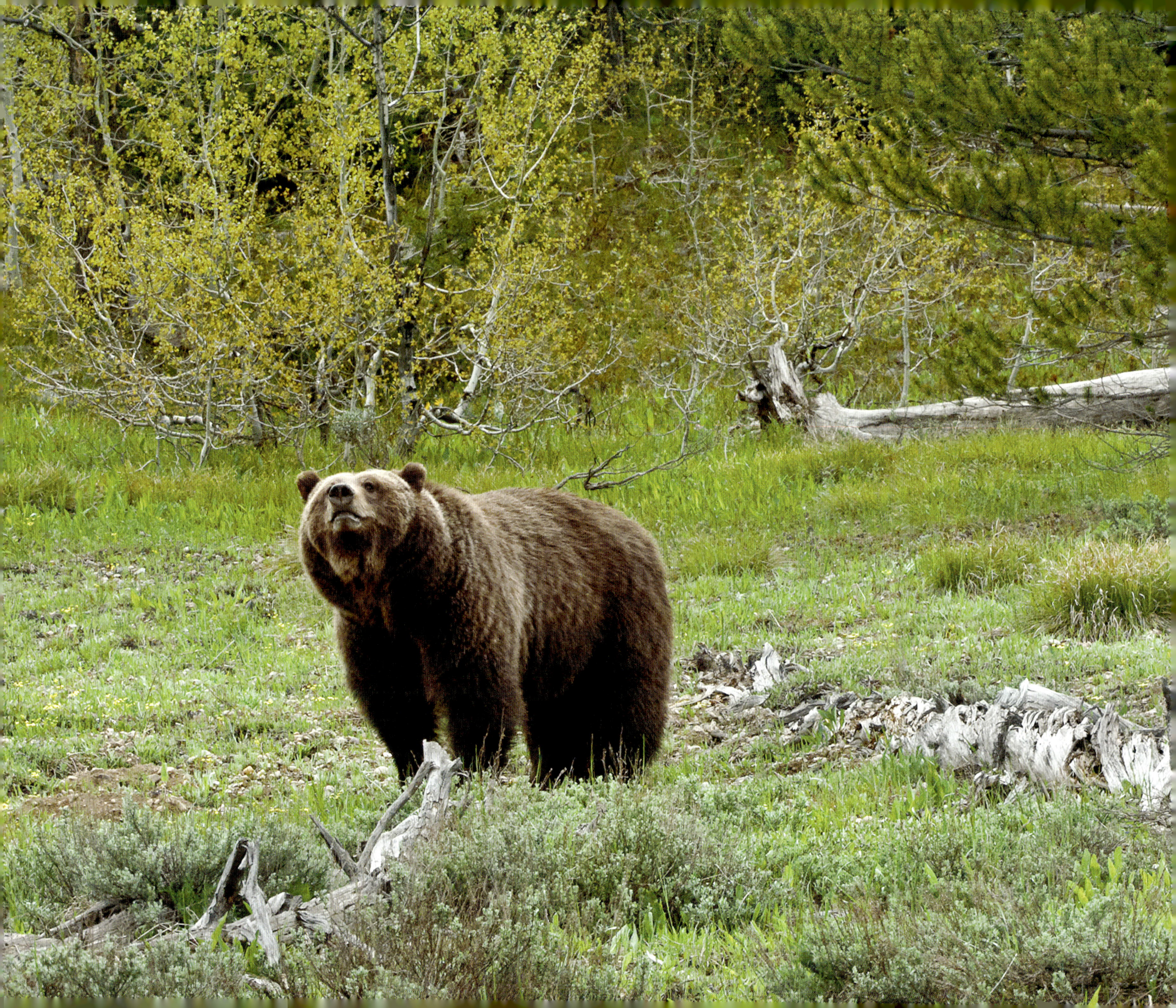

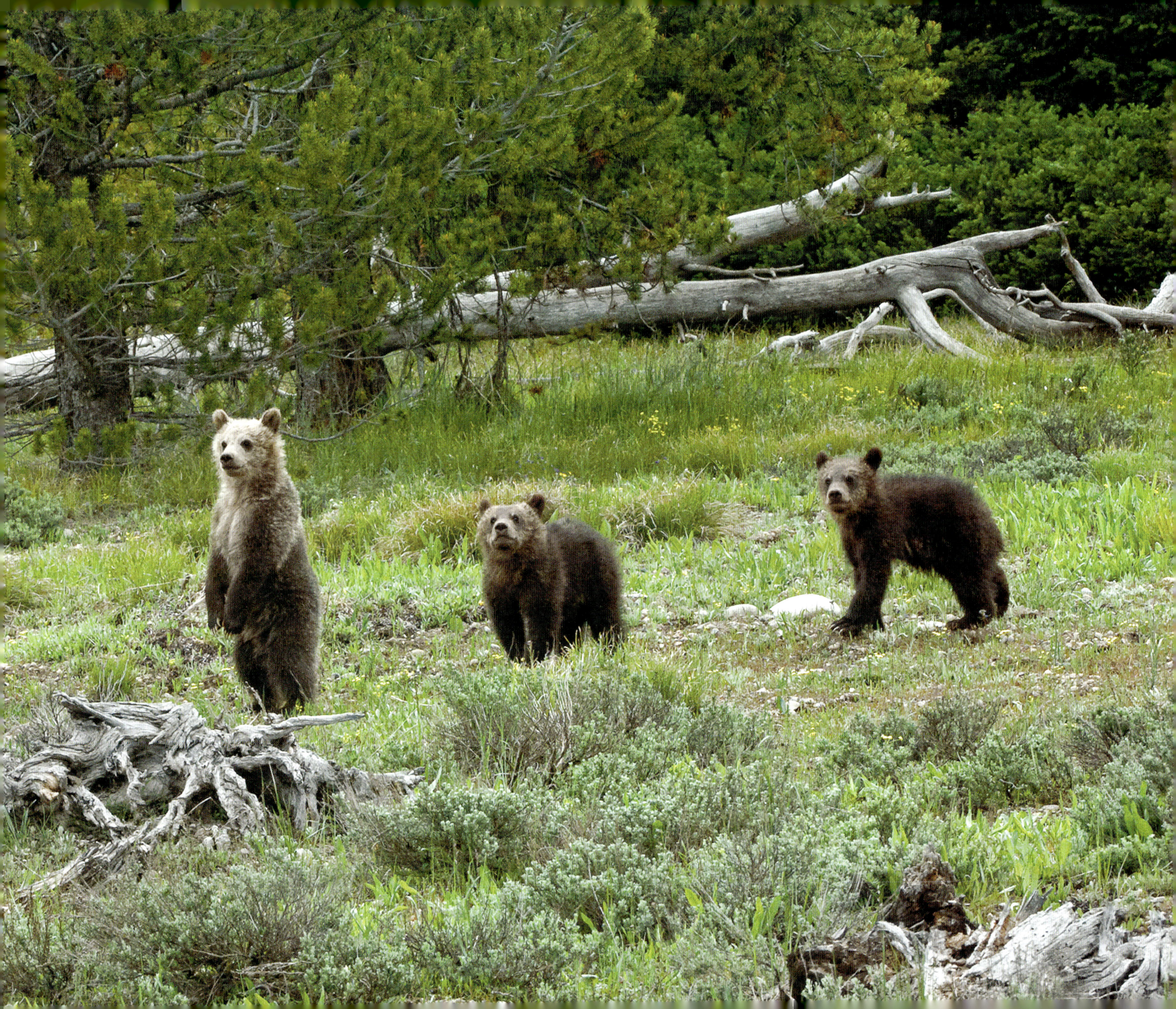

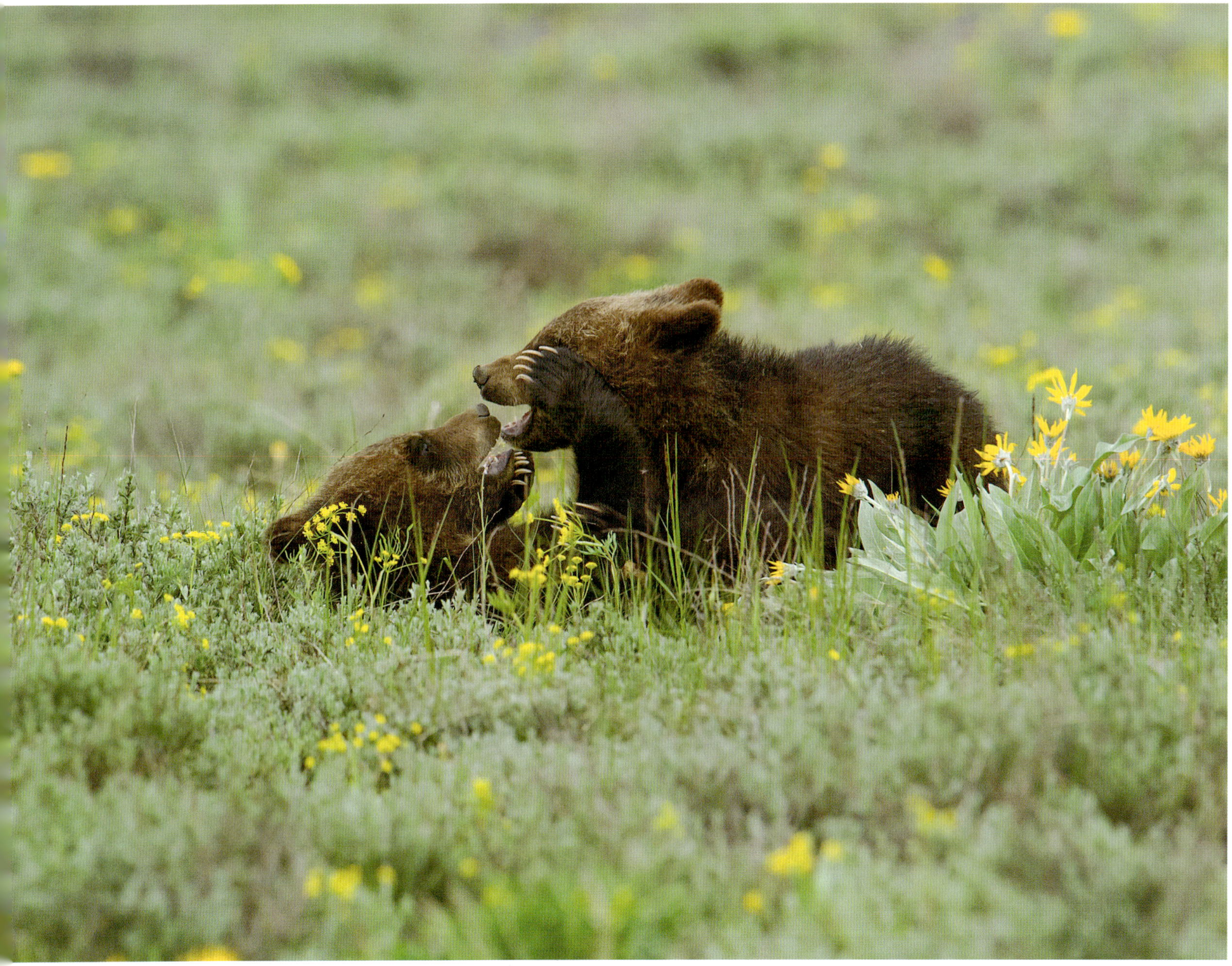

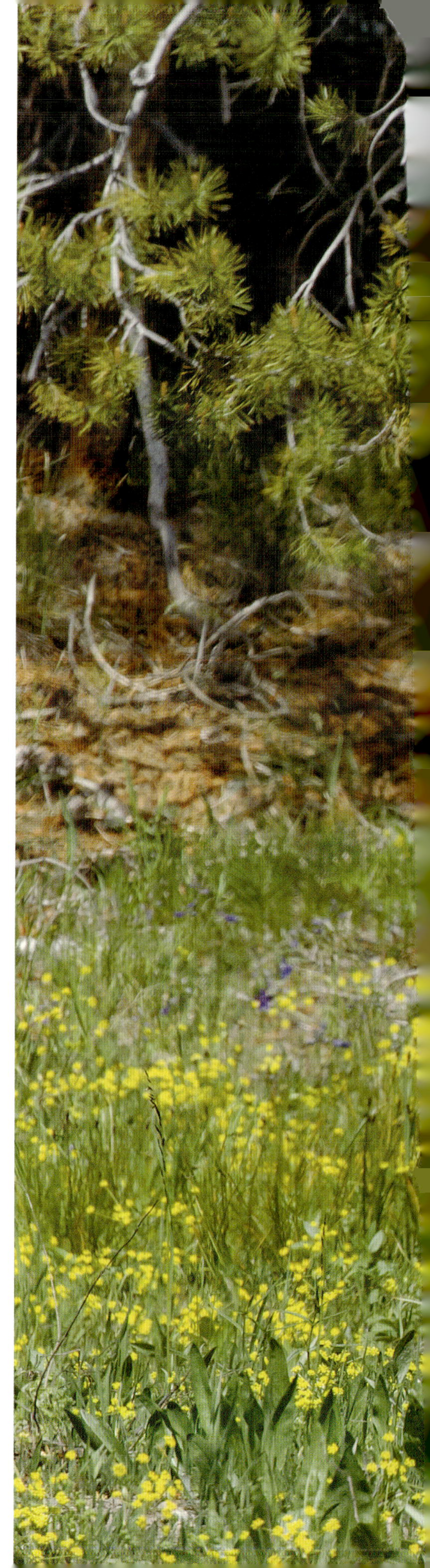

Two of 399's cubs, Brownie and his dark-colored male sibling, play in a meadow of biscuitroot near Pilgrim Creek.

OPPOSITE: Ash, with a mouth full of biscuitroot, grazes with her mother and siblings.

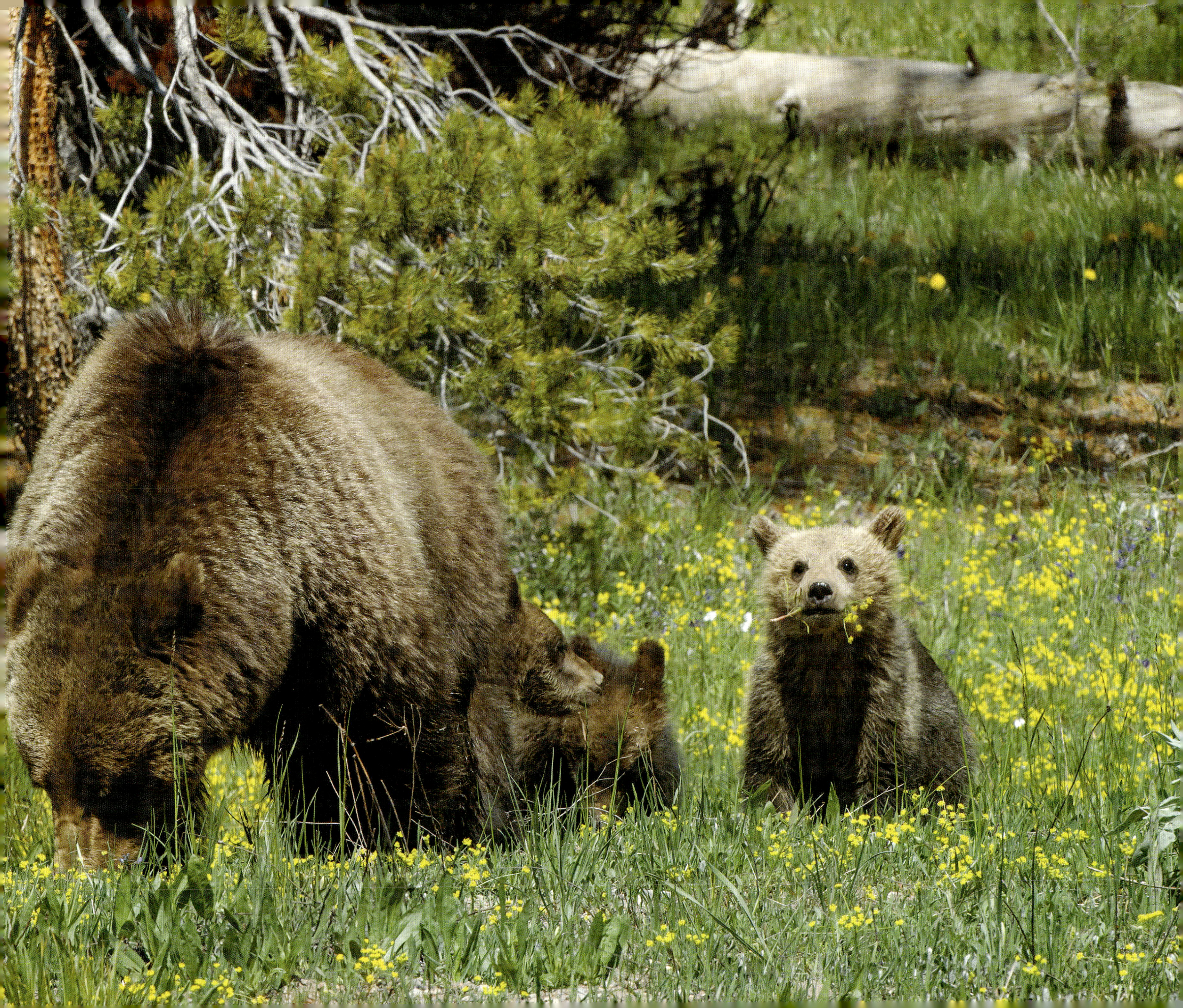

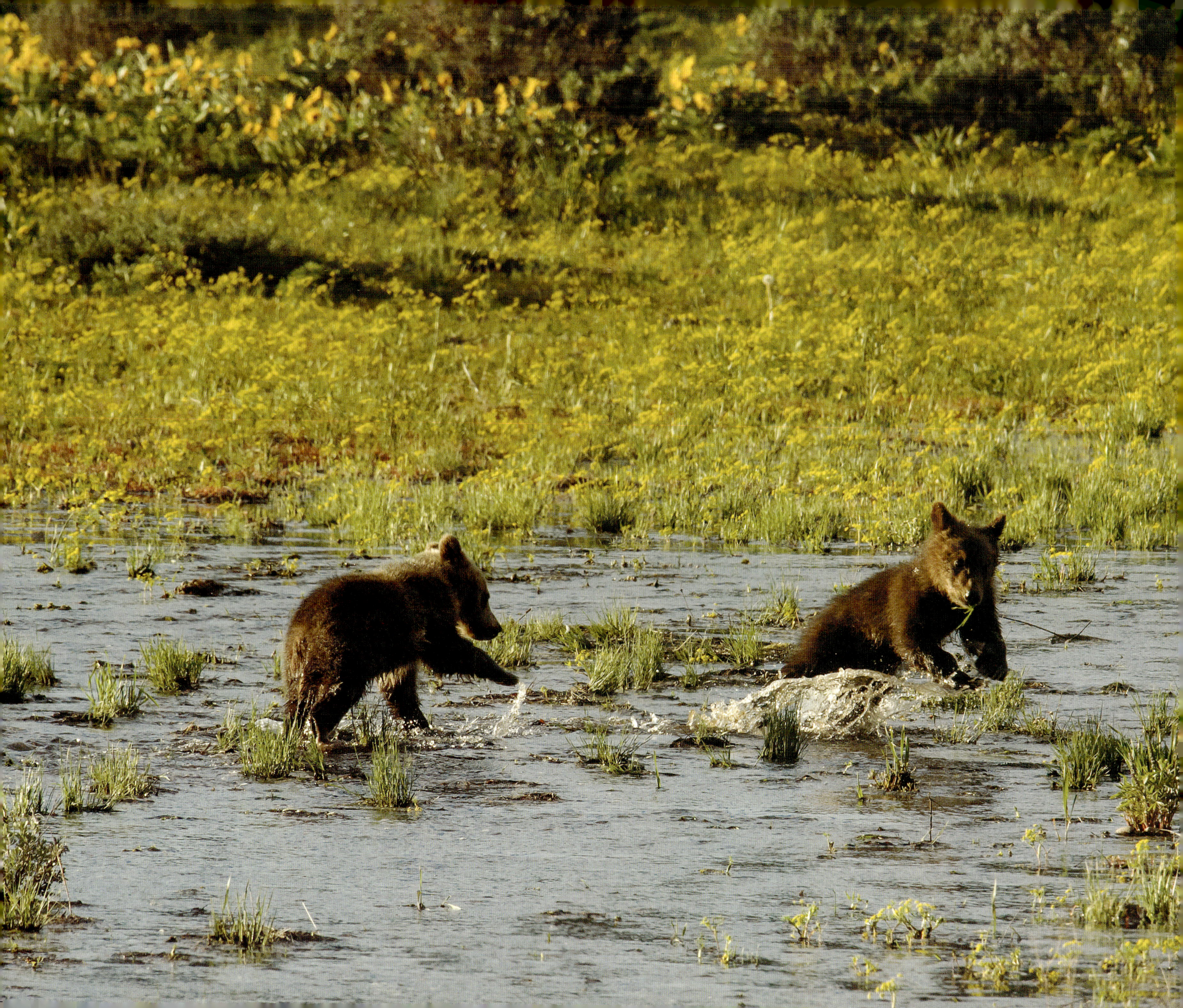

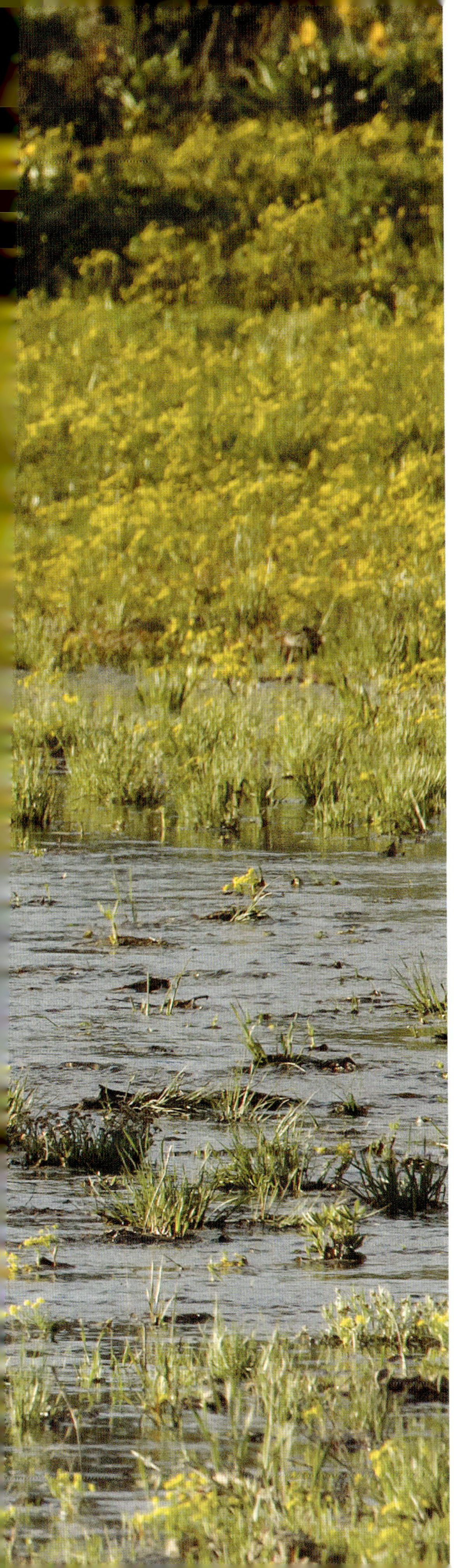

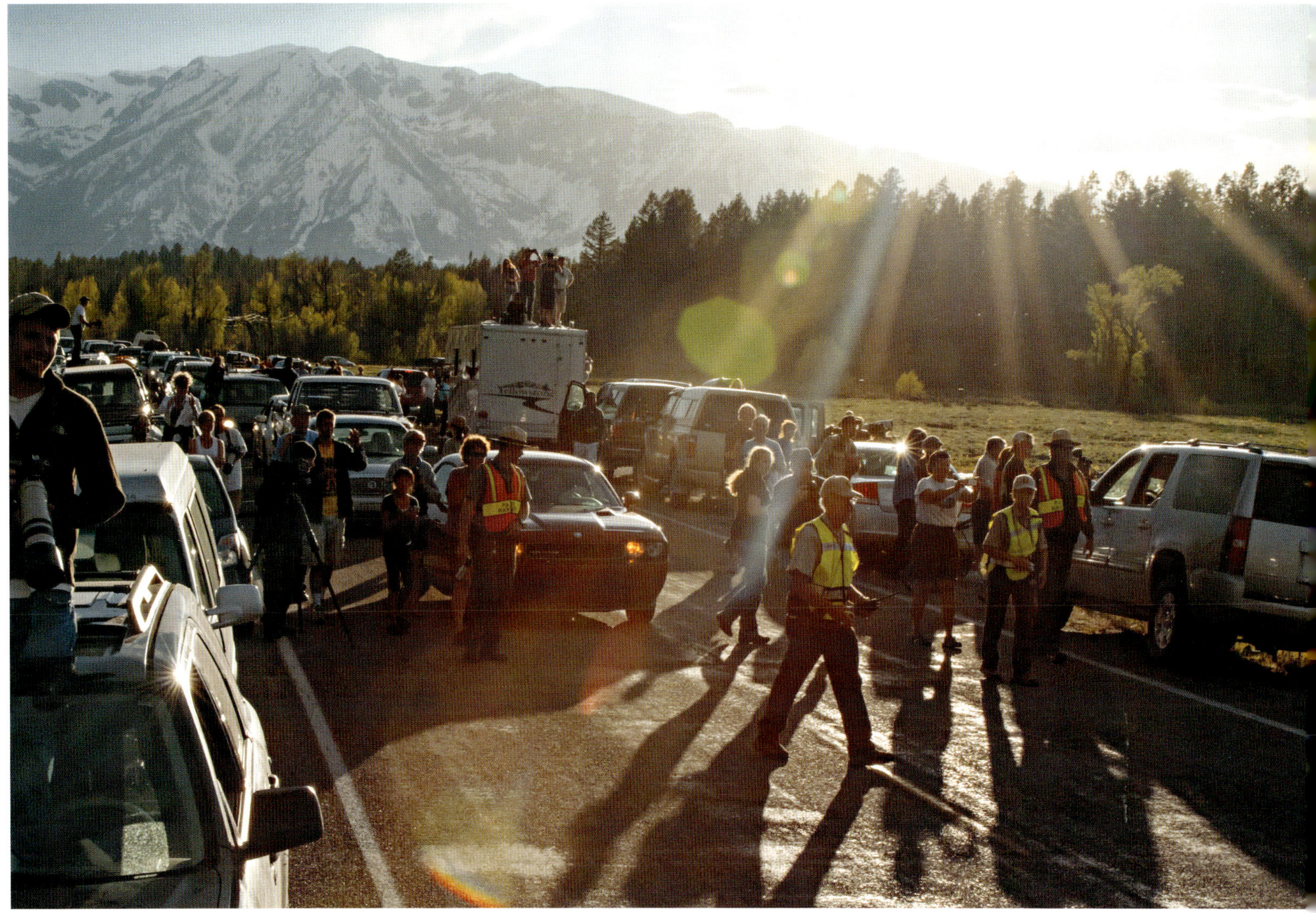

OPPOSITE: Surrounded by flowering balsam and biscuitroot, 399's cubs frolic and splash in a meltwater creek that feeds into Pilgrim Creek.

While grizzlies have an inestimable existence value, changes in society and the increasing rareness of large carnivores have made them far more valuable alive, for the tourist economy, than dead, when they are hunted and stuffed as trophies.

FOLLOWING SPREAD: Female grizzlies of reproductive age are the most important part of a bear population. Whether they live or die can determine whether a population rises or falls.

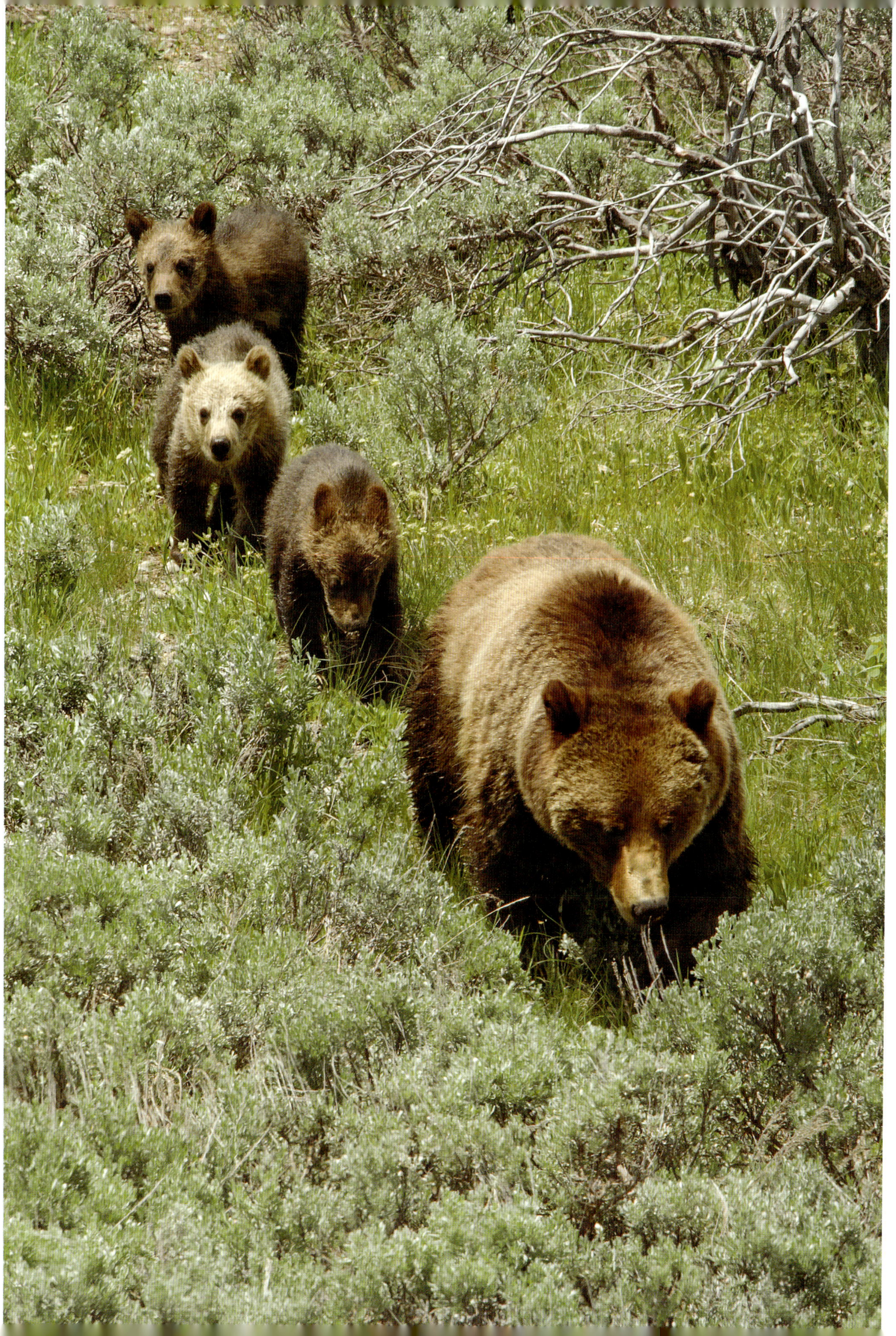

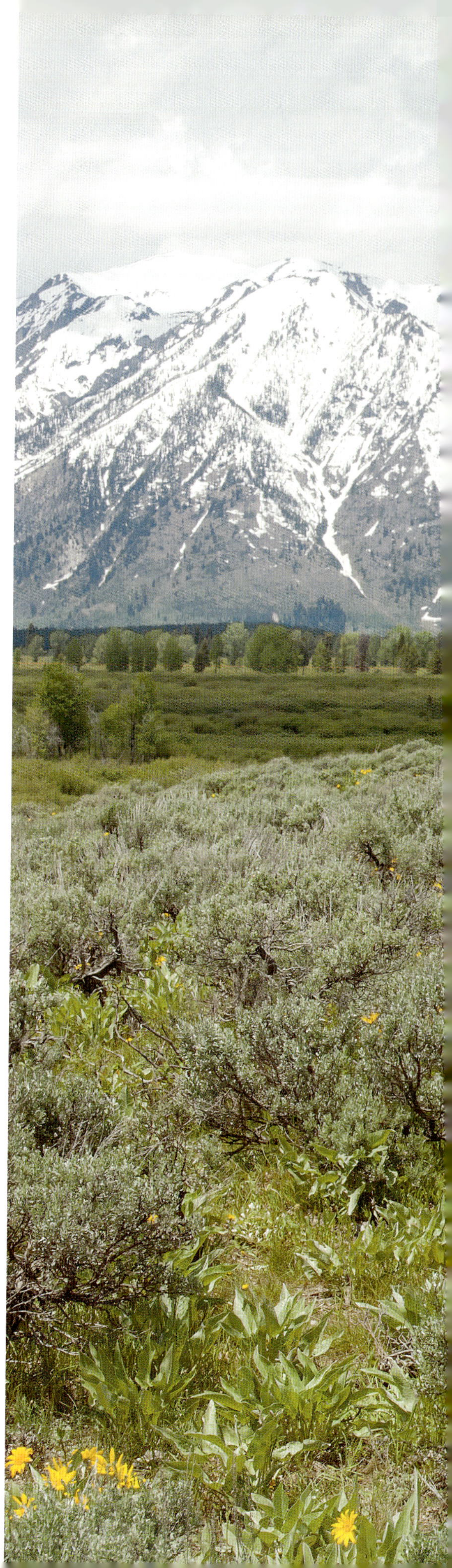

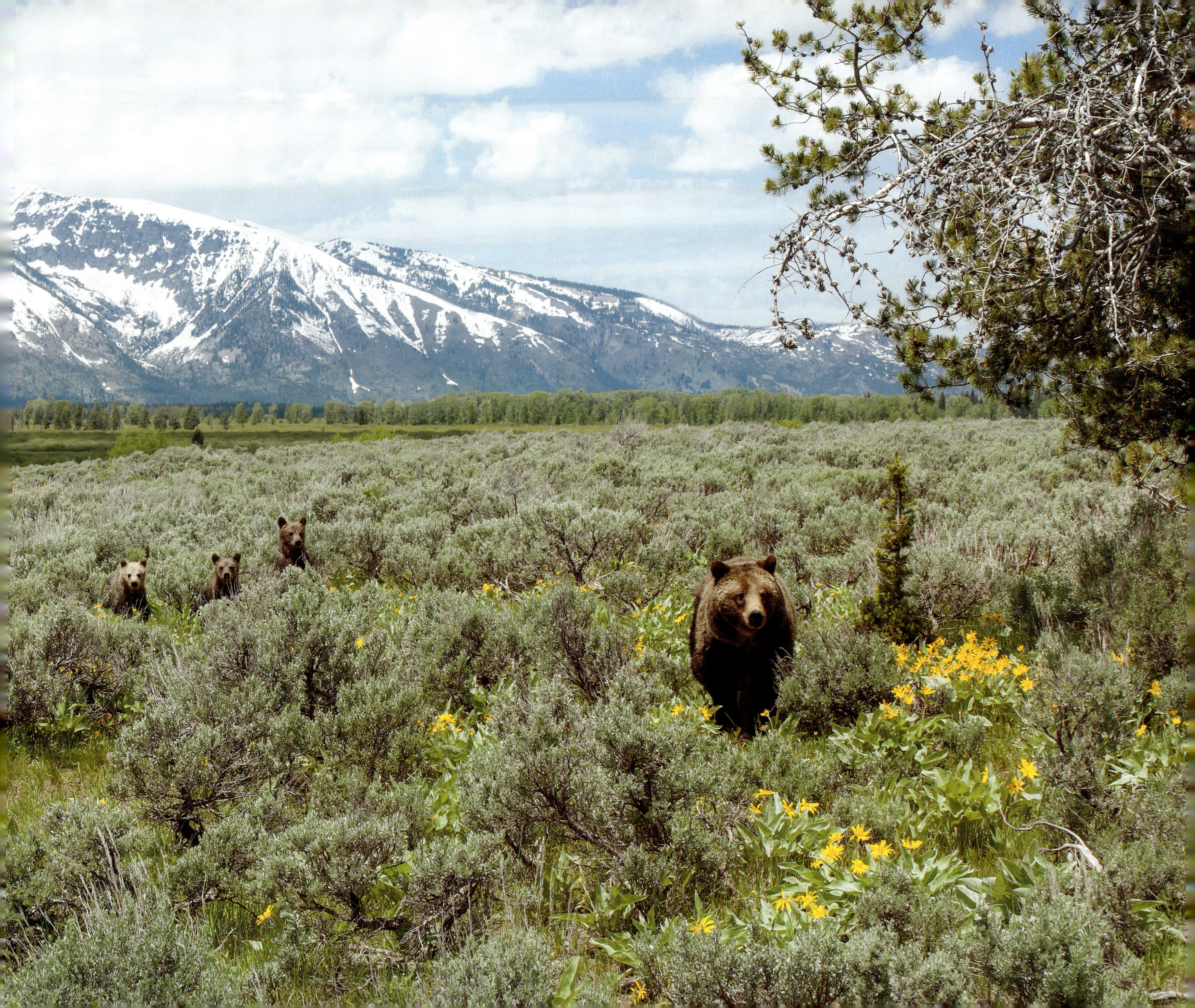

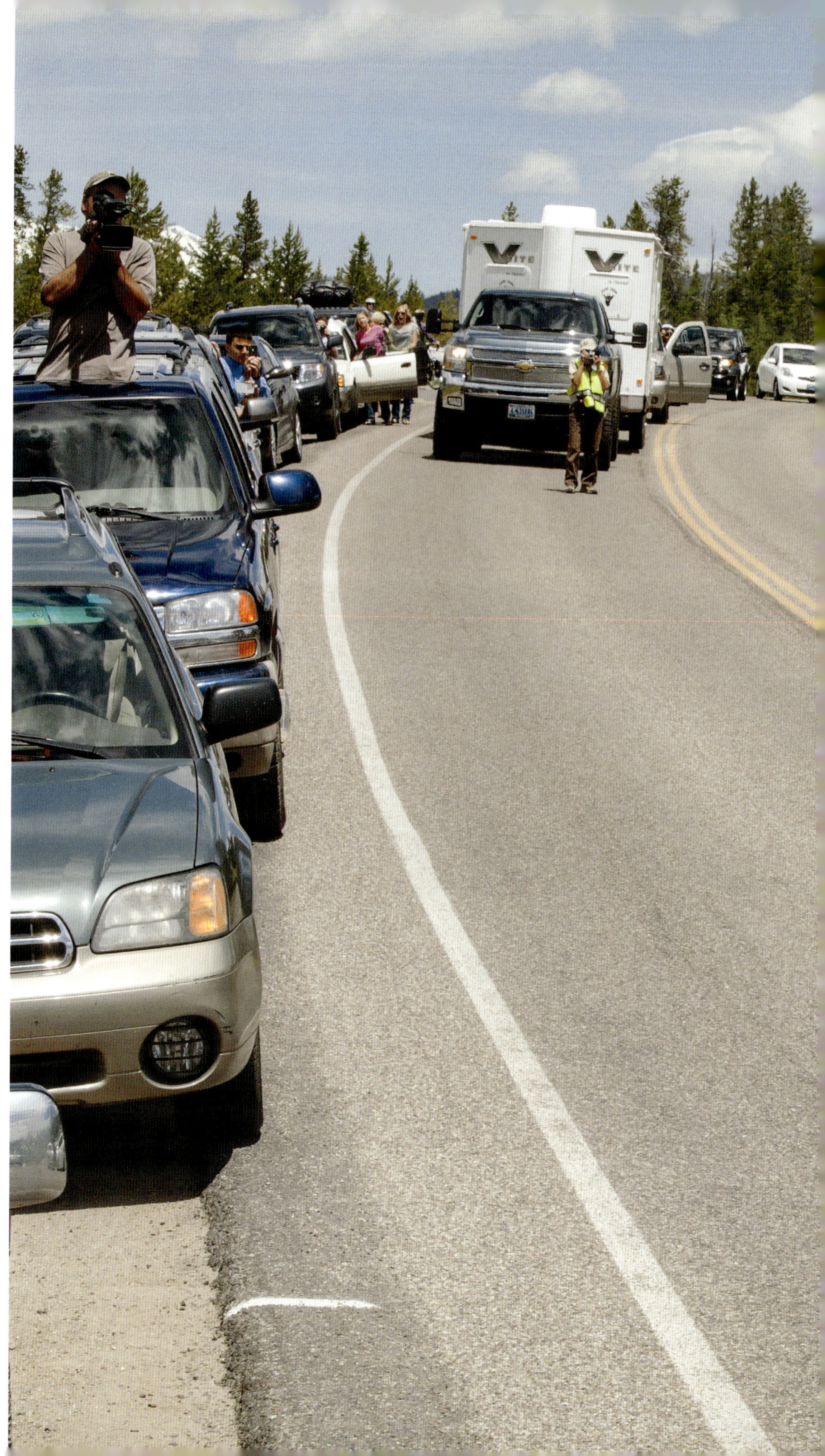

Greater Yellowstone is both a sanctuary and a land mine for grizzlies. In any given year, the bears must navigate a gauntlet of millions of tourists, hunters, inattentive drivers, cattle and sheep, loose pets, human foods, recreationists, and other bears.

FOLLOWING SPREAD, LEFT: 399 keeps a watchful eye for any unexpected dangers that might befall her offspring in a meadow near Pilgrim Creek.

FOLLOWING SPREAD, RIGHT: Walking the balance beam of a fallen log, this young cub explores its surroundings.

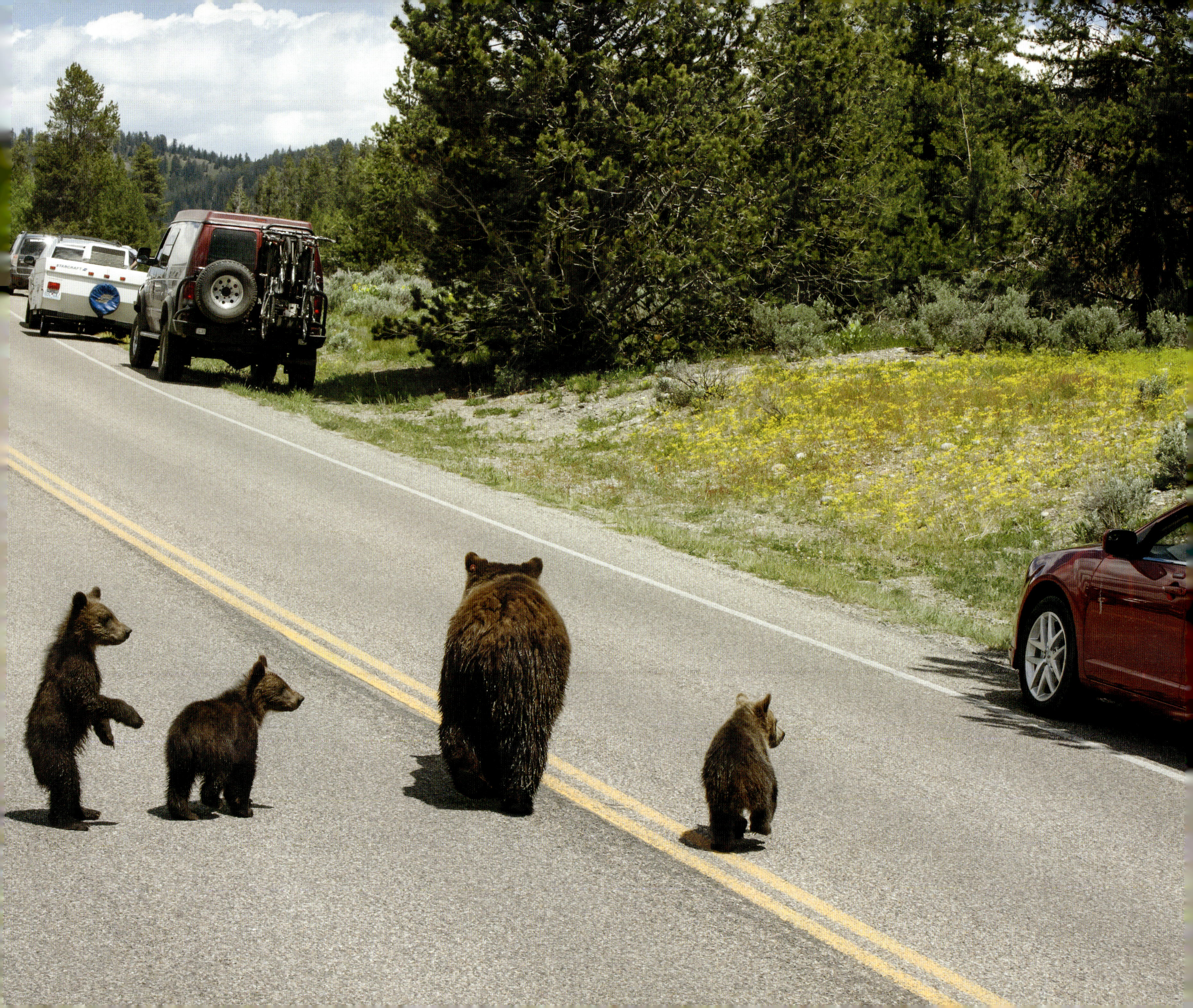

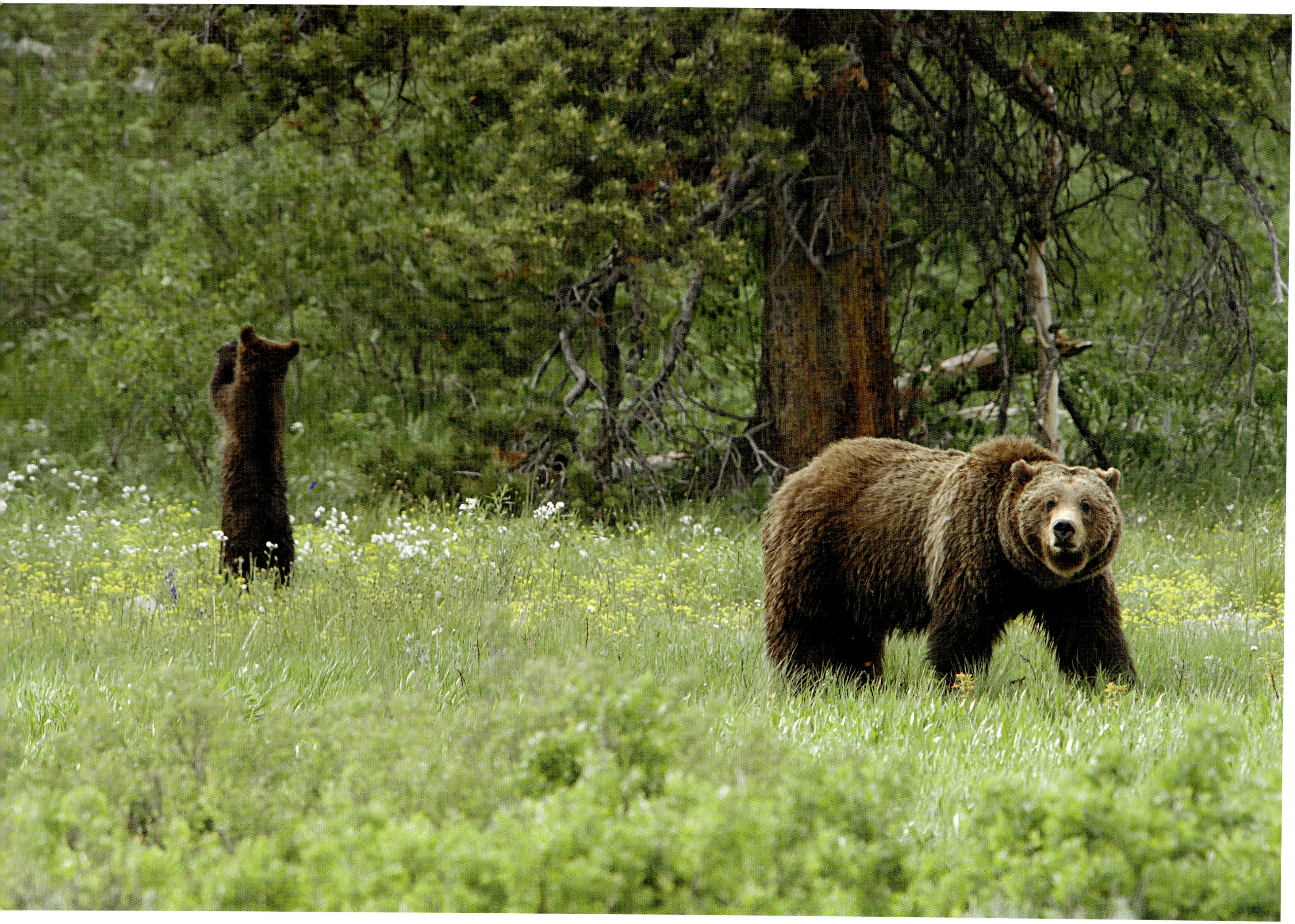

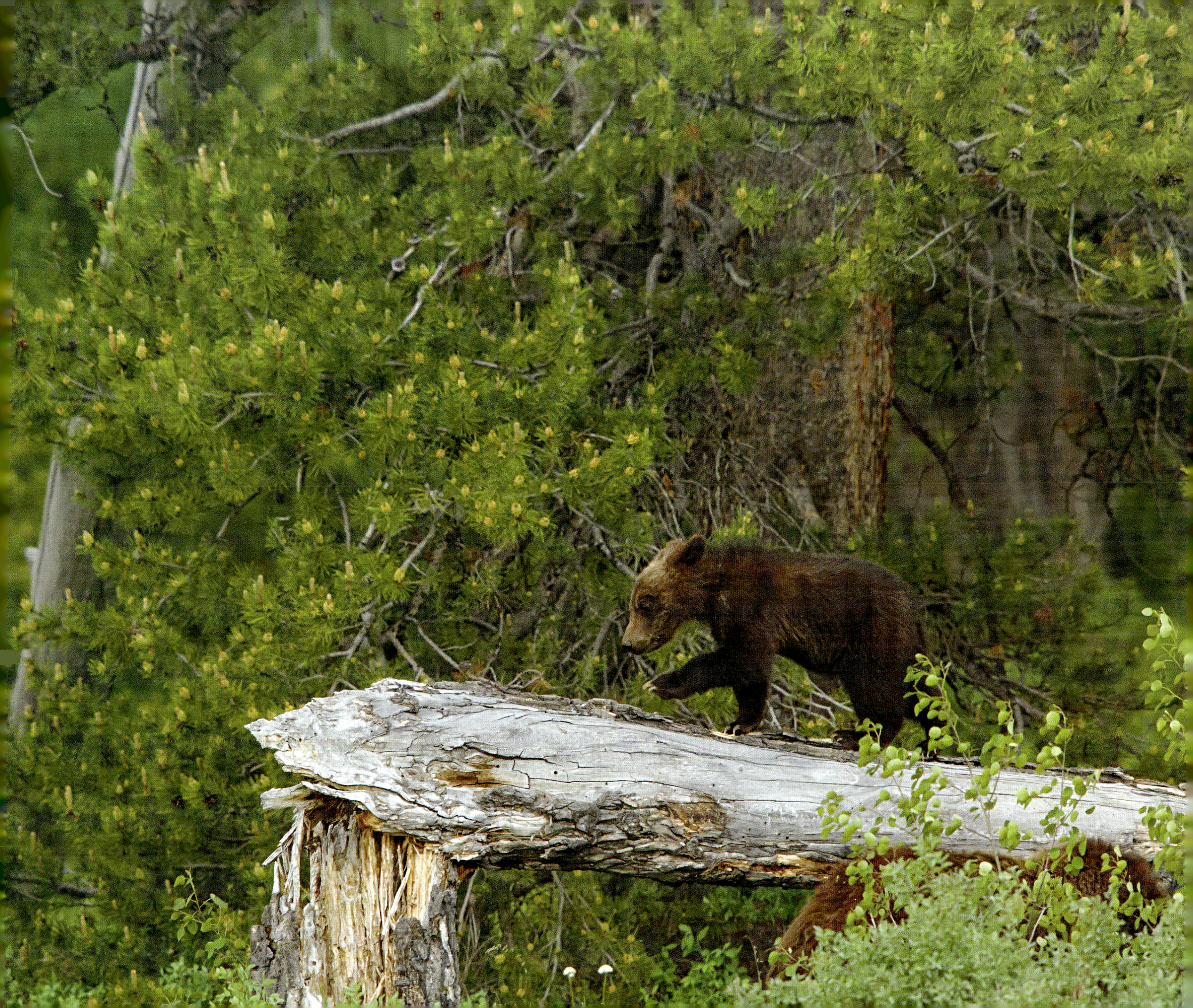

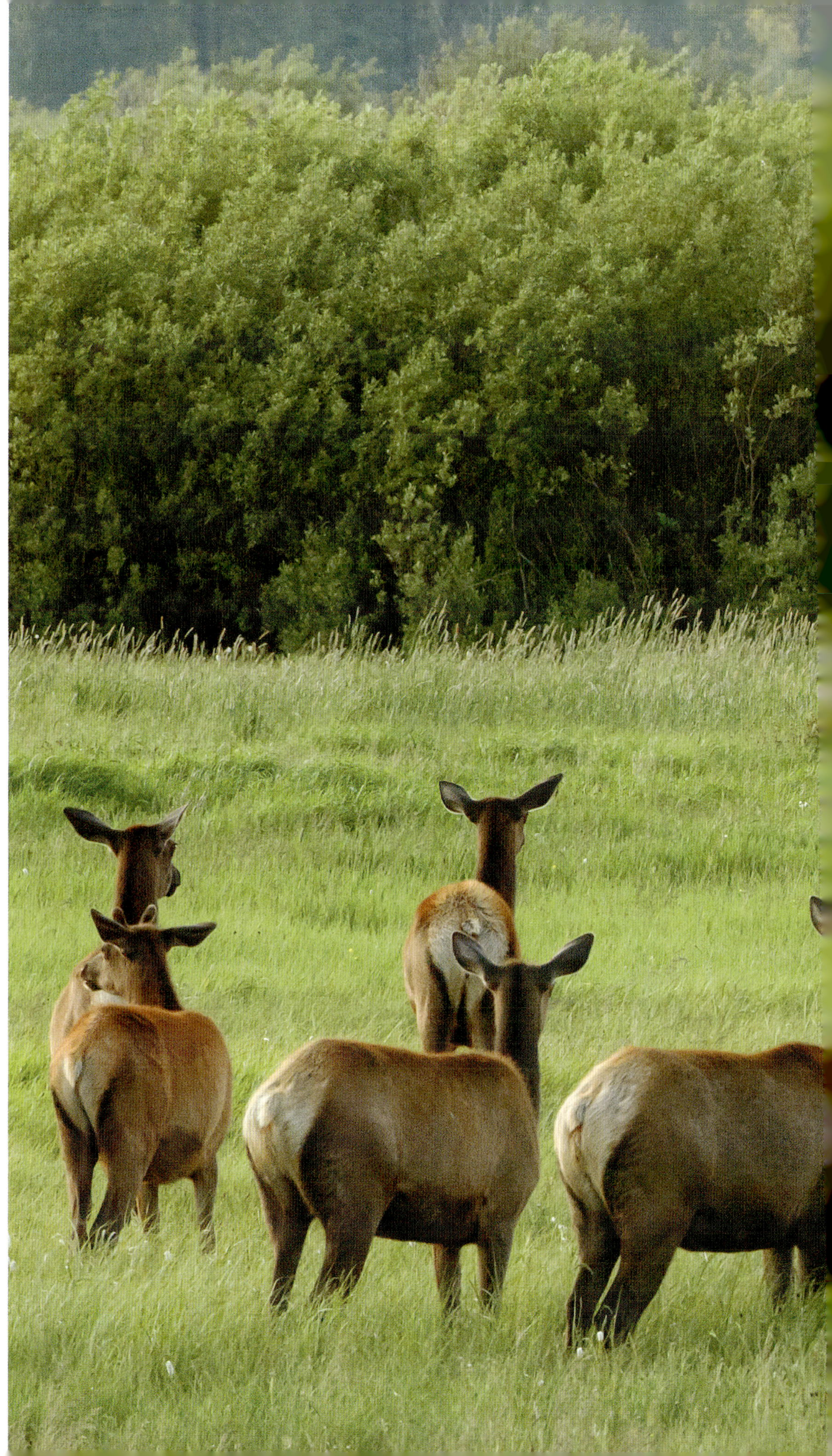

Look who is coming to dinner: at dusk in Willow Flats, cow elk draw closer to 610 and her cub in order to keep a close eye on their every move.

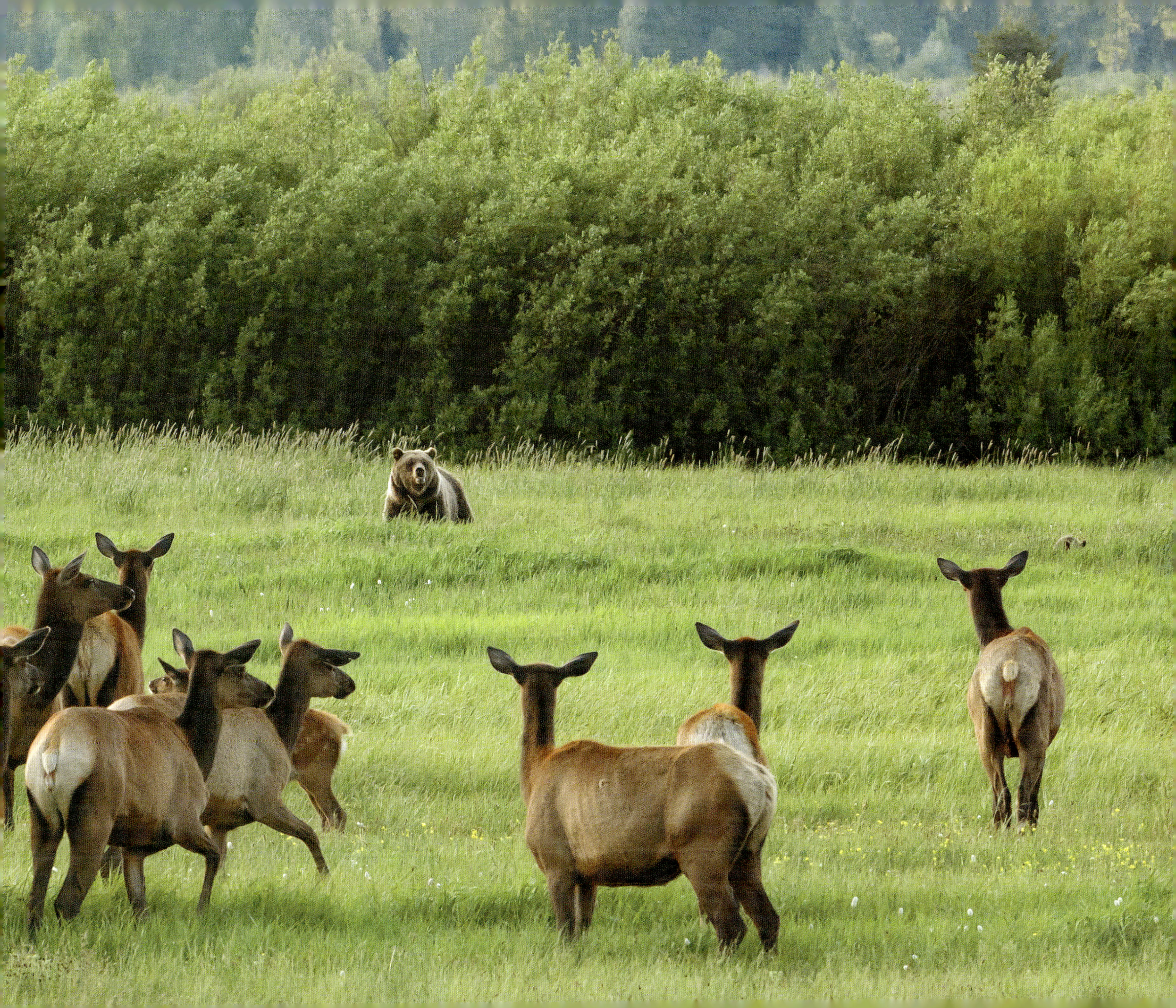

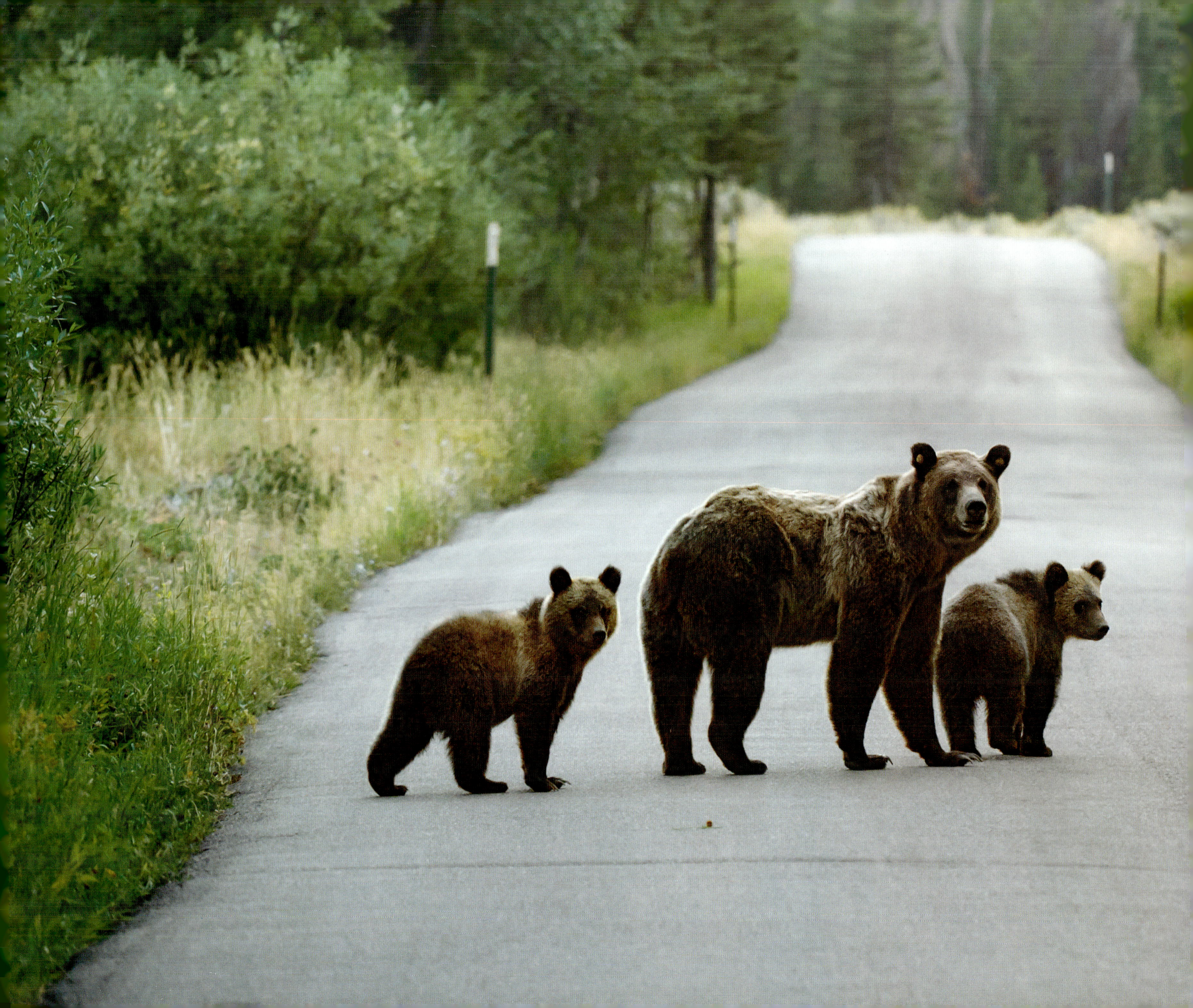

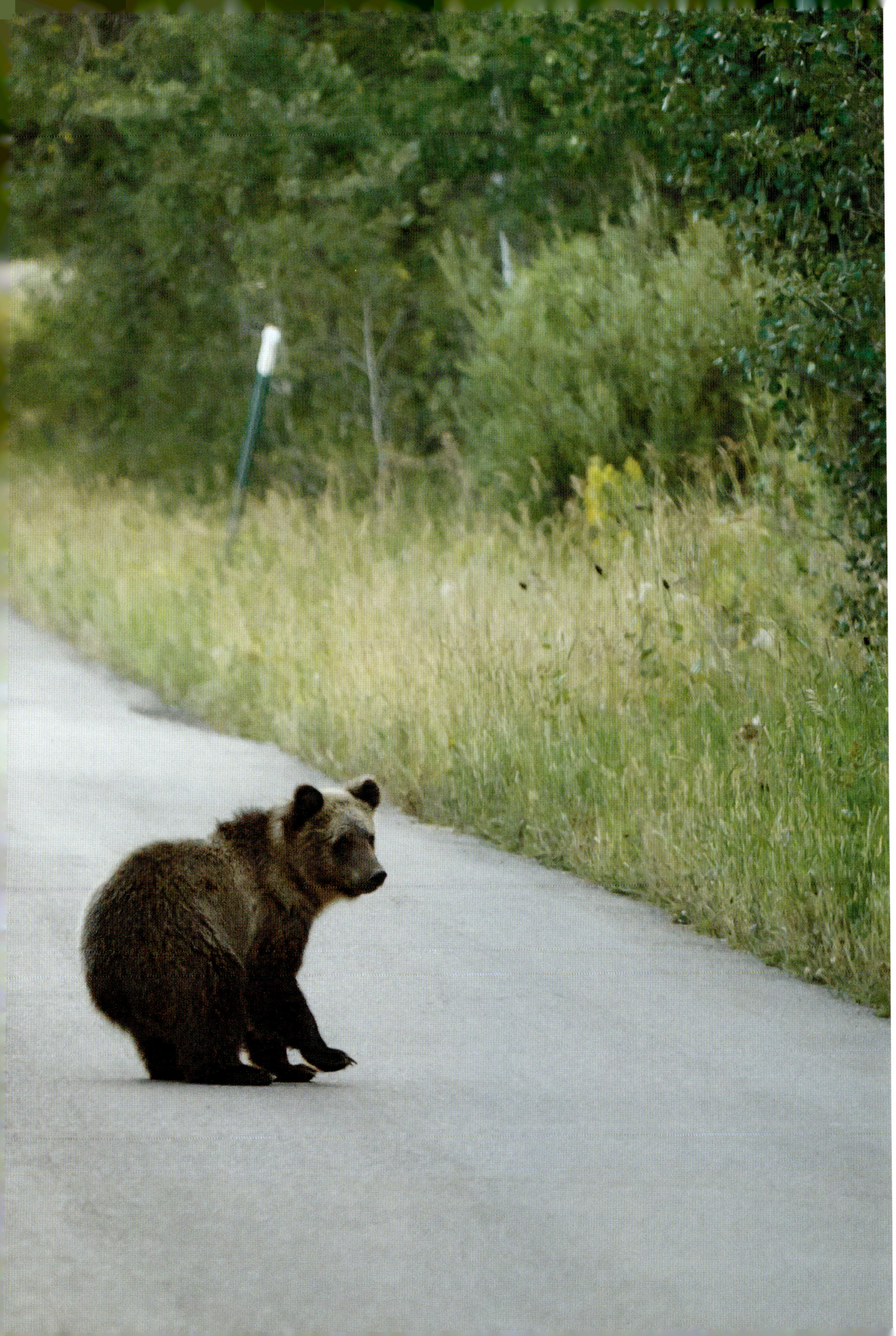

After a strange mix-up, 610 and her two cubs, as well as one cub adopted from her mother, 399, walk along Pacific Creek Road. The reason is unknown, but the mix-up may have occurred while both families were hunting elk calves in Willow Flats. Here, 610 in her summer coat looks thin, likely due to the stress of nursing and providing for an additional cub.

FOLLOWING SPREAD, LEFT: Standing on her tiptoes, 399 stretches for berries along Pilgrim Creek.

FOLLOWING SPREAD, RIGHT: 399's cubs Ash and Brownie find a swimming hole in Pilgrim Creek.

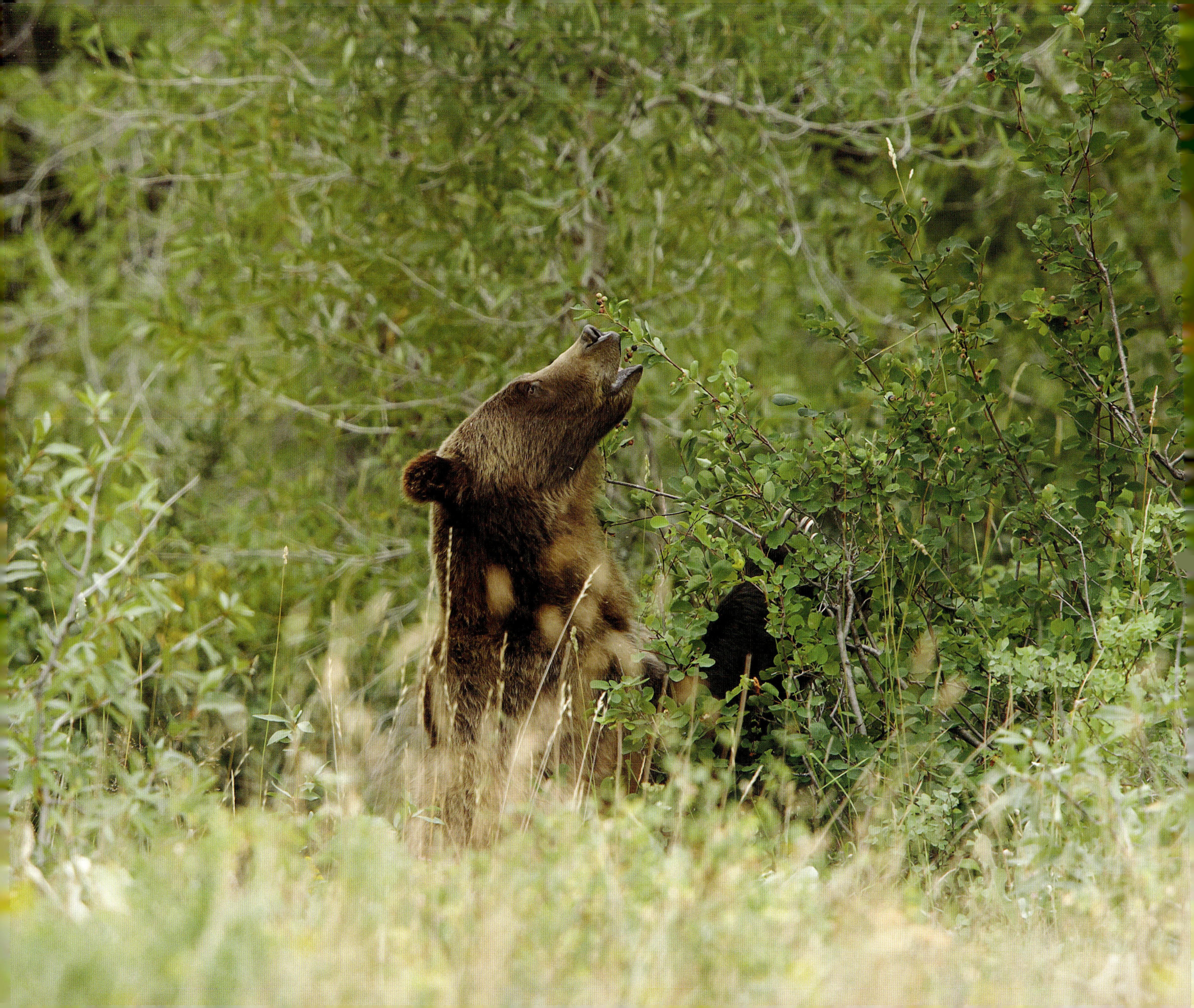

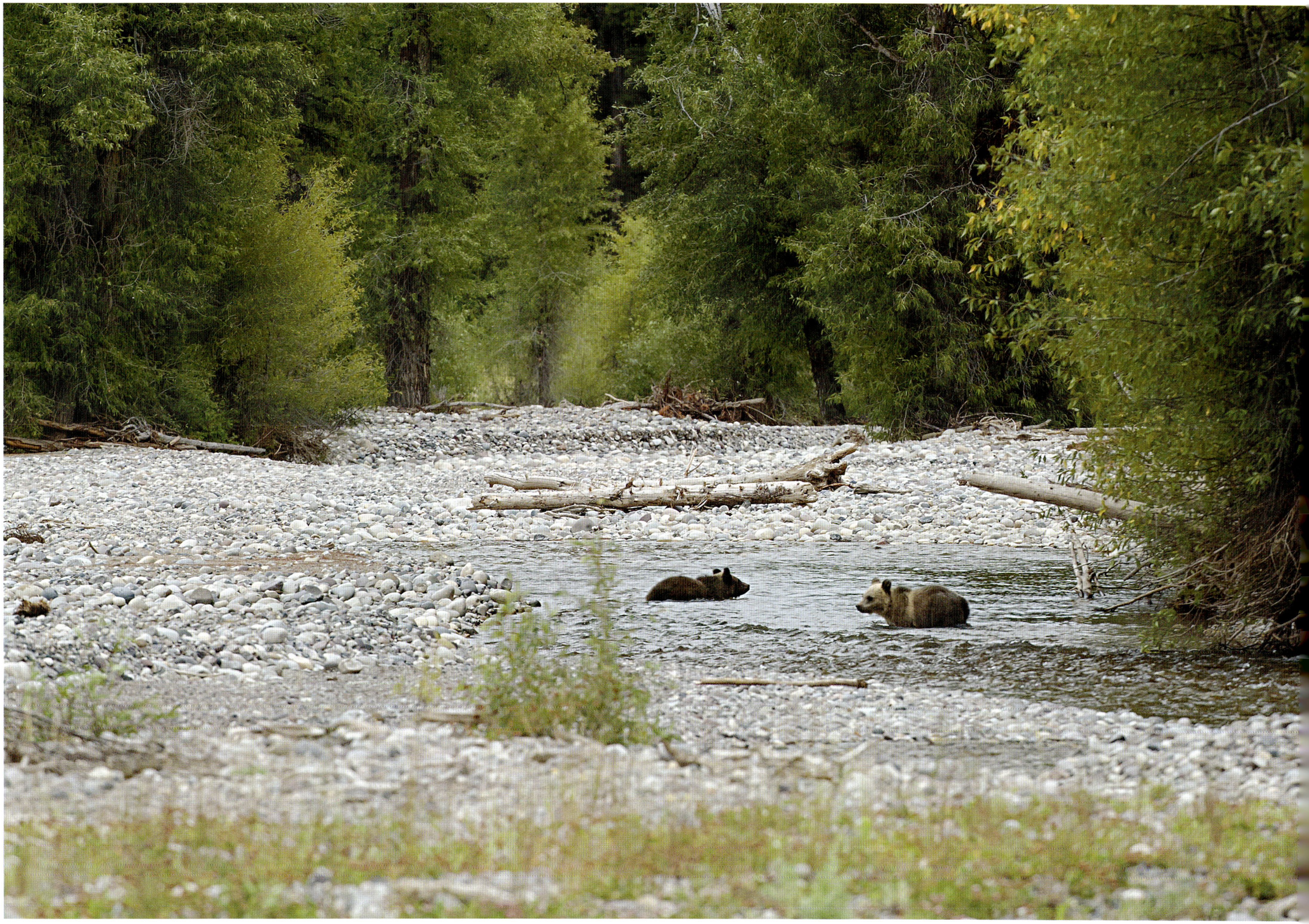

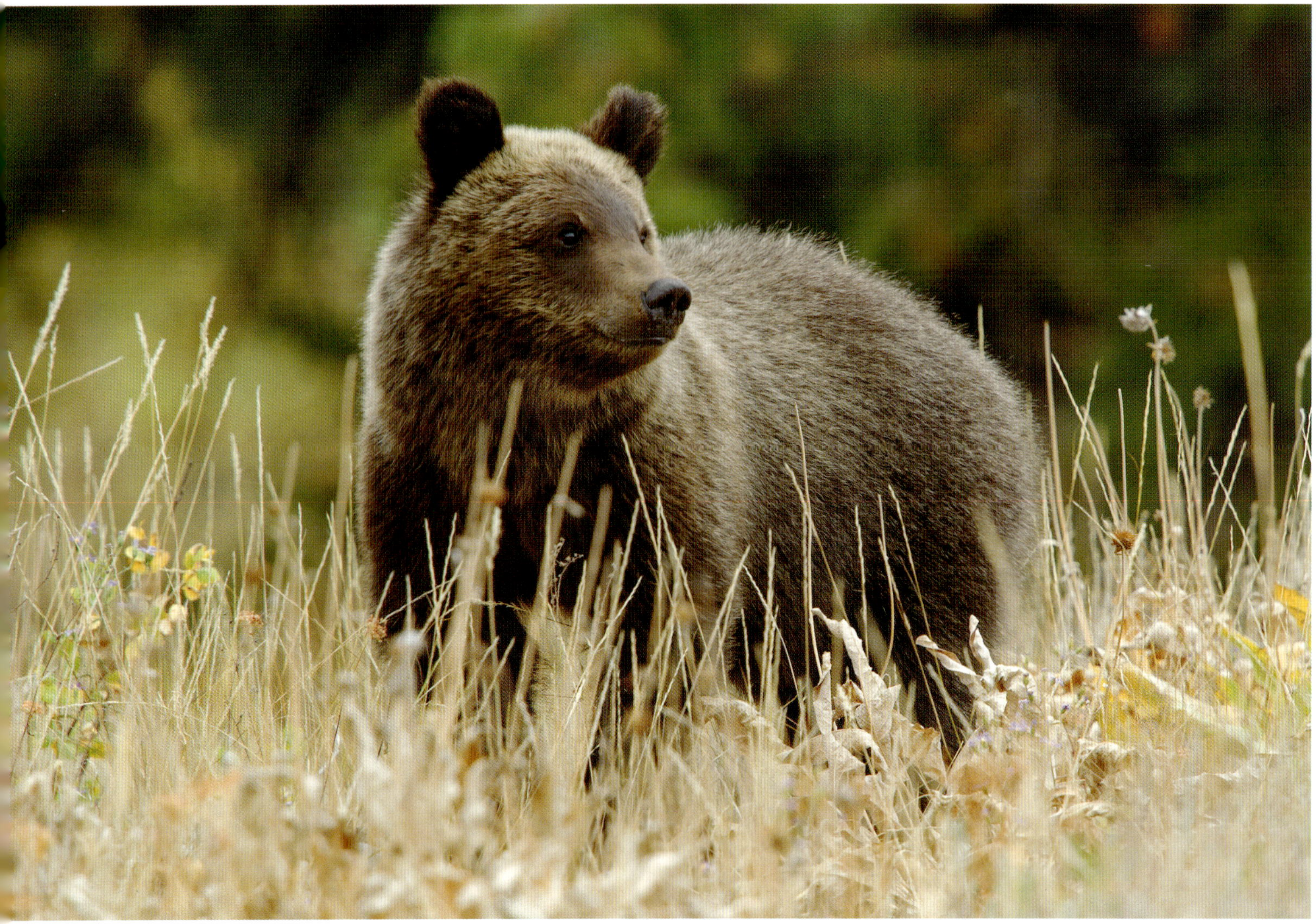

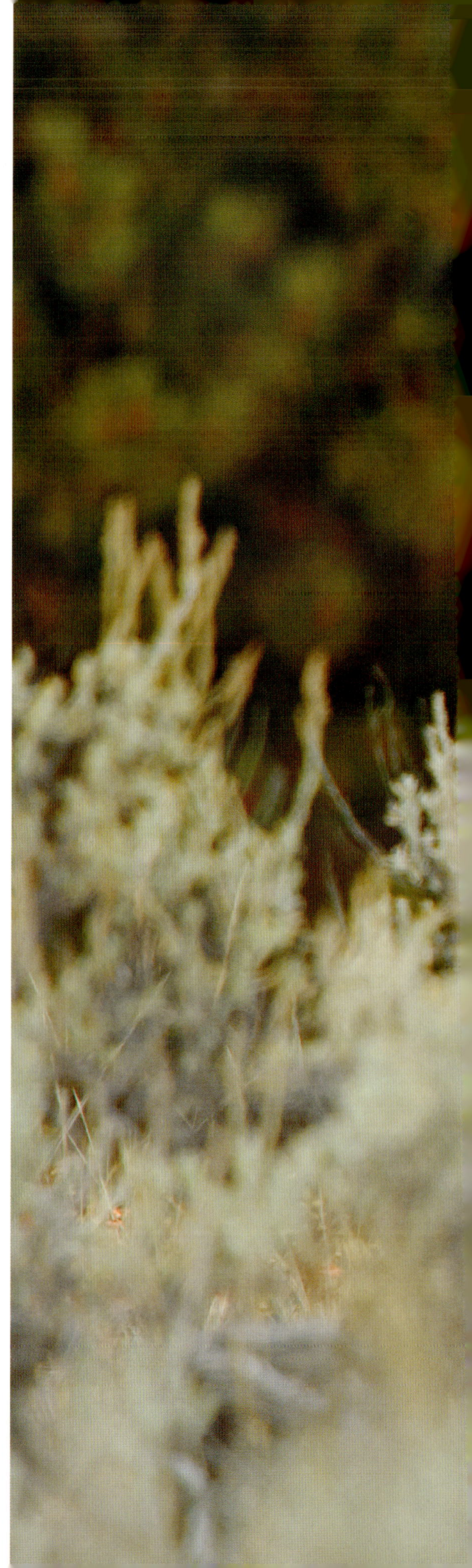

With the face of a teddy bear, one of 399's cubs mugs among autumn grasses.

OPPOSITE: Easily identified by her yellow ear tags, 610 follows a game trail paralleling Pilgrim Creek.

FOLLOWING SPREAD, LEFT: Edible couch? 399 lounges against a carrion sofa while her cubs have their meal.

FOLLOWING SPREAD, RIGHT: 399 lashes out at an aggressive gang of ravens that threaten to ravage her buffet.

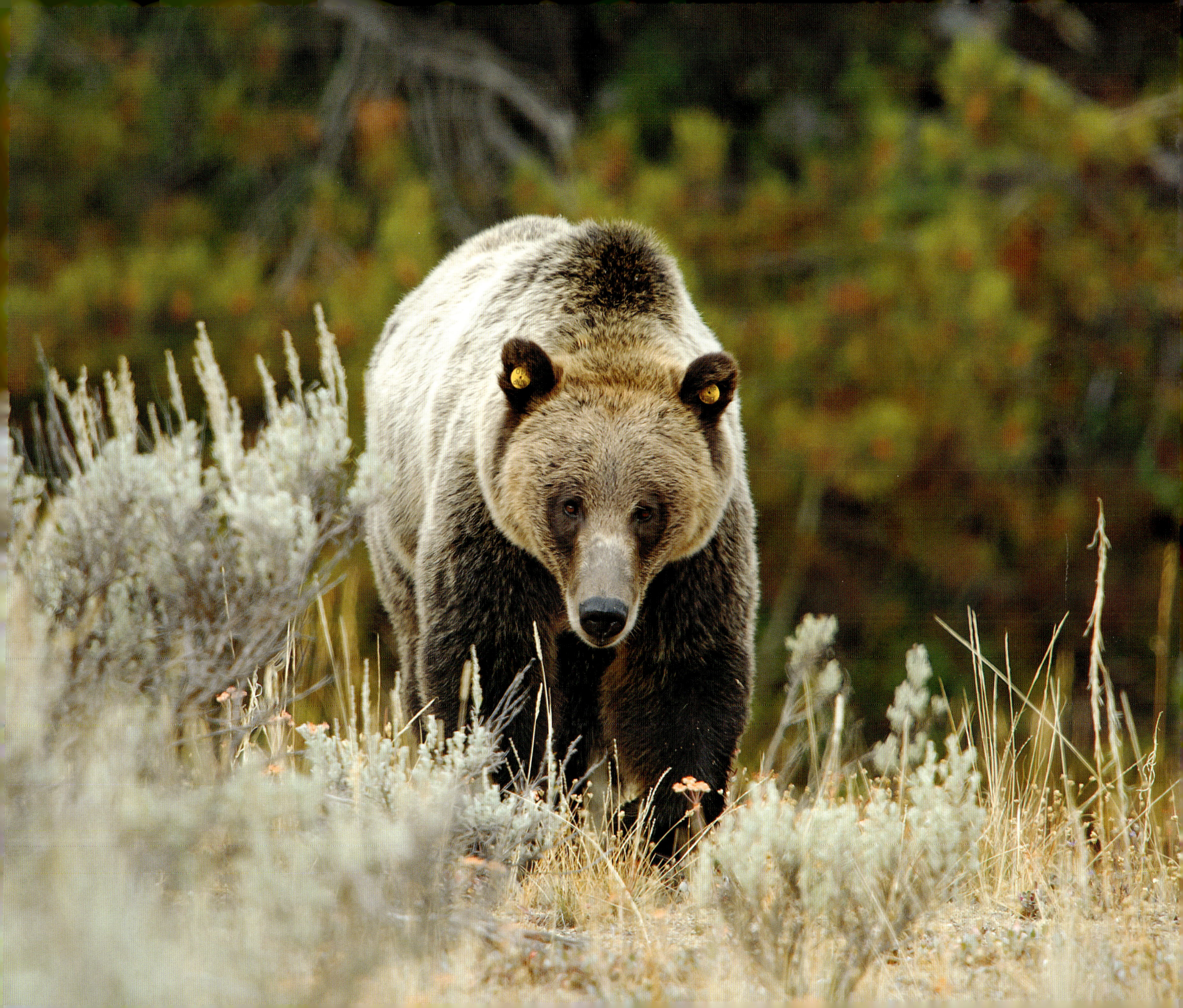

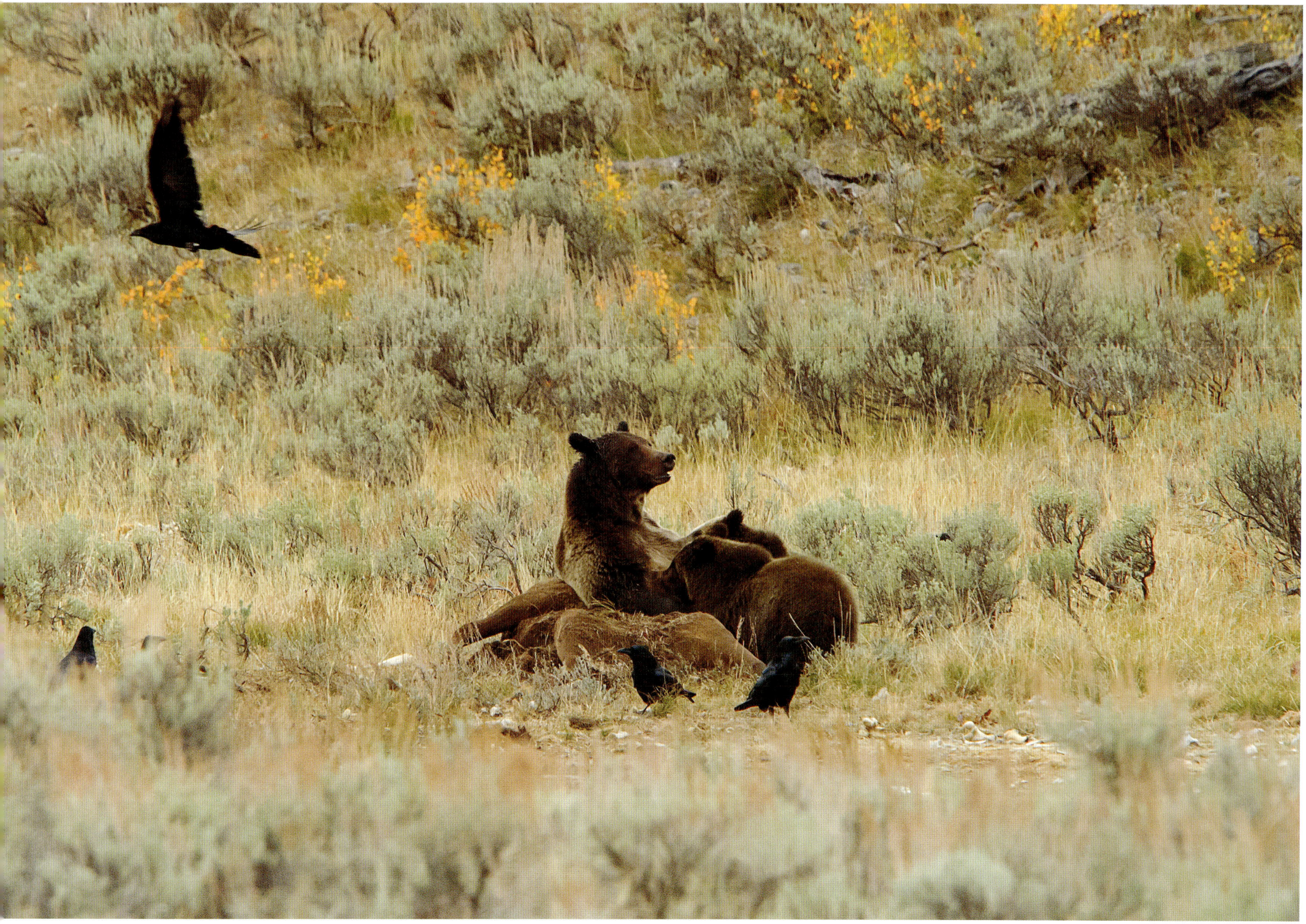

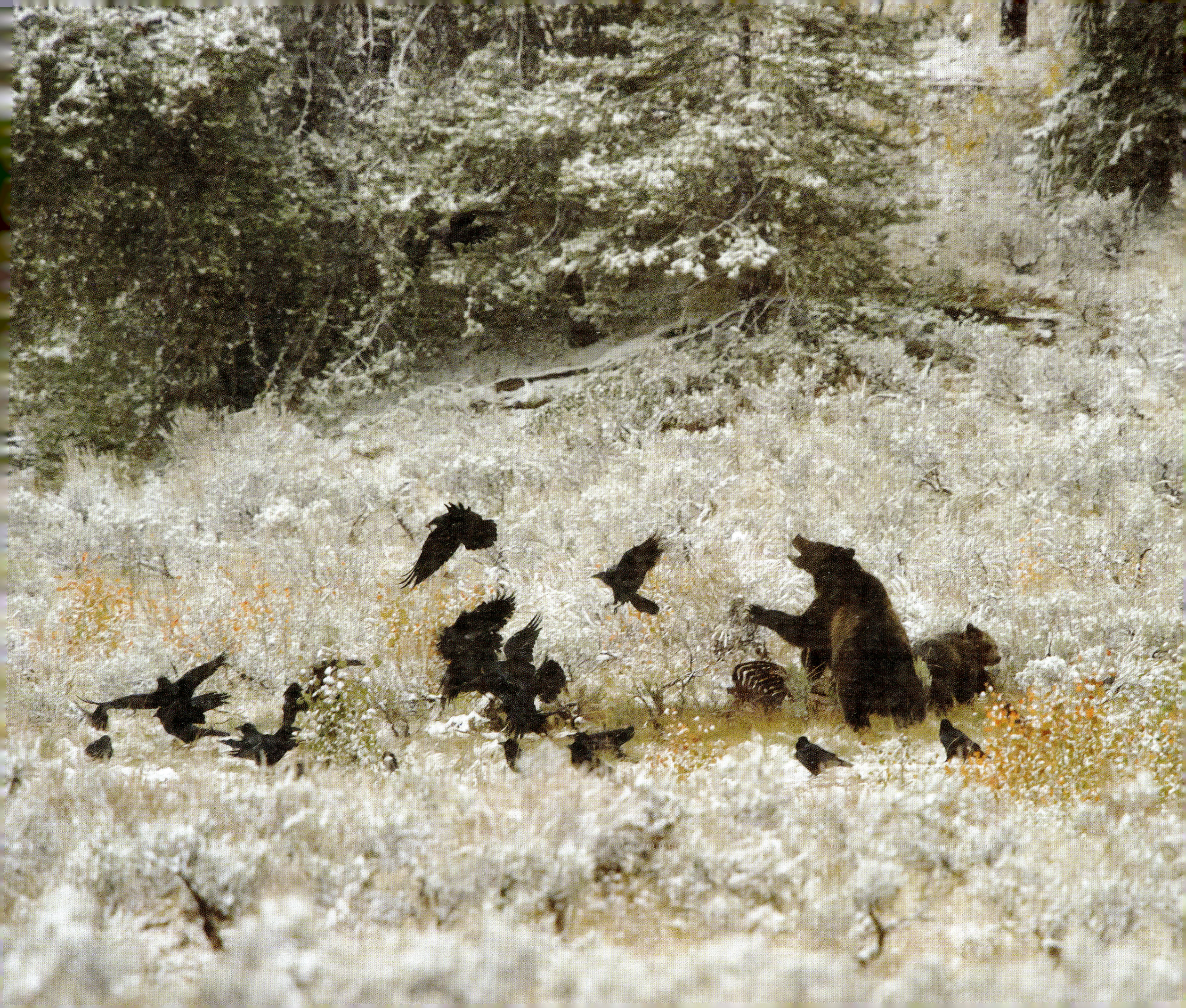

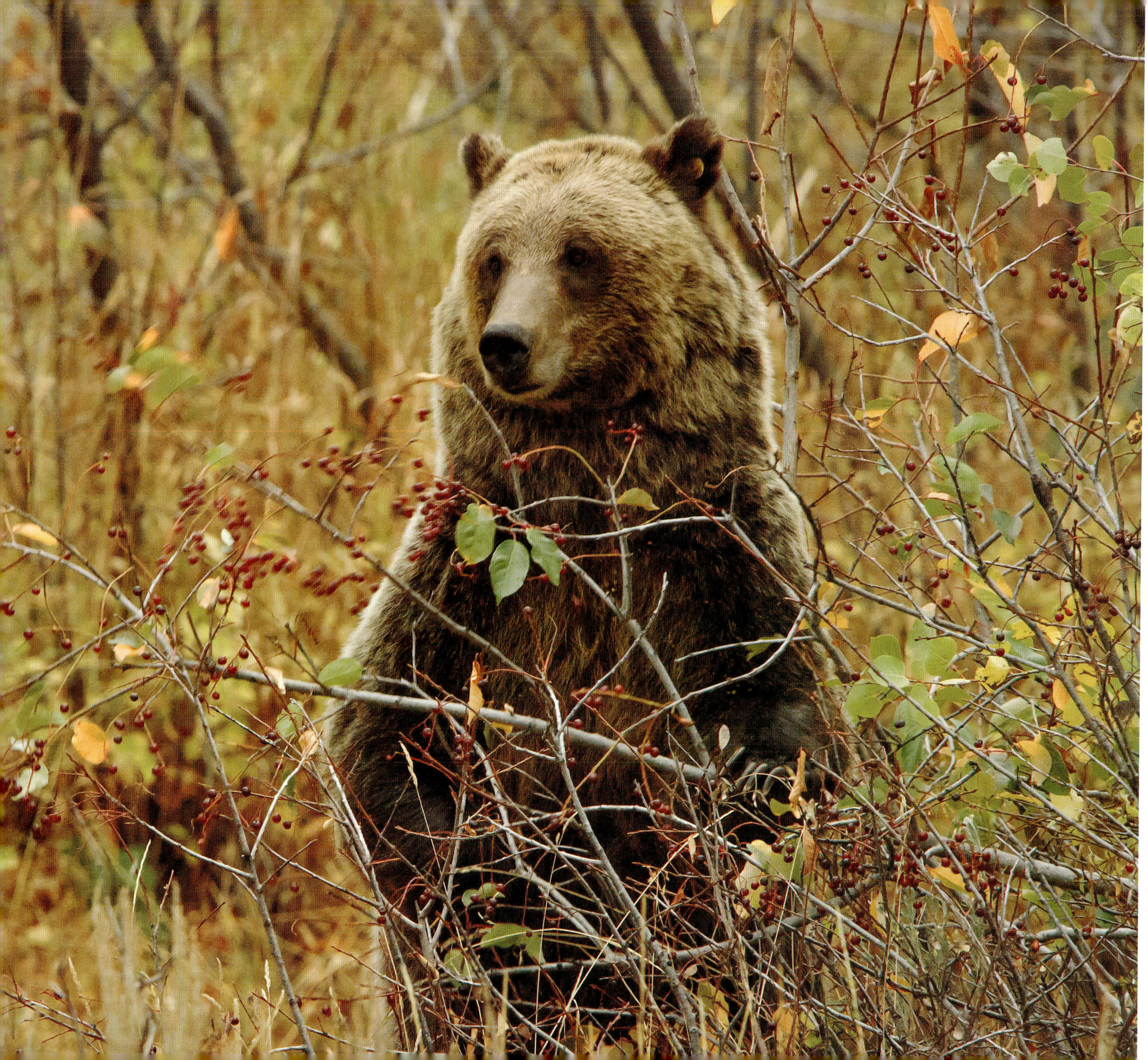

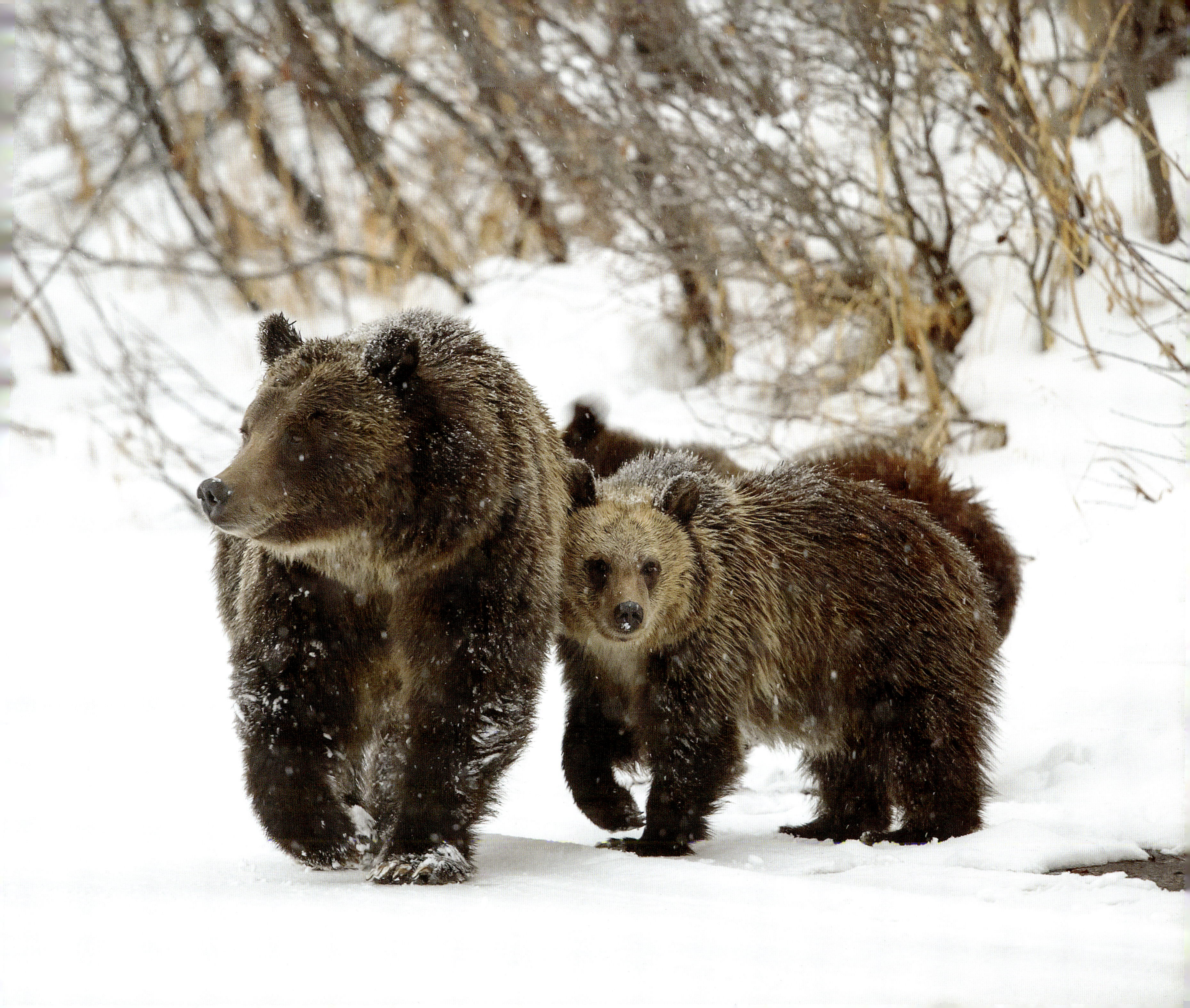

PREVIOUS SPREAD, LEFT: Known for its numbers of hawthorns and chokecherries, the Moose-Wilson corridor is a magnet for bears in the fall. Previously it was only frequented by black bears until 399 and her lineage discovered the cornucopia of berries. They were the first grizzlies to wander this far south in many decades.

PREVIOUS SPREAD, RIGHT: During an evening snowfall, 399 and her cubs lumber along Moose-Wilson Road in search of the last remaining hawthorn berries.

Near Windy Point, 399, with Ash and Brownie, surveys approaching traffic.

FOLLOWING SPREAD: Traveling through deep snow, 399, Ash, and Brownie head toward the Snake River bottom to find elk carcasses and gut piles left by hunters.

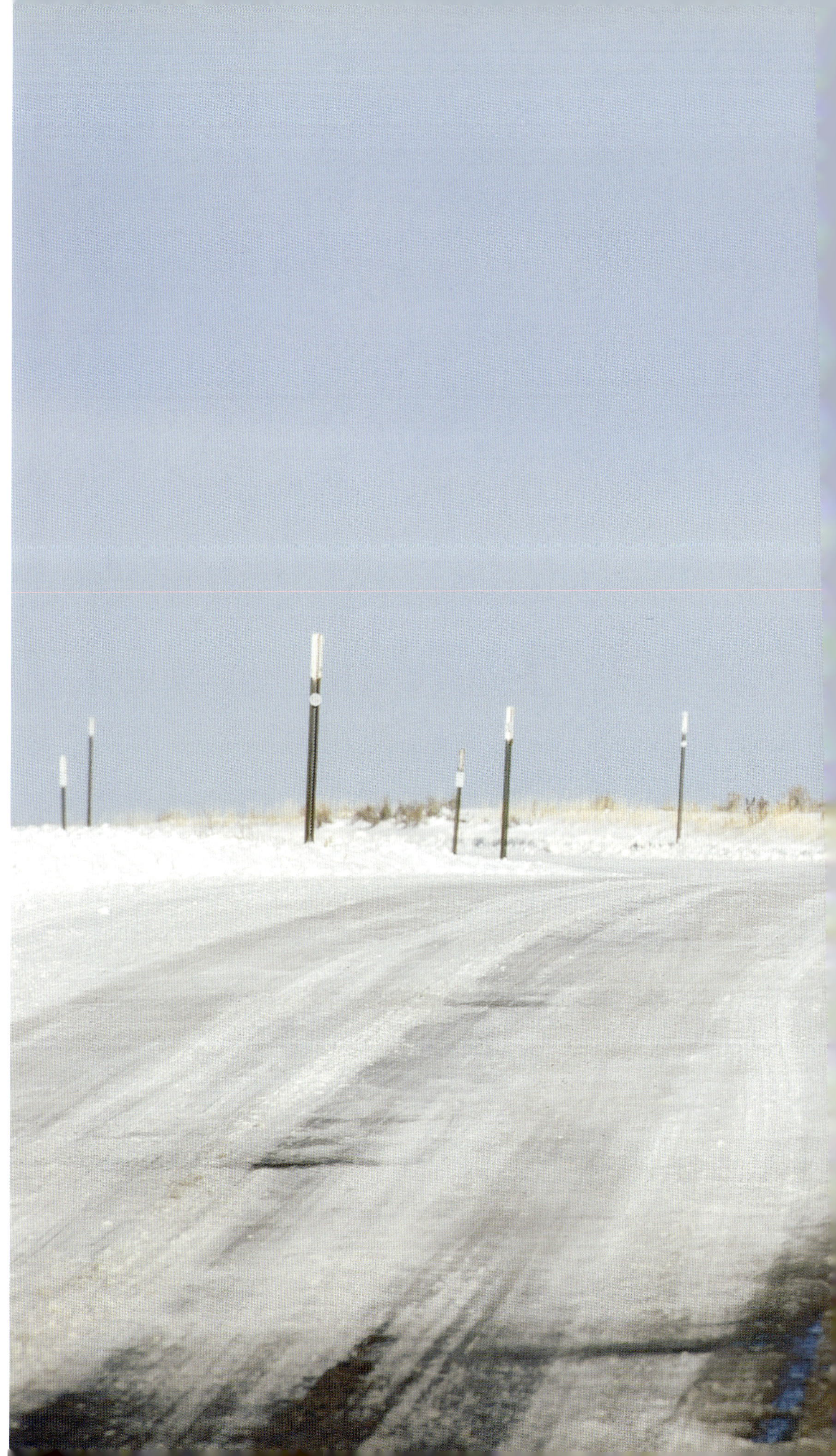

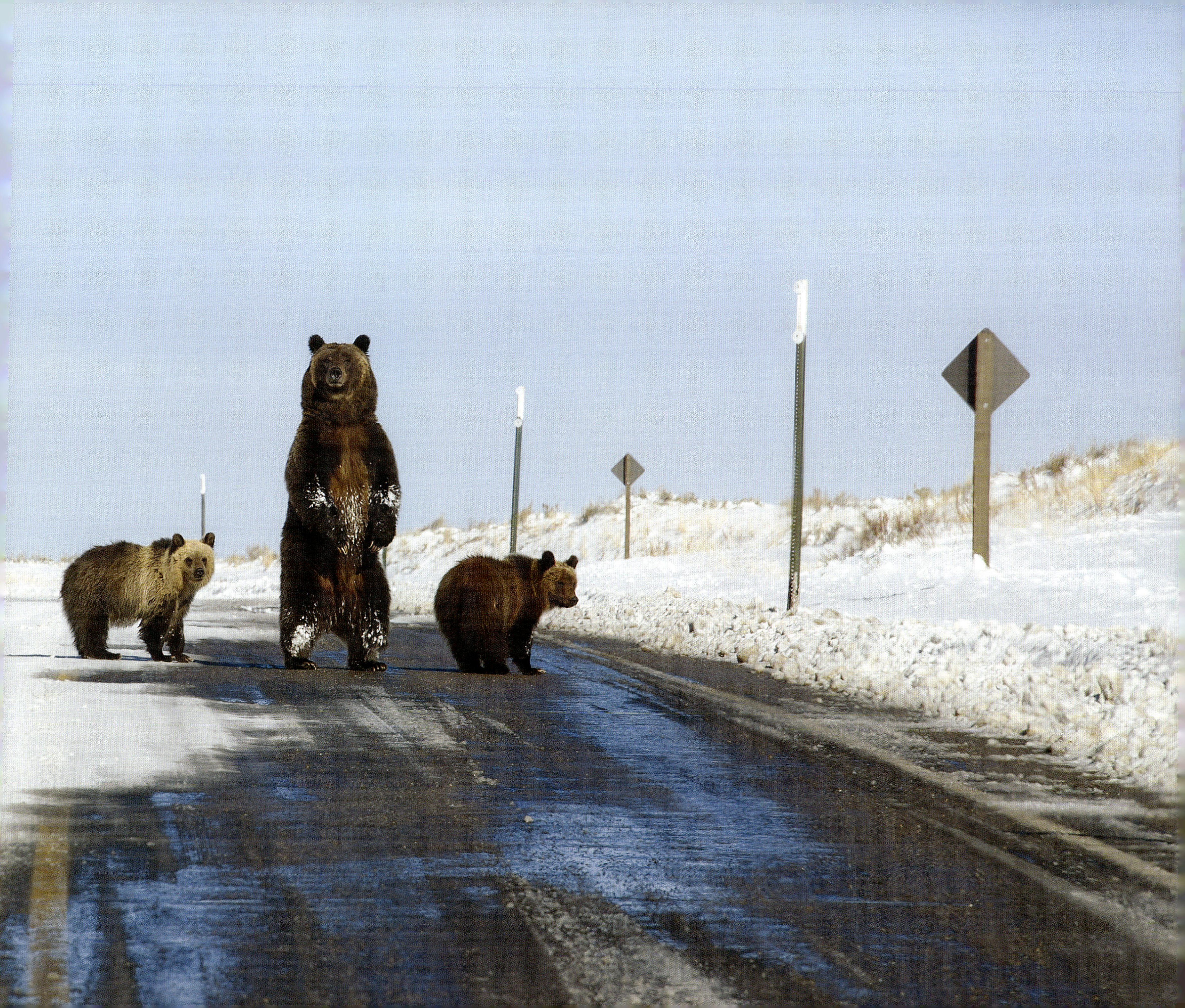

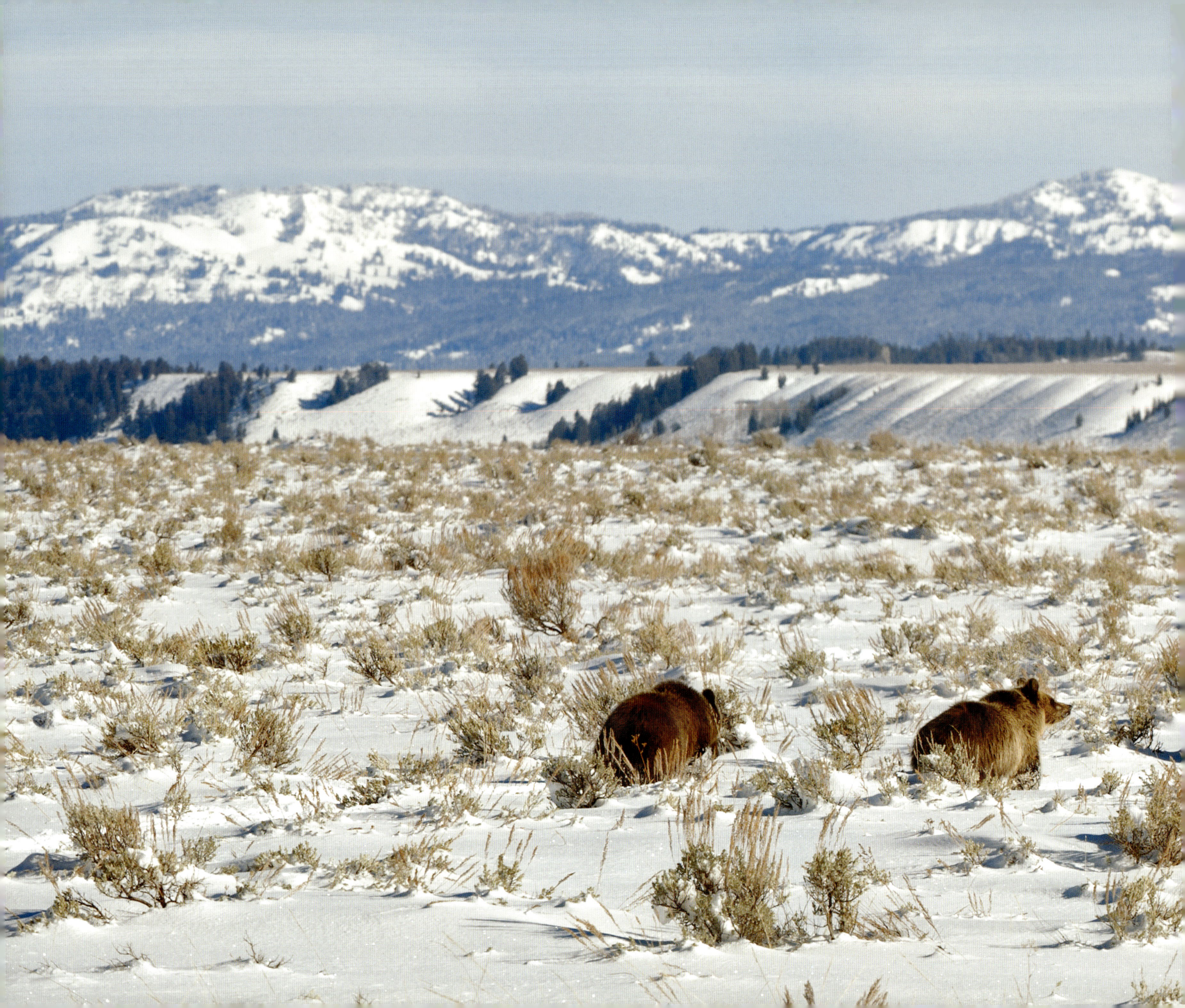

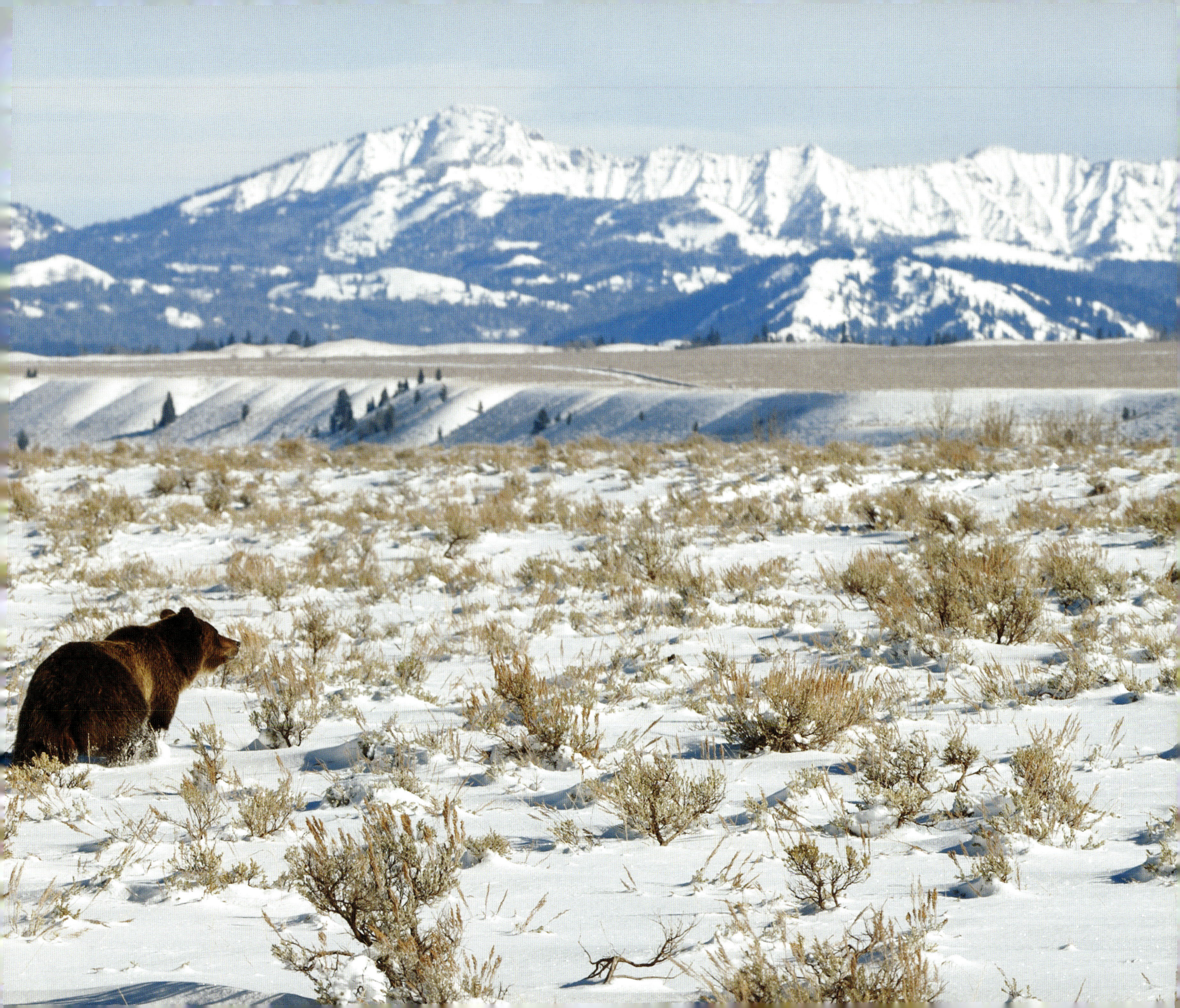

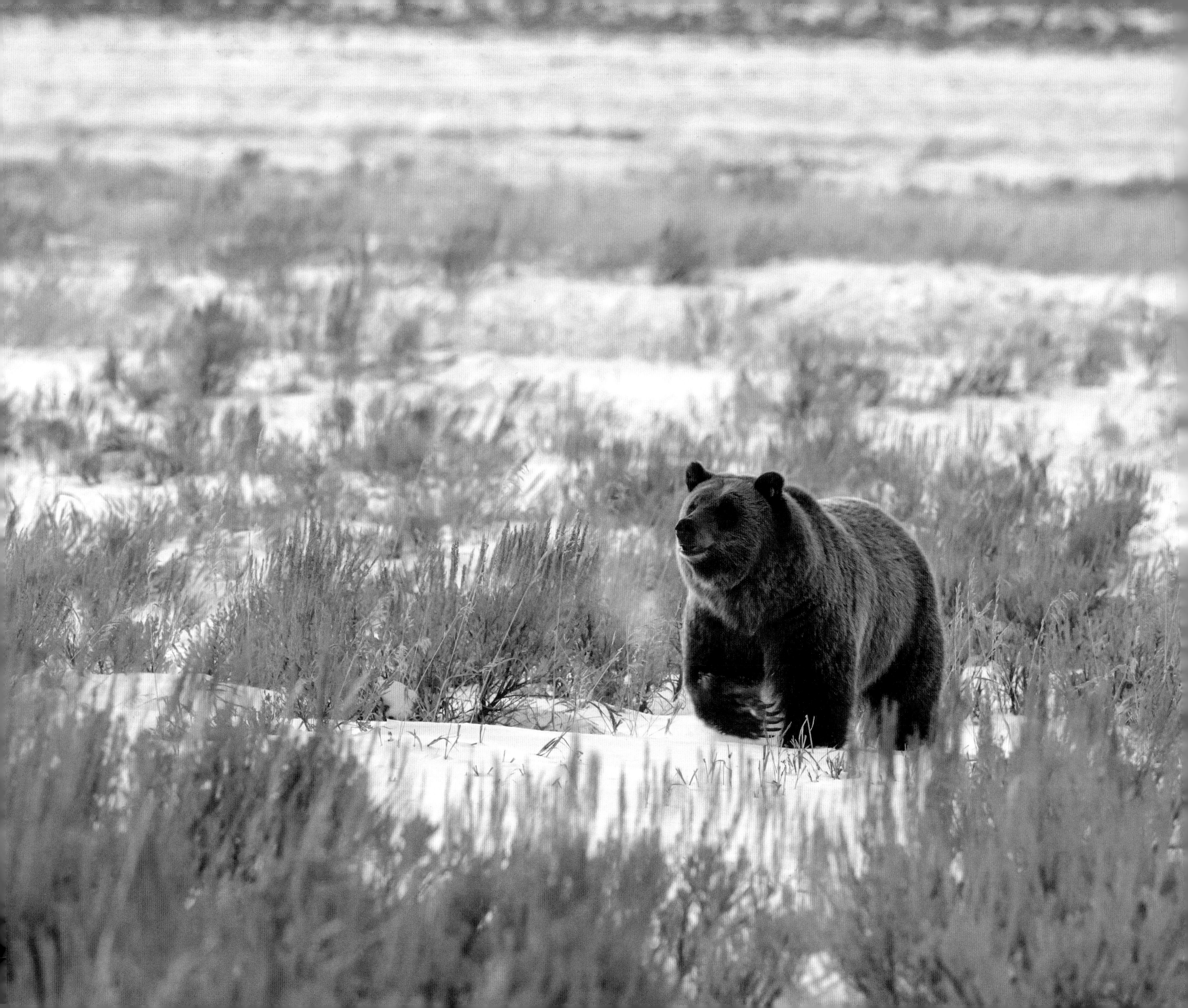

OPPOSITE: As evening falls and temperatures drop, 610's breath can be seen as she walks through the snow, likely toward the sanctuary of her den.

Covered in snow from a wet blizzard, 610 rises from feeding on an elk gut pile left by hunters near Antelope Flats.

FOLLOWING SPREAD, LEFT: 610 and her yearling cubs, shortly after emerging from their underground sleeping chamber, walk along the shoreline of Jackson Lake's Catholic Bay.

FOLLOWING SPREAD, RIGHT: After leaving the shoreline, 610 and her cubs cut uphill through the woods and cross the Brinkerhoff property, a place where human dignitaries often stay.

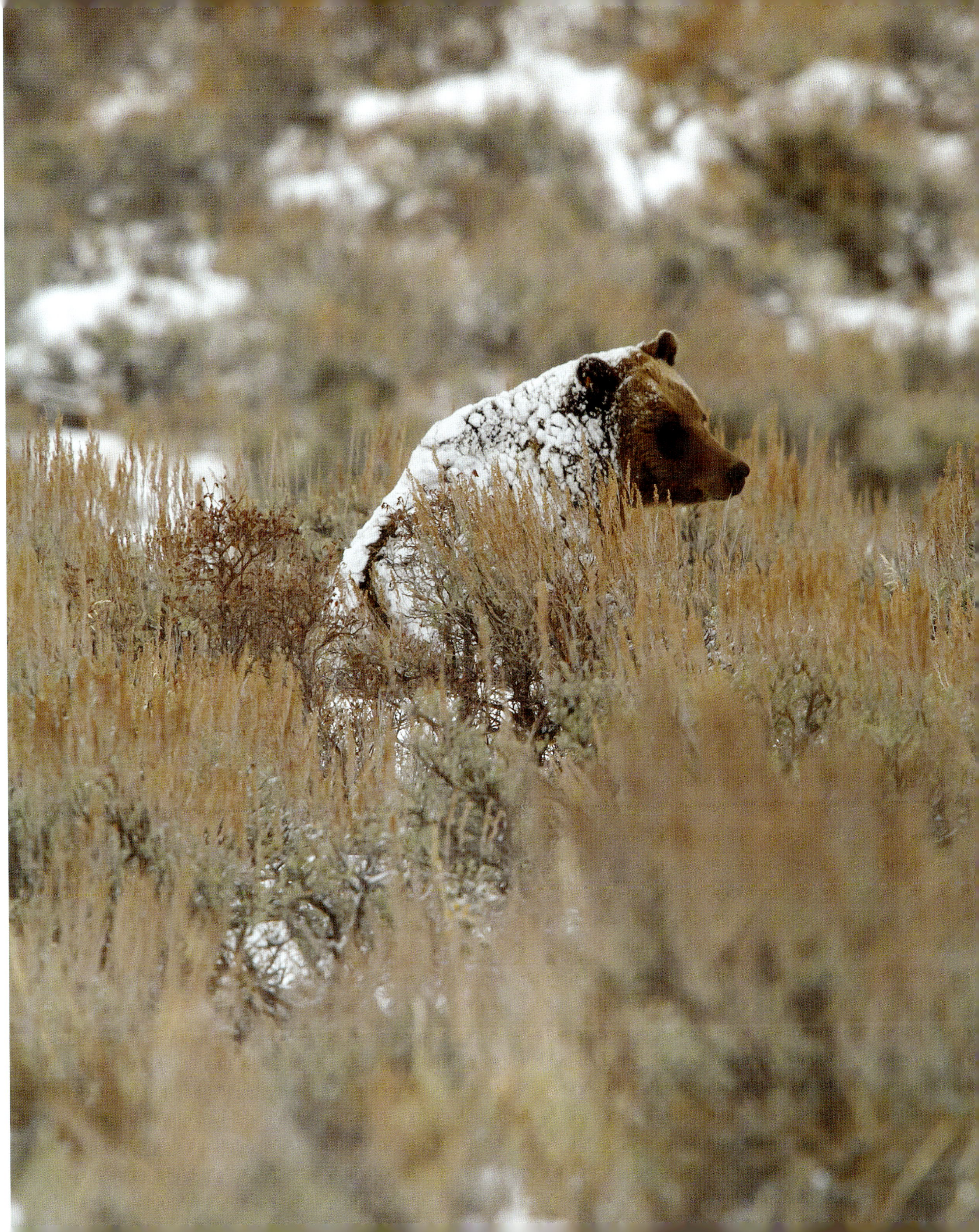

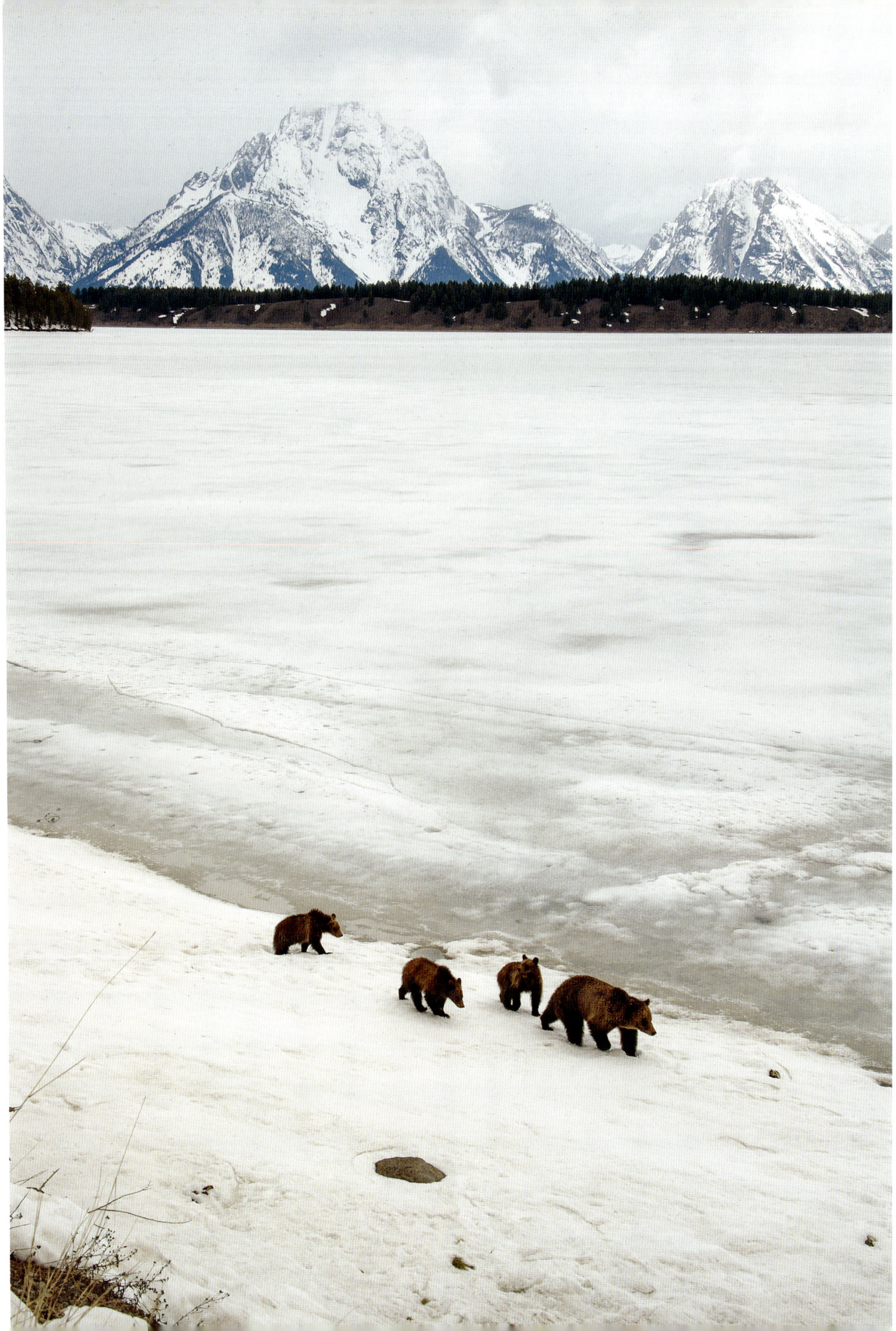

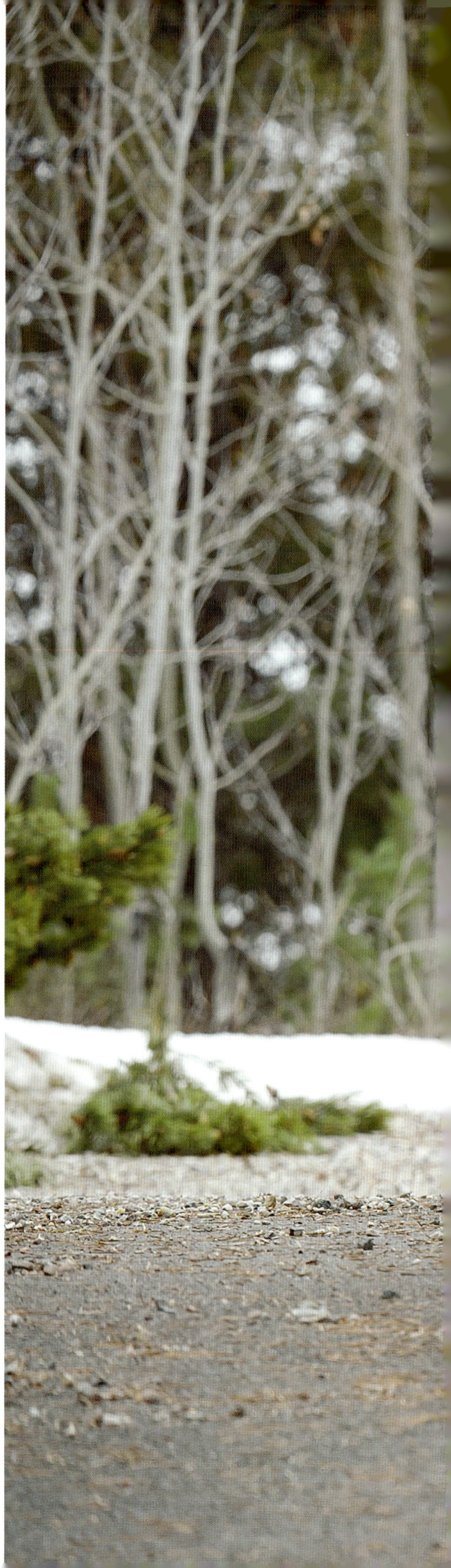

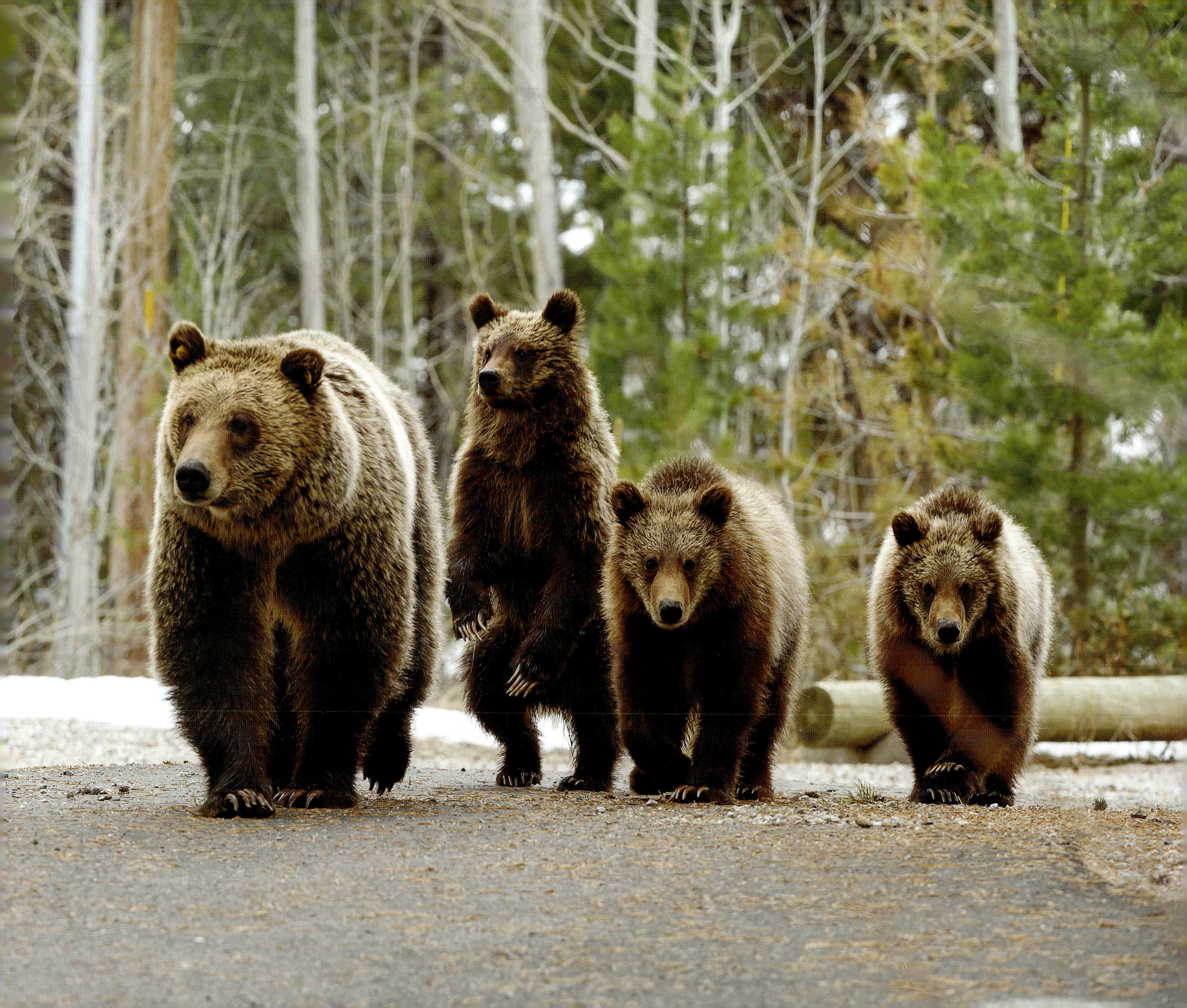

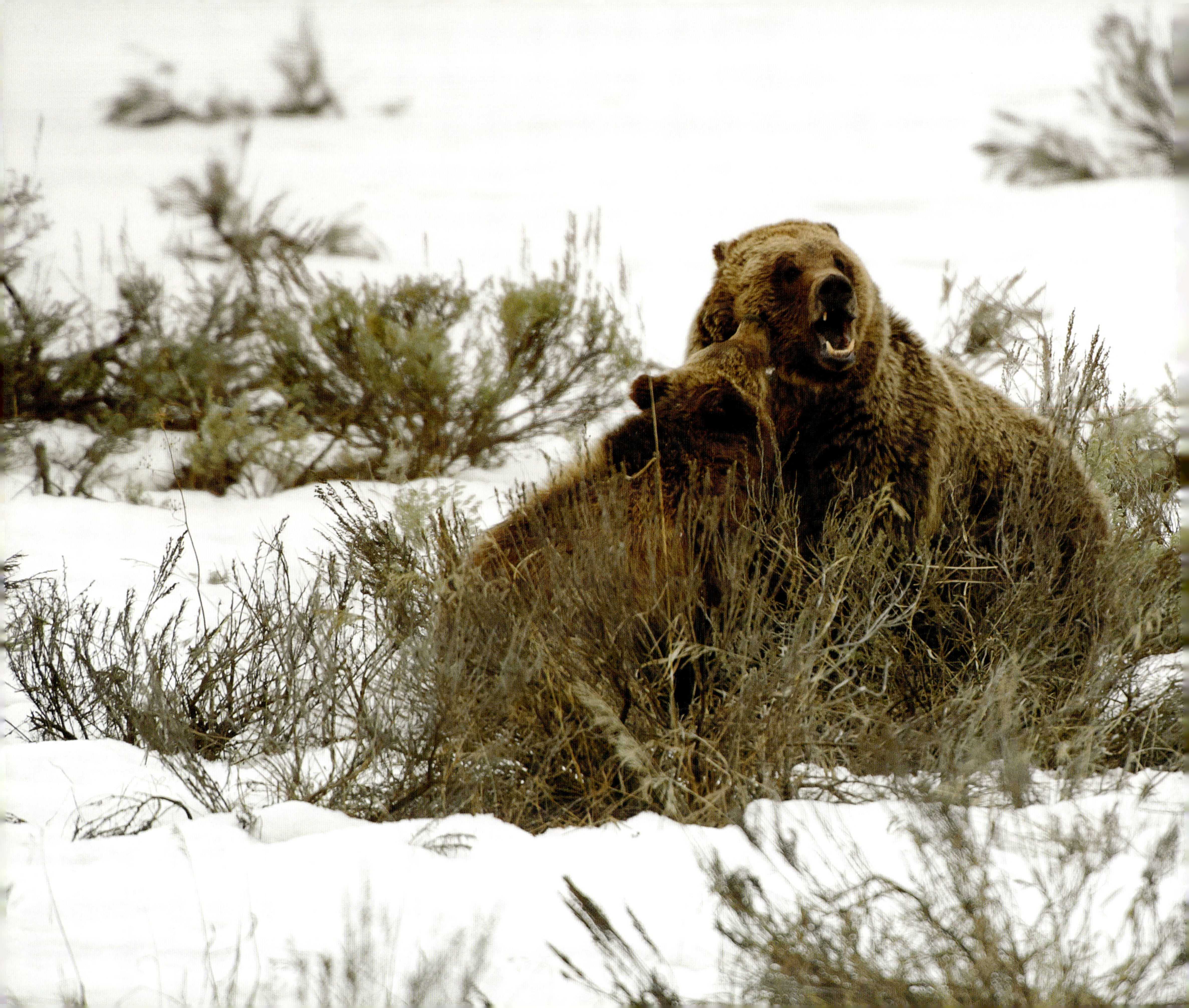

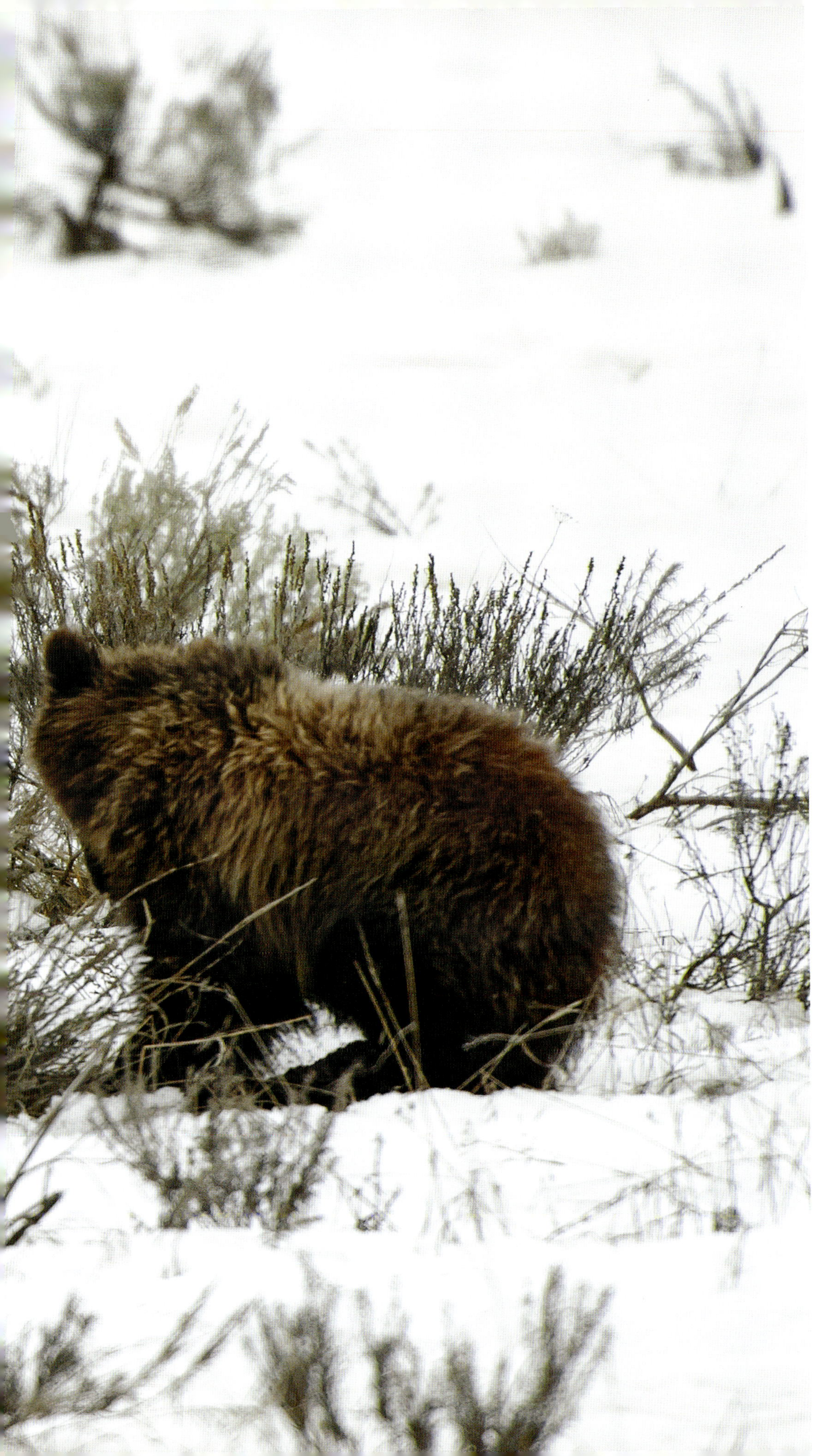

Driven by hunger, 610 and three yearling cubs scrap over a pocket gopher in Willow Flats.

FOLLOWING SPREAD, LEFT: In early spring, the meadows of Willow Flats are bursting with skittering pocket gophers and their stashes, making it a favorite snacking ground for 610 and her cubs.

FOLLOWING SPREAD, RIGHT: To view her surroundings, 610 stands in Willow Flats.

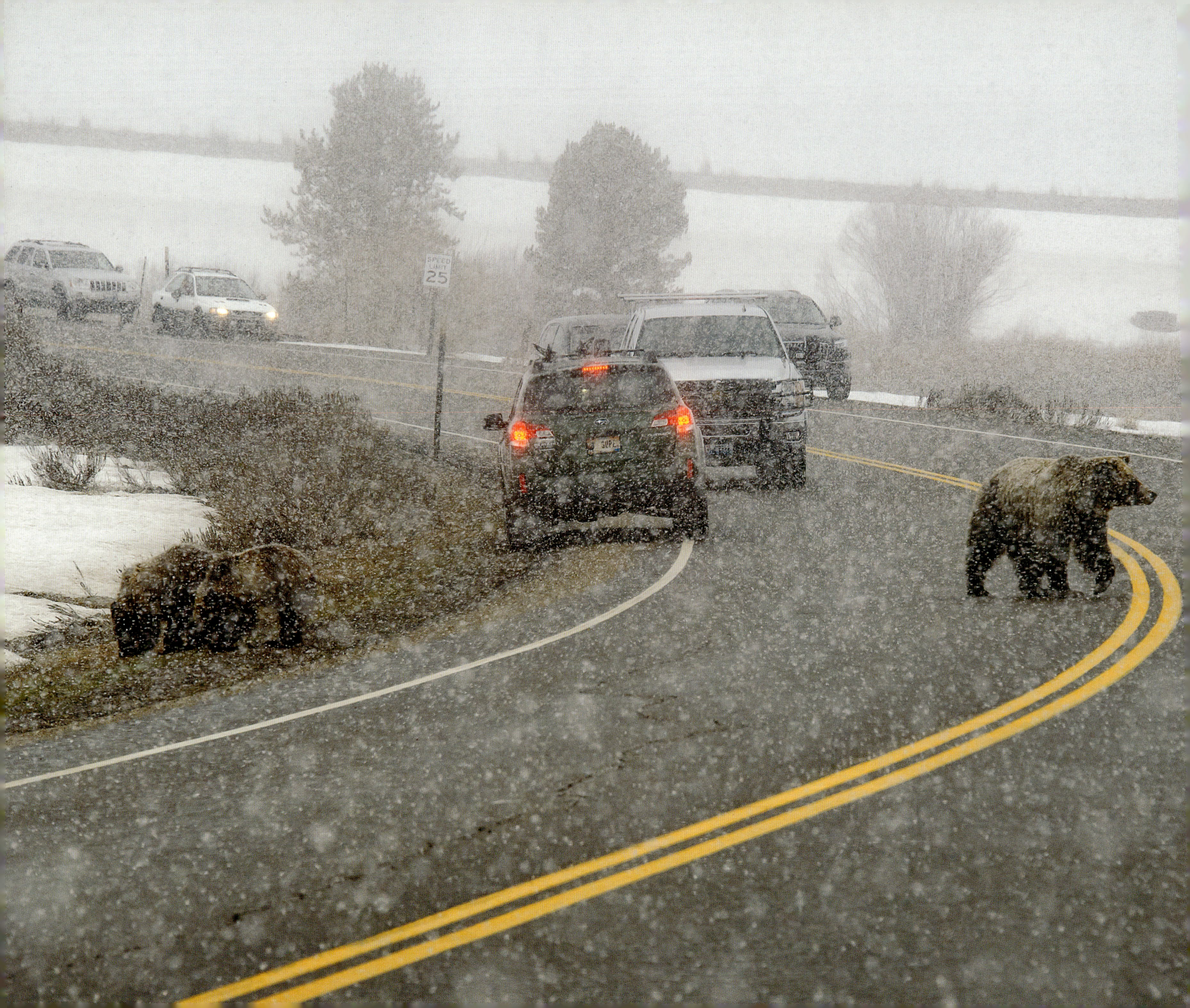
SPEED
25

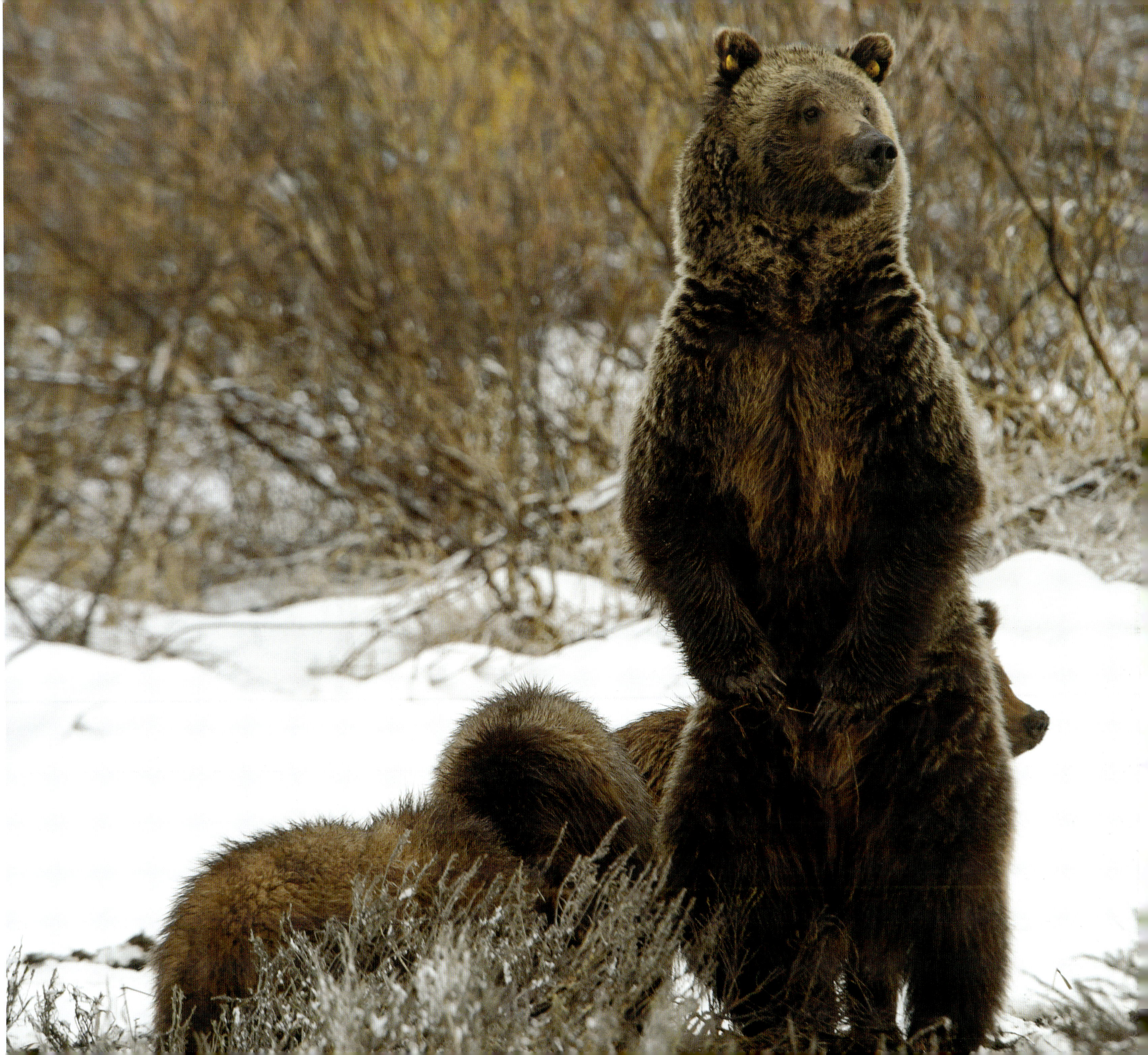

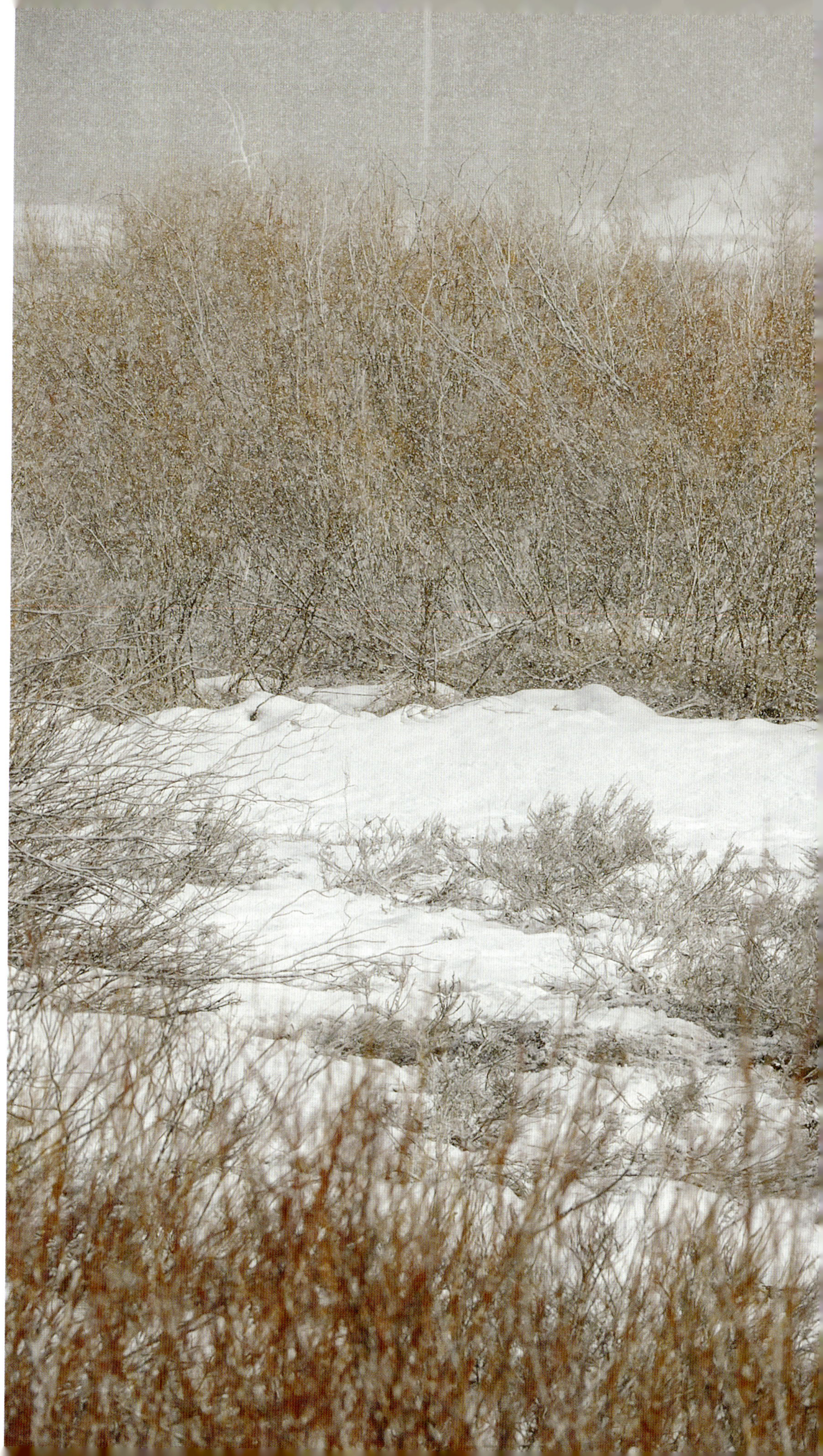

Pocket gophers and their stashes of roots, tubers, and other vegetation are an important early season food cache for grizzlies.

FOLLOWING SPREAD: The first sightings of grizzlies in the Tetons signal the true beginning of the long-awaited spring.

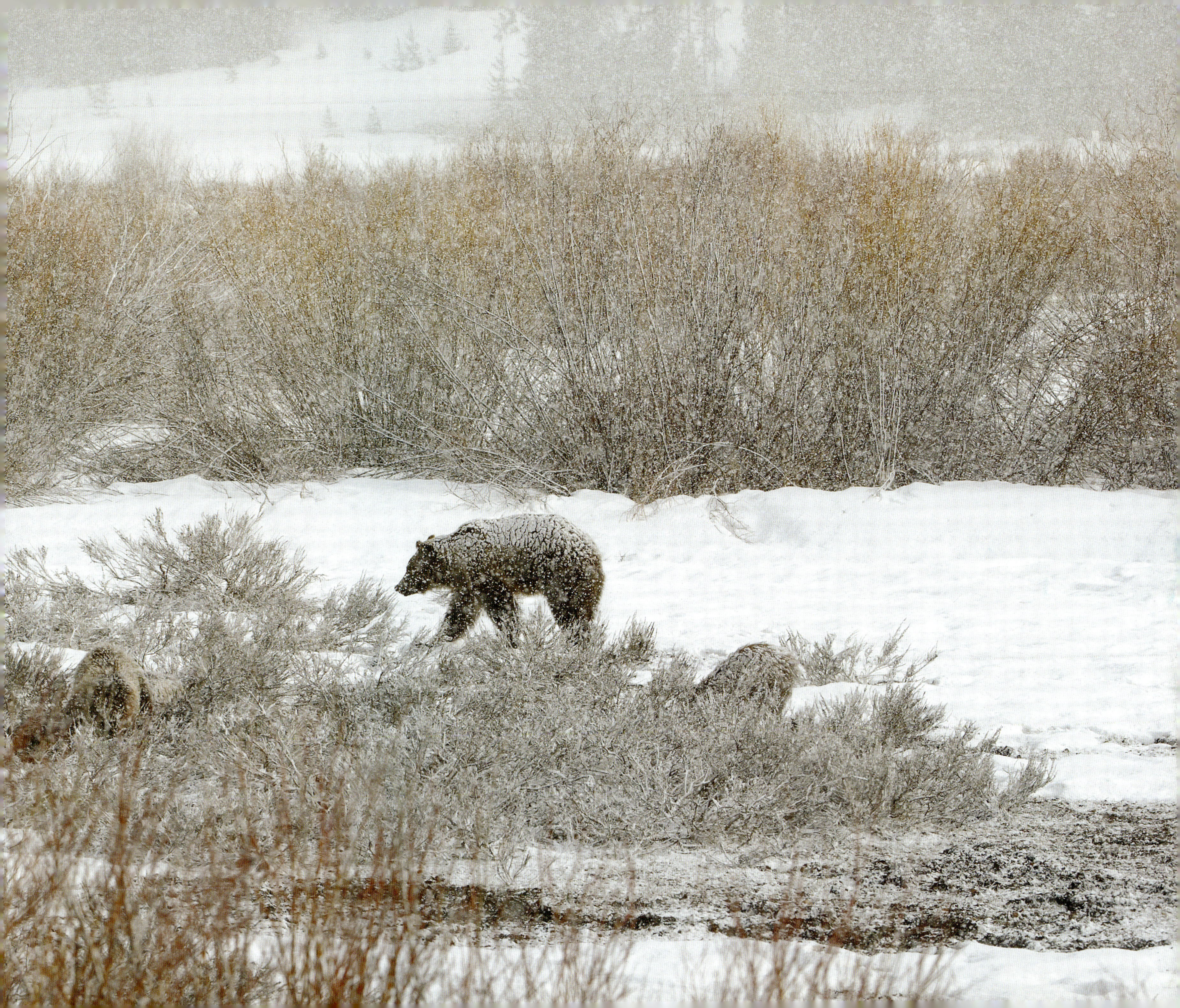

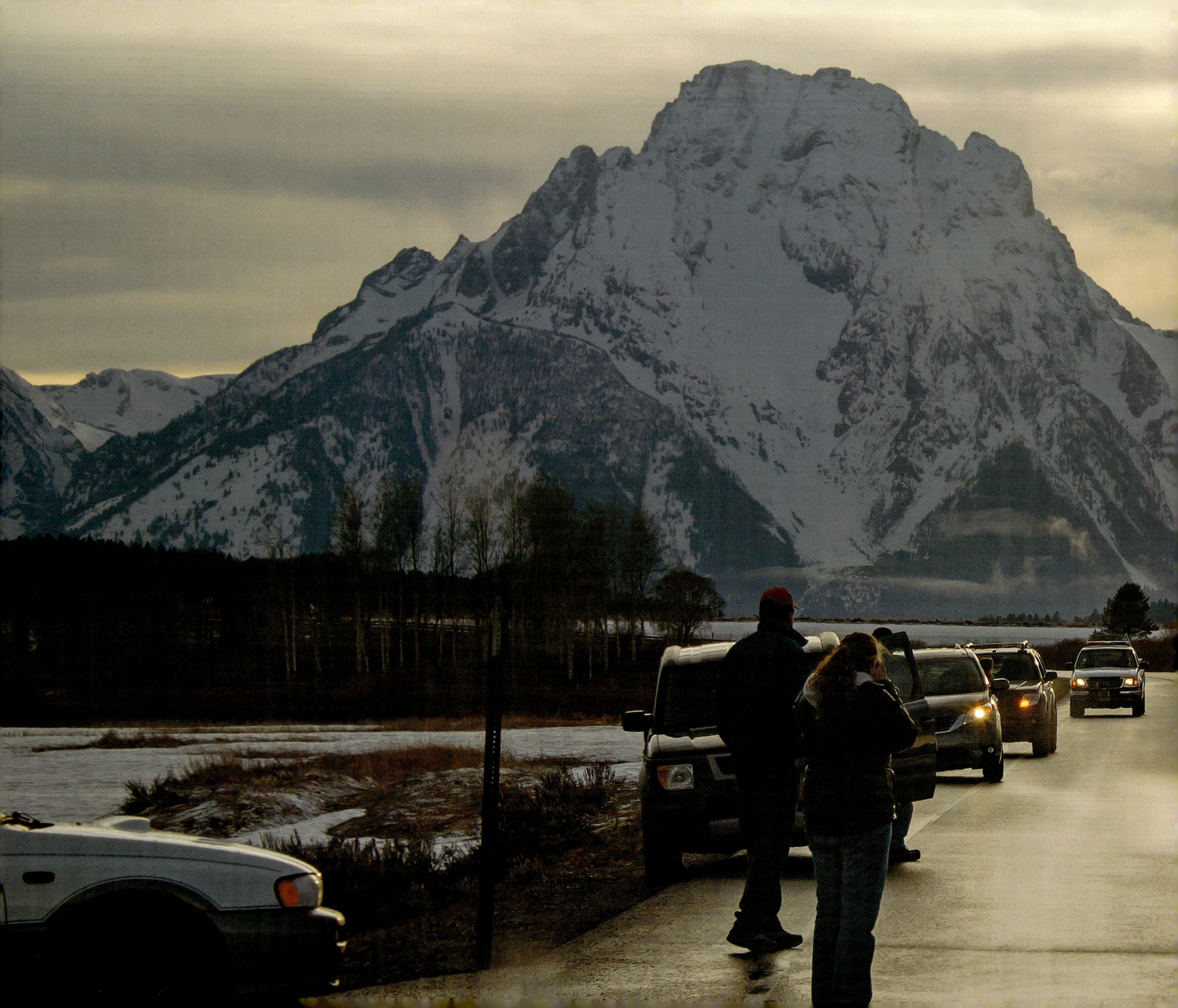

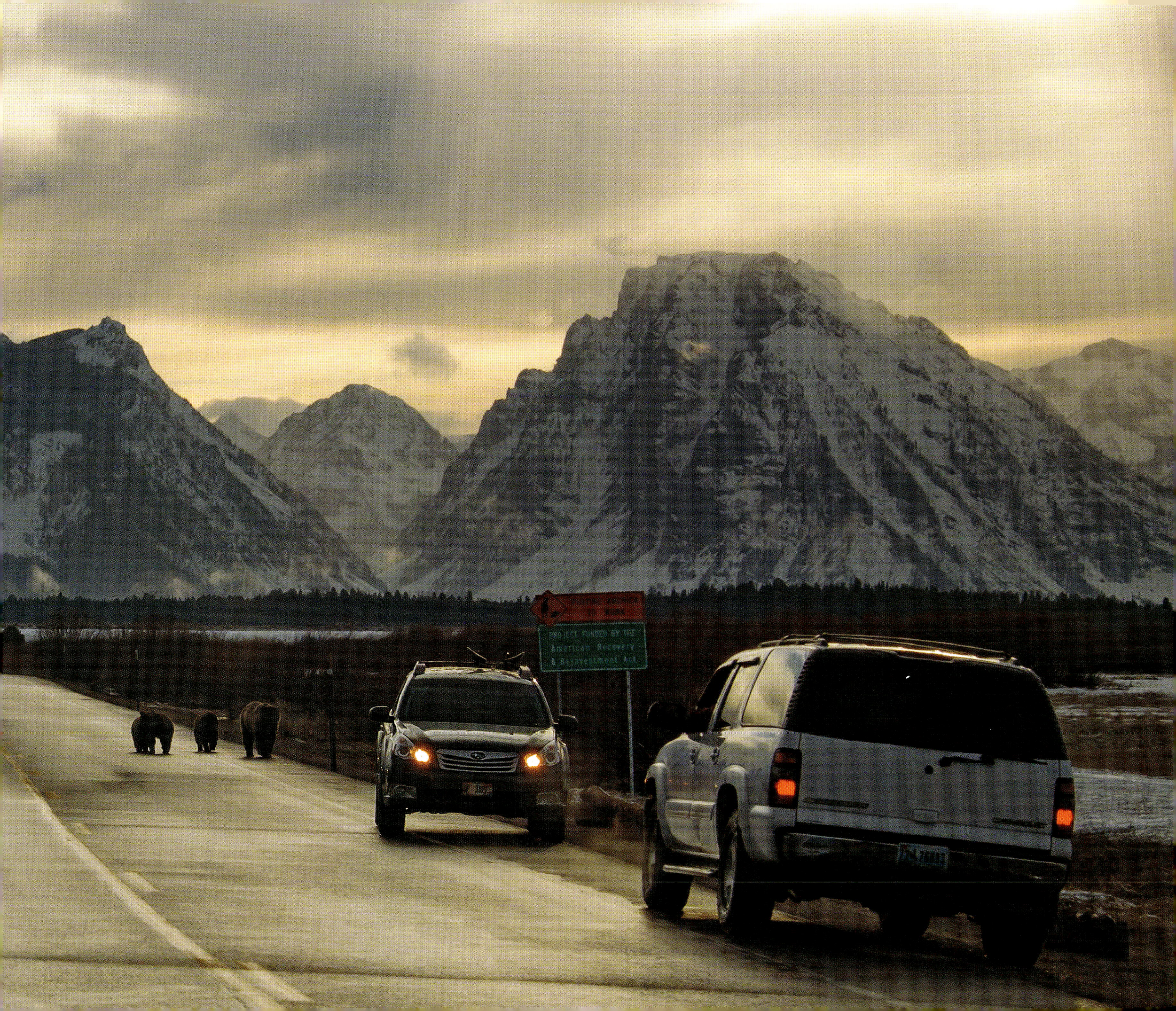
PROJECT FUNDED BY THE
American Recovery
& Reinvestment Act

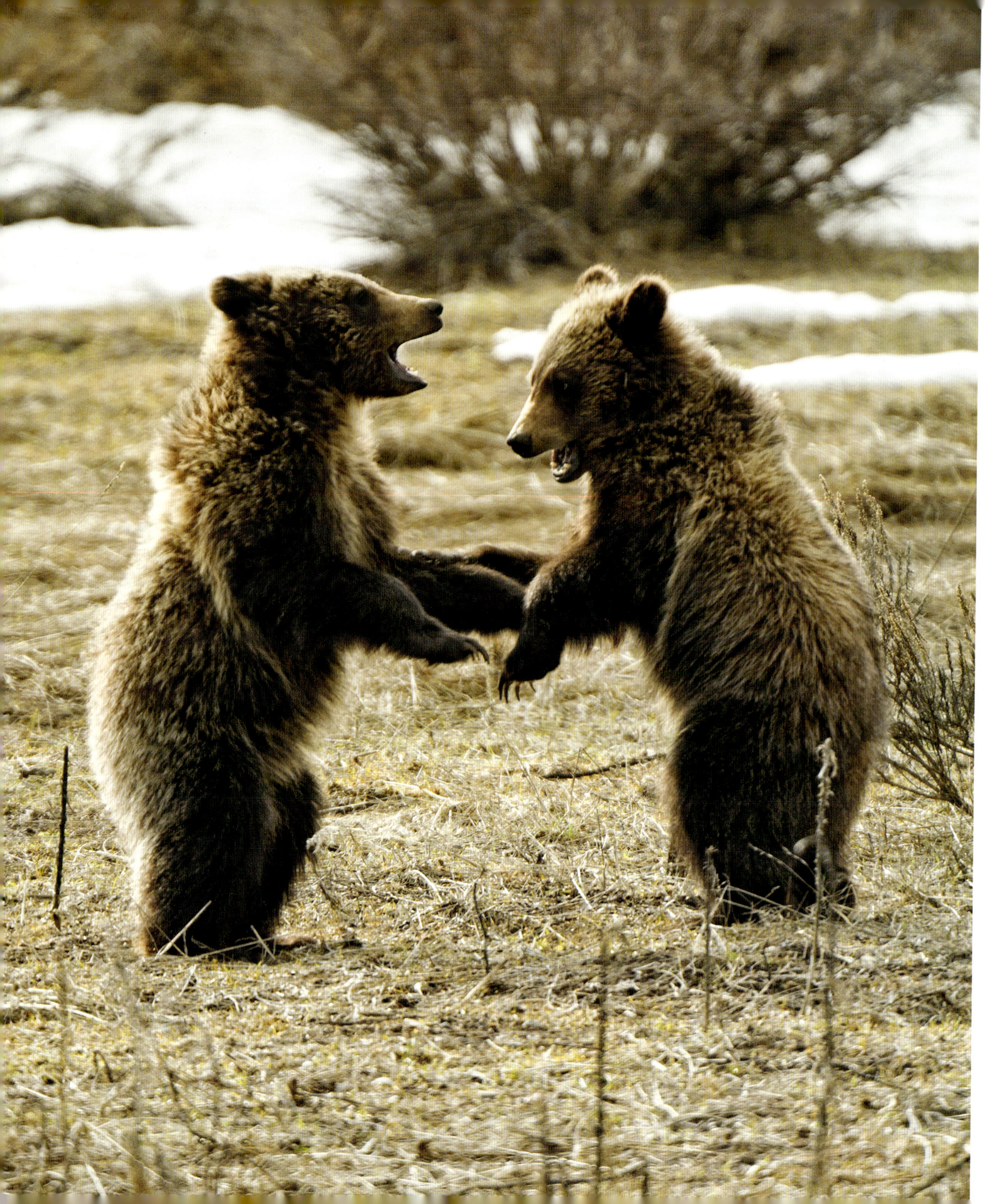

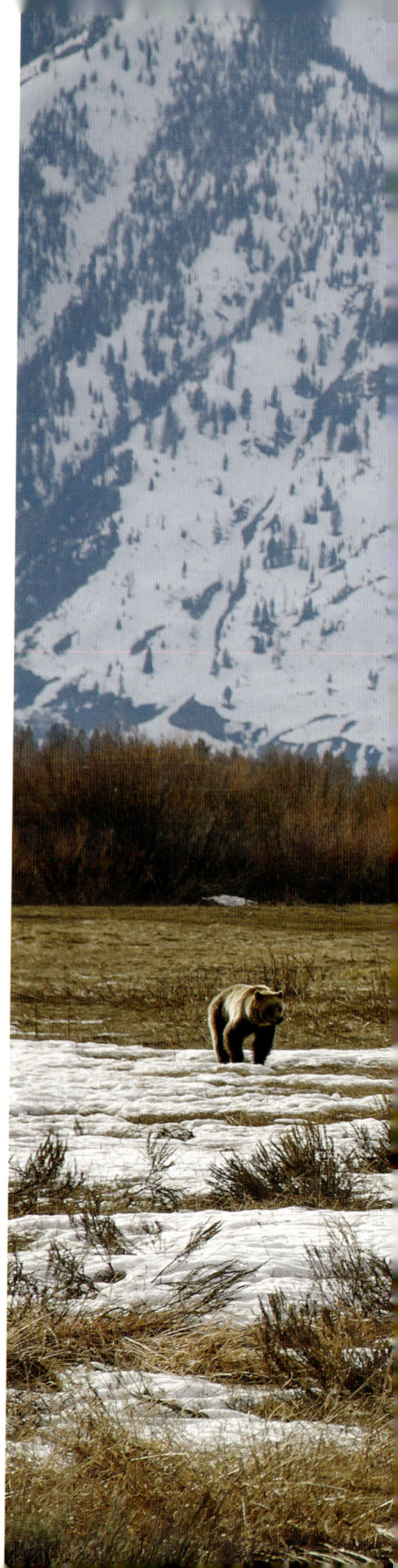

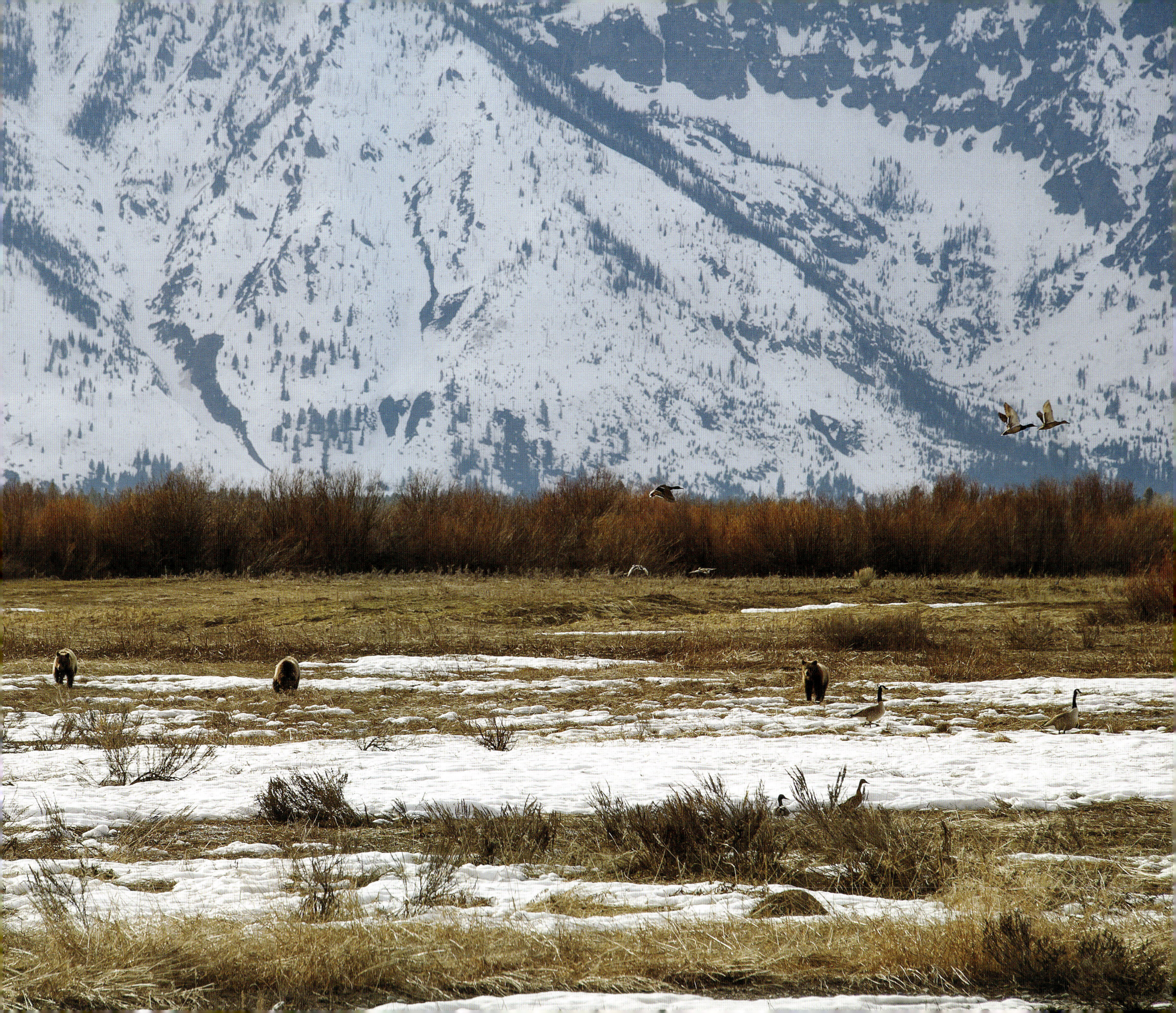

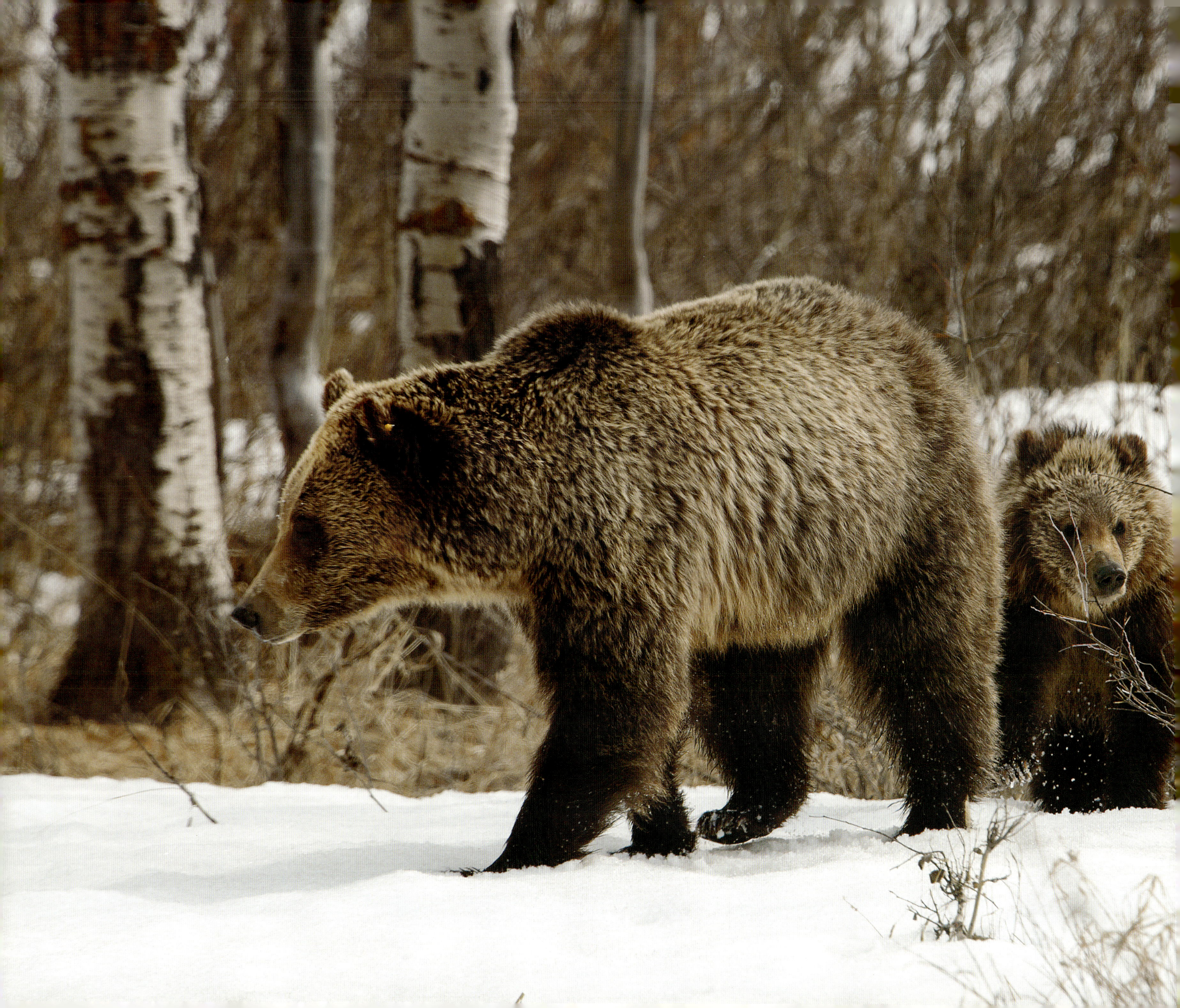

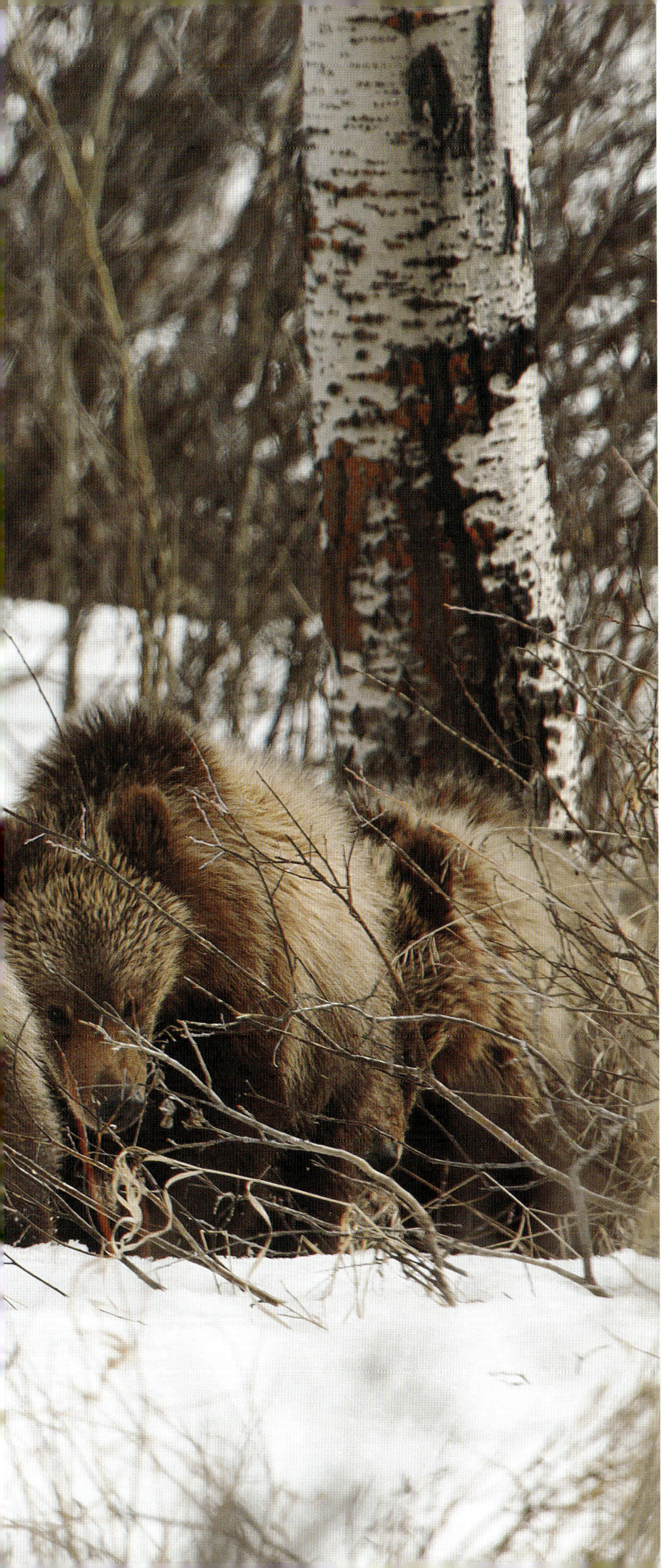

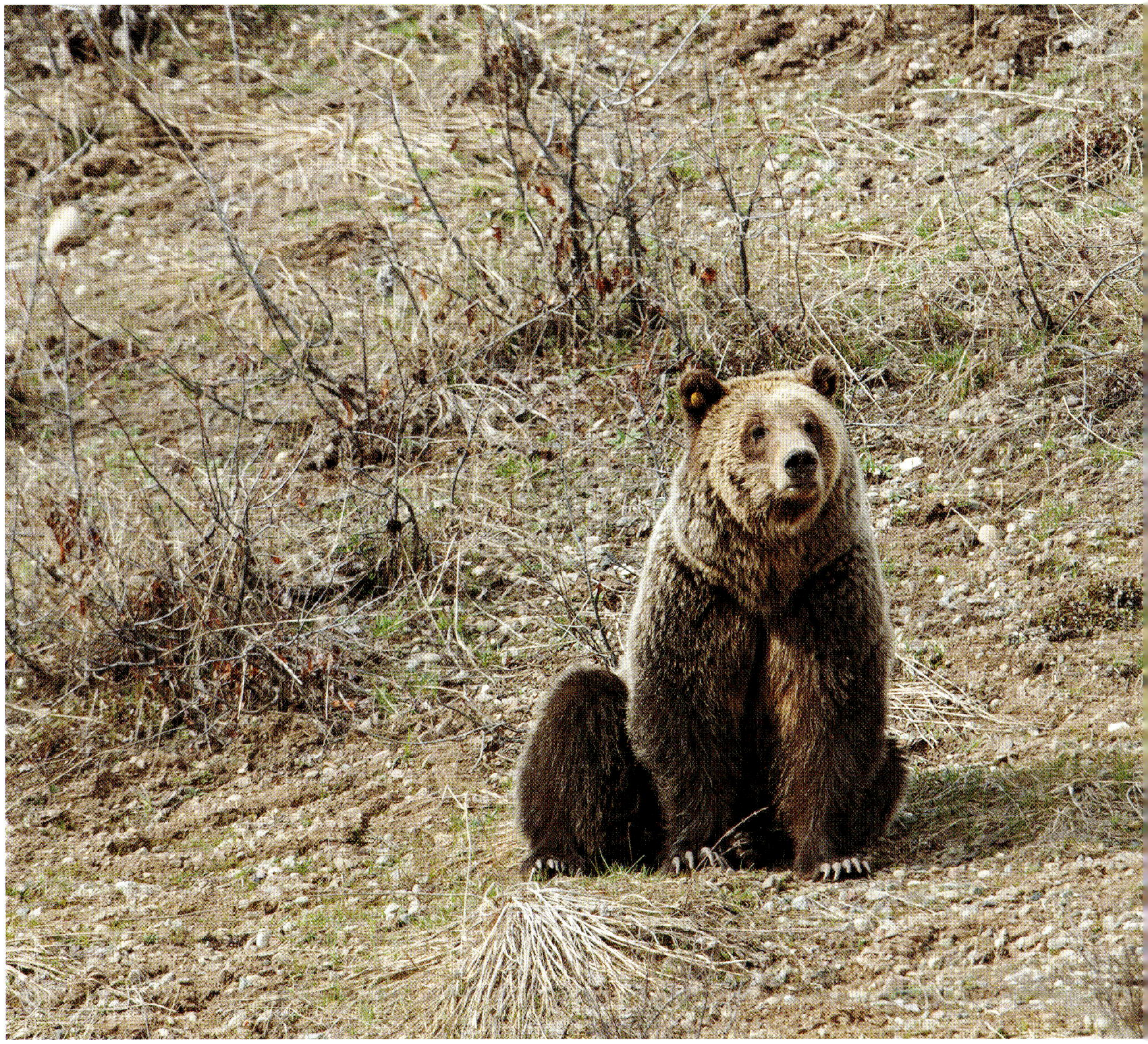

PREVIOUS SPREAD, LEFT: Feeling frisky, two of 610's cubs play fight in Willow Flats.

PREVIOUS SPREAD, RIGHT: Hoping to add fowl to their smorgasbord, 610 and her cubs chase ducks and geese in Willow Flats.

OPPOSITE: Heading toward Lozier Hill, where elk are grazing on new grasses, 610 and her cubs meander through an aspen grove.

Taking a break from digging pocket gophers, 610 basks in the warming sunshine.

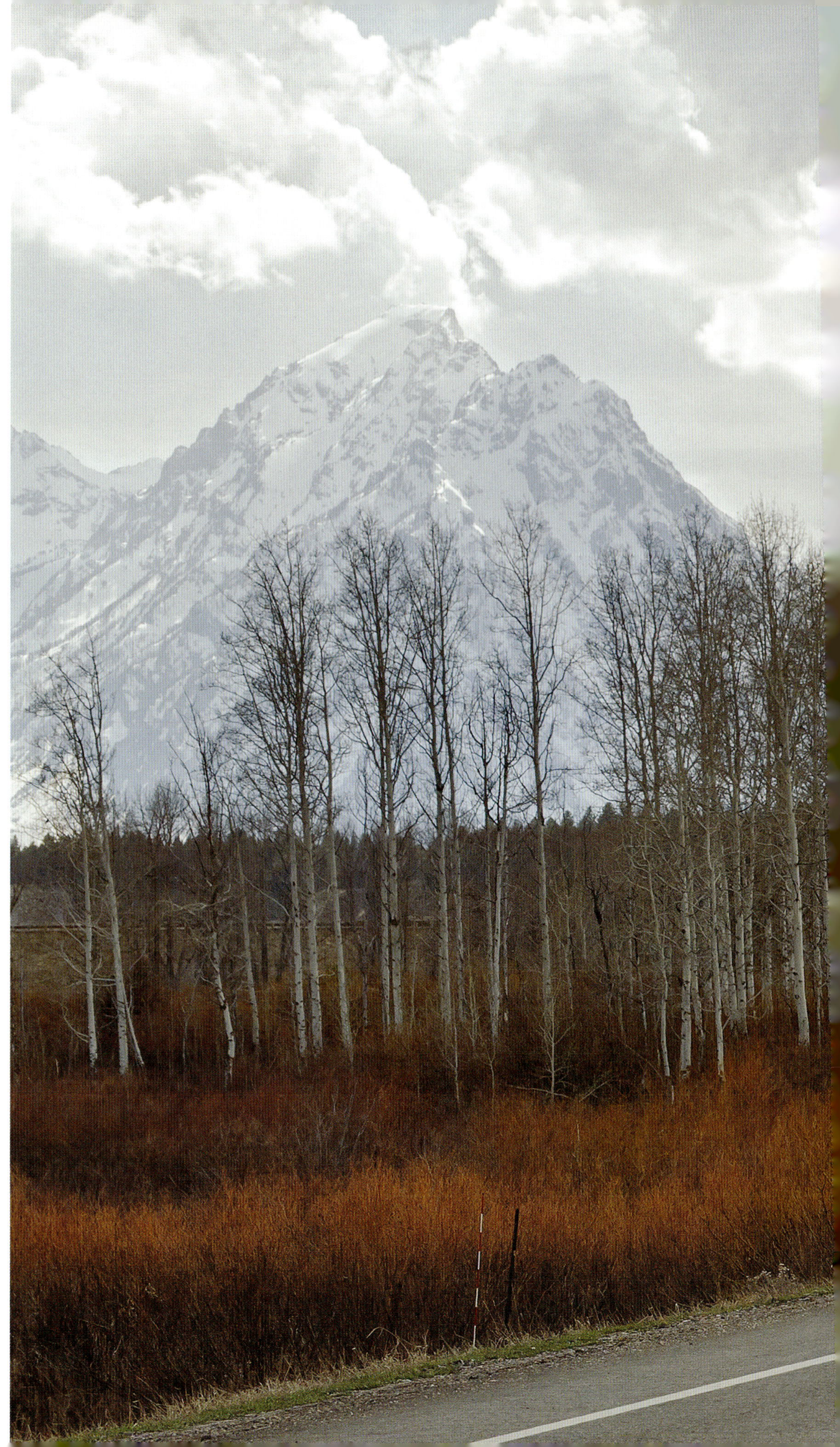

610 and her three yearling cubs cross the Willow Flats highway as Mount Moran rises behind them.

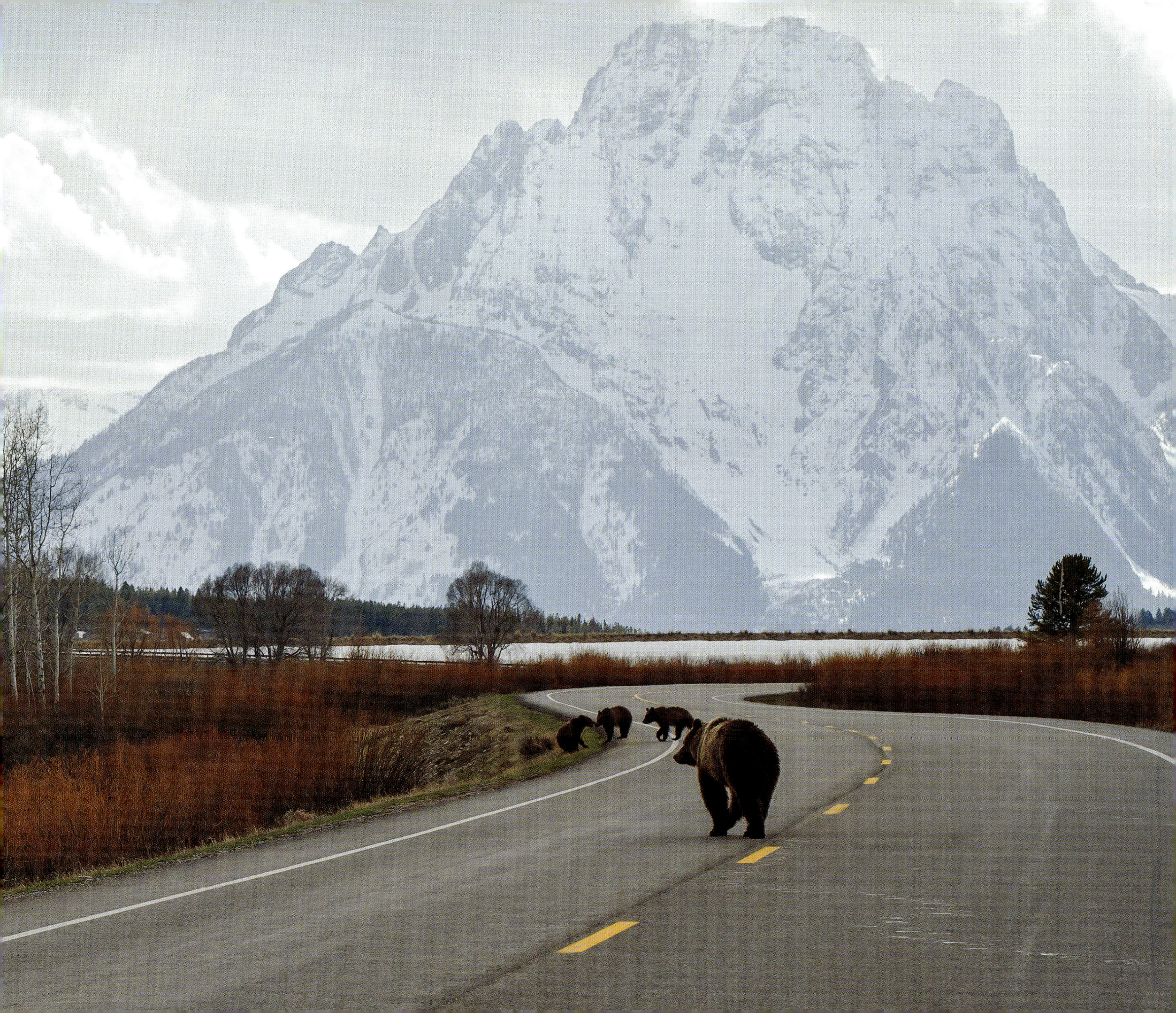

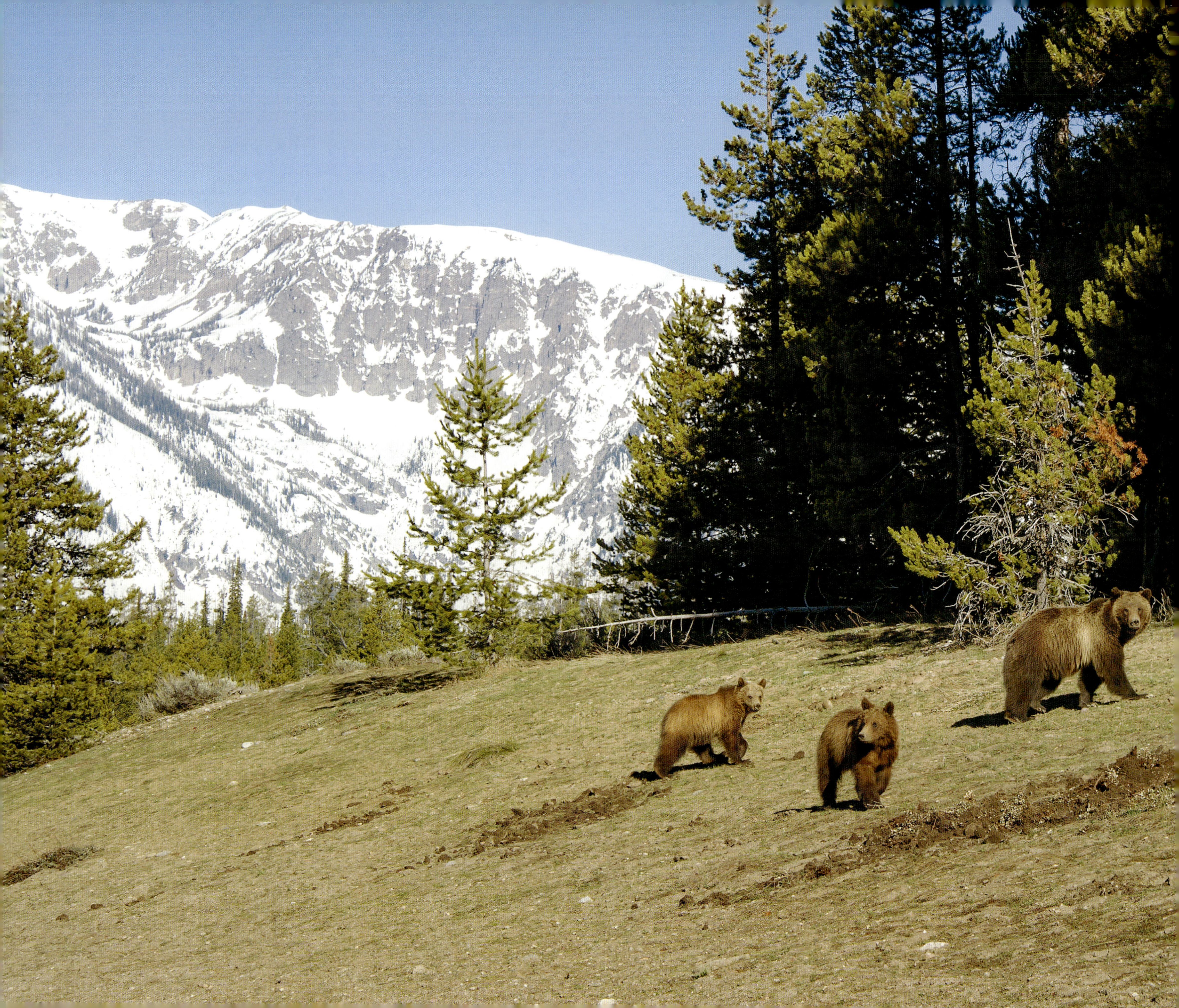

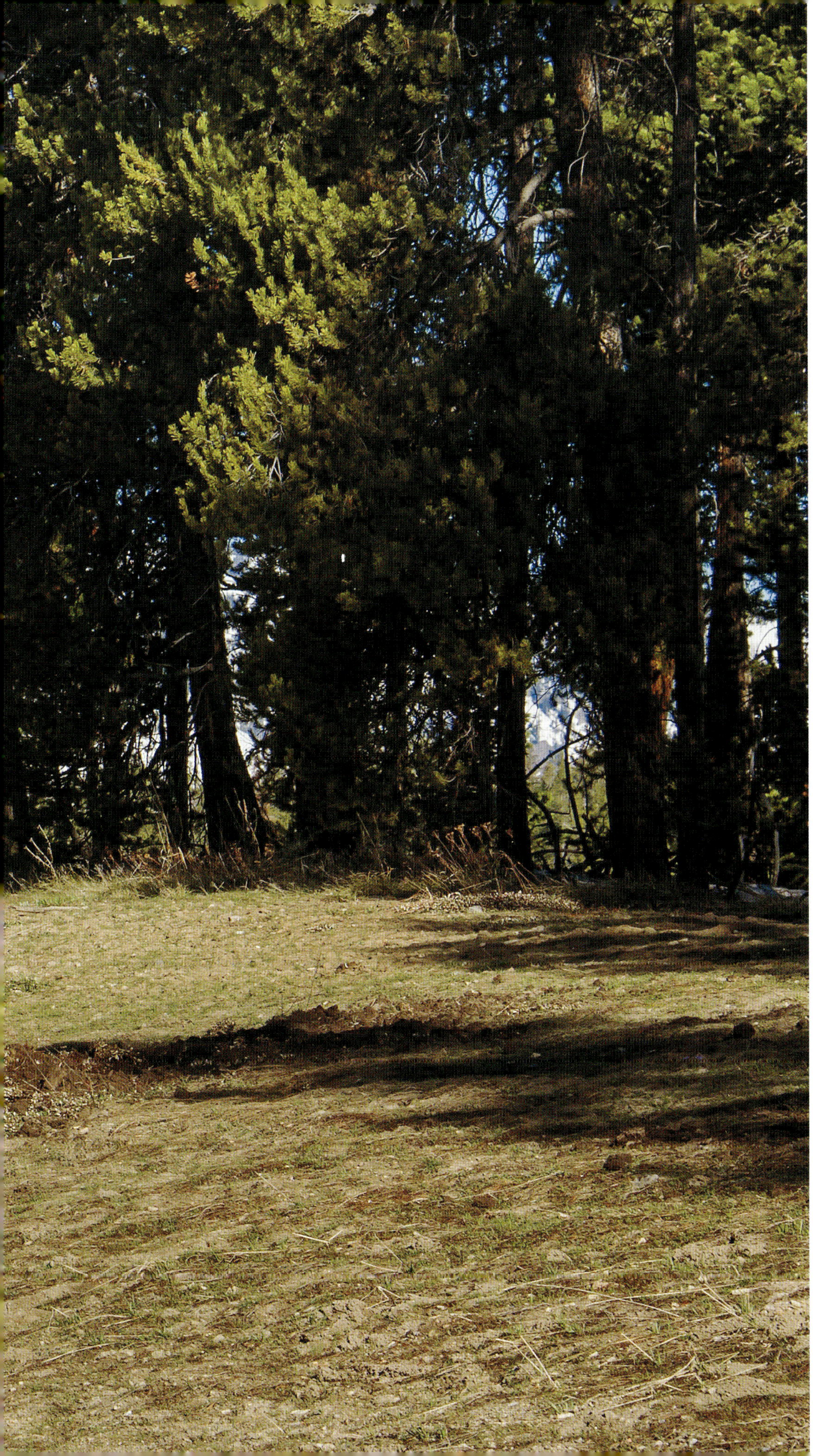

Leaving evidence of their diggings on the greening hillside nicknamed the "grassy knoll," 399 and her cubs head into the shade for a well-deserved nap.

FOLLOWING SPREAD, LEFT: After a light snowfall, Brownie, 399's yearling cub, crosses a meadow at Colter Bay.

FOLLOWING SPREAD, RIGHT: Ash and Brownie play in the Pilgrim Creek Meadow.

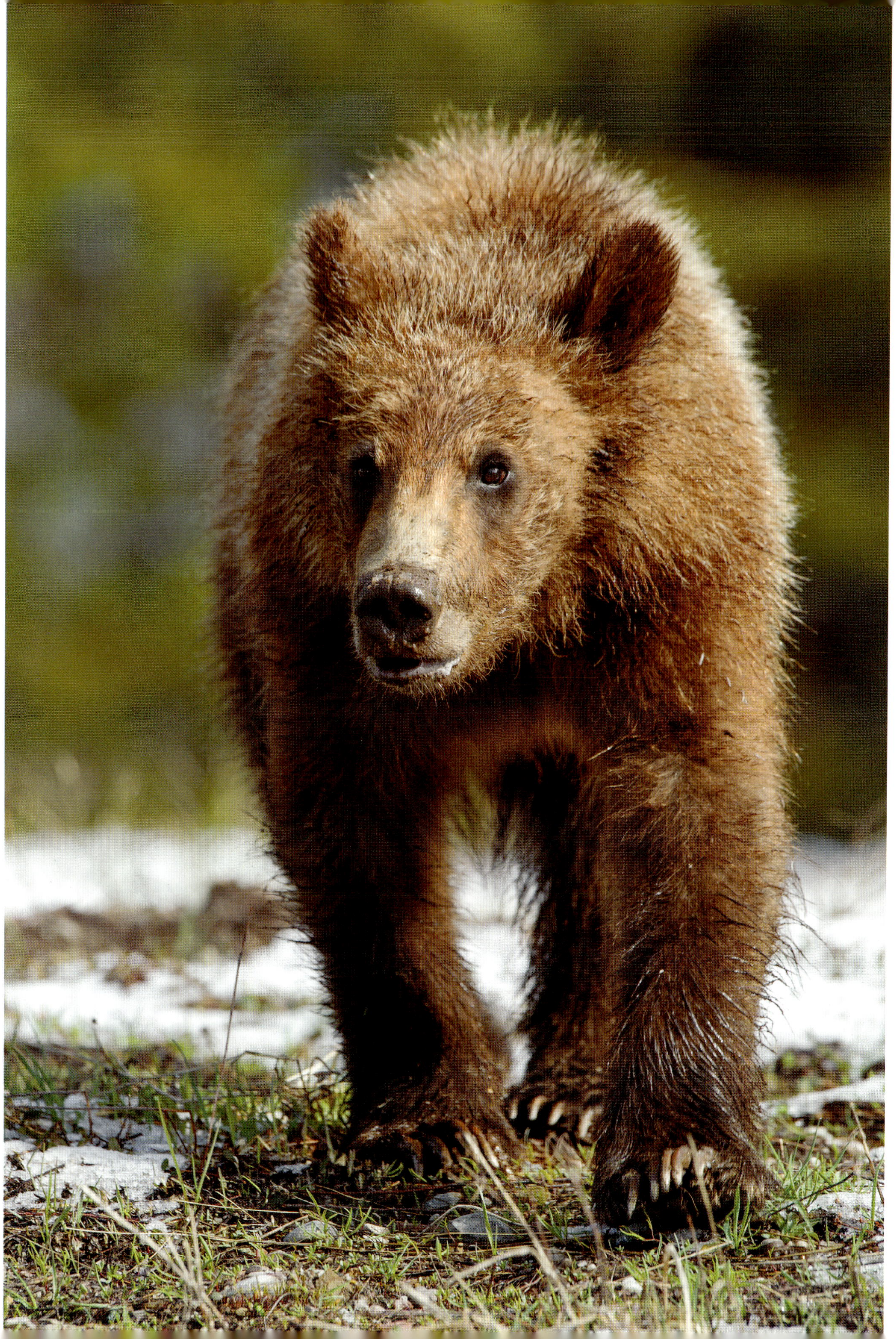
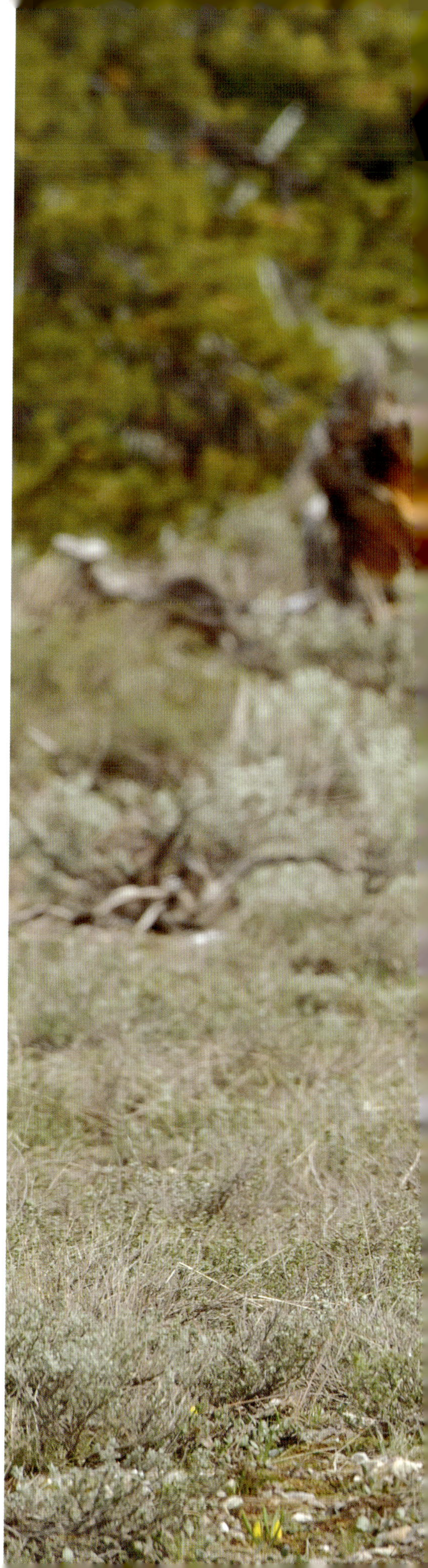

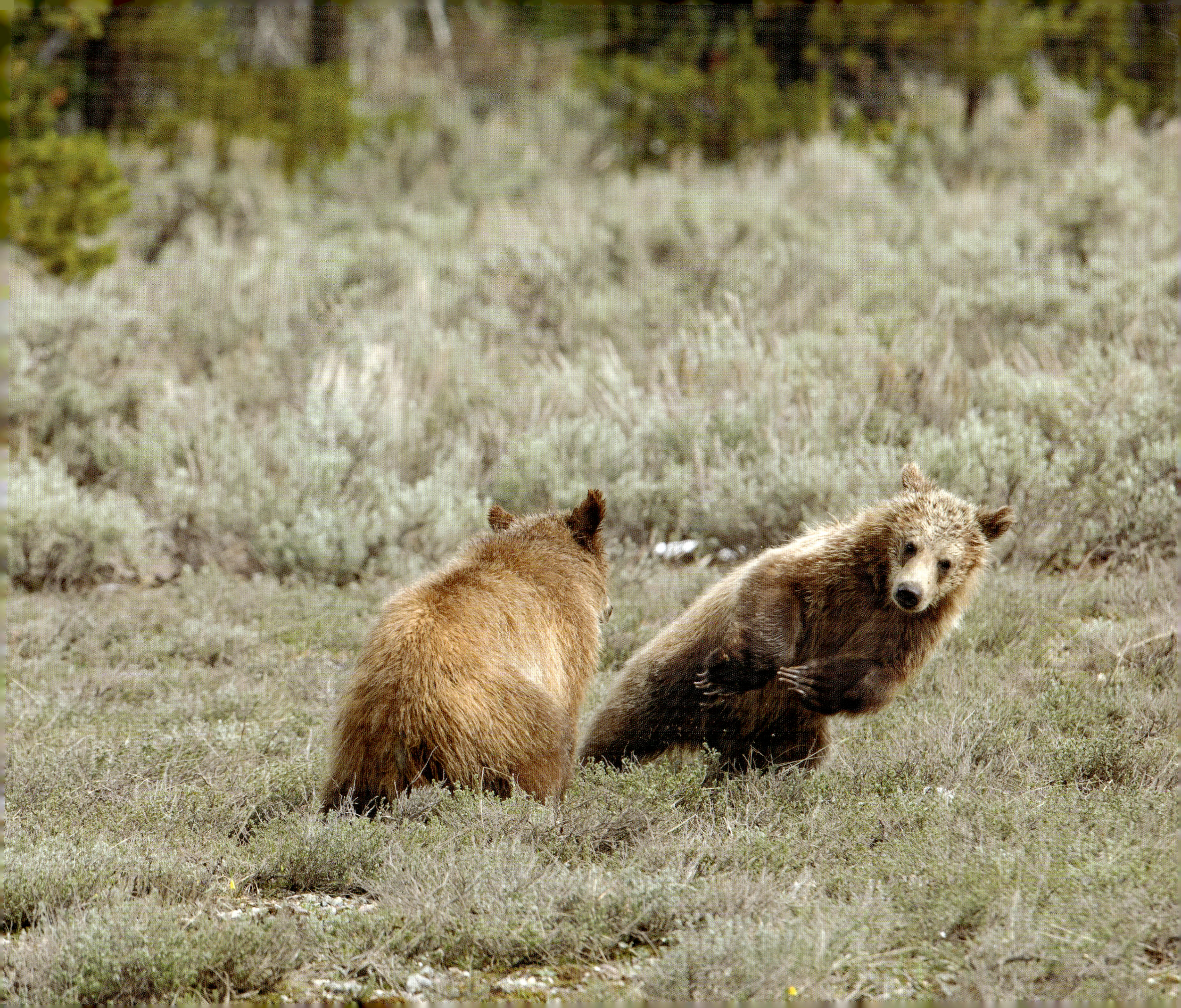

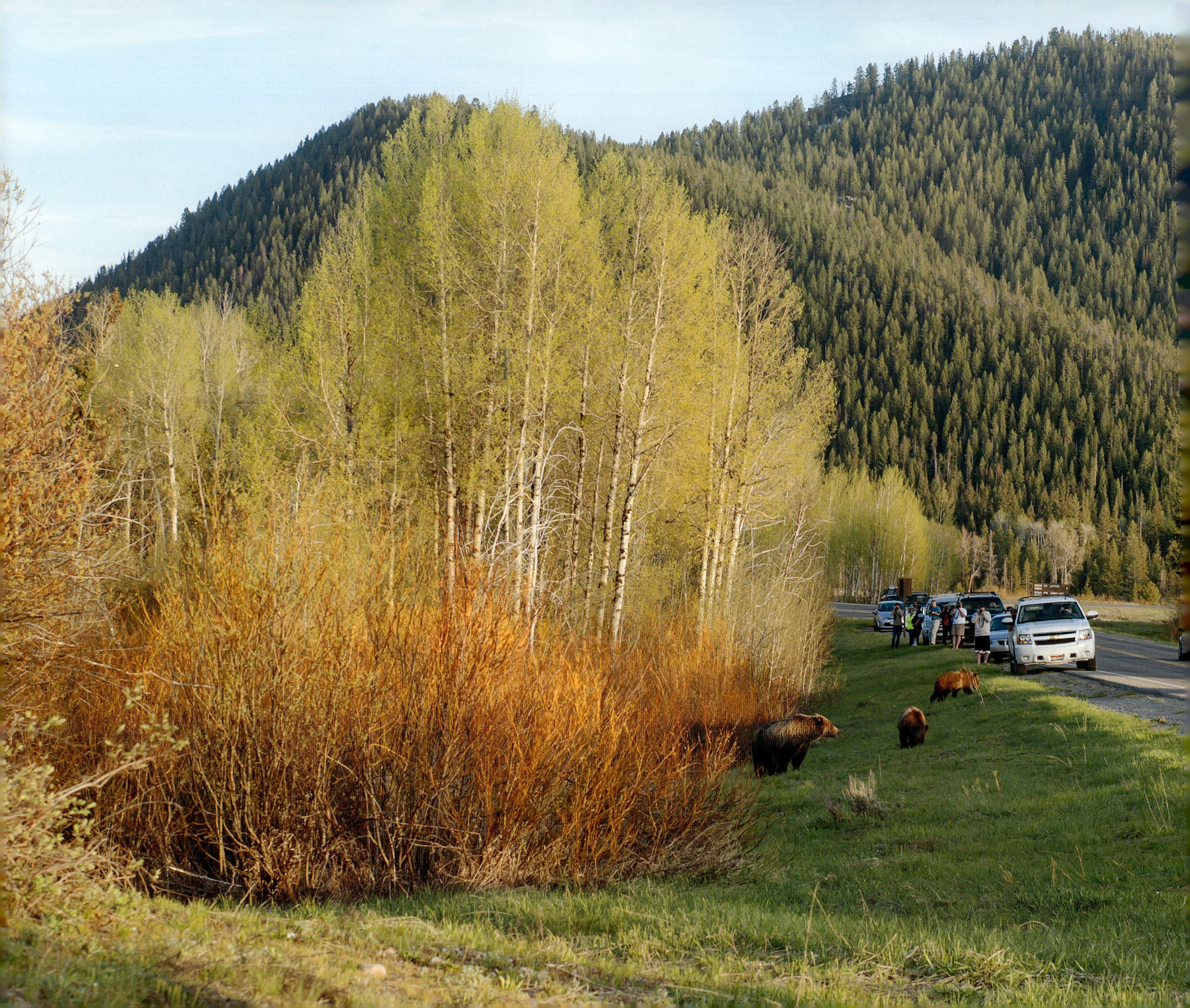

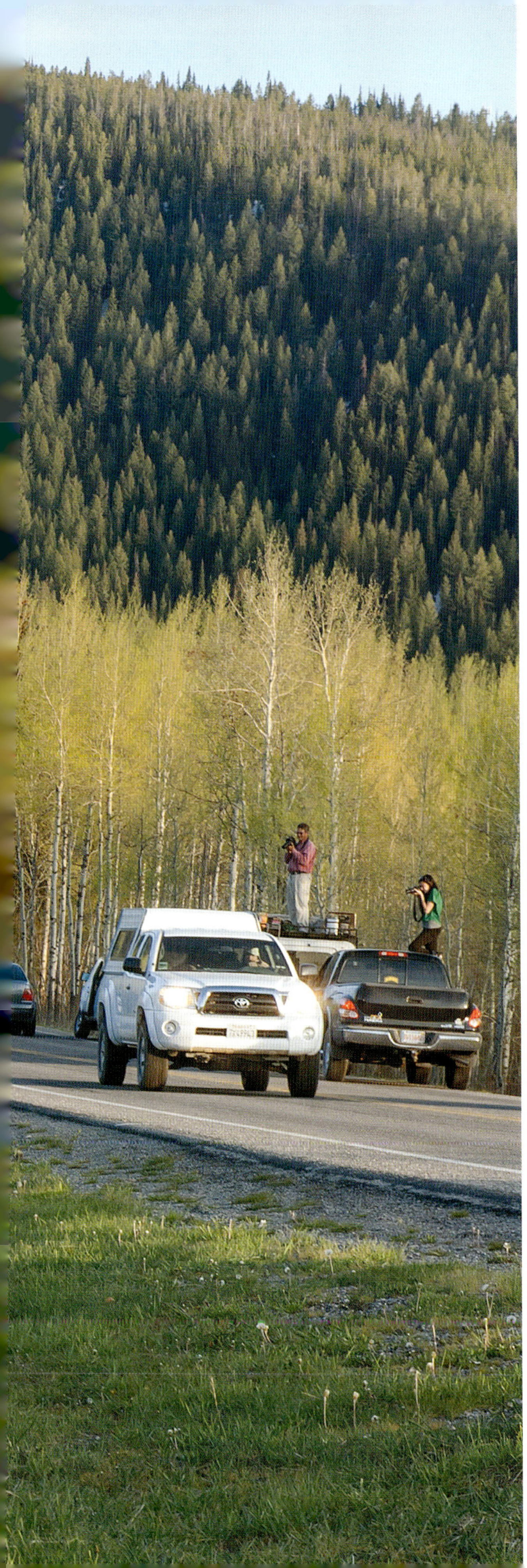

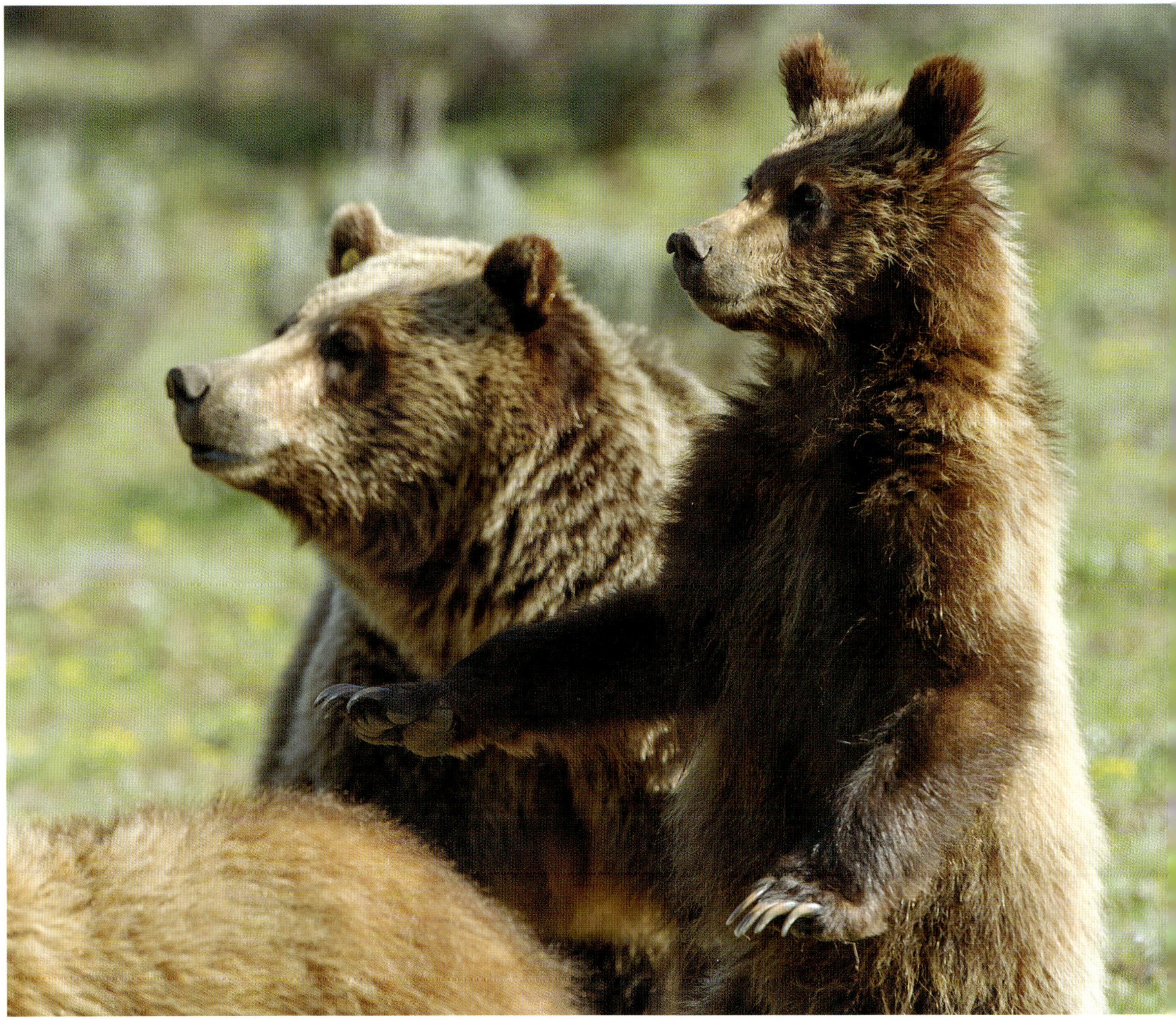

OPPOSITE: Attracting a crowd, 610 and her cubs munch at a salad bar along the highway near Jackson Lake Junction, paying little attention to two-legged onlookers.

One of 610's yearling cubs stands to get a better look at its surroundings near Oxbow Bend.

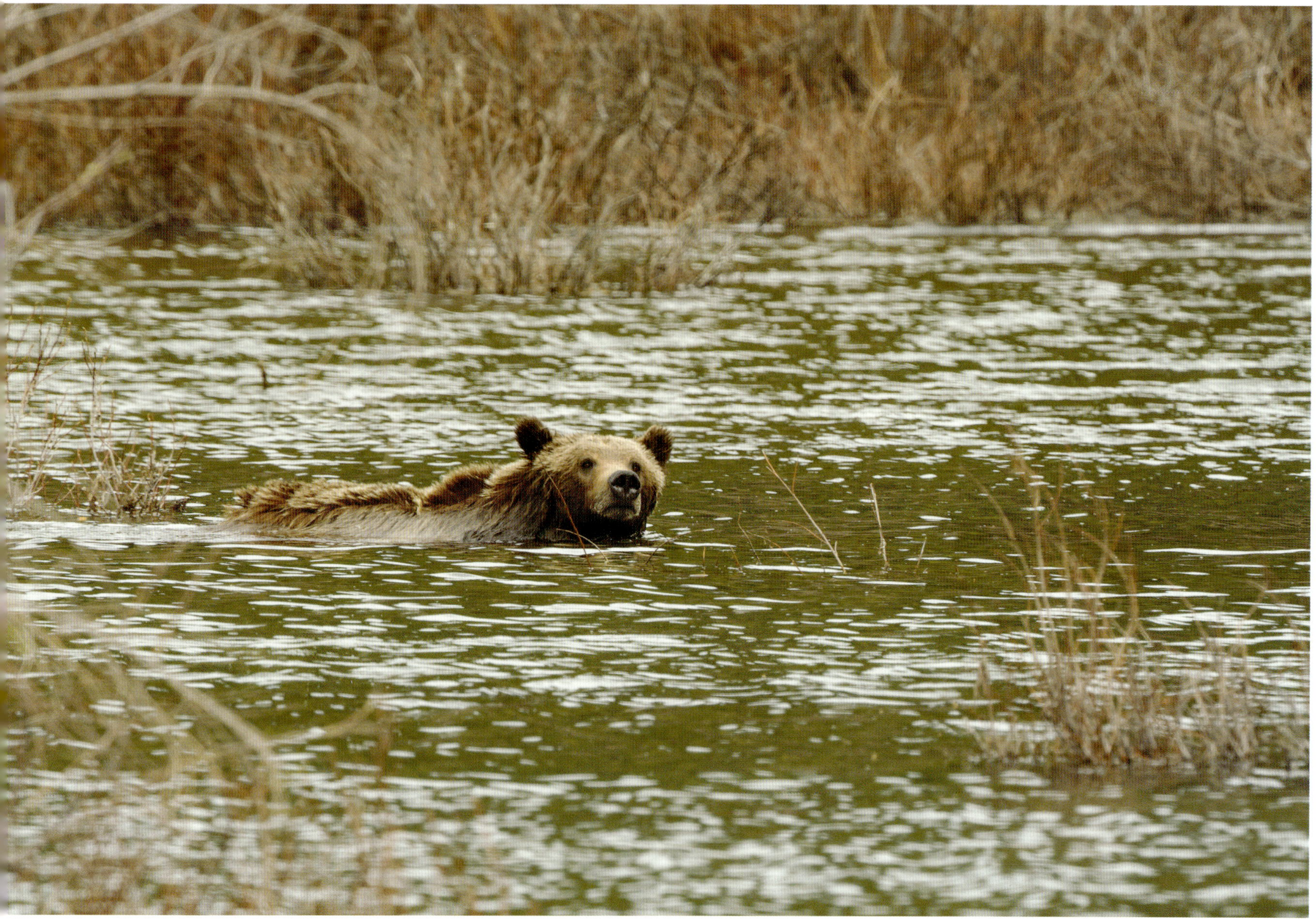

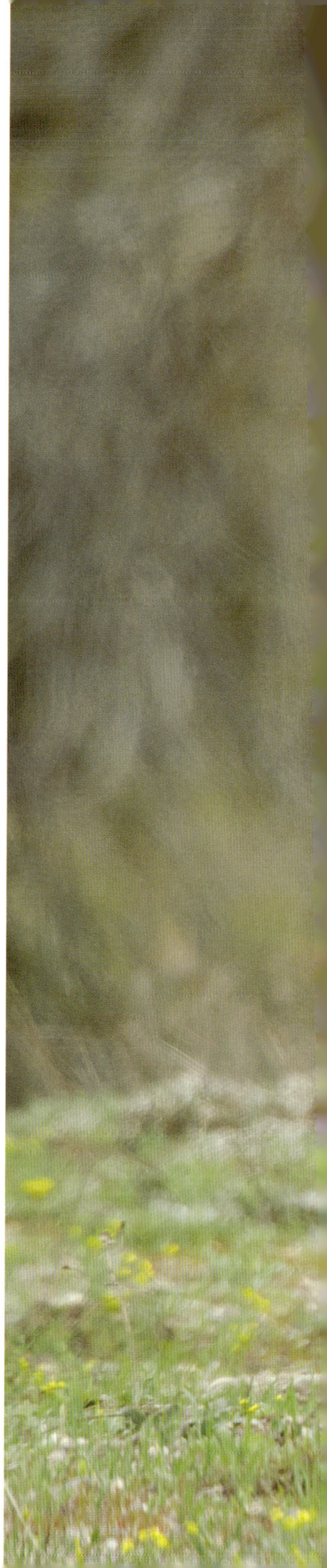

Near Pilgrim Creek, 399 swims across the "Moose Pond" after being separated from her cubs for at least two weeks. This was one of the first documented sightings after they were lost, bringing joy to many of her fans.

OPPOSITE: 399 was never observed in the company of Brownie and Ash again after the separation two weeks earlier.

FOLLOWING SPREAD: Traveling up Pilgrim Creek alone, 399 was likely searching for her missing cubs, Ash and Brownie.

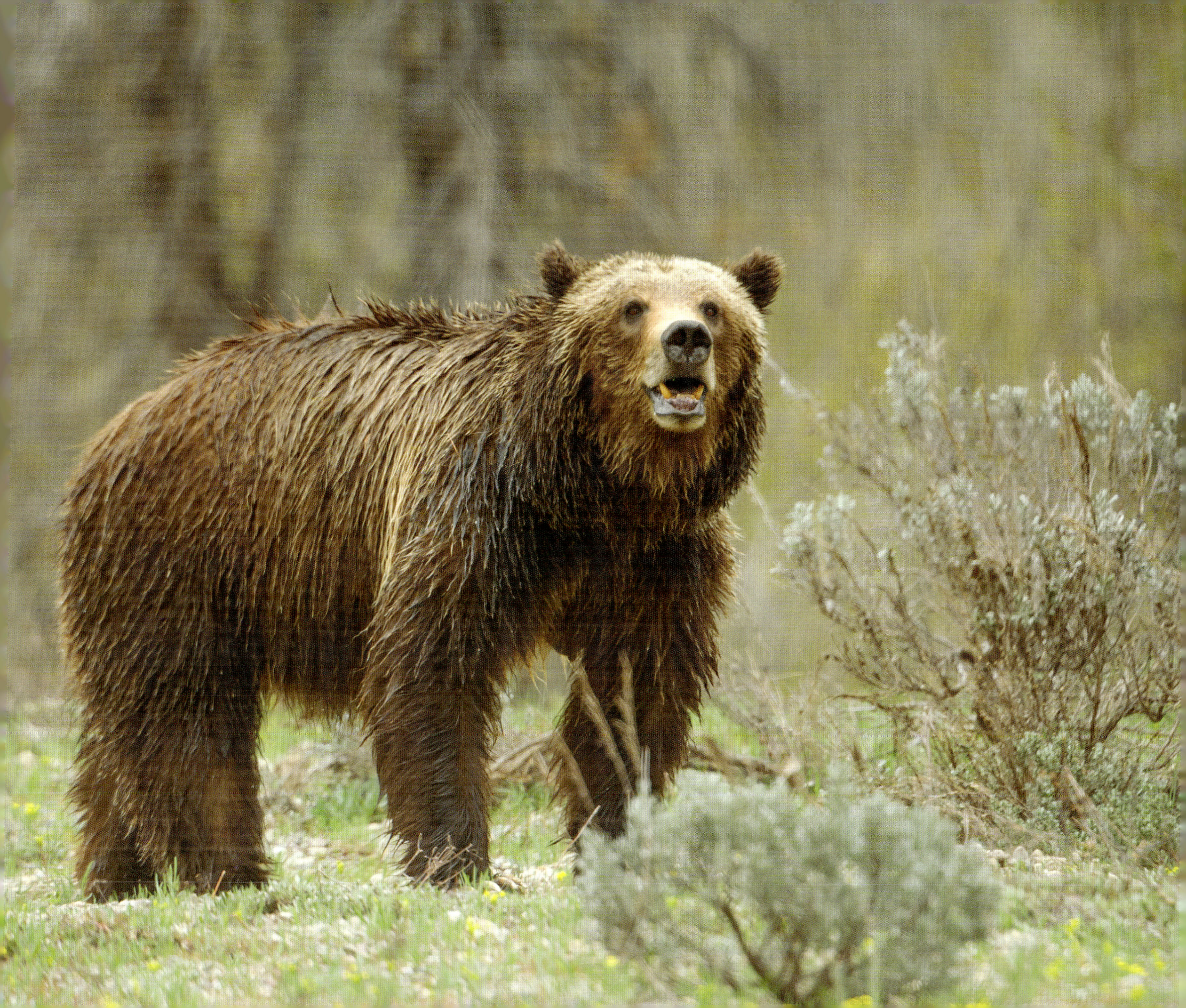

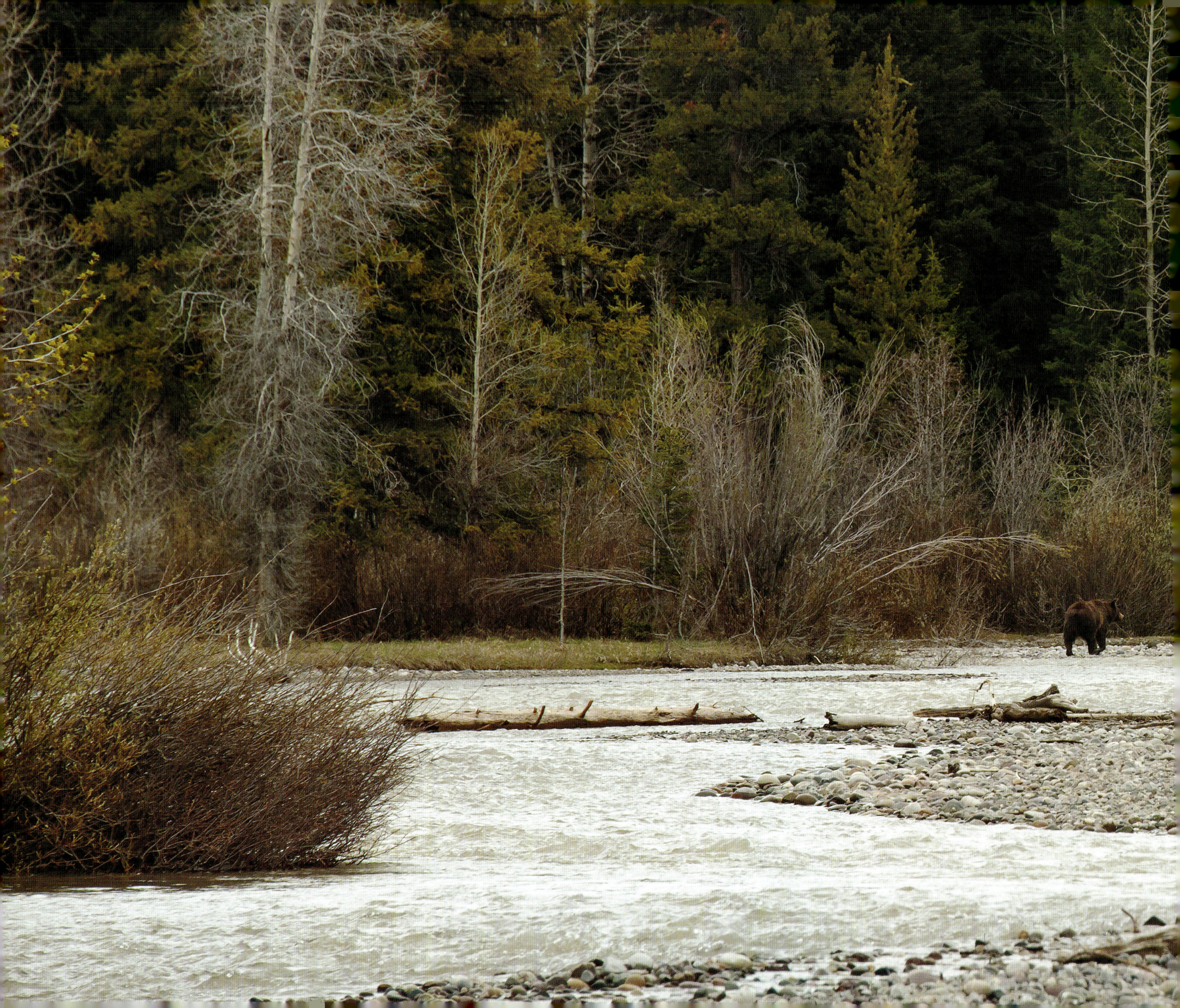

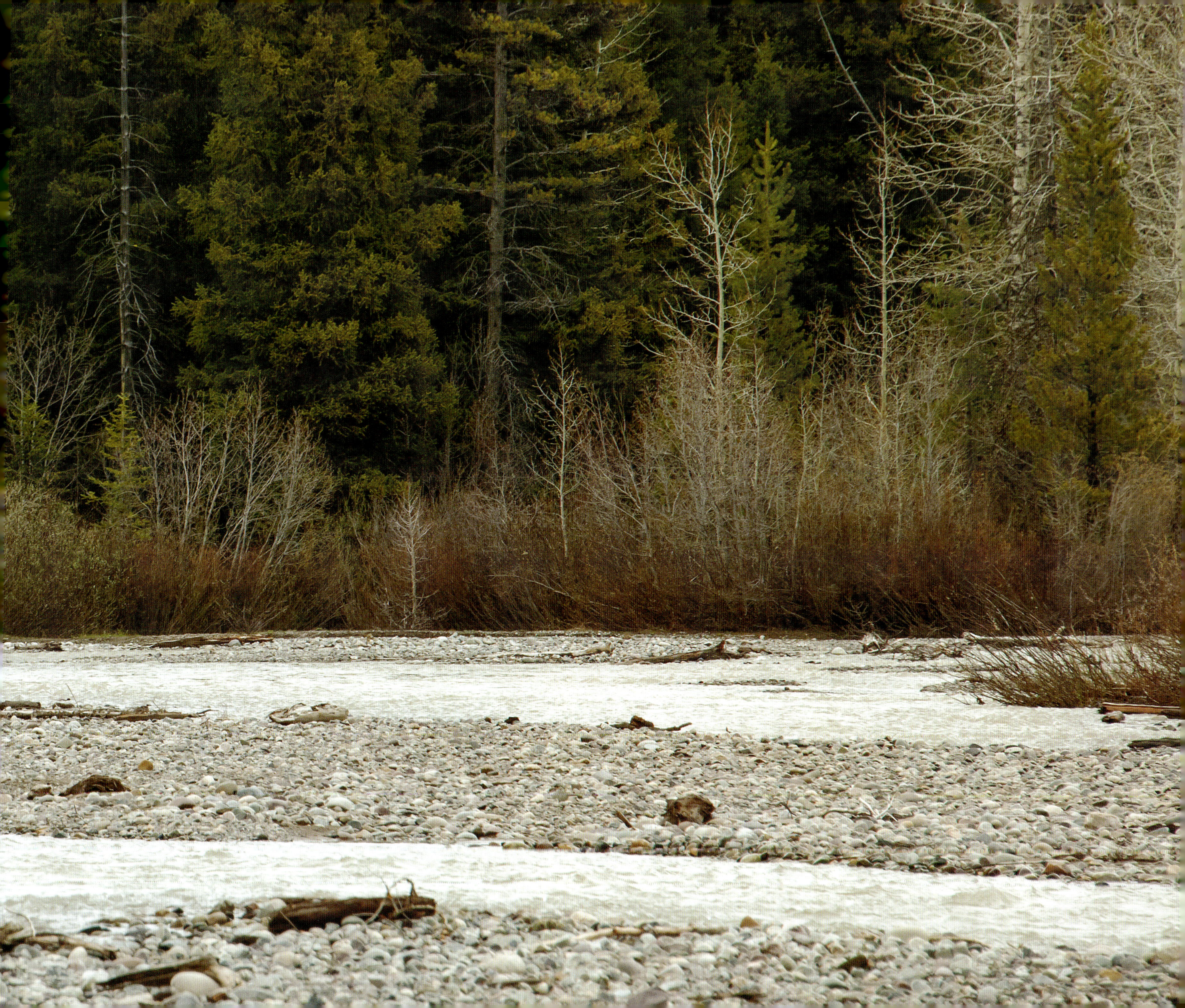

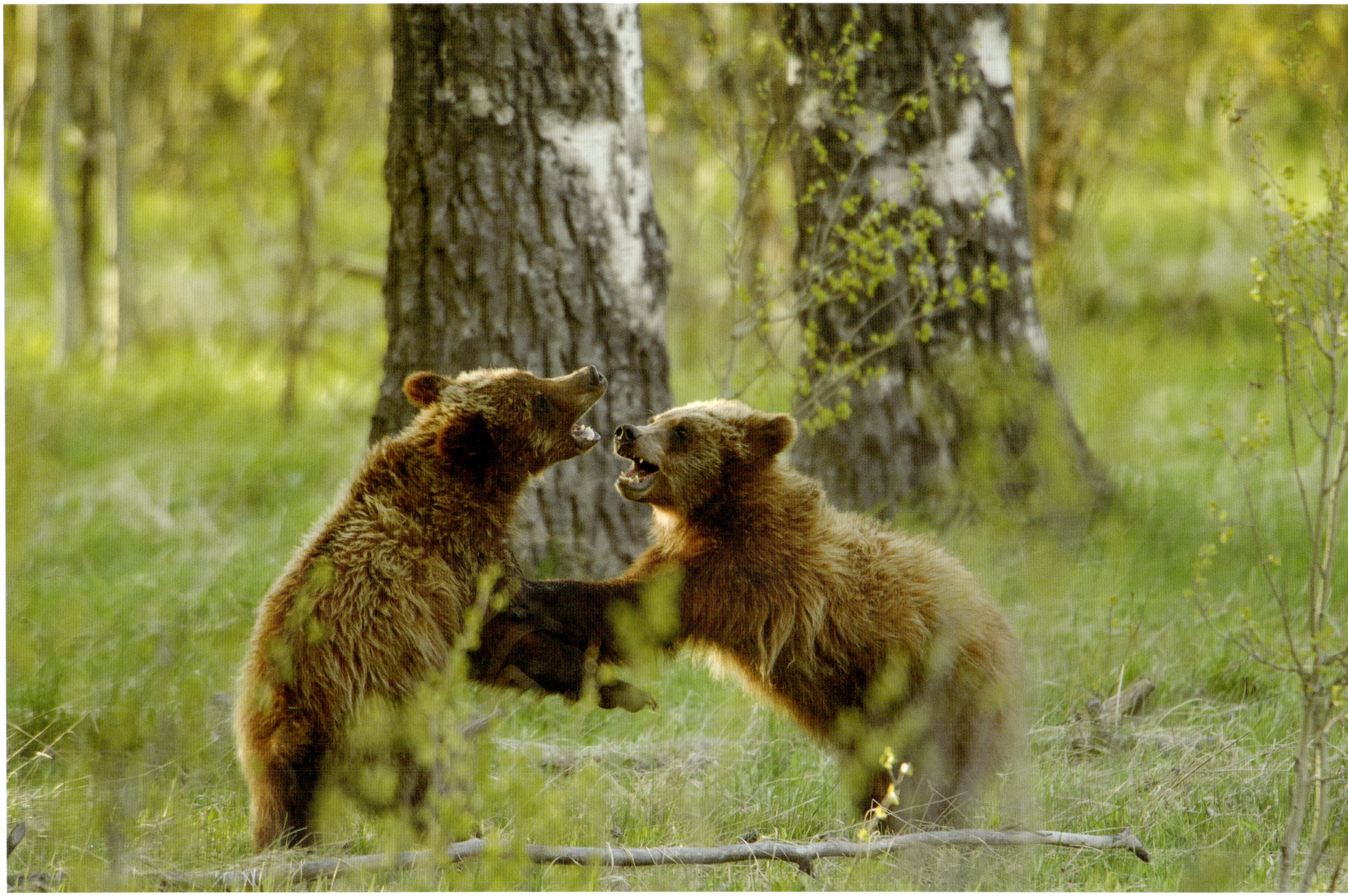

A pair of 610's cubs enjoy the carefree days of youth at Oxbow Bend.

OPPOSITE: 610 and one of her cubs amble between Oxbow Bend, one of the best wildlife-viewing spots in the Lower 48, and Jackson Lake Junction. Her yellow ear tags are marked with GTNP (for Grand Teton National Park).

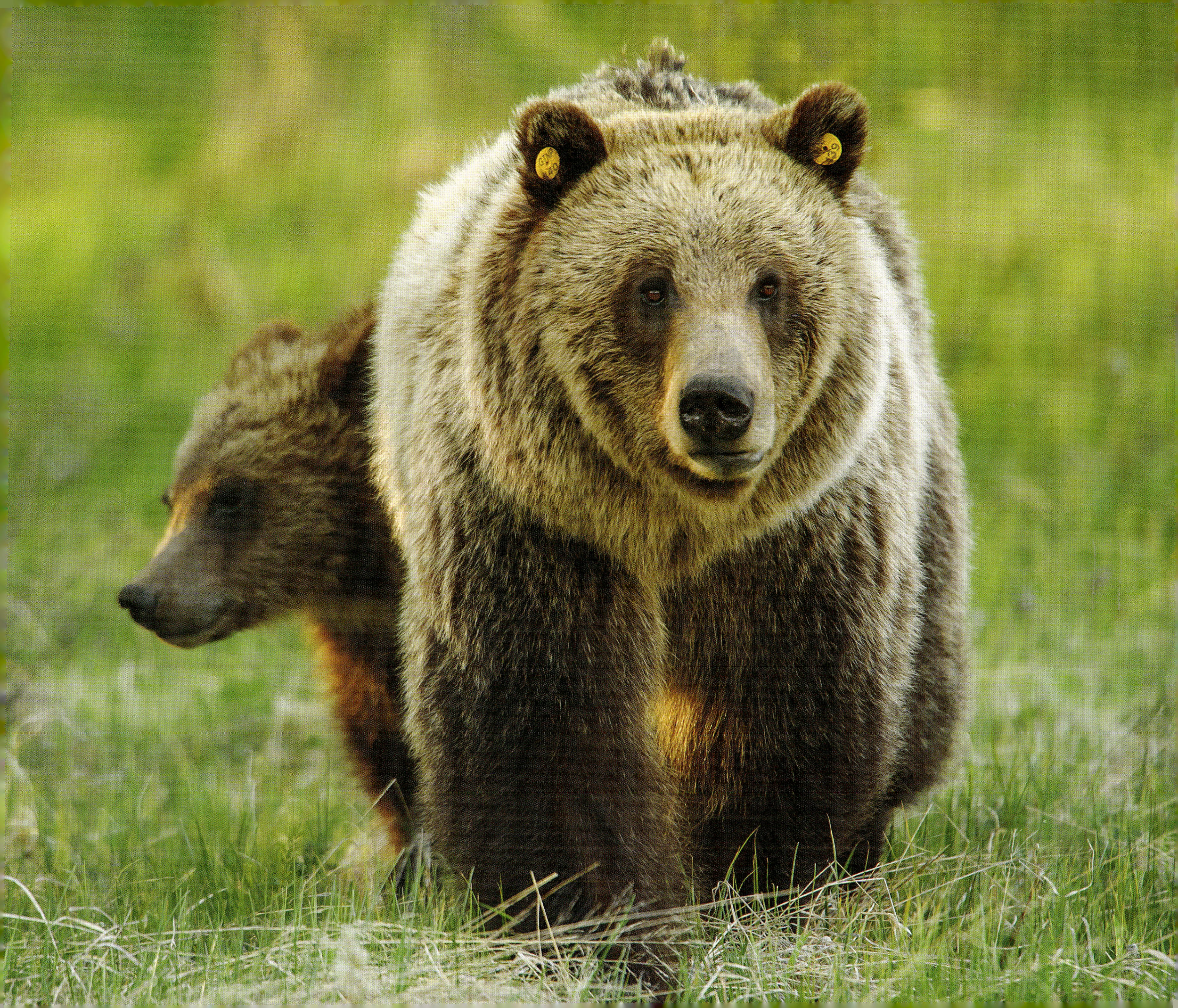

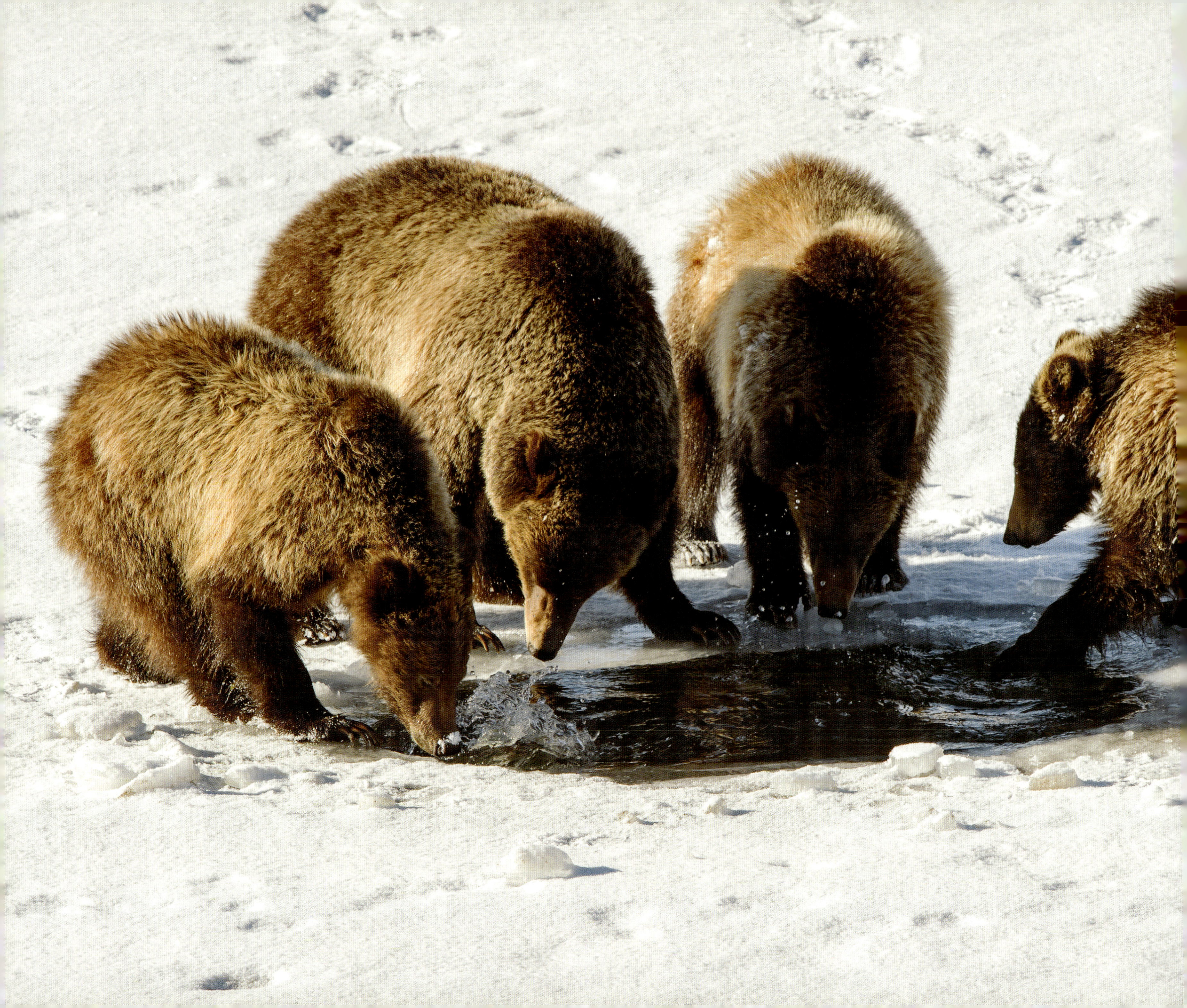

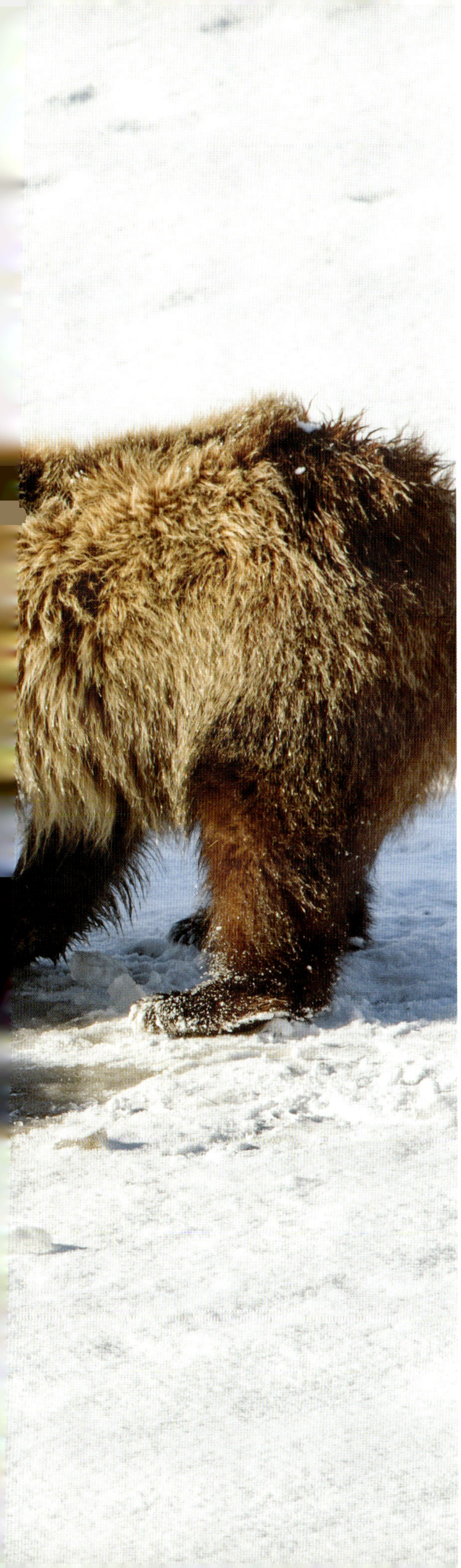

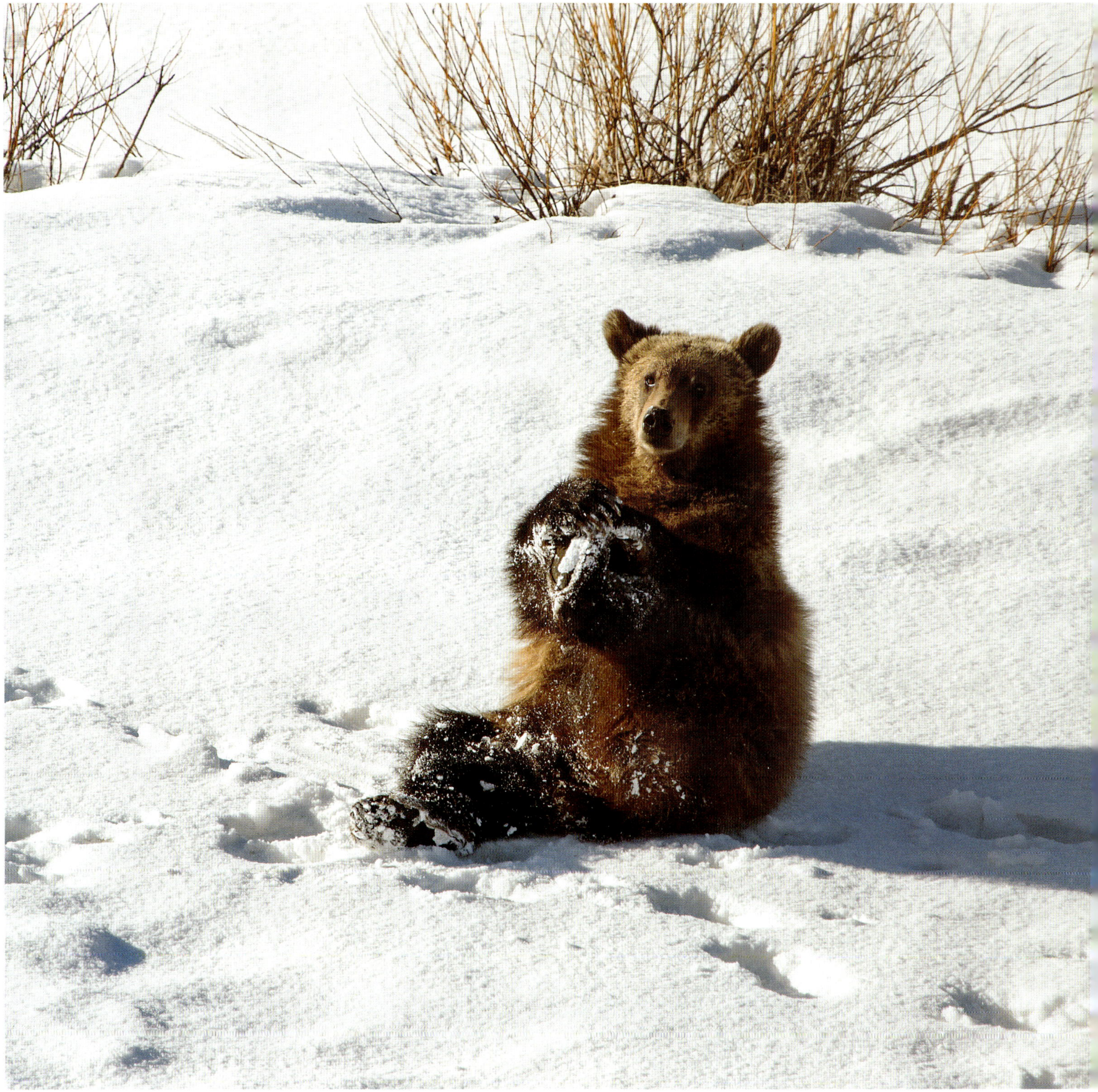

PREVIOUS SPREAD, LEFT: 610 and her cubs paw at the water exposed by this hole in the ice of the Snake River.

PREVIOUS SPREAD, RIGHT: Being limber is a virtue: one of 610's cubs stretches in the snow at Oxbow Bend.

After failing in an attempt to retrieve a beaver carcass and breaking through the thin ice, one of 610's cubs struggles to get a paw hold.

FOLLOWING SPREAD: Fresh out of the den, 610 leads her hungry cubs to Oxbow Bend, a traditional place for early season scavenging. The open waters are a magnet for a plethora of other wildlife, including beavers, otters, muskrat, and a variety of waterfowl, including trumpeter swans, Canada geese, and many ducks.

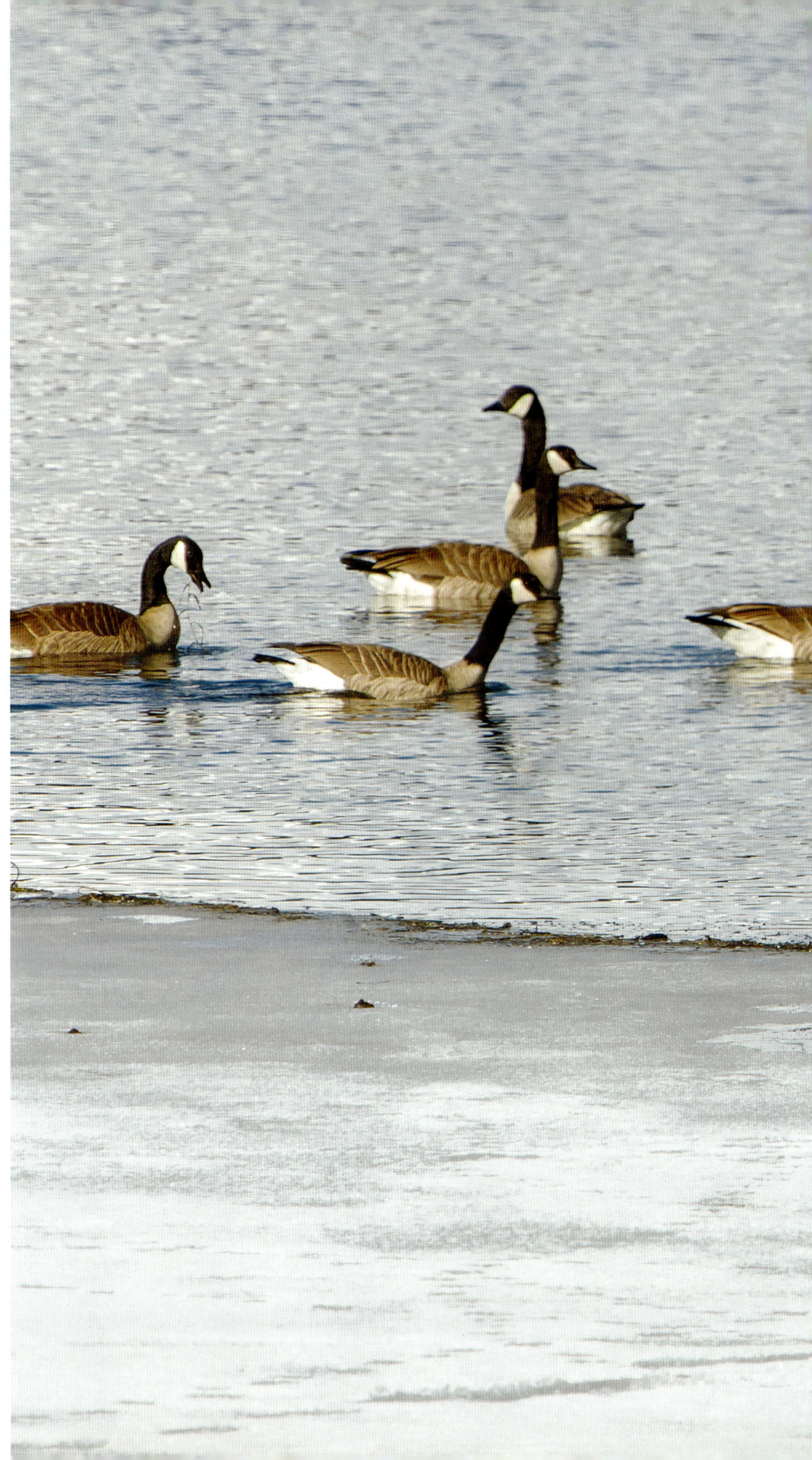

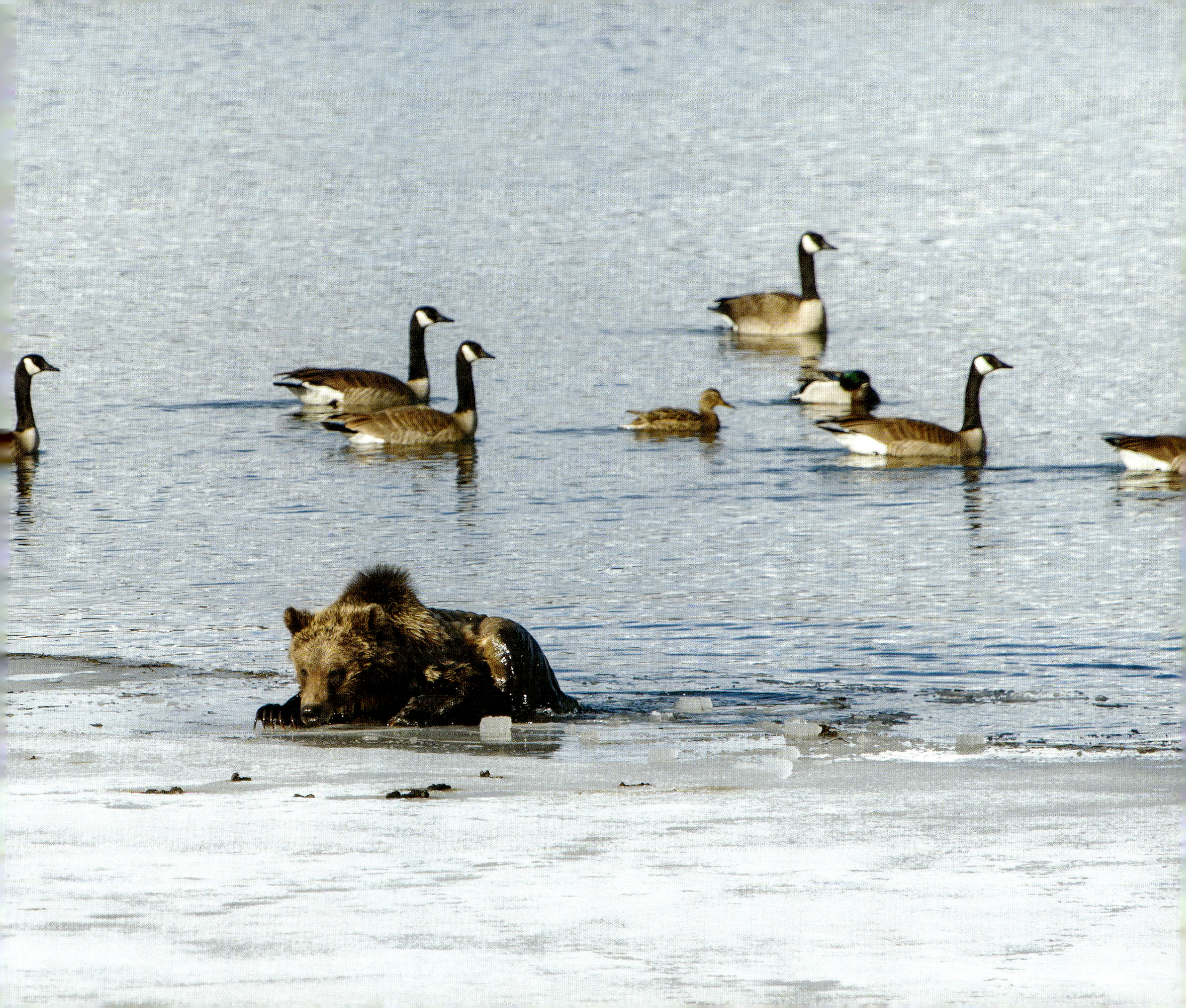

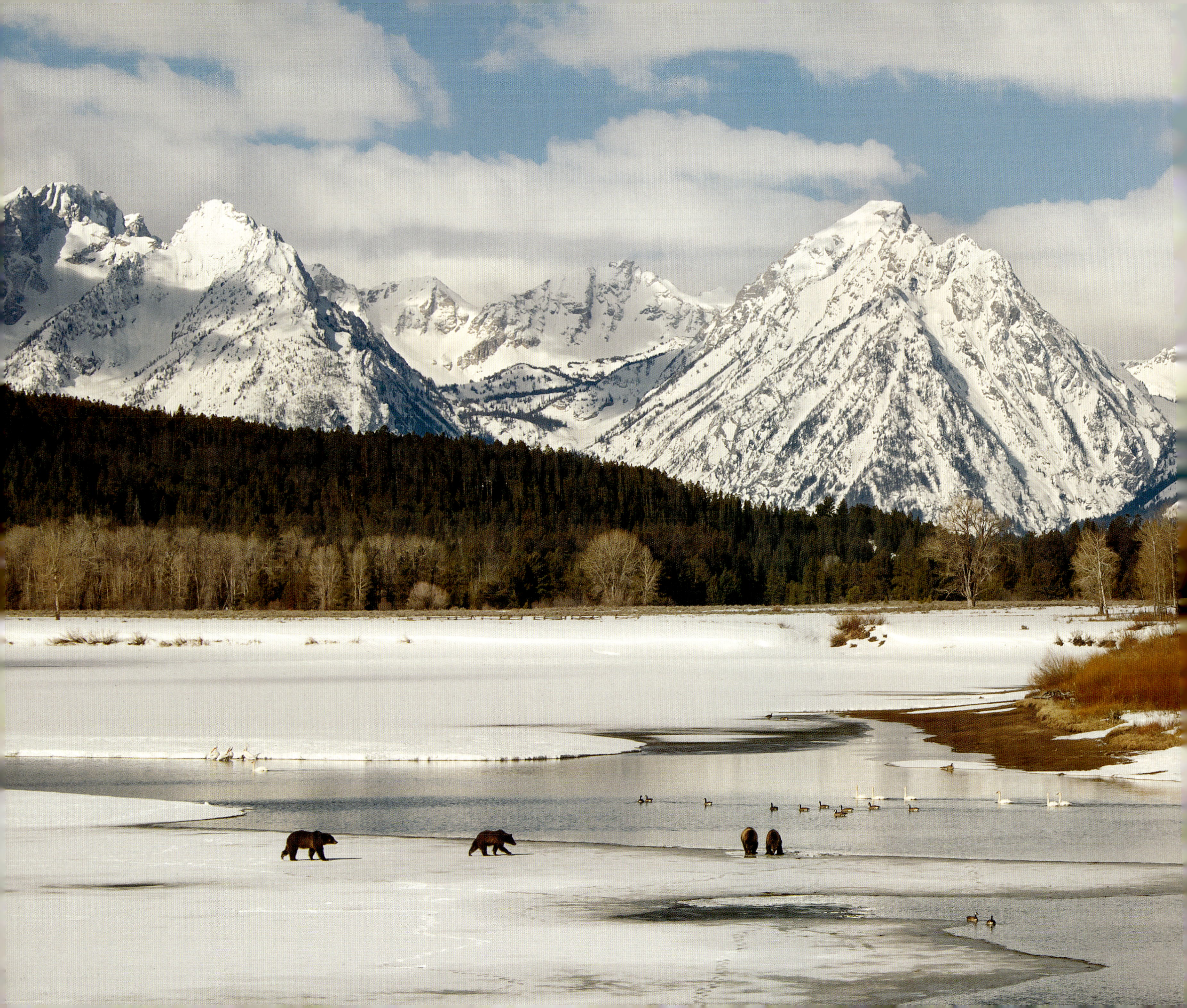

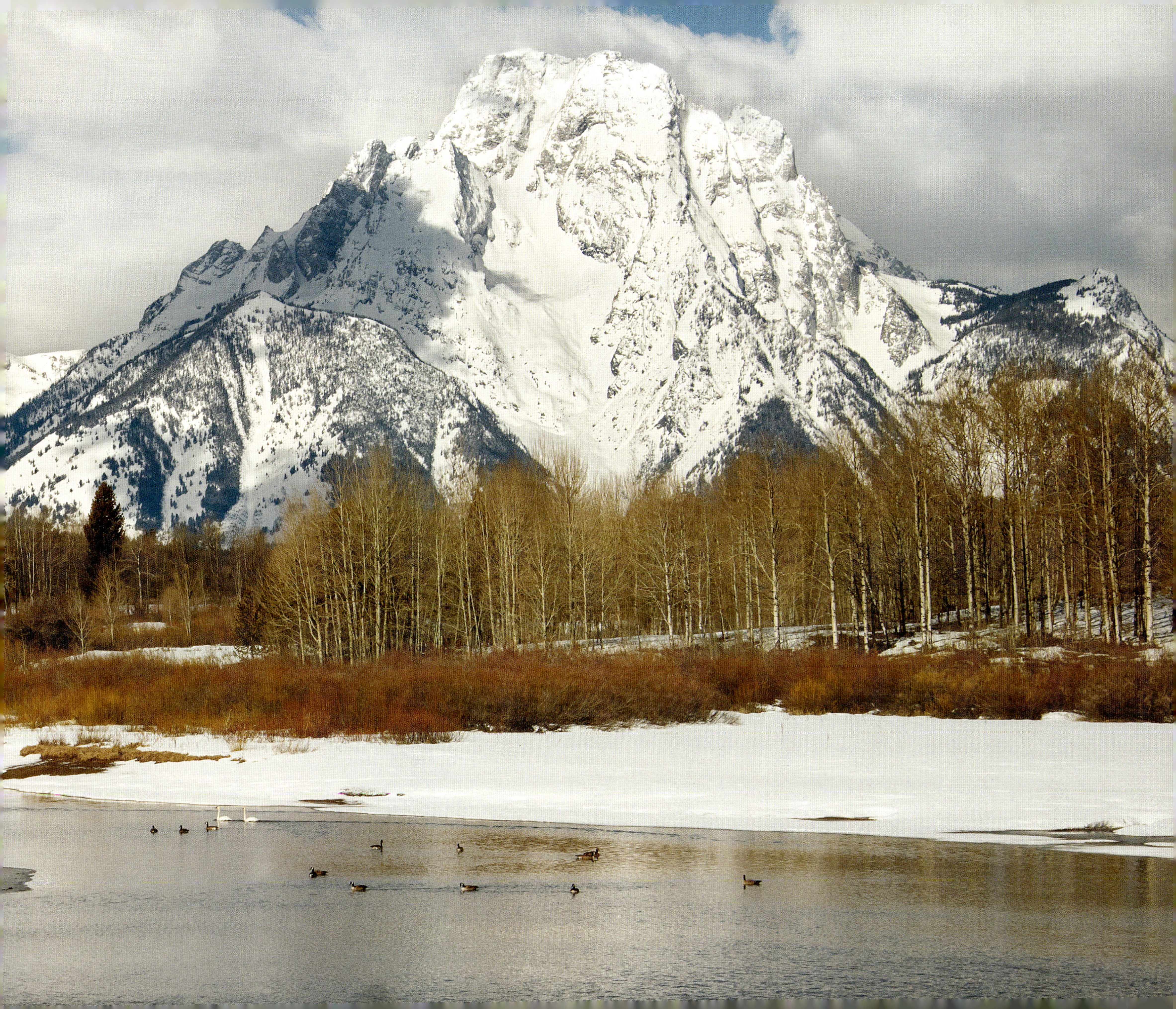

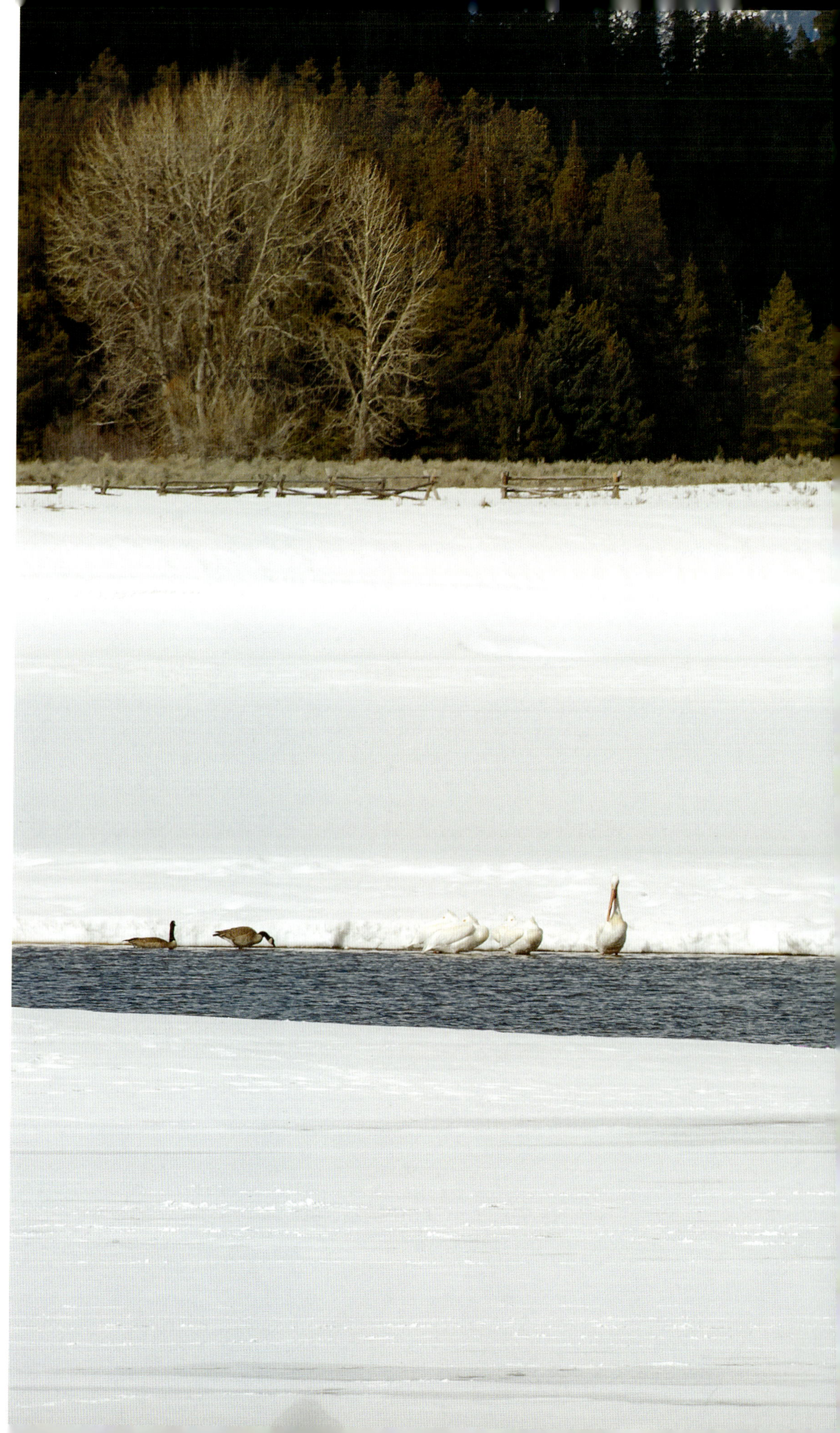

Let mama show you: after her cubs failed in their retrieve, 610 broke through the ice to grab the prize and carried it back to her cubs to share.

FOLLOWING SPREAD: 610 and her brood emerge from the willows along the Snake River, continuing on their never-ending search for food.

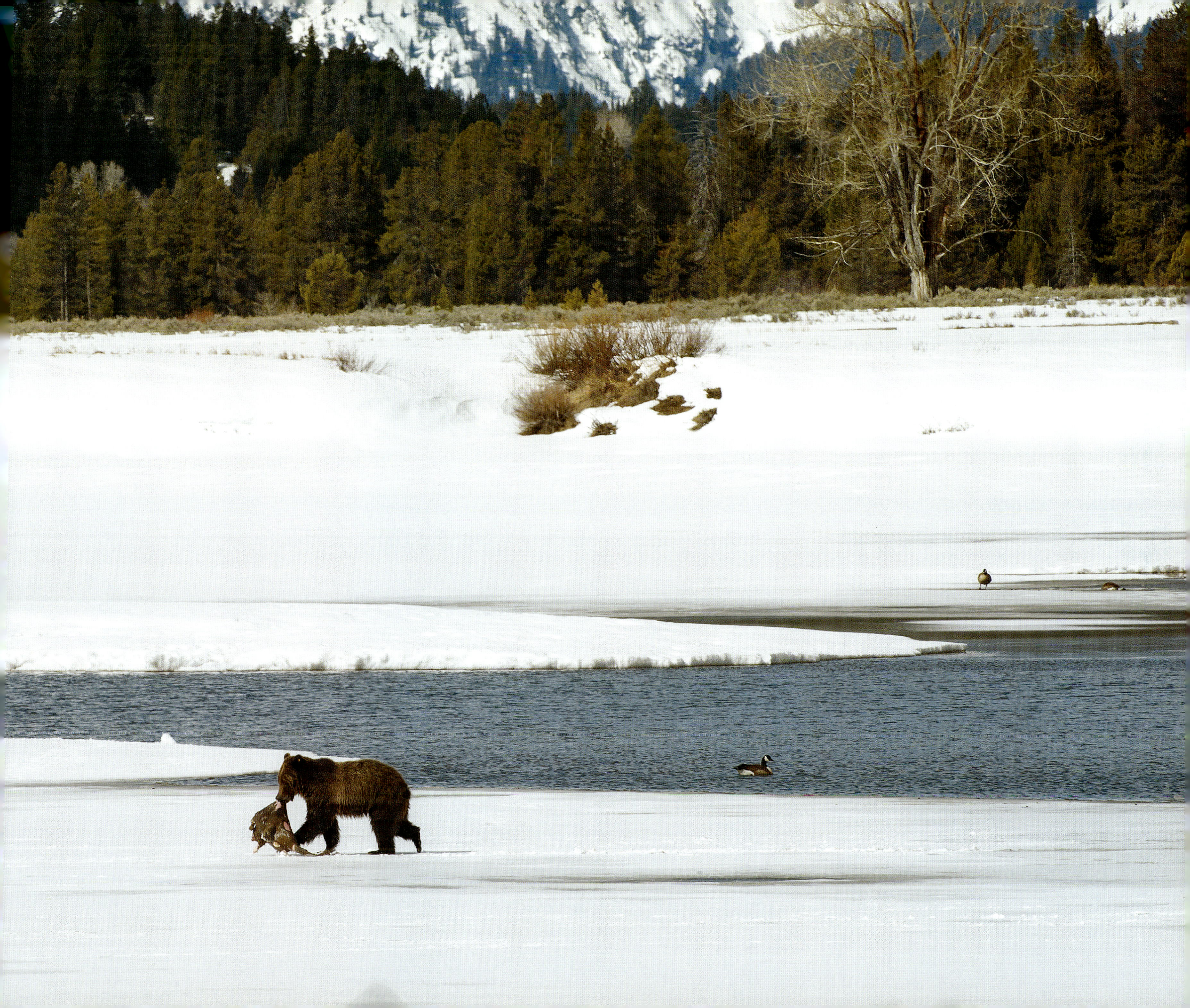

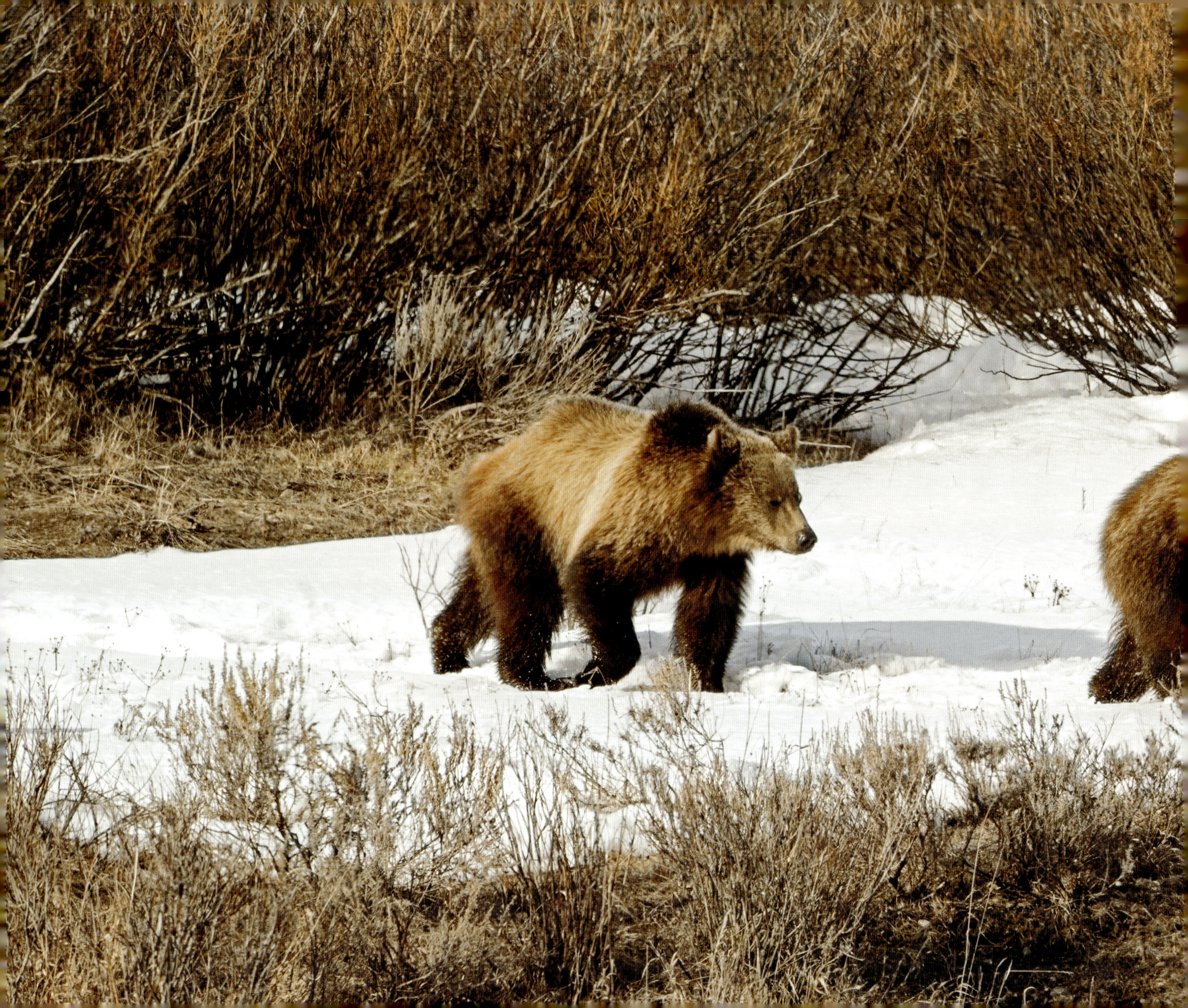

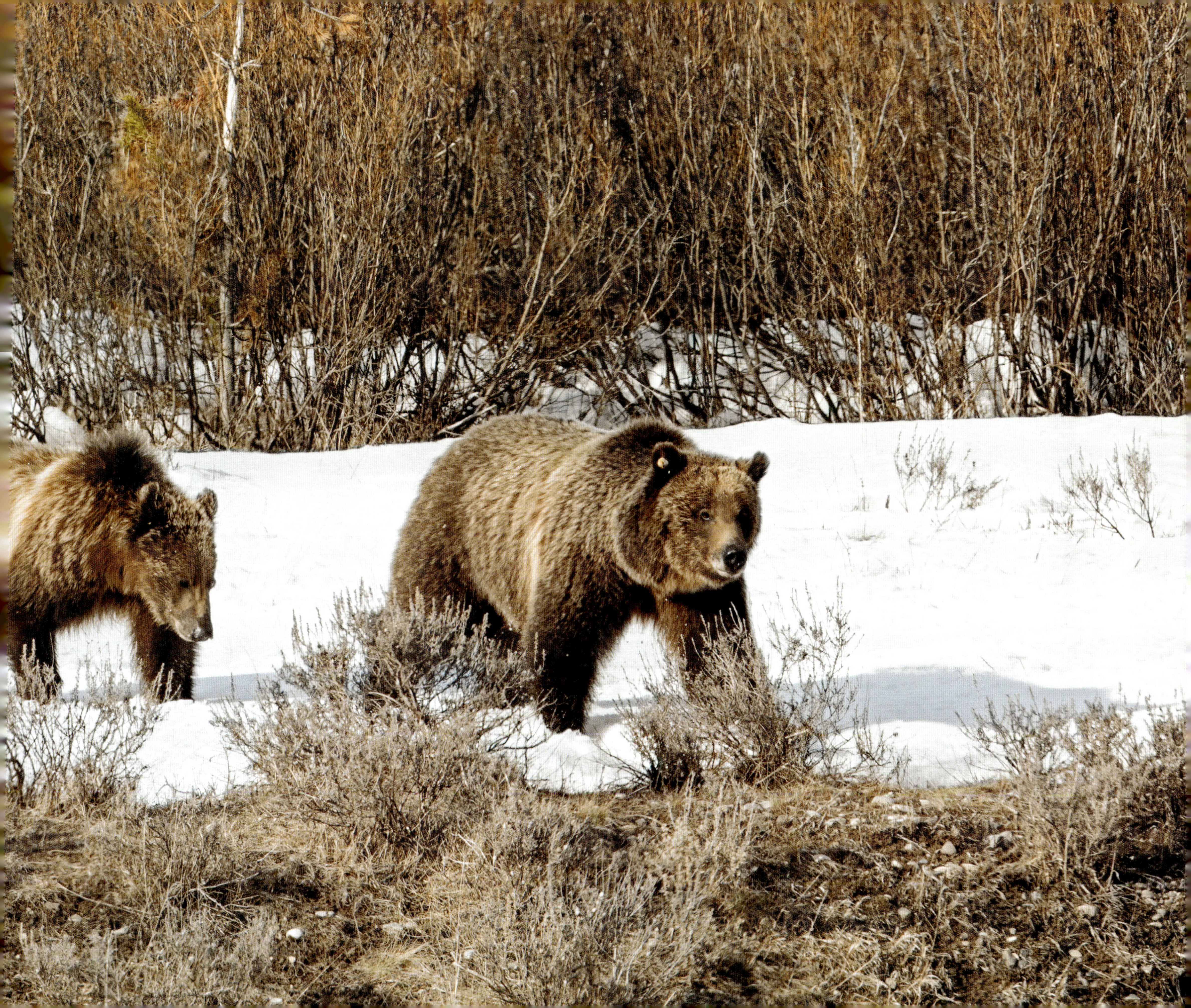

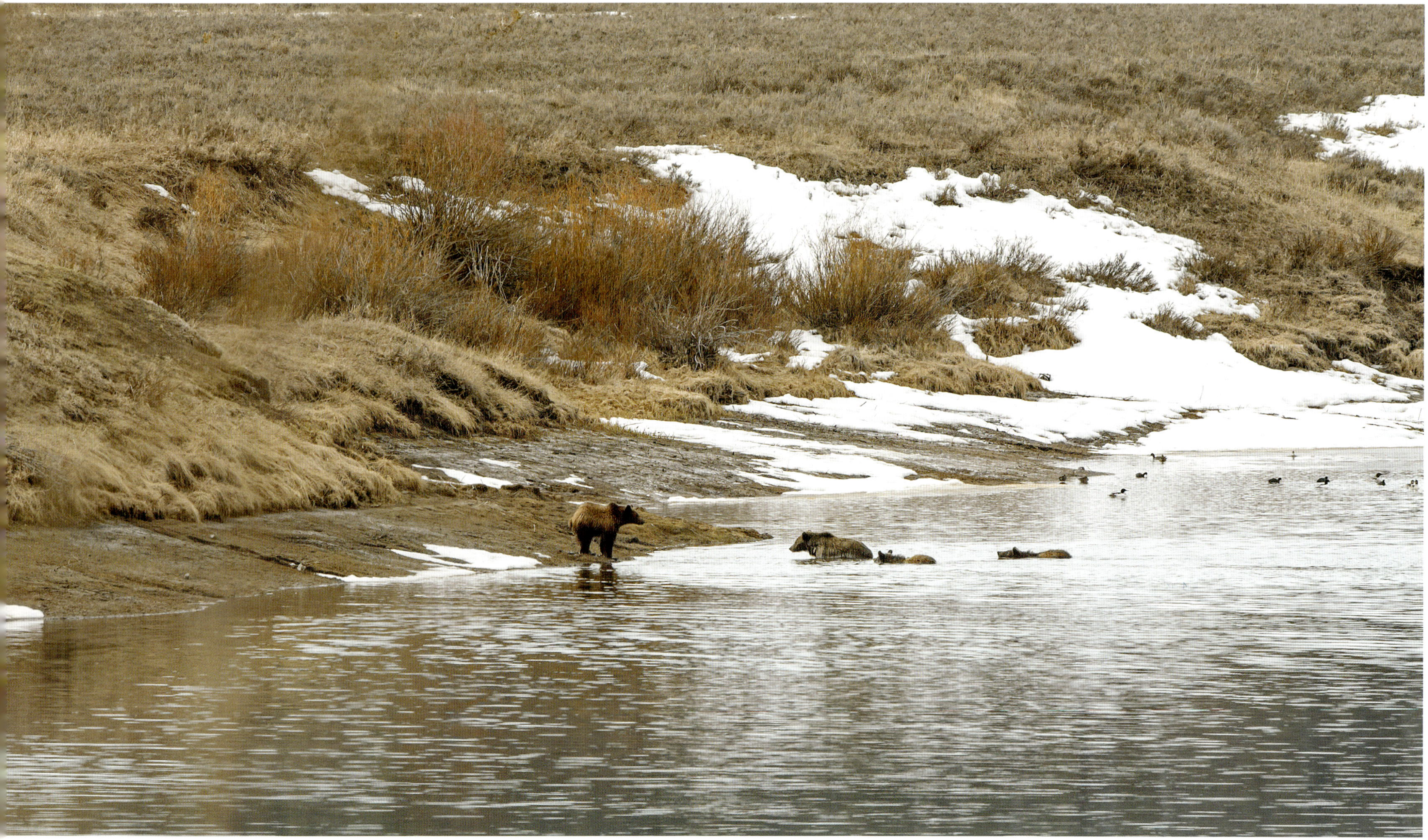

One of 610's juvenile cubs waits on the bank of Oxbow Bend for the rest of the family to finish crossing the river.

OPPOSITE: 399 sits on a hillside in the Teton Wilderness after nursing her third set of triplets. Small and vulnerable, her tiny cubs will soon experience life in the lower elevations and all that comes with being a roadside bear.

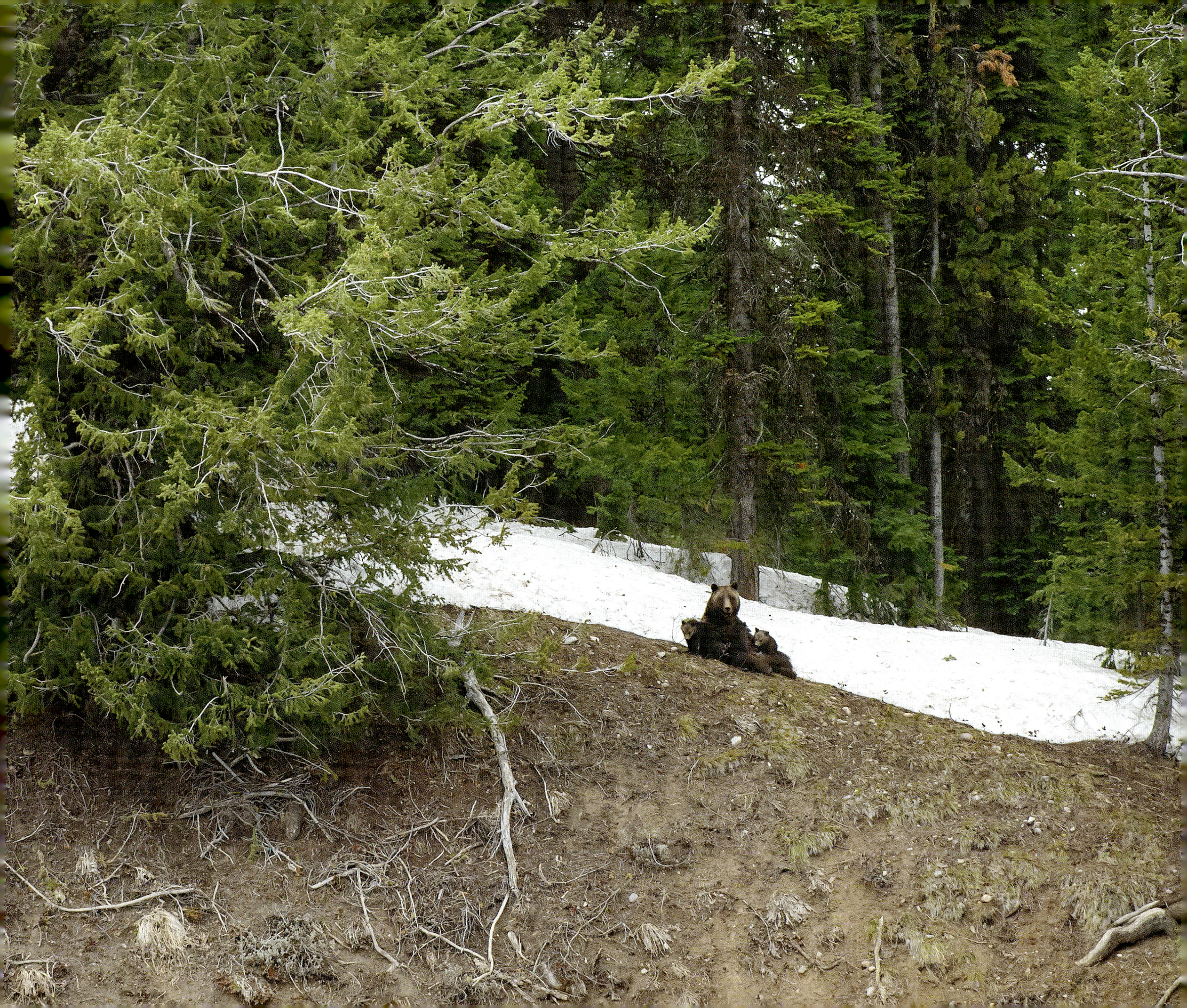

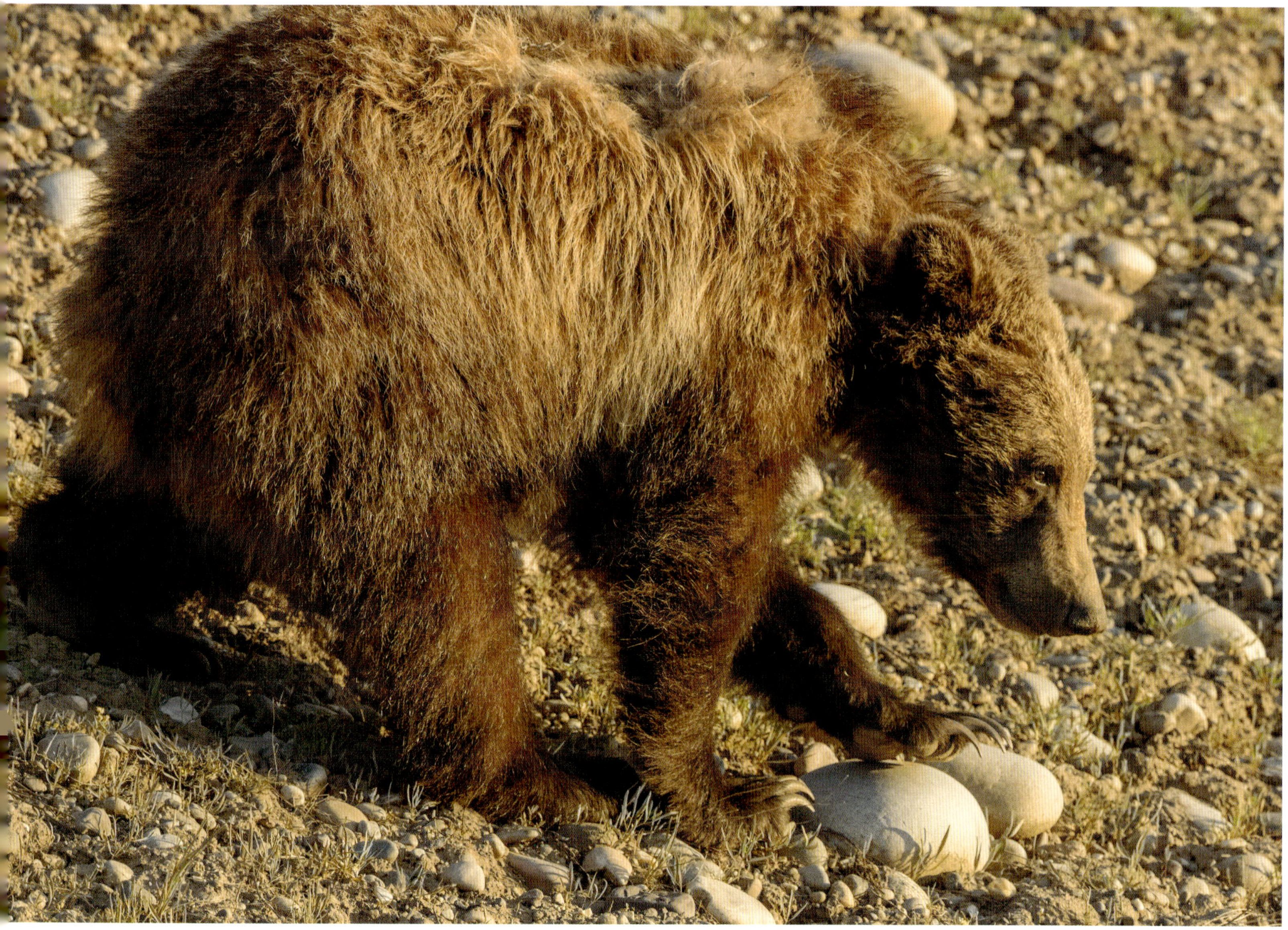

By autumn, the cubs will separate, beginning their lives as independent bears. Although they have been taught much by their mother, the next few years will be crucial to their survival into adulthood.

OPPOSITE: 610's cubs remained together soon after their mother pushed them out on their own. Here, they stay close while feeding on fresh grasses along the hillside near Windy Point.

FOLLOWING SPREAD, LEFT: 399's new triplet cubs stay close to one another in a meadow near Pilgrim Creek.

FOLLOWING SPREAD, RIGHT: Much of a young grizzly's life is spent playing with siblings, climbing trees, and exploring.

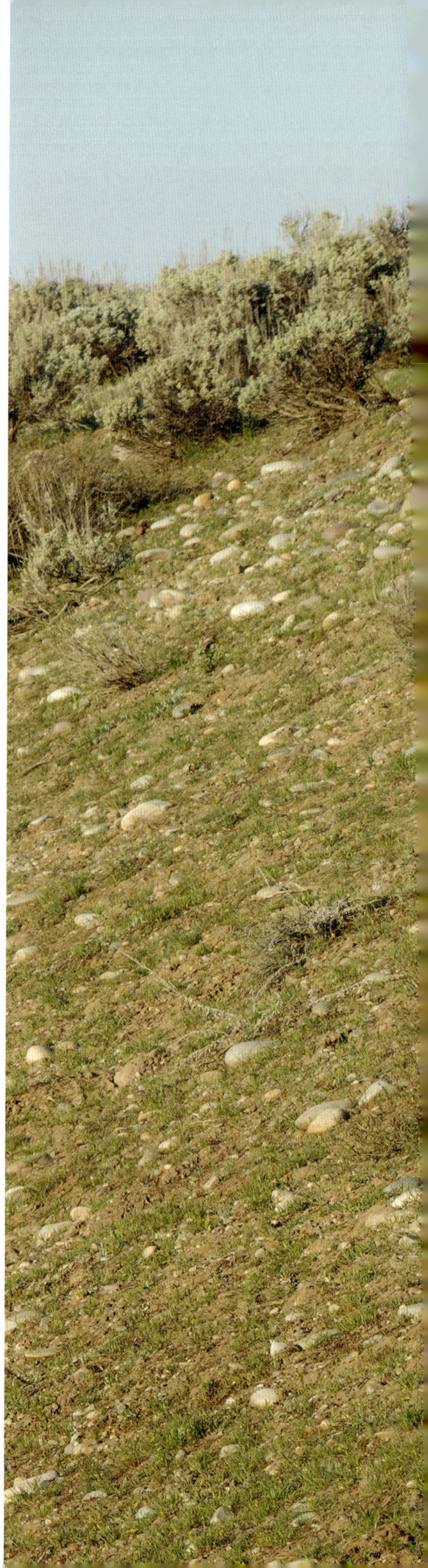

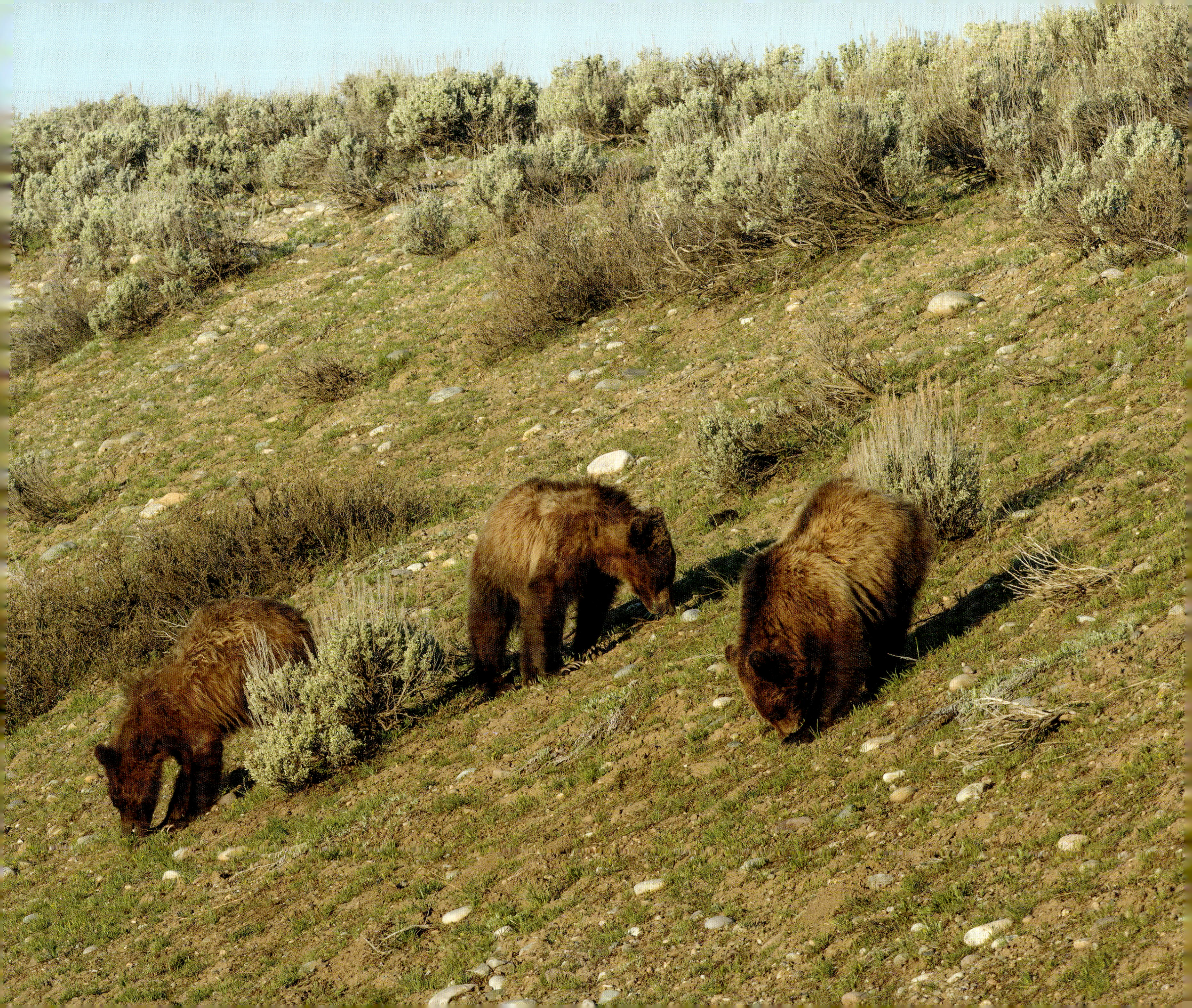

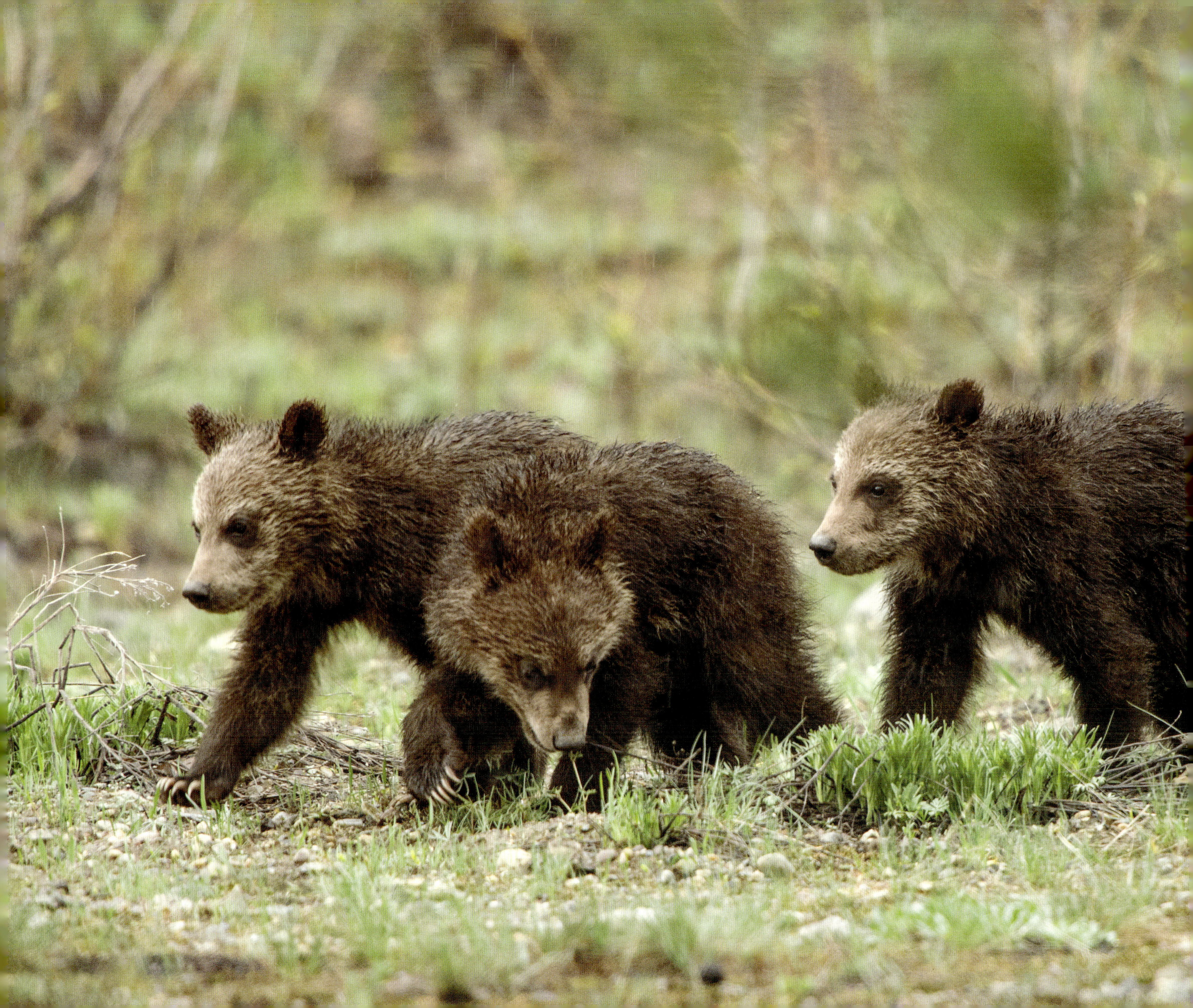

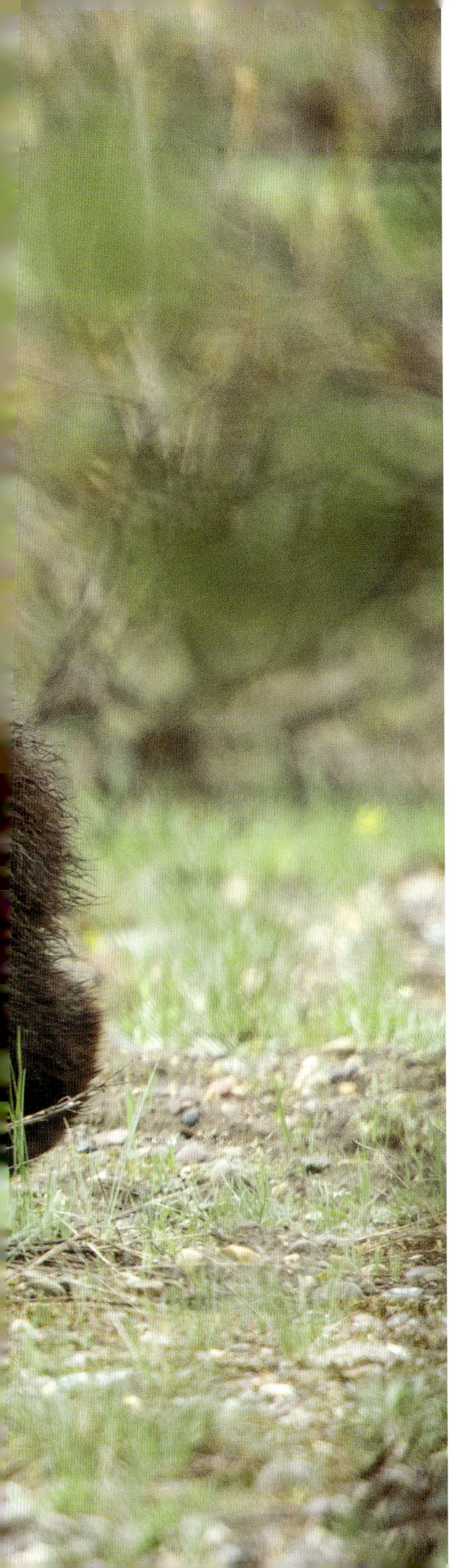
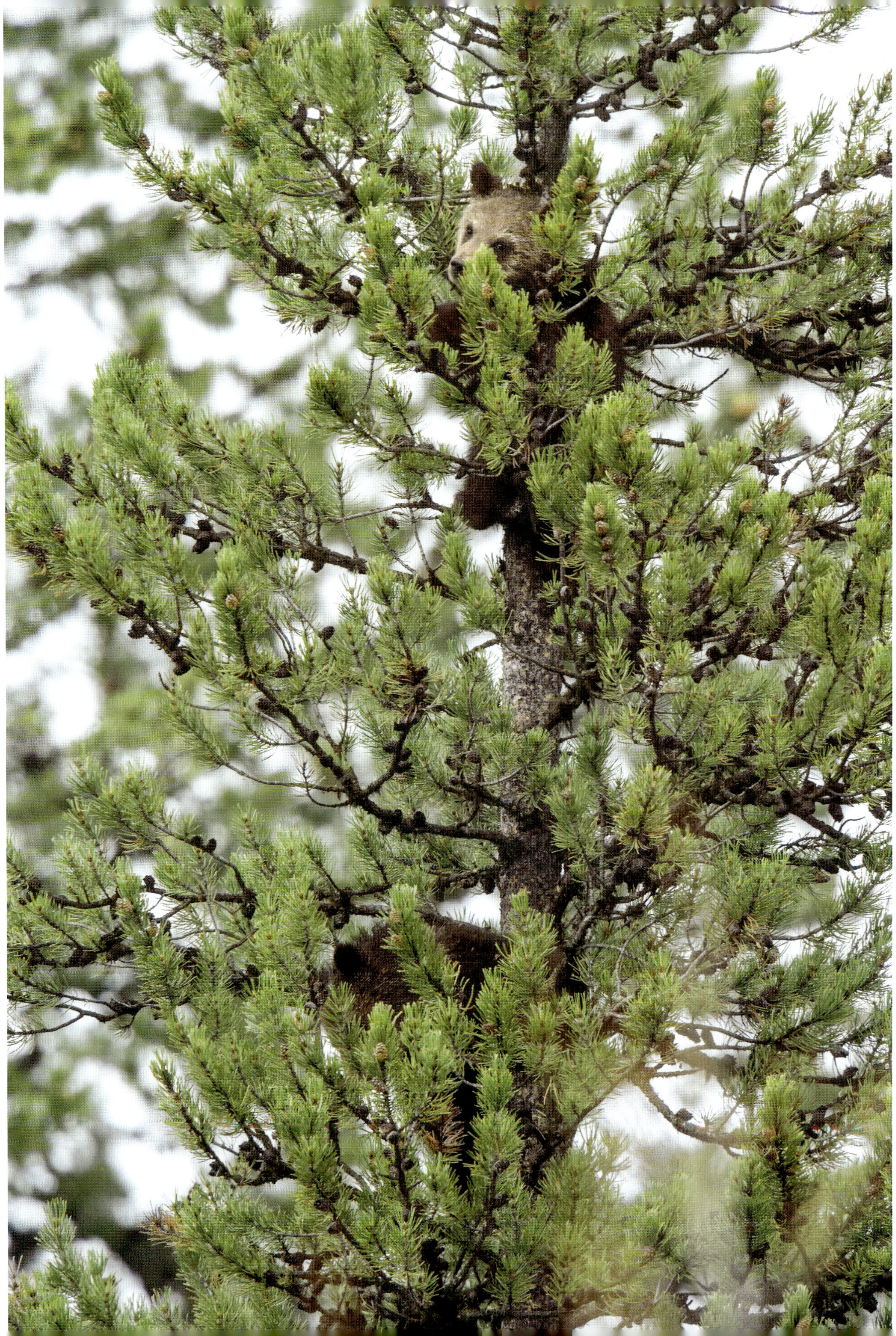

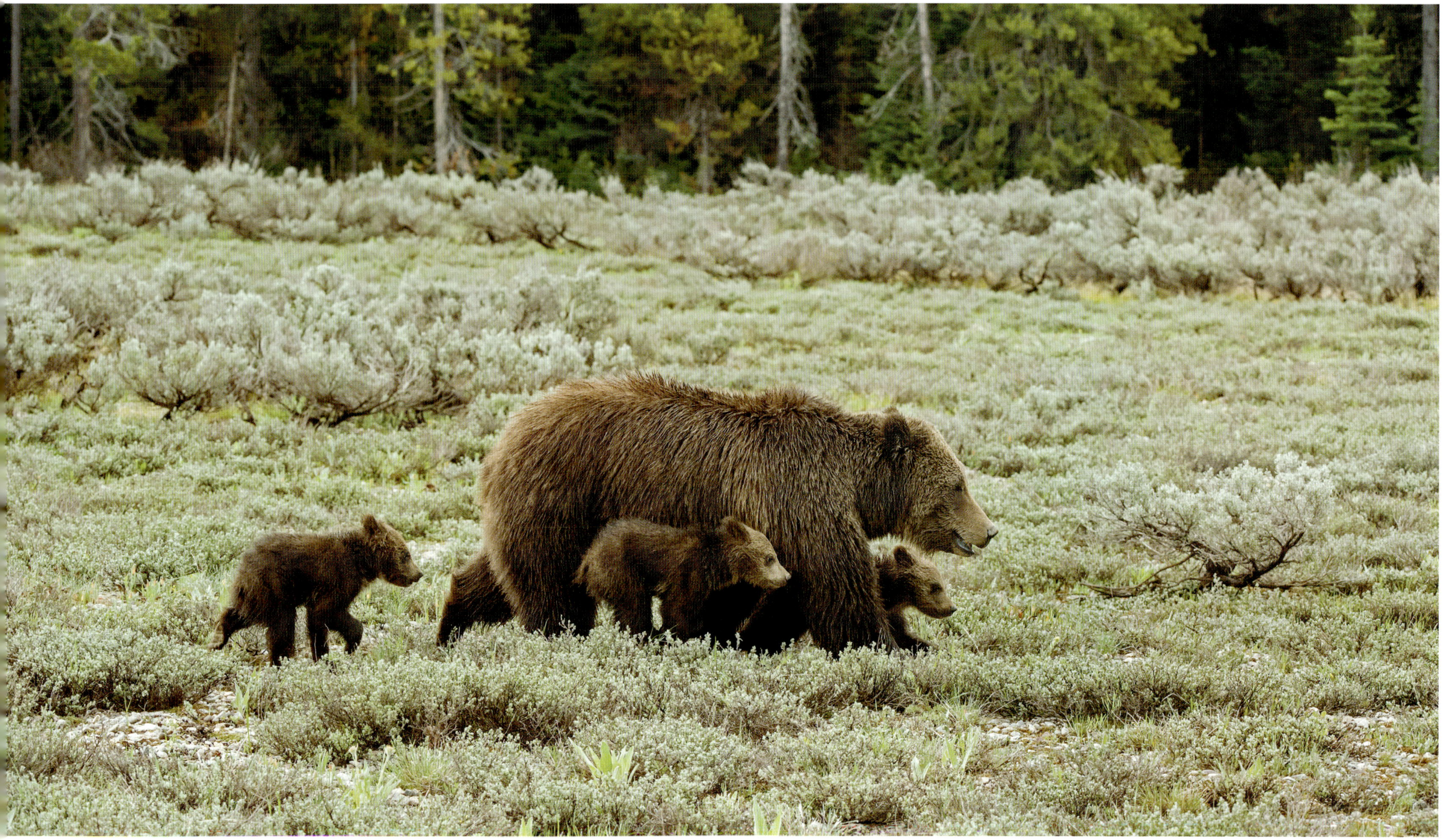

399's cubs stay close to mom while exploring the entrance to their neonatal birthplace, the winding drainage of Pilgrim Creek.

OPPOSITE: Bear necessities: new adventures await at every turn in Pilgrim Creek, connecting the high country with the Snake, a major headwater to the mighty Columbia River.

FOLLOWING SPREAD: After taking a drink, 399 surveys the far shore of Colter Bay while her three cubs wander the shoreline prowling for fun.

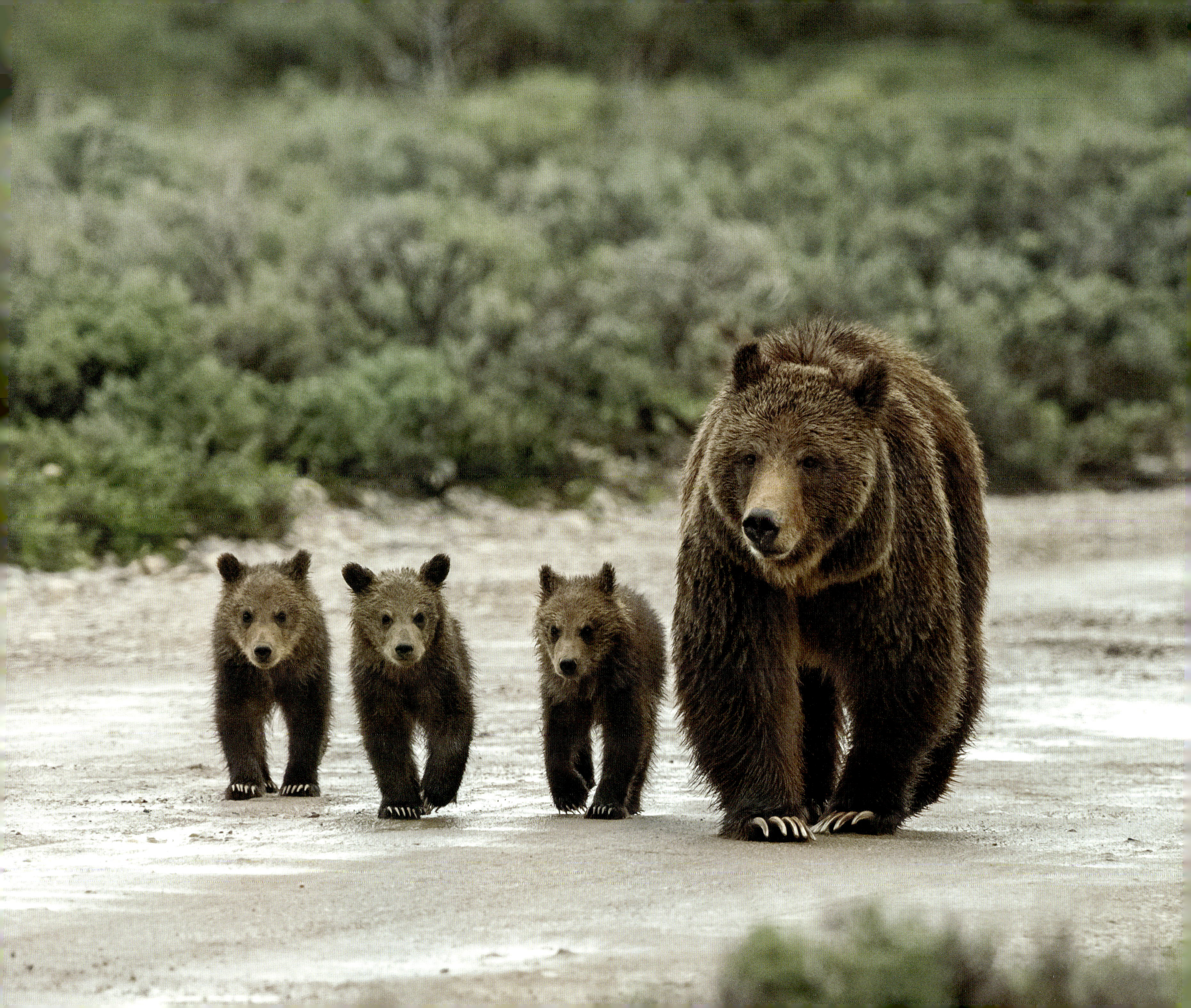

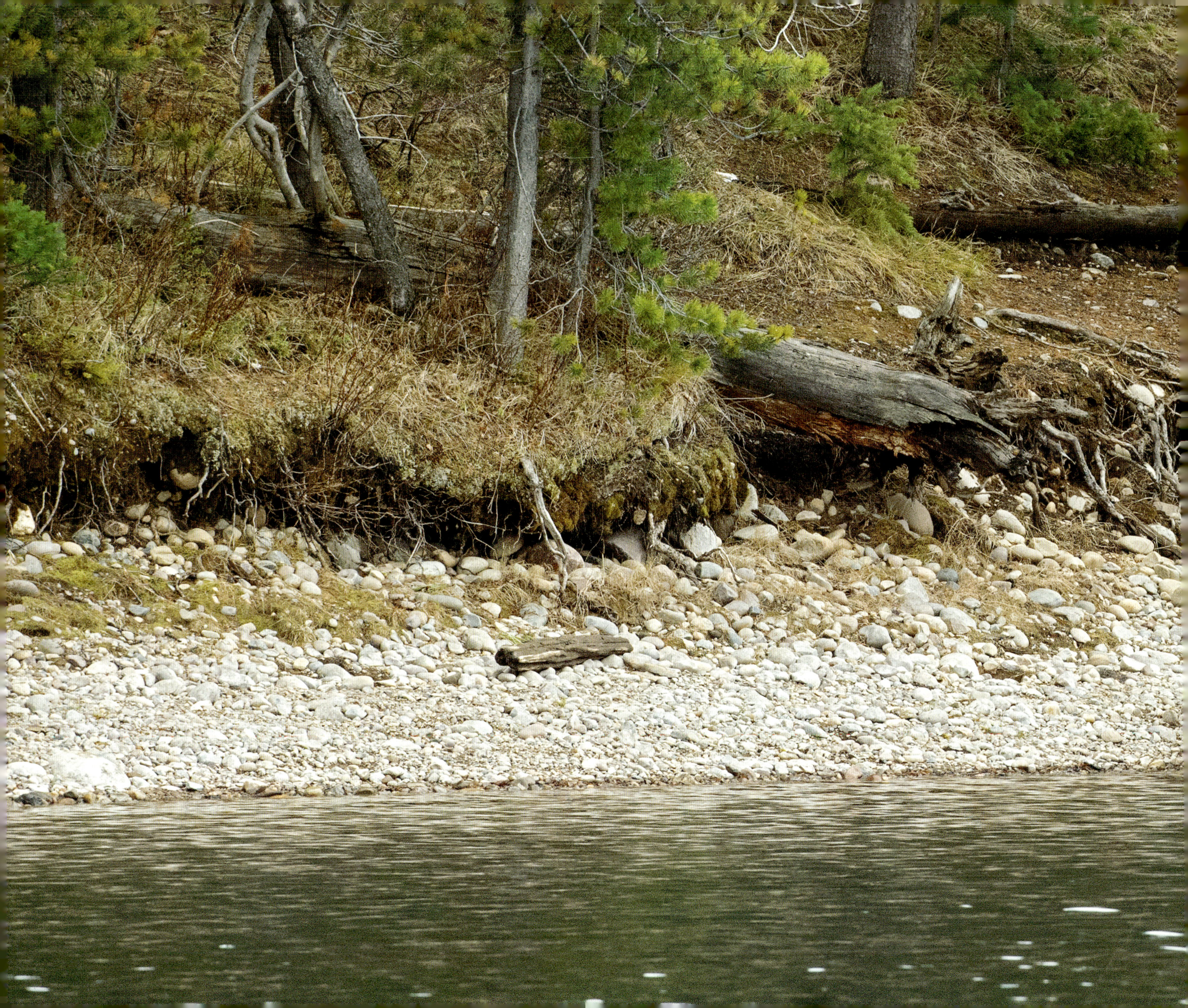

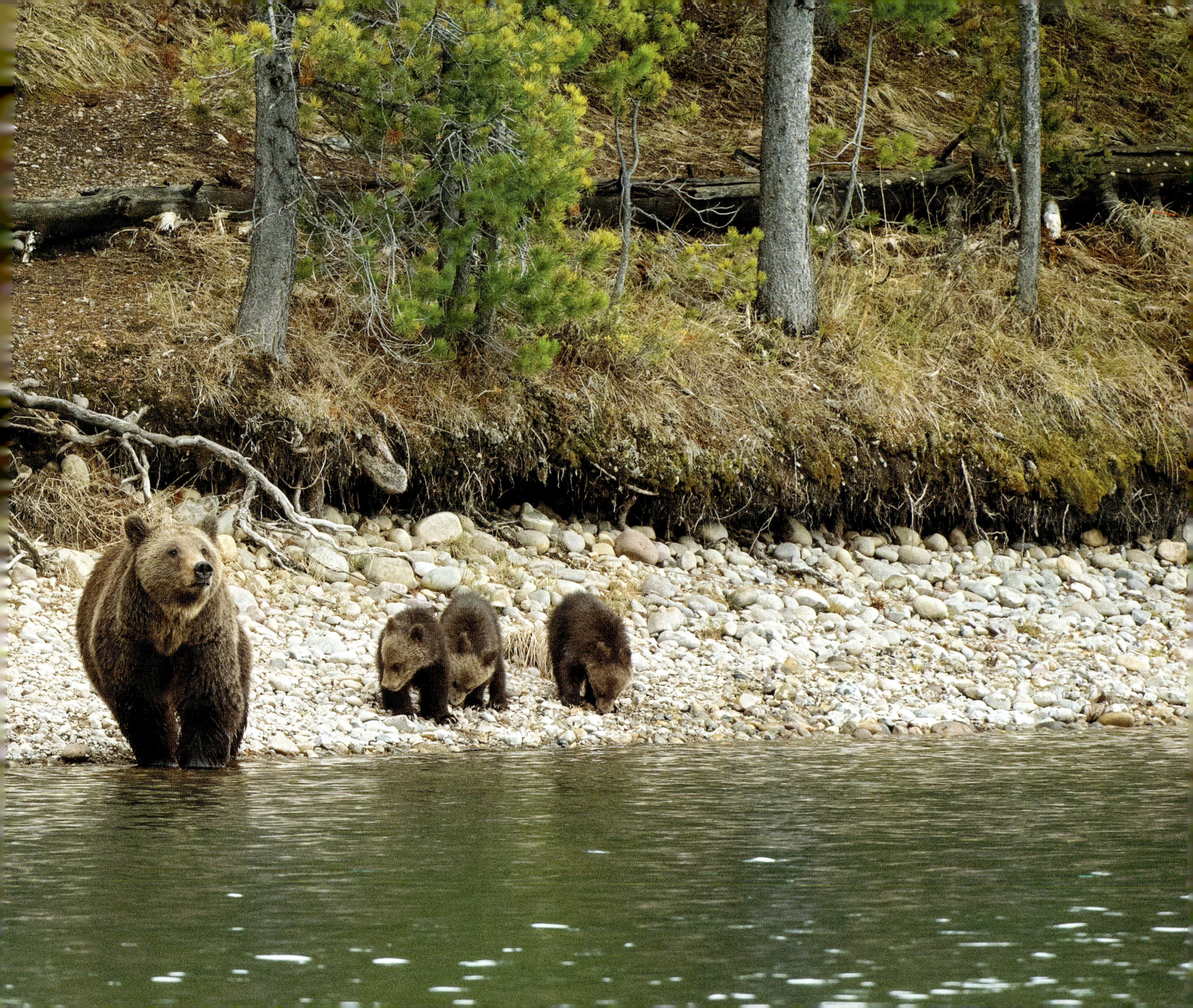

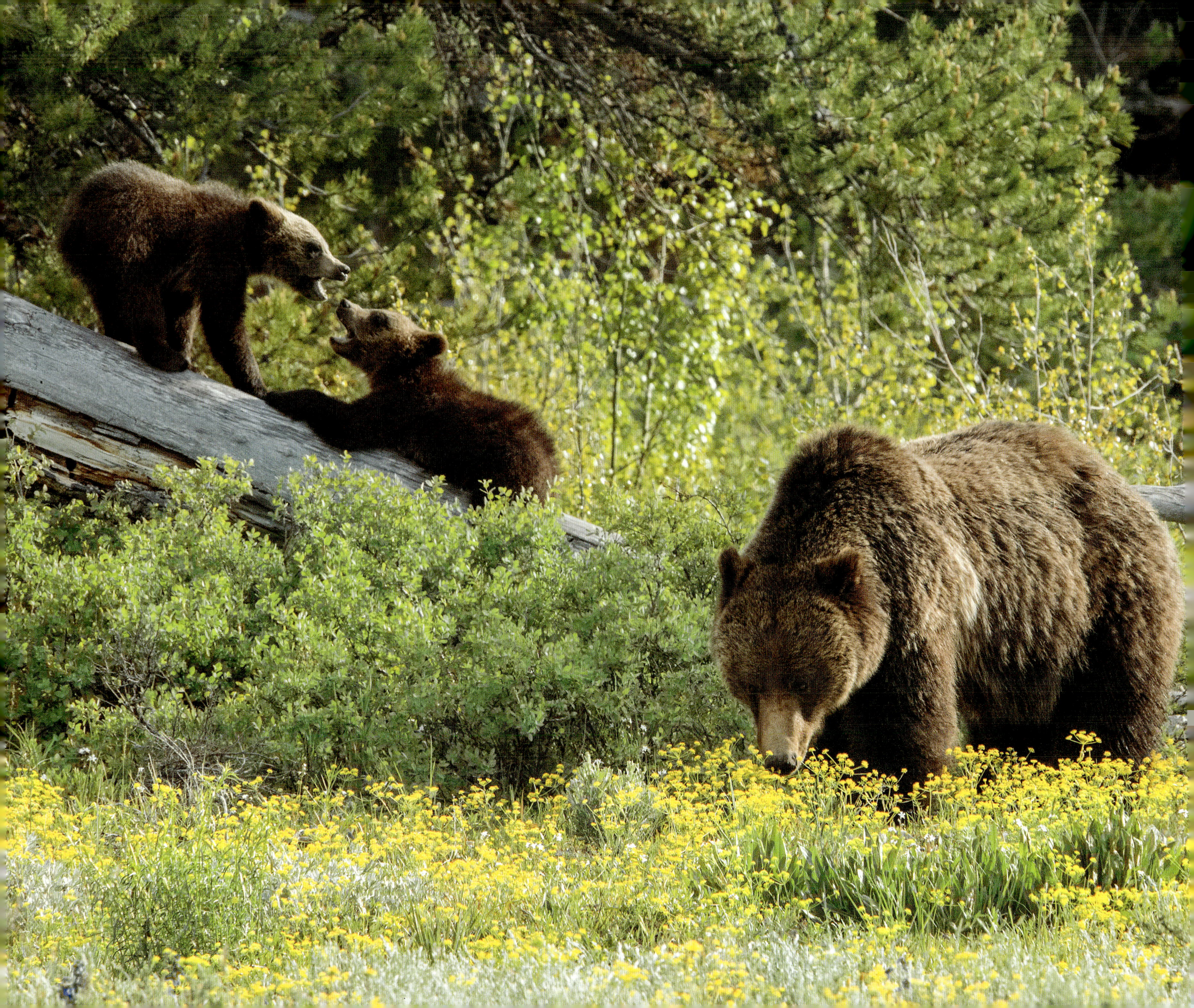

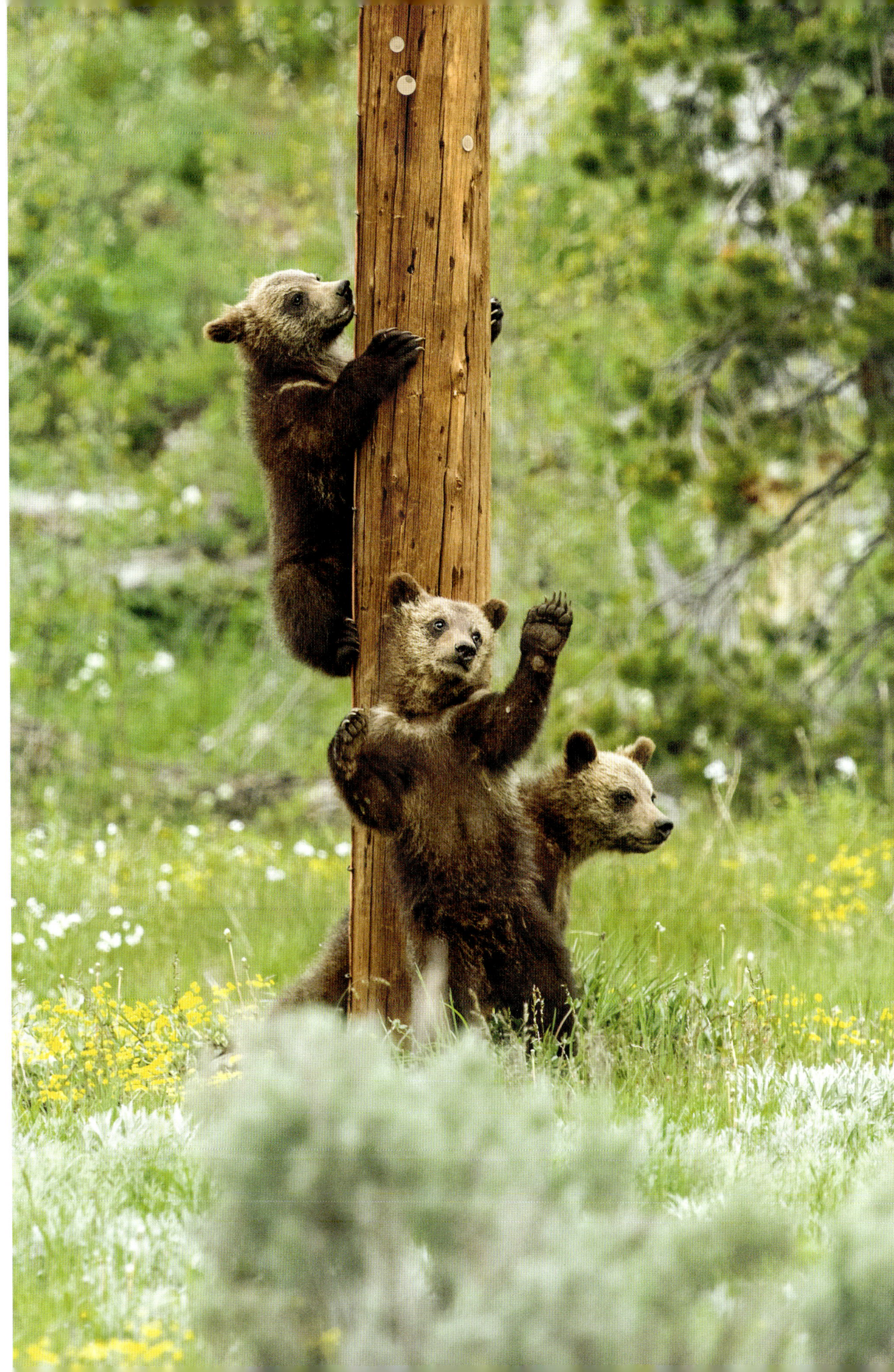

OPPOSITE: 399 gets her roughage while dining on biscuitroot as two cubs play on a fallen tree behind her near Pilgrim Creek.

Following their mother's lead, 399's triplets climb and scratch against a utility pole, leaving behind their scent.

FOLLOWING SPREAD: Of all the wildlife in his portfolio—hundreds of species encompassing millions of images—Mangelsen has felt the strongest connection to 399 and her family. "We all love what we know and I've spent more time in the field watching them than any other animals," he says.

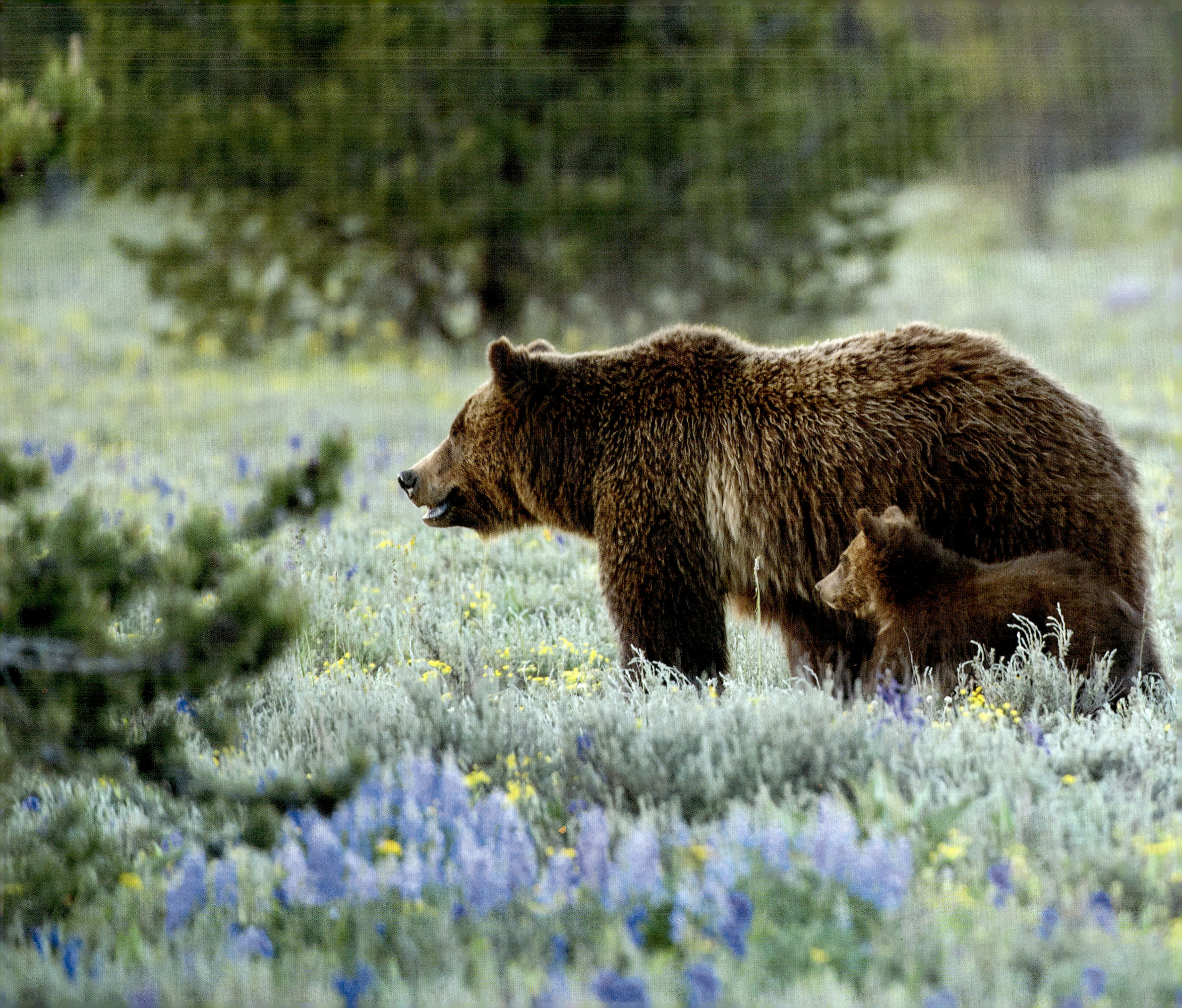

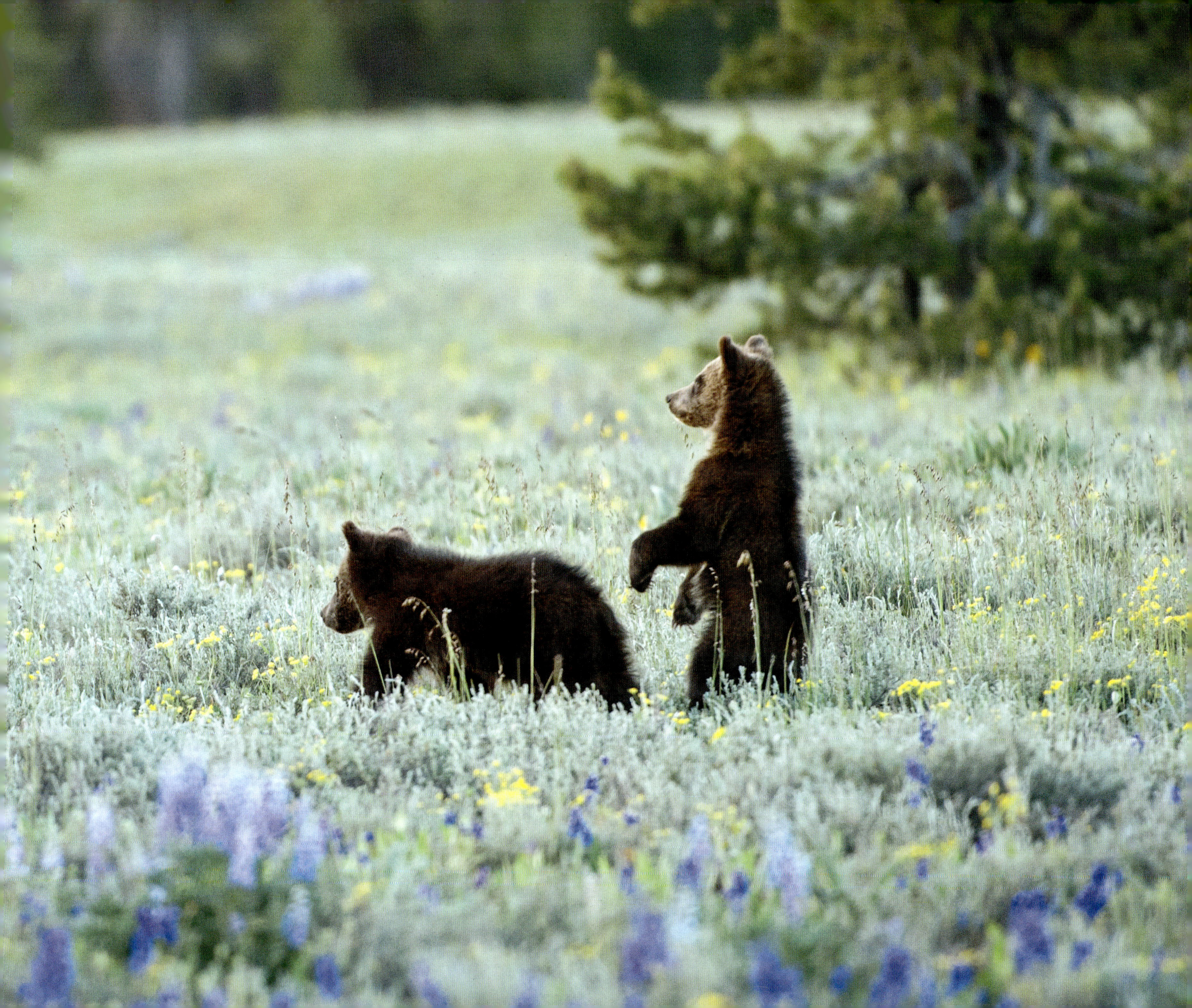

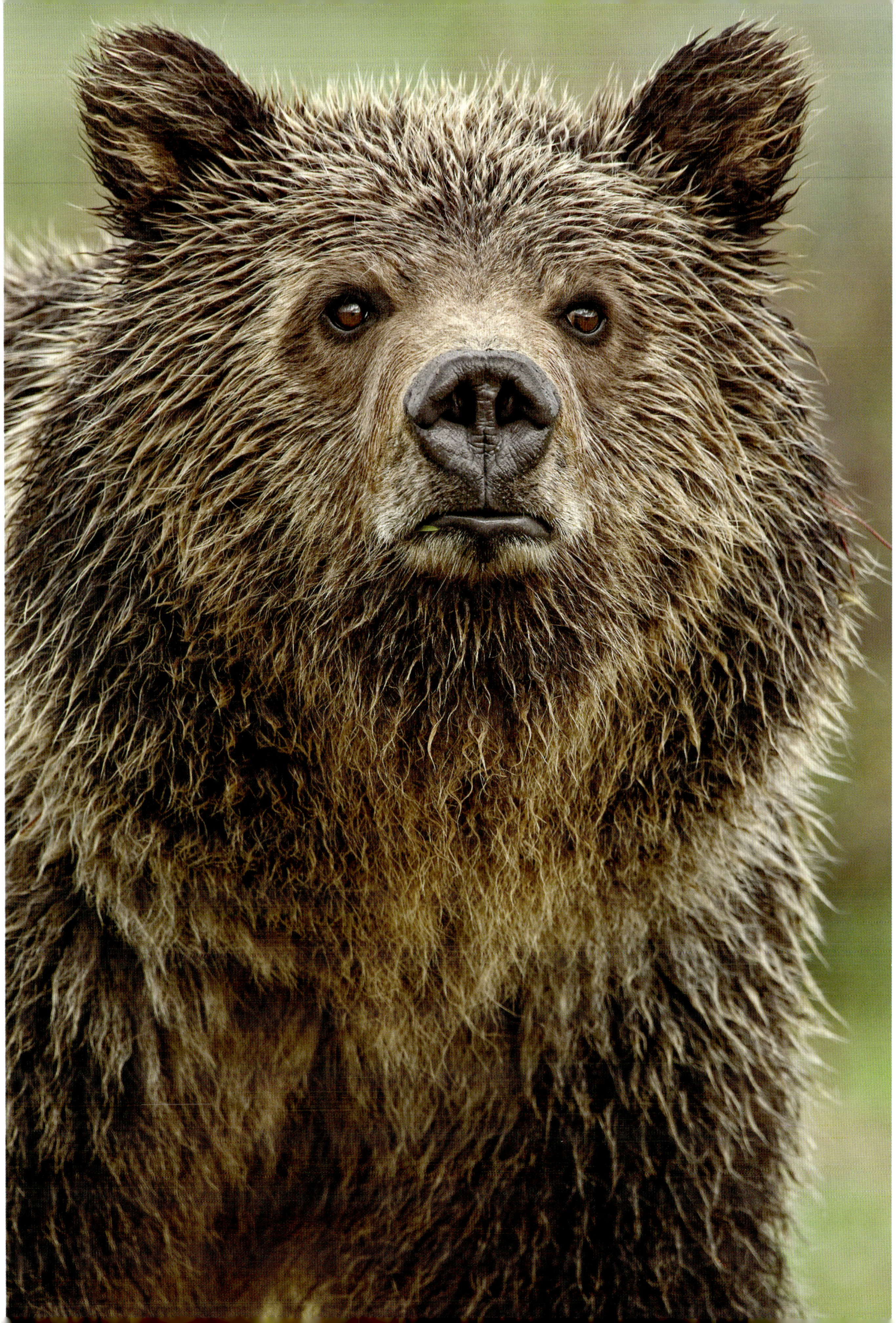

With a small blade of grass in the corner of her mouth, this young grizzly, nicknamed Blondie, takes a break from being a vegetarian to survey the meadow along Pilgrim Creek.

OPPOSITE: In midsummer, one of 610's subadult cubs wades through shoulder-high grasses and wildflowers near Signal Mountain.

FOLLOWING SPREAD: Always alert, 399 looks up to make sure her cubs aren't wandering far from their wildflower meadow play area near Oxbow Bend. Among the predators of young grizzlies are wolves.

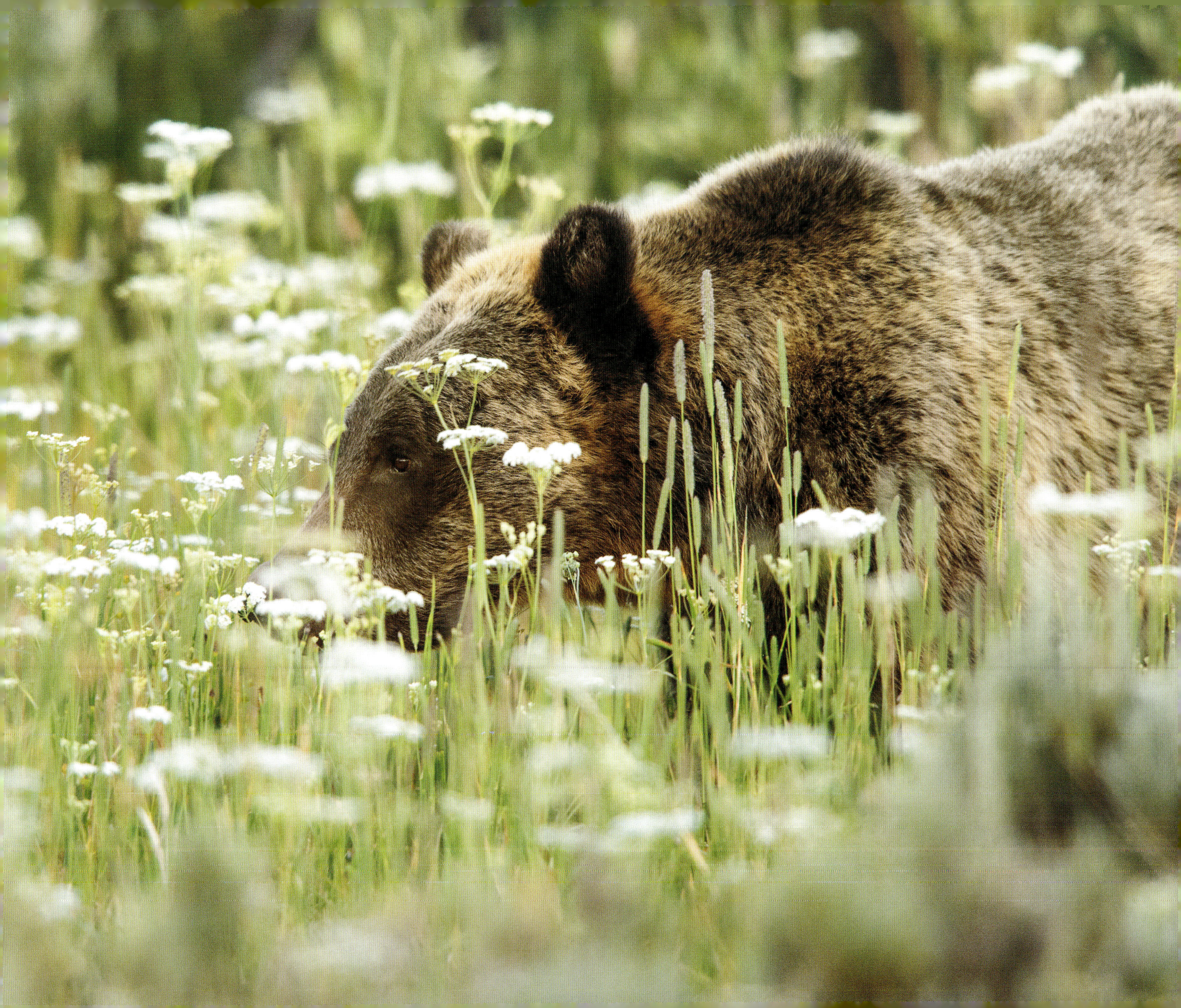

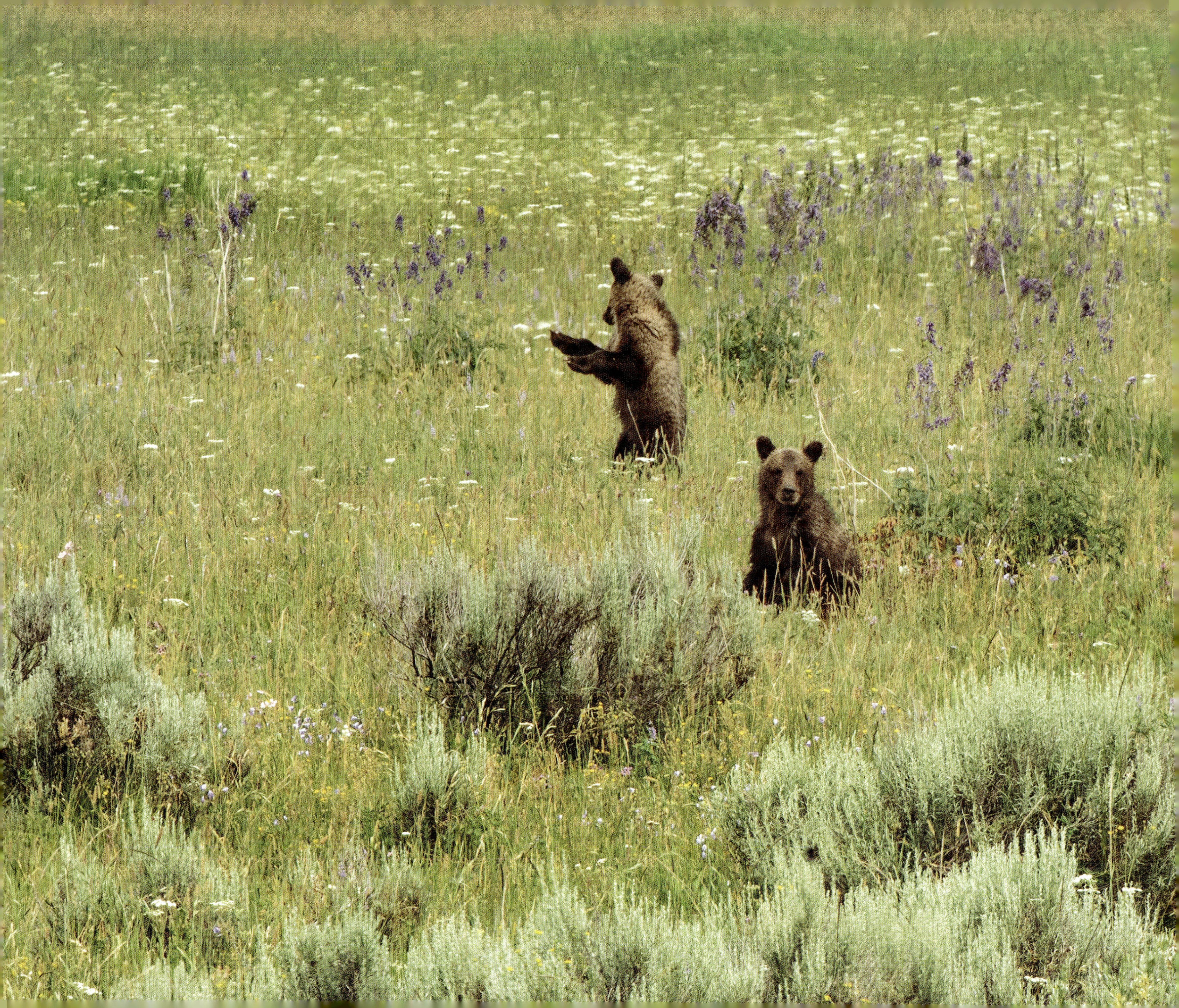

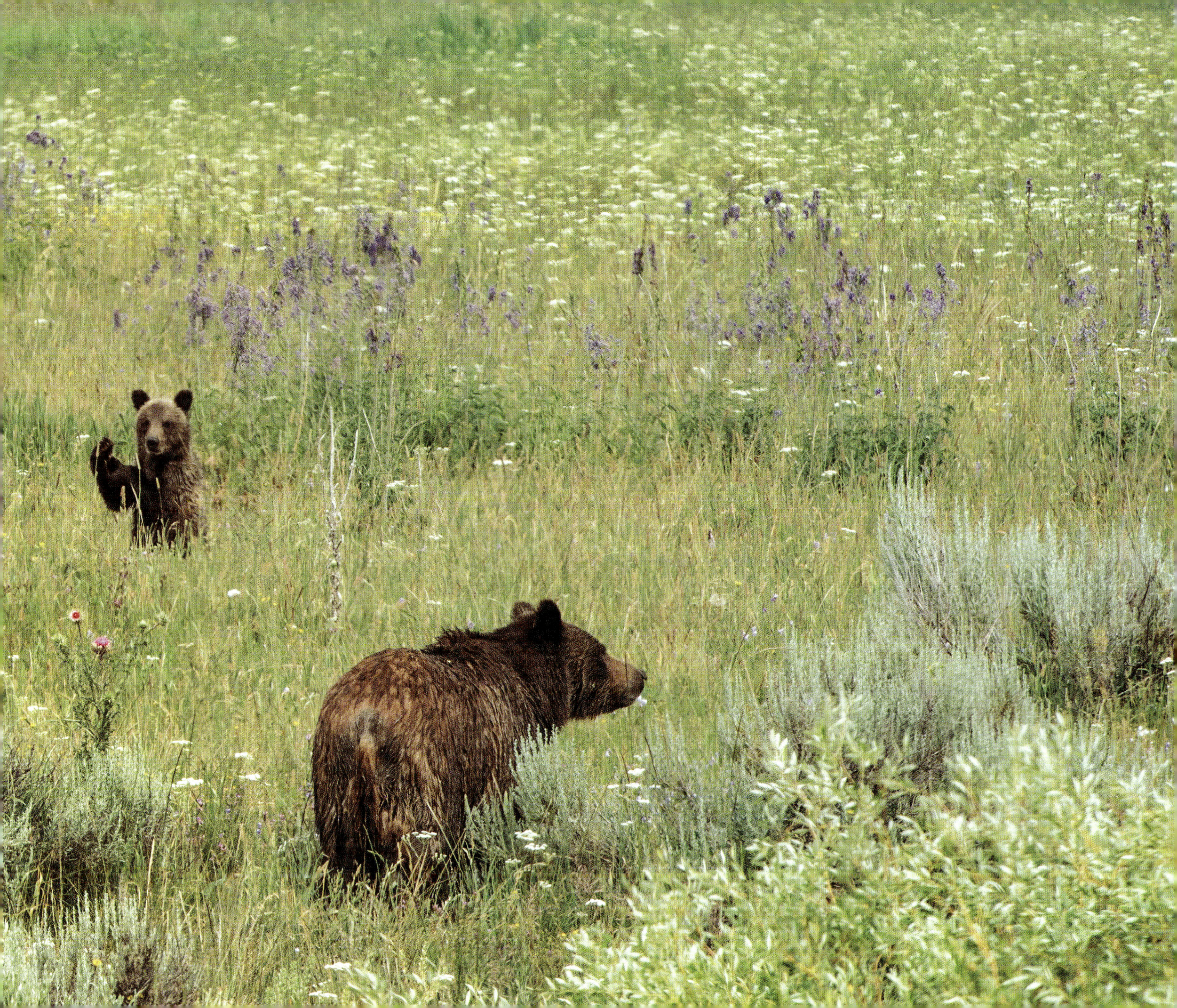

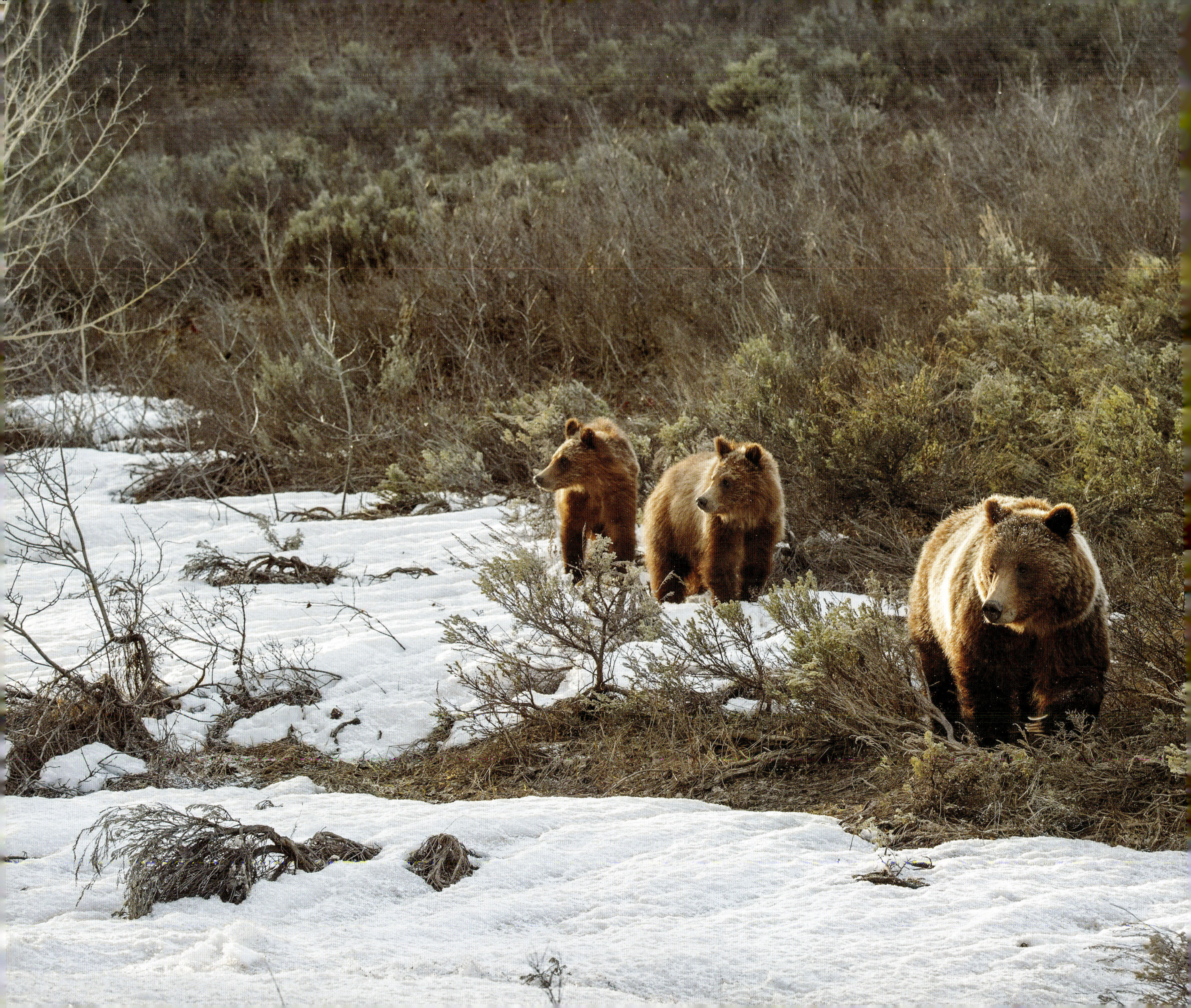

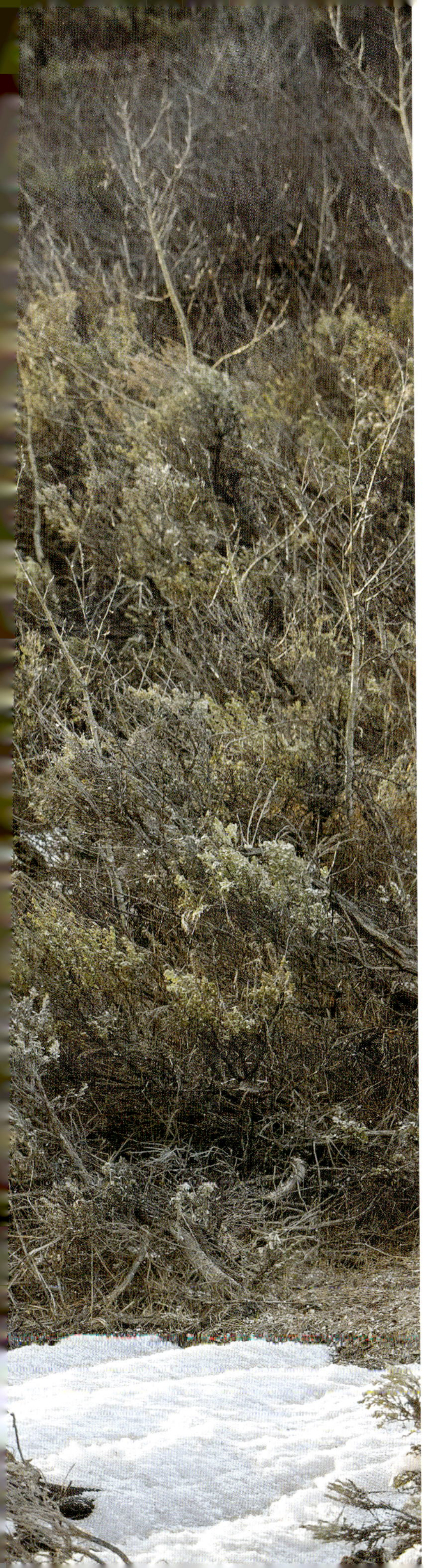

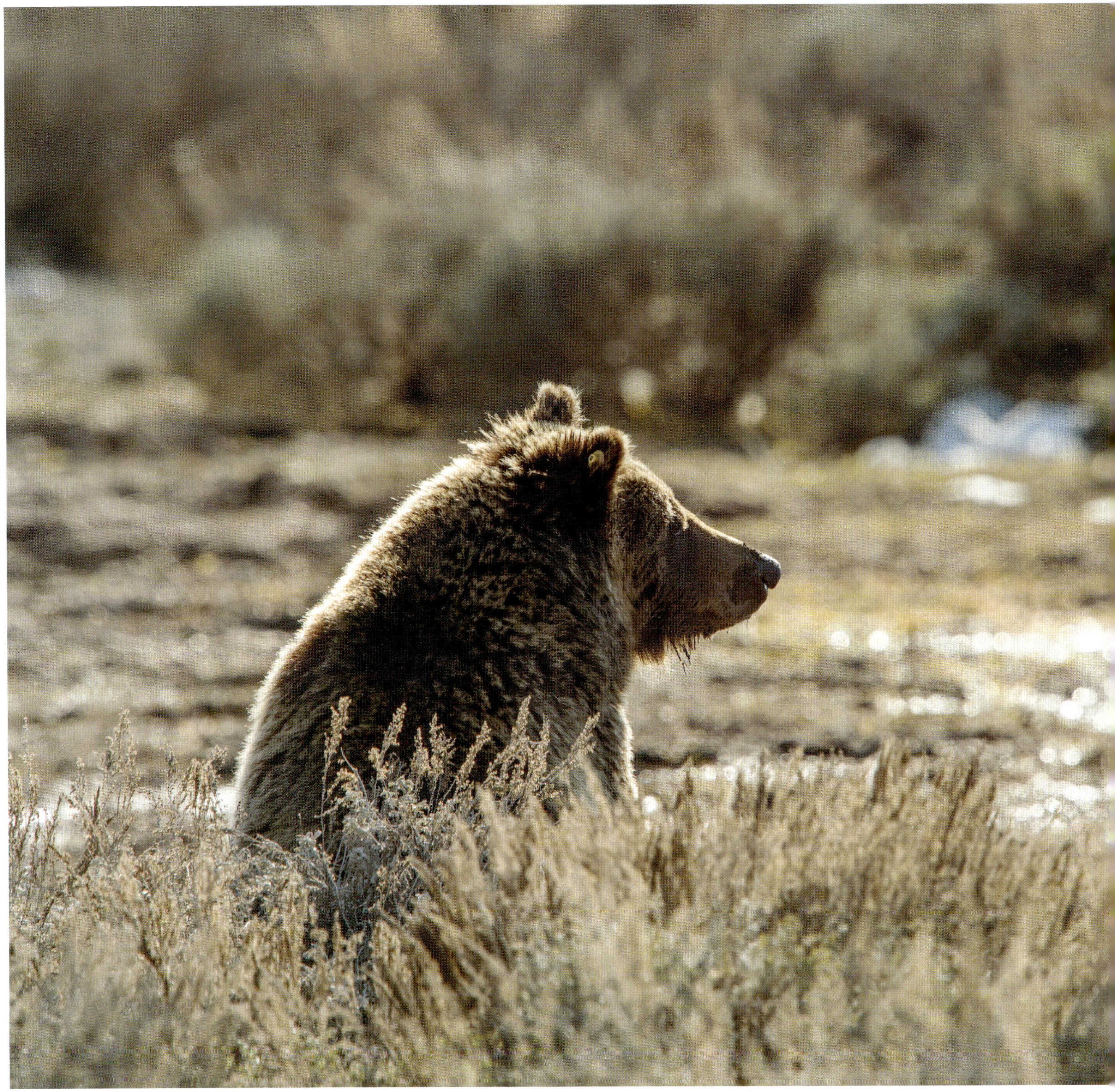

PREVIOUS SPREAD, LEFT: Soon, Mama, a bedtime story? As the sun sets, 399 and her two cubs look for a place on Lozier Hill to settle down for the night.

PREVIOUS SPREAD, RIGHT: 760 takes a break from digging for pocket gophers and their stashes in Willow Flats.

610 checks for a scent on the morning breeze, while her single cub sits close by.

FOLLOWING SPREAD, LEFT: In 2014, 610 had a single cub. Young, tiny cubs are the most vulnerable in the population.

FOLLOWING SPREAD, RIGHT: Unfortunately, the evening following this photo would be the last time 610 and her tiny charge were documented together.

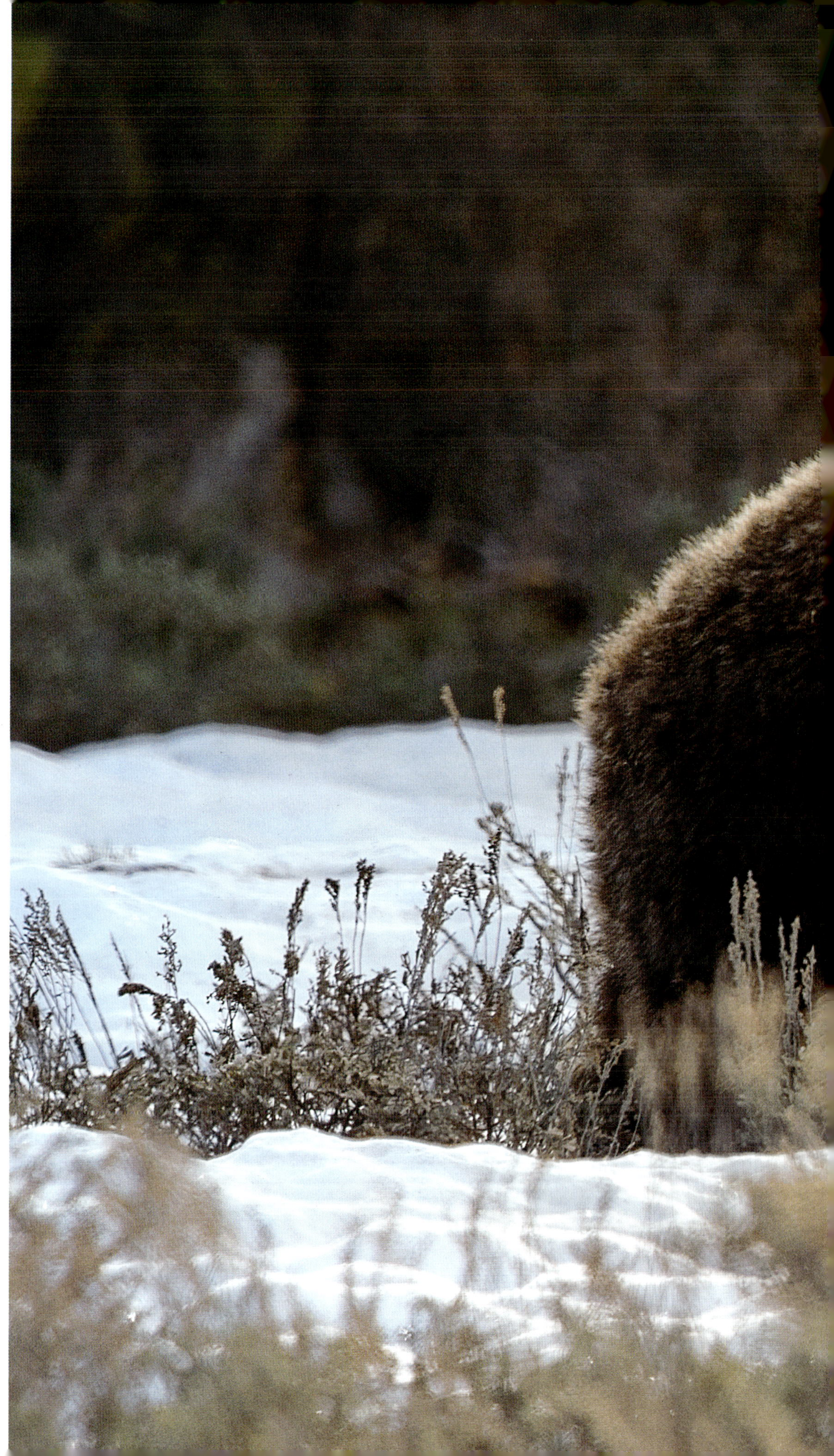

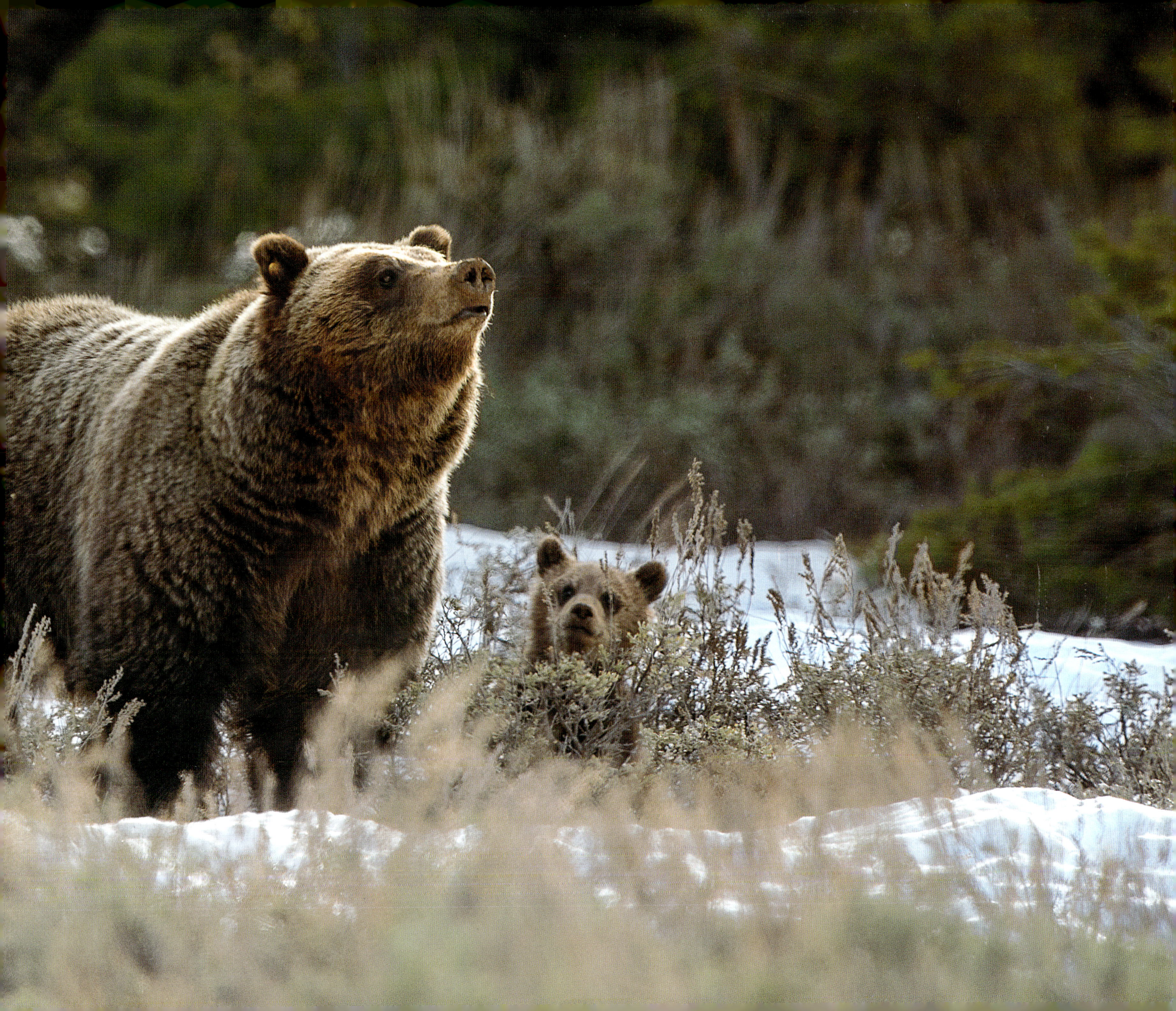

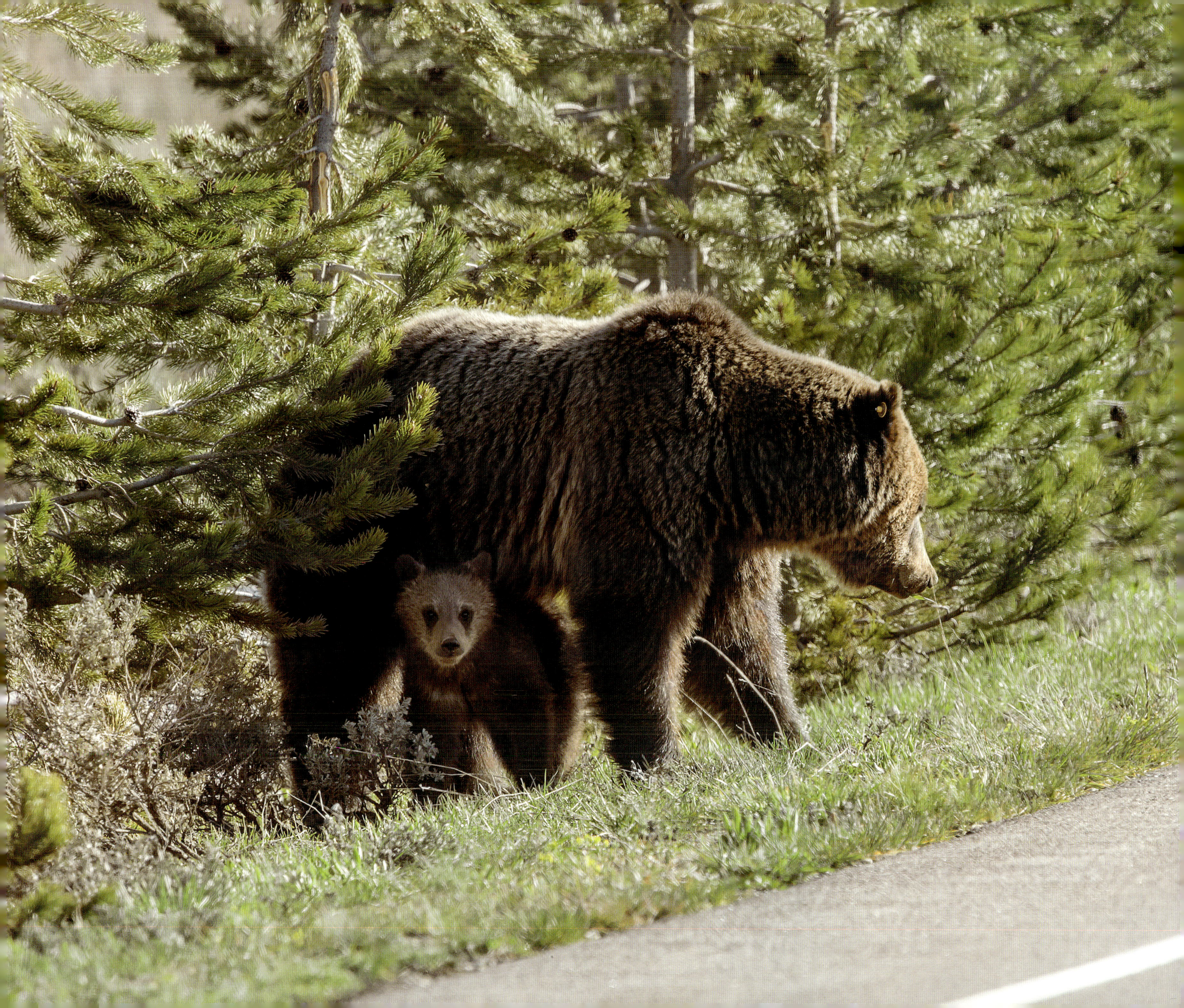

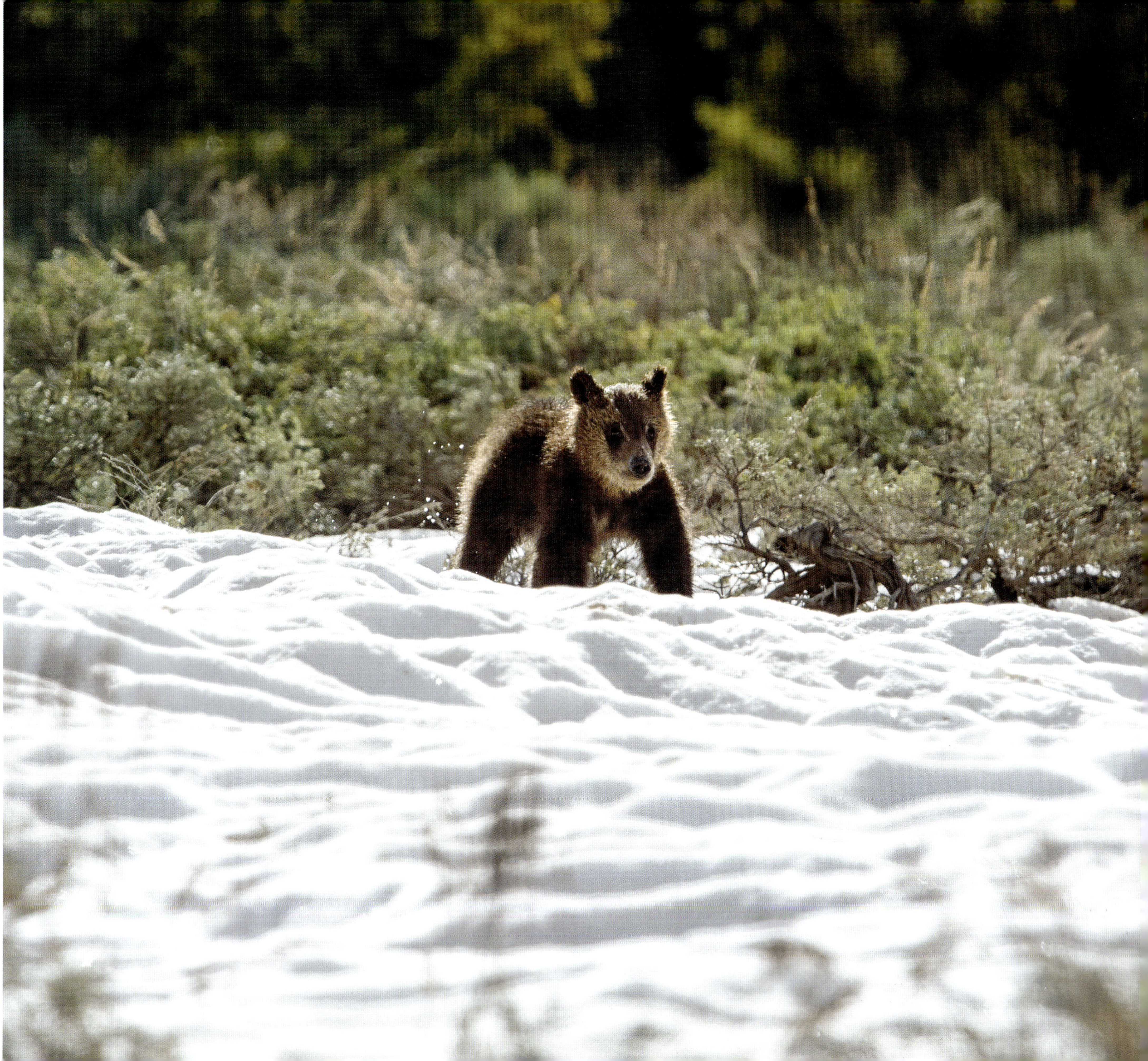

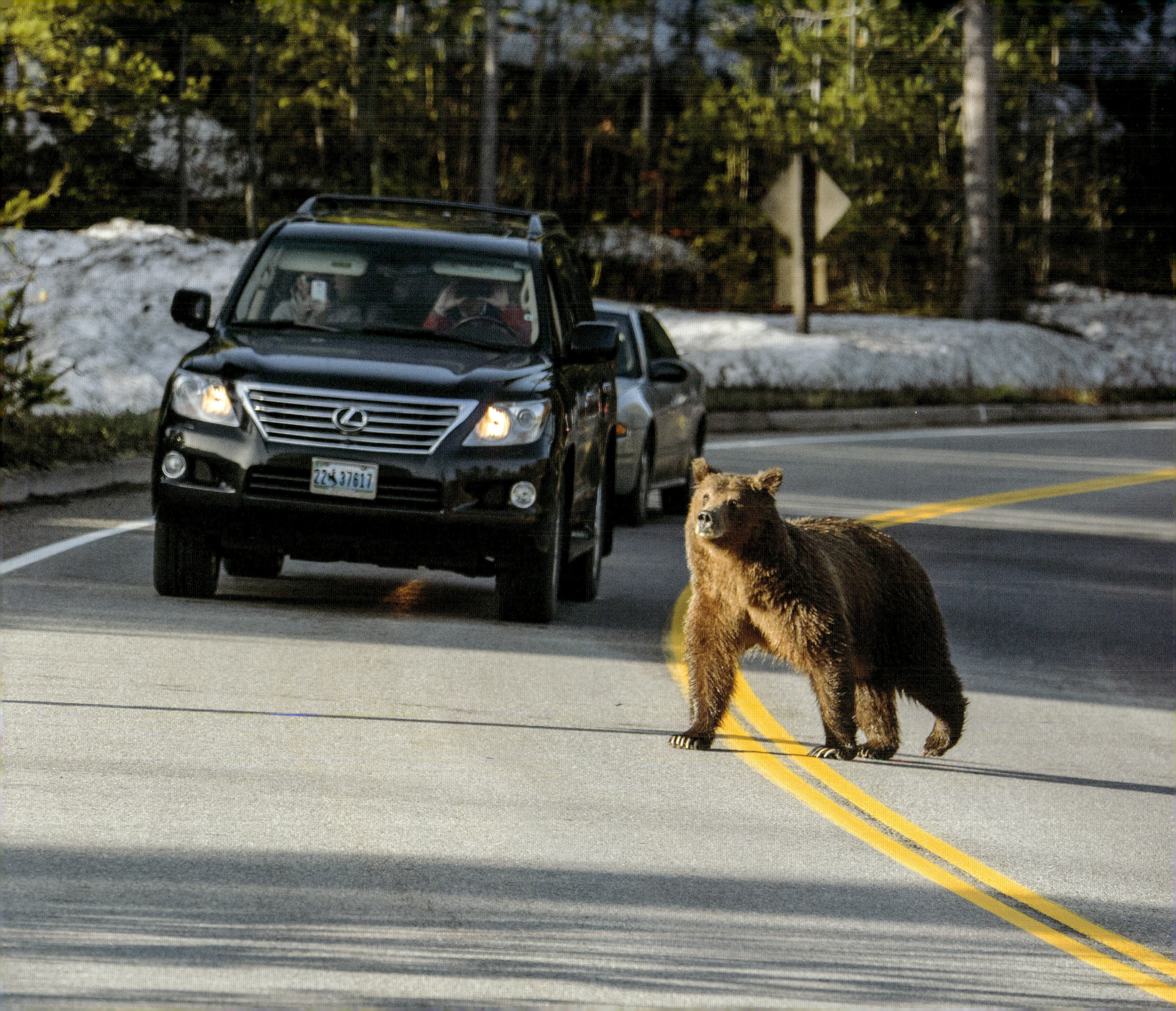
22 37617

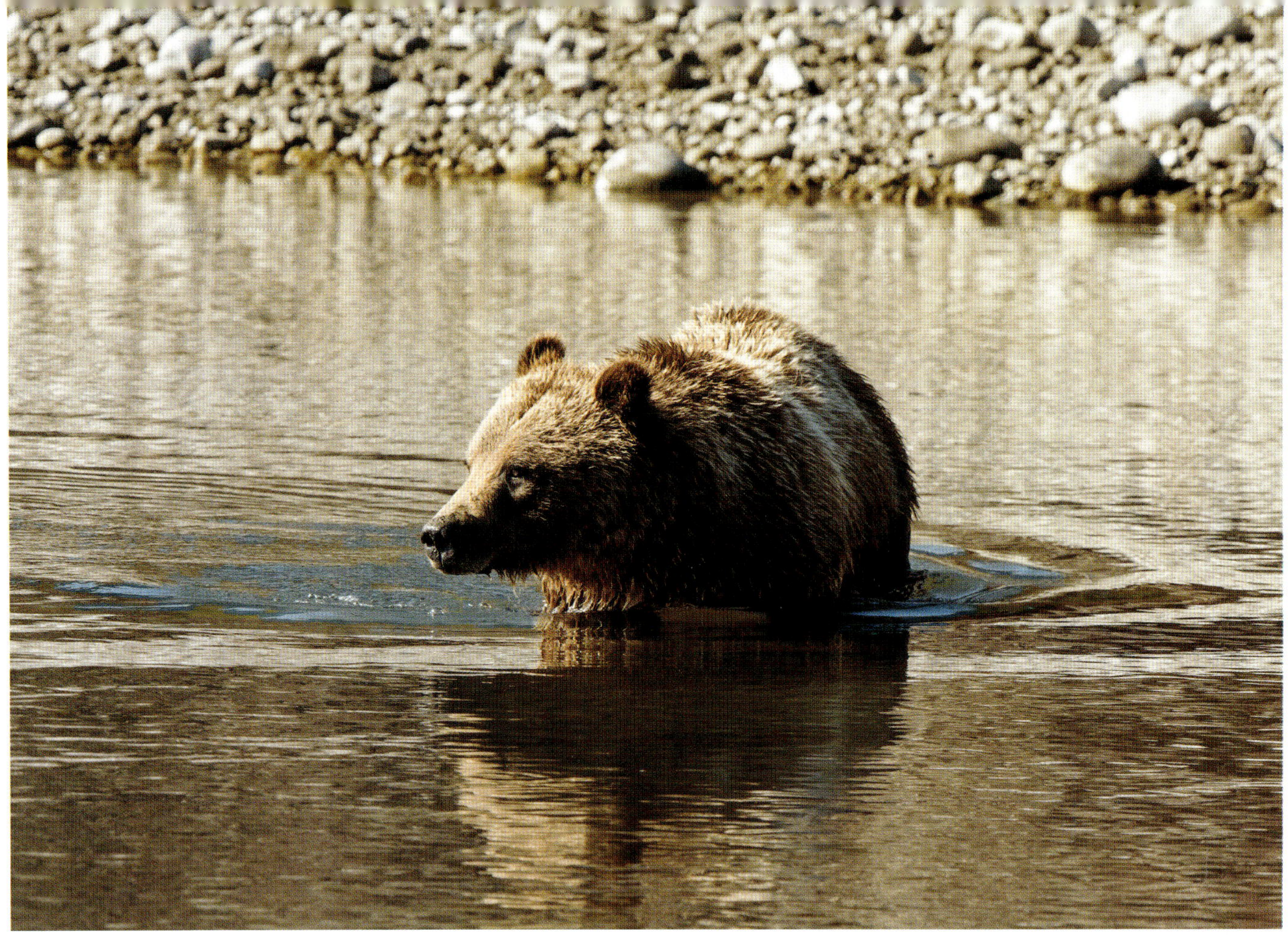

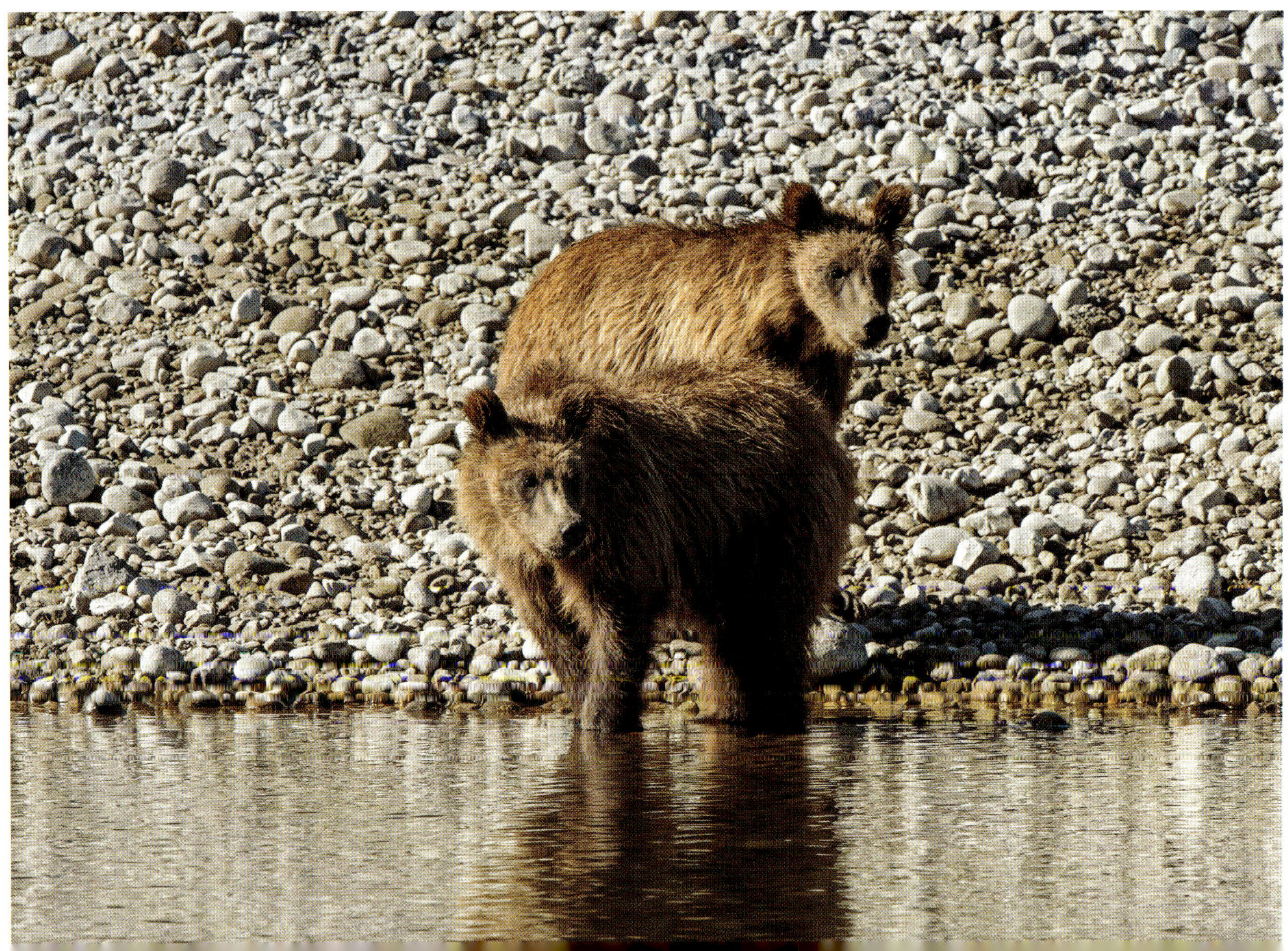

OPPOSITE: Many travelers are delighted to pause on their journey, allowing 399 to safely traverse the highway.

TOP: Is it cold? 399 splashes across the shallow Snake River downstream from Jackson Lake Dam while her two yearling cubs pause on the shoreline before entering the water.

BOTTOM: The previous fall (2013), 399 lost one of her triplets to unknown causes.

FOLLOWING SPREAD: Grizzly bears are known to travel many miles daily in their search for food. Here, 399 leads her two yearling cubs north toward Arizona Creek.

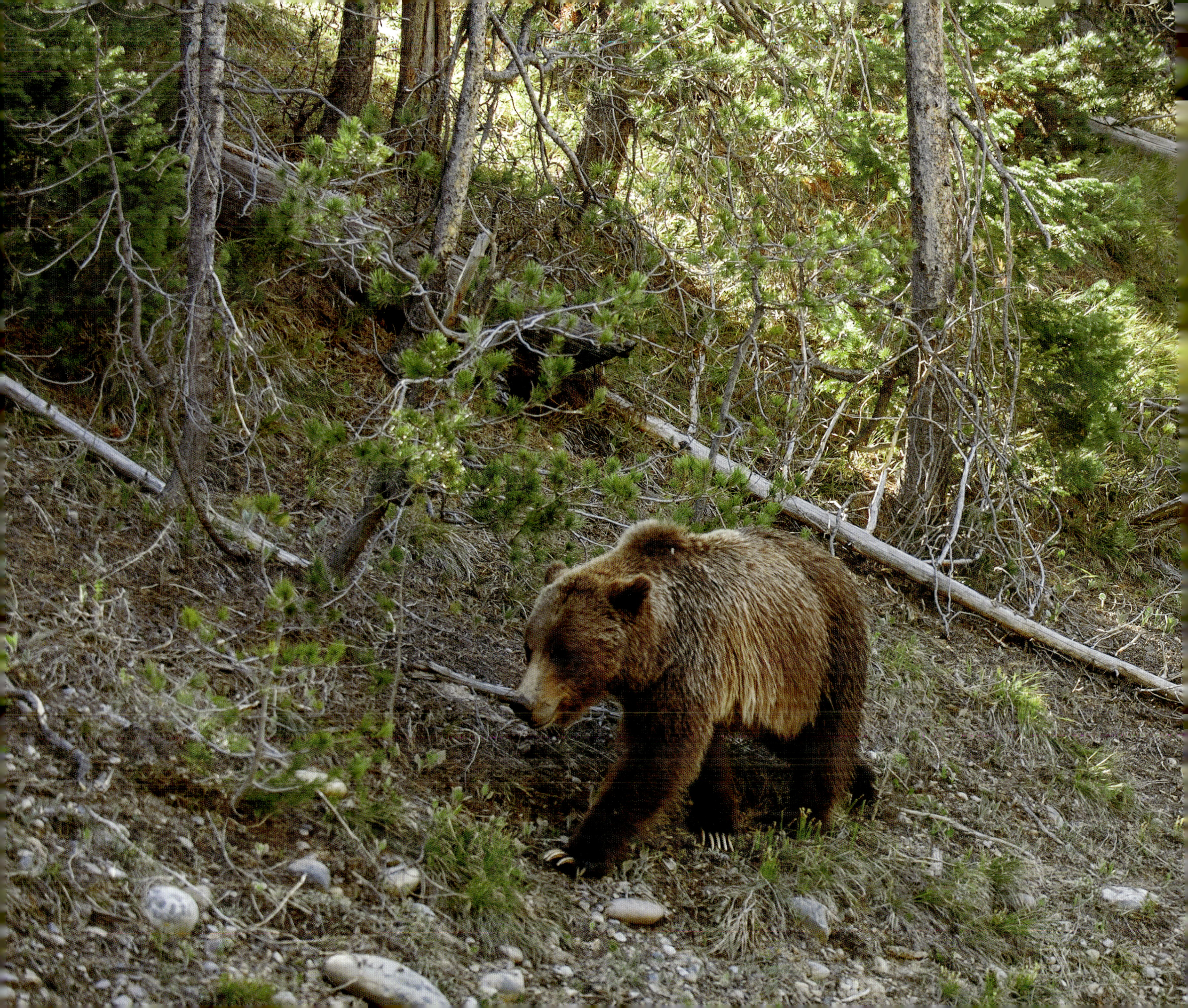

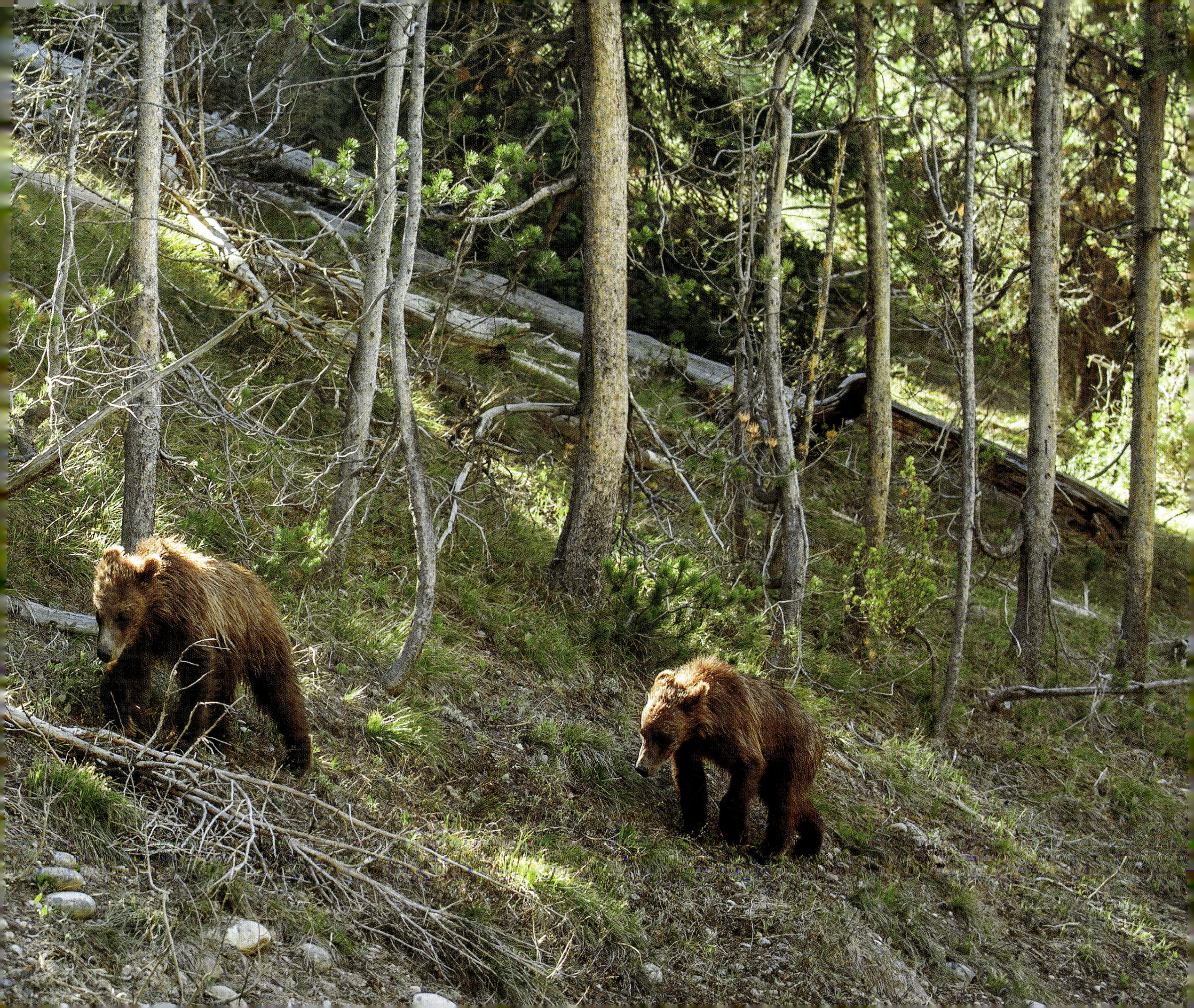

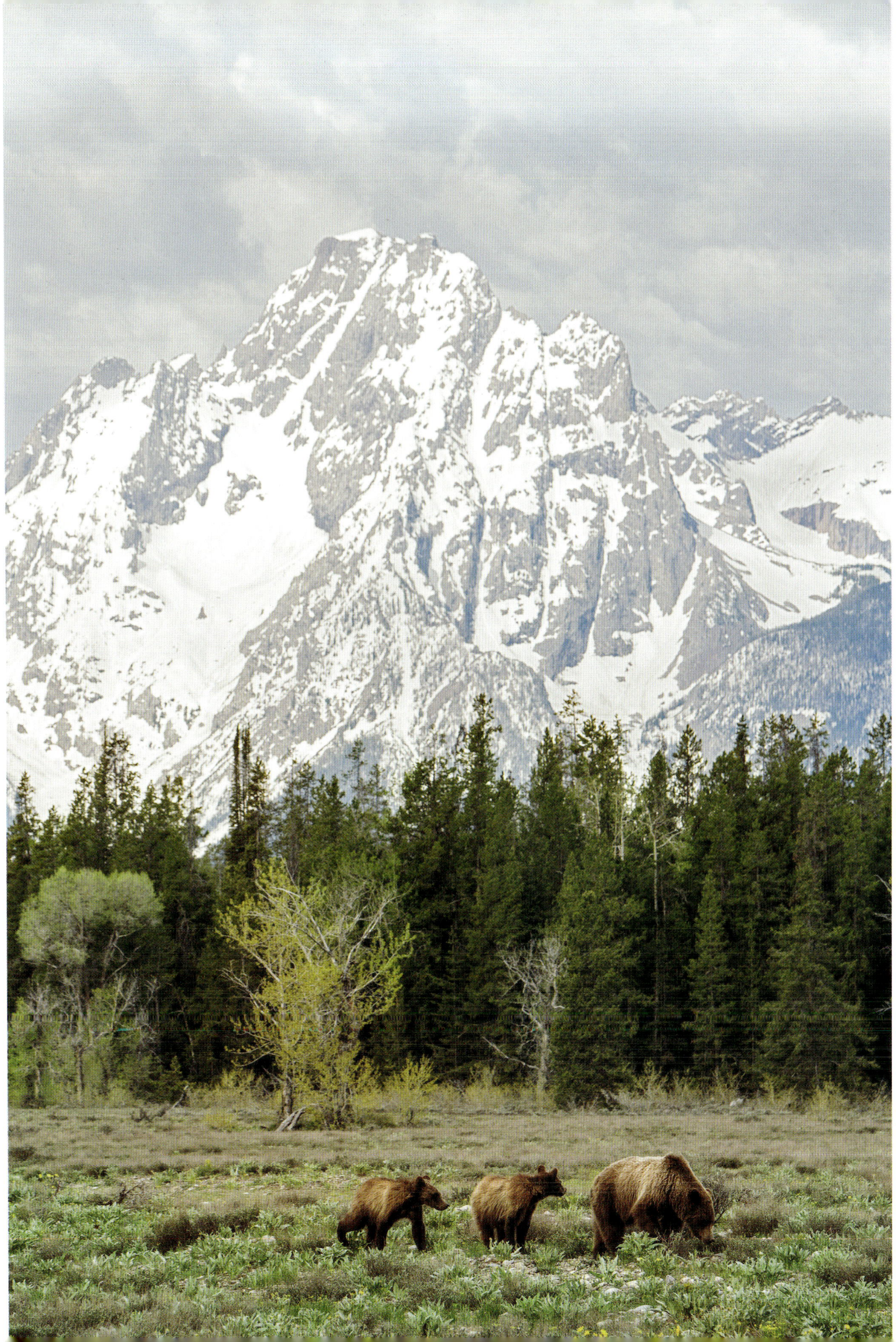
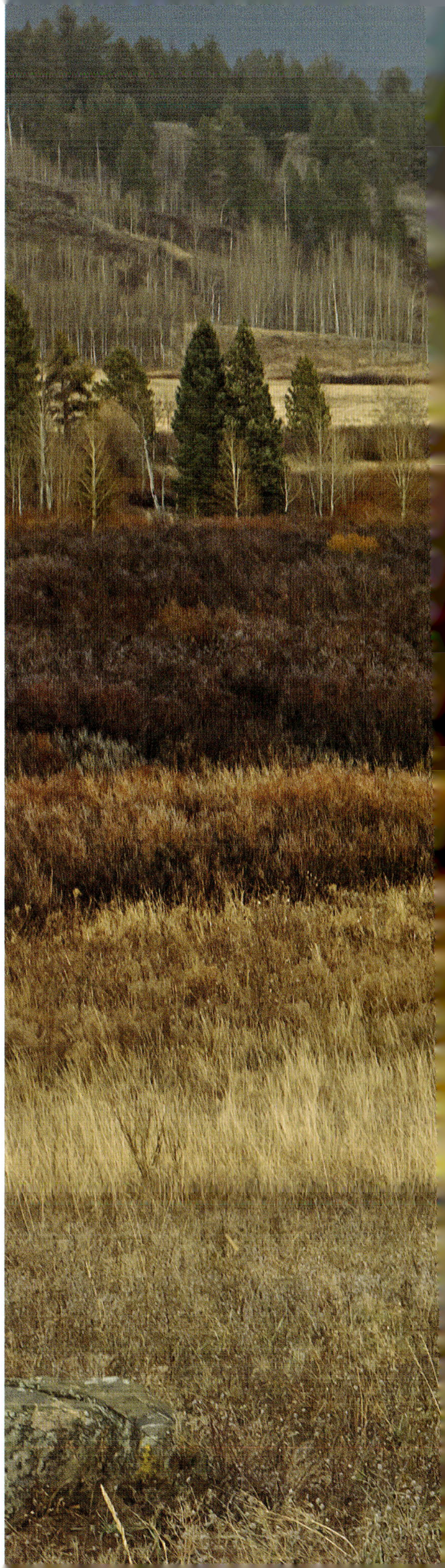

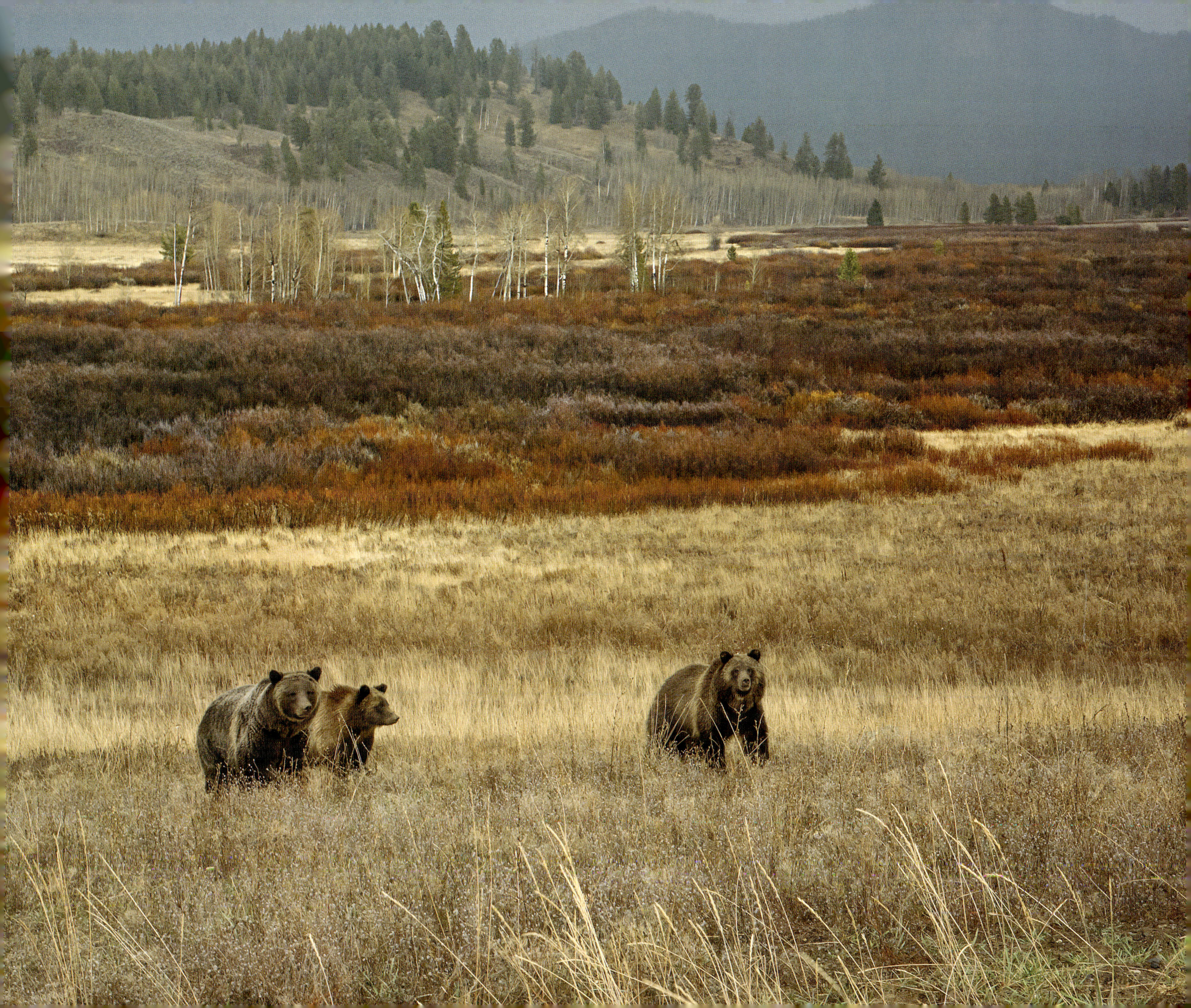

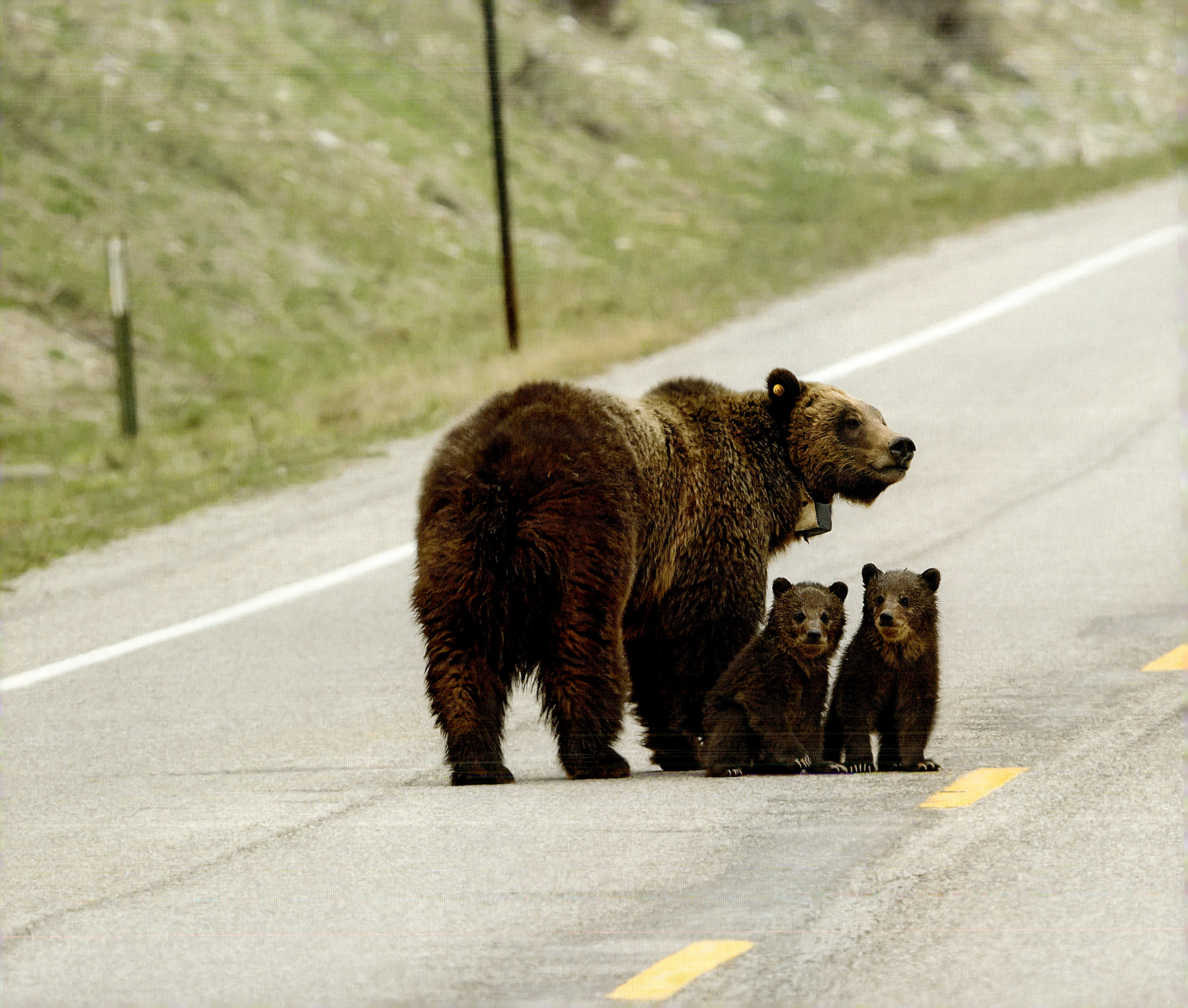

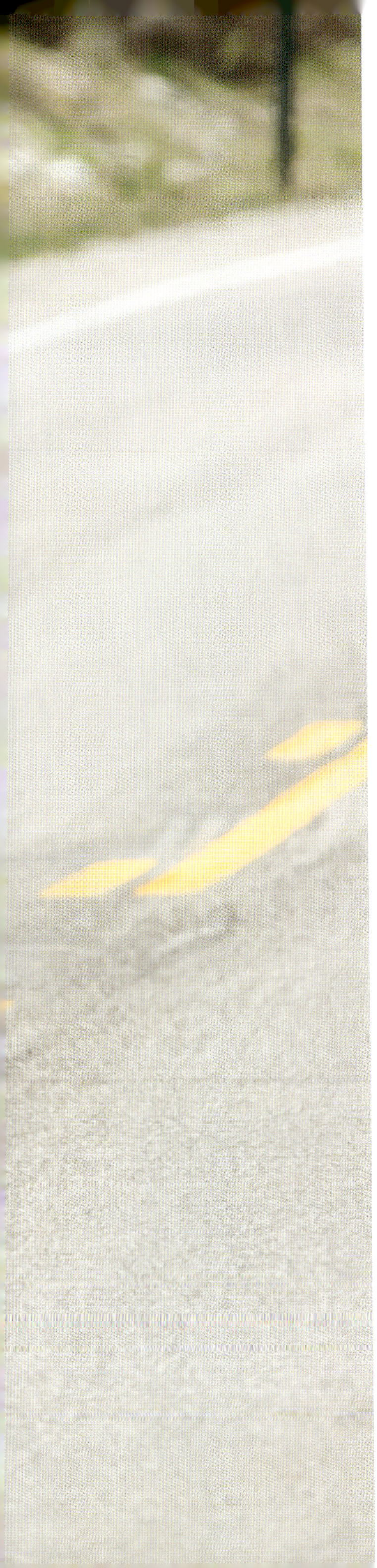

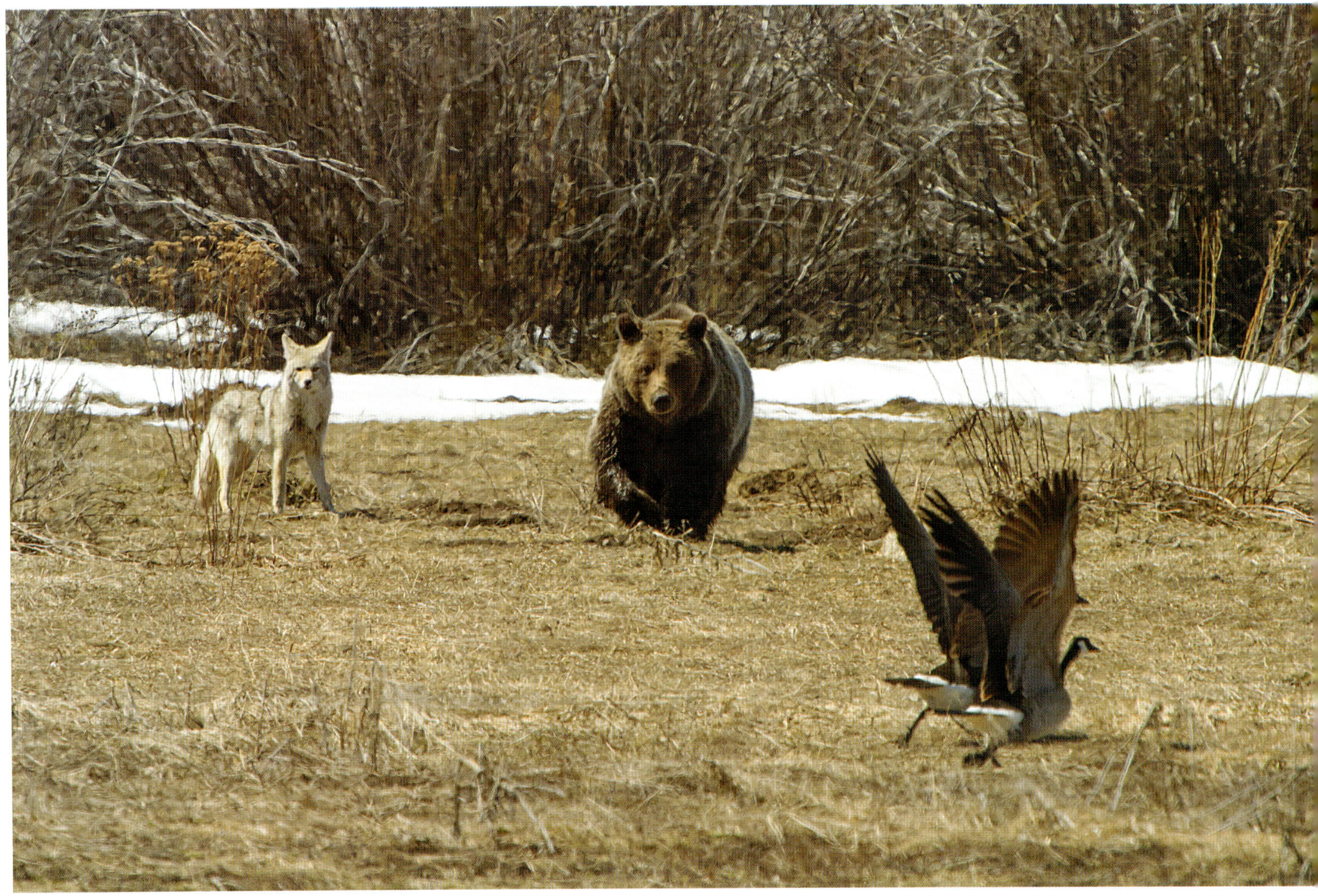

PREVIOUS SPREAD, LEFT: Grizzly 399 teaching her yearling cubs what's in their natural larder while visiting a meadow beneath Mount Moran.

PREVIOUS SPREAD, RIGHT: The recovery of grizzlies in Greater Yellowstone is a conservation success story touted around the globe, a hopeful sign for those trying to rescue imperiled populations of African lions, elephants, rhinos, Bengal and Siberian tigers, snow leopards, and jaguars.

2015 had an exciting start. On April 6, Mangelsen saw 399 and her two juvenile cubs on the Island of Oxbow Bend, at a distance too far to photograph. This was the exact spot where he first saw her in 2006. A week later, on April 13, one of the subadults raised by 610, either her own or the one she adopted from 399, was seen in Willow Flats being harassed by a pair of coyotes.

OPPOSITE: On April 23, 10 days later, 610, wearing a large radio collar, was seen with two tiny new cubs.

AFTERWORD | THOMAS D. MANGELSEN

MAYBE ONCE IN A LIFETIME an animal like this appears and her mere presence seems to galvanize everything. An icon of motherhood. An emblem of wildness. A sentient creature who has taught us new ways to think about others of her kind. 399 summons us together at a time when so much of the natural world is at loose ends and falling apart.

When I see her and other grizzlies in my home valley of Jackson Hole, Wyoming, and as I look into my camera viewfinder, gleaning an understanding that I've never had with other animals in my 40 years as a wildlife photographer, I try to remind myself how much their continued existence in the backyard of our country is a twenty-first-century *miracle*.

I have never forgotten the first grizzly I ever saw. Nor will you. For me, it happened during an epic youthful road trip long ago when a friend and I were driving from Boulder, Colorado, to Alaska. We were angling through the Canadian Rockies just west of Jasper National Park, and there it was, a gigantic mass of brown fur 50 yards off the highway with only a chasm of space separating us.

The grizzly was huge, and though only a few lousy photographs were taken, the encounter remains even larger in my imagination—this near-mythical animal with a giant head, silver-tipped fur, and six-inch-long, ivory-colored claws. This monarch was lying on his stomach, facing us, but with eyes shut as if asleep. He almost appeared dead. I wondered why he was there, so perilously close to the road. The possibility crossed my mind that maybe he had been hit by a vehicle and simply crawled to a place on the hillside. But then he reconfigured himself upright, head on his front paws. He had a few scars on his muzzle, the marks that came with the distinction of achieving a long life, against long odds, to be an elder bruin; an old sage on the mountain.

My friend, Bob Perath, and I decided to move closer; I was packing a 400mm lens. My legs were shaking. My preconception had been formed by tales of bear attacks, probably 95 percent of them untrue. Even from a distance, we were prepared to drop everything and flee for the car. Maybe I was foolish. I was certainly naive, having never seen a grizzly before and having believed every word I had ever heard about these supposedly menacing beasts.

Seeing the bear with my own eyes made the fiction fade away. He revealed himself as a gentle giant, napping in the woods. He opened his smallish amber eyes briefly to let us know he was indeed not dead and very aware of our presence. That experience, as I am writing this, almost 40 years ago to the day, changed my view of "the horrible one" forever.

At a conference on US national parks in 2015, E. O. Wilson said, "A species becomes rare and extinct usually because its support system in a natural environment is taken away." When we protect habitat for grizzlies, when we give them room to roam and protect them from needless human malice, we are safeguarding homelands for hundreds of other species. The fact is that we—you and I—are the best support system a grizzly bear in Greater Yellowstone has. We must give the Great Bear a voice. It would not delight Todd and me more than for us to learn that this book provided you with inspiration to declare that bears like matriarch 399 matter.

Maybe the most powerful thing we can do is celebrate grizzlies and bring them into the imaginations of young people, showing that they are worthy of our respect and admiration. We can only hope that human tolerance prevails and leads to a wider acceptance of wild things—especially predators like bears, wolves, and cougars—that are not always convenient to live with. As 399 and 610 show, they are important, sentient beings, with intellect, feelings, and emotions just like us. They are not only crucial in a truly healthy ecosystem but also belong among us in their own right. Their acceptance into our midst is what makes us human. Join me in telling everyone you know that grizzlies belong out there in the wild places. Our children and grandchildren will thank us.

On May 18, 2013, 399 and her third set of triplets (one is behind her) sit together in a meadow. Here, the cubs of the year had their earliest experiences with humans near Pilgrim Creek Bridge, continuing 399's legacy.

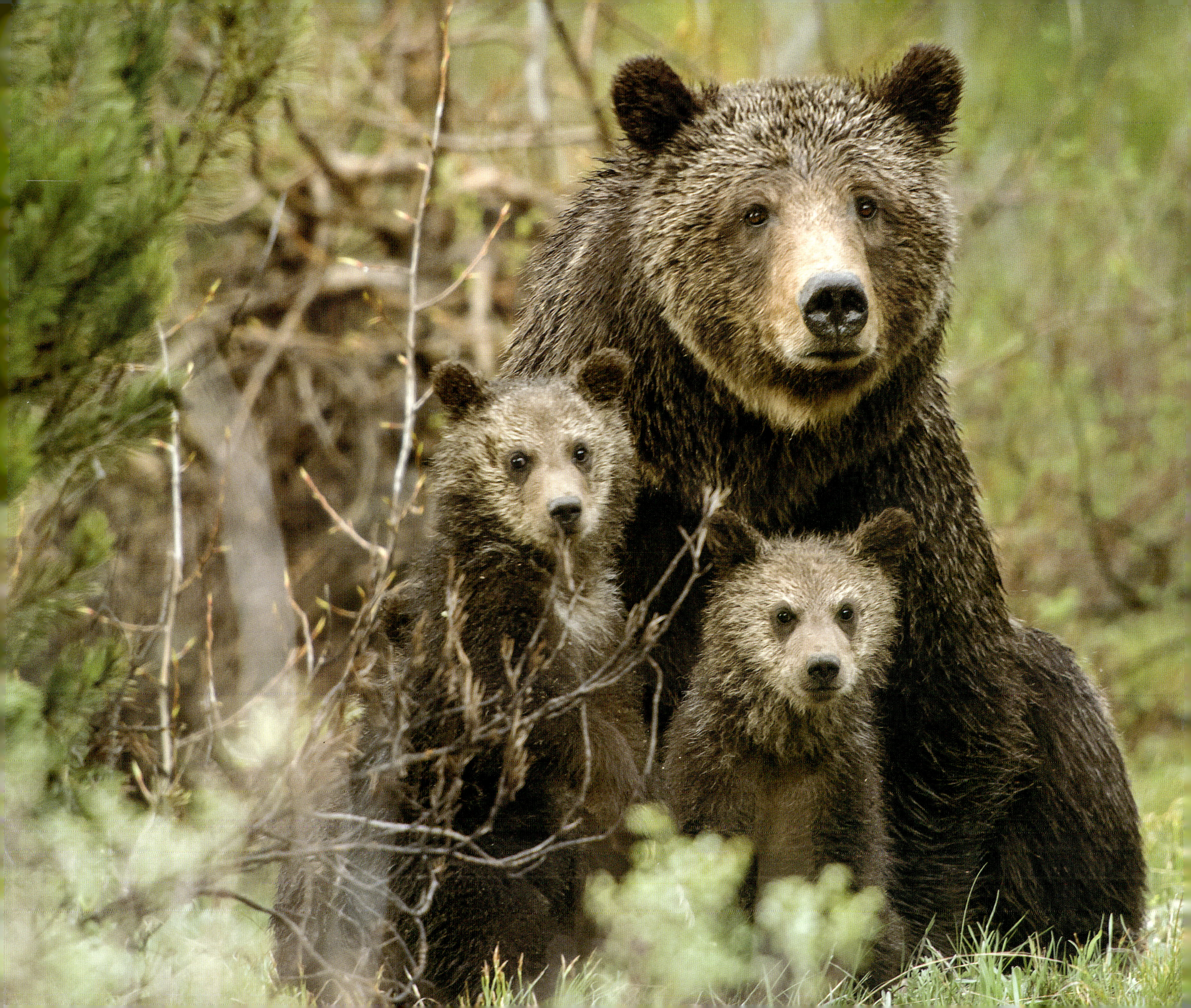

ACKNOWLEDGMENTS

GOOD FRIENDS, we (Tom and Todd) have been contemplating this ambitious collaboration for nearly a decade, trying to figure out how to make the story of 399 speak to the larger arc of wildlife conservation. We hope we have succeeded. Bringing this book to life has been a significant undertaking. It has involved a core group of very passionate, talented, inspired individuals and they deserve special recognition for helping to assemble the book before you now.

Standing on the summit of our gratitude is Sue Cedarholm, who has spent long hours in the field assisting Tom to make these images possible, as well as Andrew Bennett, Jessica Baker, Shalese Hill, and Victoria Blumberg in the Mangelsen Jackson Hole office. Right next to them is the very gifted publishing team at Rizzoli led by Jim Muschett and colleagues Candice Fehrman, editor; Susi Oberhelman, designer; and James Daley, graphic artist and cartographer. Tom also thanks his family, especially brothers David and Hal, and all of his Mangelsen—Images of Nature associates who not only champion his images and keep him out in the field but also help further his mission by speaking out about the importance of protecting wild places and wildlife to hundreds of thousands of people who annually visit his galleries. A very special thanks to Jane Goodall, a dear friend, and the one who more than anyone continues to inspire Tom. Todd thanks Jeanne, Carter, Natalie, and Rosie. Further, we salute the entire Craighead clan for embedding the importance of grizzly bear conservation deep in the conscious awareness of citizens everywhere. Beyond them, the list of people who have shared this special family of bears, and influenced and enlightened Tom and Todd over the decades about bears, is simply too long to cite in one space. We value you more than you know.

Finally, we want to profoundly thank all of the committed "Bear People" out there—millions upon millions of you who *love* grizzlies and are dedicated to their long-term persistence in the wild. You have mattered in the past and you do now, more than ever, as we go forward. This book is for you.

DEDICATED TO SUSIE WITH LOVE

First published in the United States of America in 2015
by Rizzoli International Publications, Inc.
300 Park Avenue South | New York, NY 10010 | www.rizzoliusa.com

© 2015 Thomas D. Mangelsen, Inc.
Photographs © 2015 Thomas D. Mangelsen
Text © 2015 Todd Wilkinson
Foreword © Ted Turner

All rights reserved. No part of this publication may be reproduced, stored in a retrieval system, or transmitted in any form or by any means, electronic, mechanical, photocopying, recording, or otherwise, without prior consent of the publishers.

MANGELSEN®—Images of Nature Gallery is a federally registered trademark of Thomas D. Mangelsen, Inc., Omaha, Nebraska.

Limited-edition prints and books are available through:
Thomas D. Mangelsen, Inc.
MANGELSEN®—Images of Nature Gallery
13303 F Street | Omaha, NE 68137 | 800-228-9686
www.mangelsen.com
support@mangelsen.com

Images from this book are available for licensing through:
Thomas D. Mangelsen, Inc.
MANGELSEN®—Images of Nature Stock Agency
PO Box 2935 | Jackson, WY 83001 | 307-733-6179
www.mangelsenstock.com
stock@mangelsenstock.com

Project Coordinator: Susan E. Cedarholm
Photo Editors and Color Management: Andrew Bennett, Shalese Hill, and Jessica Baker
Associate Publisher: James Muschett
Editor: Candice Fehrman
Book Design: Susi Oberhelman
Maps: James Daley

2017 2018 2019 2020 / 10 9 8 7 6 5 4 3 2

Printed in China

ISBN-13: 978-0-7893-2949-3

Library of Congress Catalog Control Number: 2015933667

Page 1: The vaunted matriarch in her prime: 399 stands above Oxbow Bend keeping an eye on her juvenile cubs. Even with her trio being nearly three years old, her teats show she is, at times, still nursing her bruin offspring.
Pages 2–3: 610 leads her three cubs through a spring snowstorm in a marsh on Pacific Creek Road.
Pages 4–5: 399 eats fresh green grass while one of her three yearling cubs stays close.

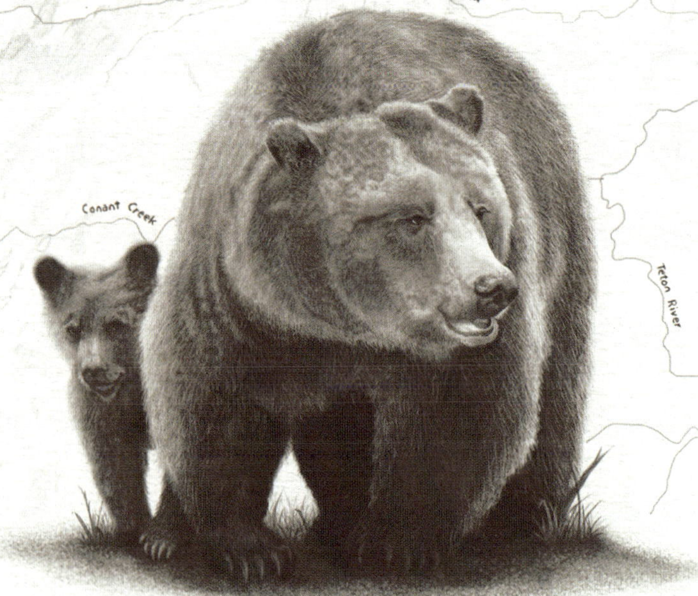

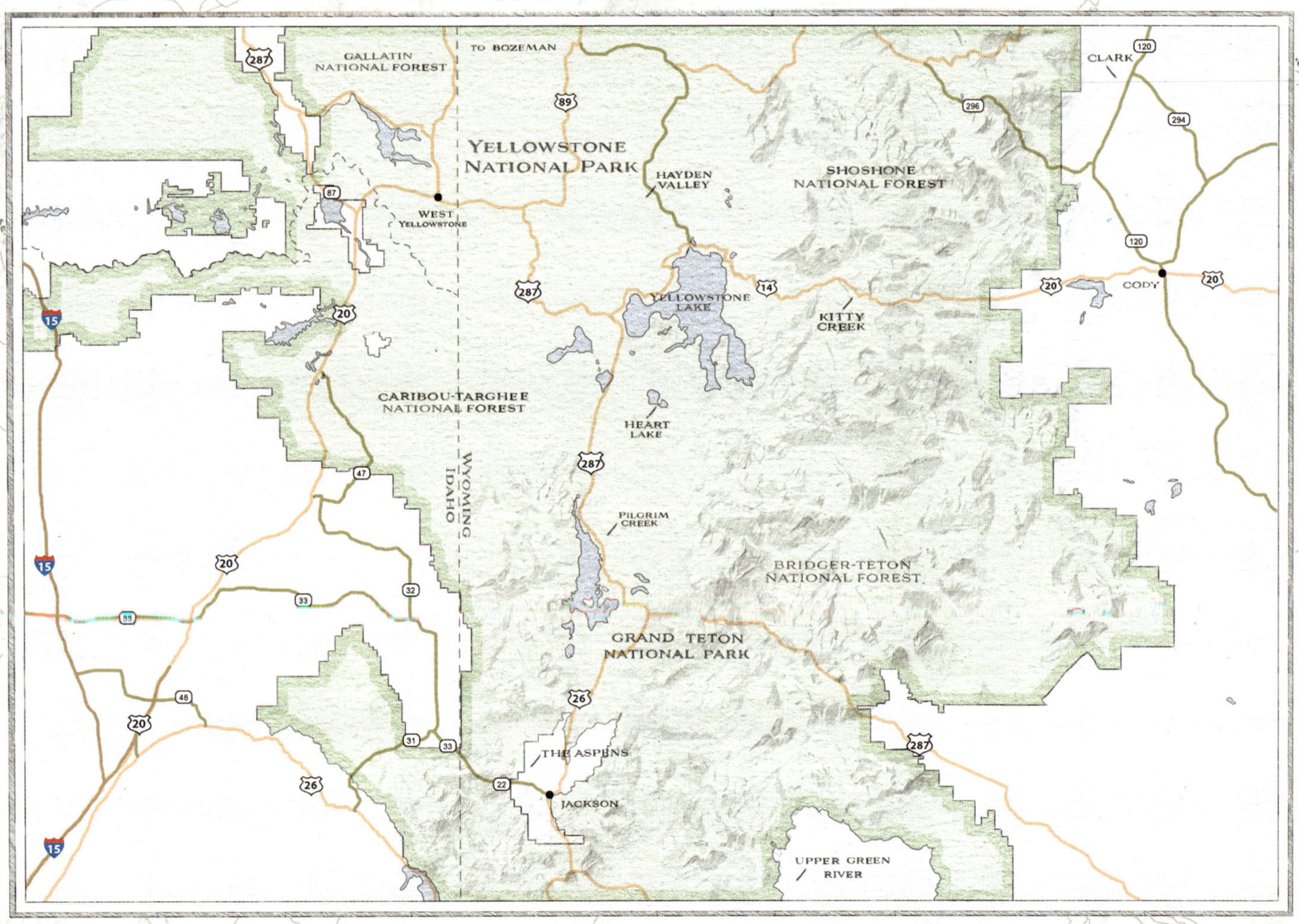

GREATER YELLOWSTONE ECOSYSTEM

AF710854

Elad Lassry

Alessandro Rabottini (ed.)

Mousse Publishing

ELAD LASSRY

Preface

by Alessandro Rabottini

We know we live surrounded by images, but what may escape us at times is the fact that we live in state of constant perceptual tension. It is the tension inherent in the production of our memory, caught between the physicality of analog formats and the uninterrupted flow of digital immediacy. The tension between visual pleasure as a form of commercial seduction and visual pleasure as a form of redemption, ambiguity, and freedom. The tension we experience in the very act of seeing, which is both consumption and autobiography.

Elad Lassry has turned this state of tension into the nerve center and production hub of his work, exploring the ubiquitous nature of the image and the many ways it can be generated and received, through a limitless vocabulary of form that is visually exuberant and conceptually meticulous.

This book documents the solo exhibition that PAC – Padiglione d'Arte Contemporanea in Milan dedicated to the artist's work: the biggest overview of it organized to date, including films, photographs, drawings, sculptures and installations.

It has been an honor to work with Elad on constructing a show where the precision and visionary nature of his expressive universe could manifest themselves so extensively, and to continue our collaboration in the months that followed, with the preparation of this volume.

I am grateful to Jörg Heiser and Aram Moshayedi for their contributions to the catalog – respectively, a conversation with the artist, and an essay – that both explore the complexity of Elad Lassry's artistic practice from multiple angles, to render the significance and impact of his work today.

I would also like to thank the many lenders who made such an ambitious exhibition possible, expressing my gratitude to Francesco Bonami and Paola Nicolin for the fundamental support they have given this project from the outset and over the course of its complex development. The staff of Galleria Massimo De Carlo and PAC also deserve thanks for the professionalism they have shown in the organization of this show: the same dedication and enthusiasm that Antonio Scoccimarro and Marco Fasolini have put into the creation of this book.

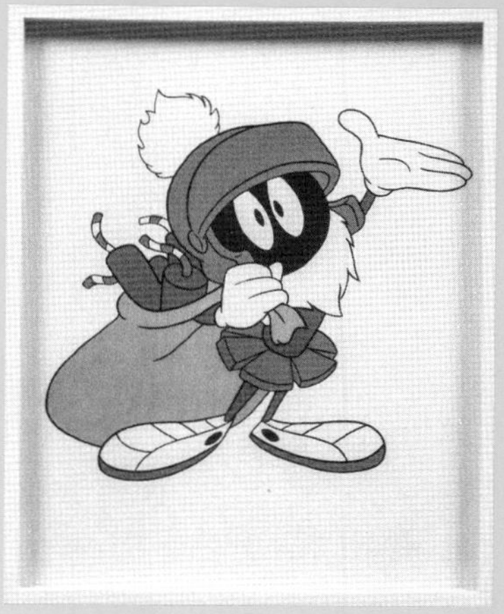

Martian, 2012
C-print, painted frame
36.8 x 29.2 x 3.8 cm (14.5 x 11.5 x 1.5 in)
Edition of 5, 2 AP
Courtesy Massimo De Carlo, Milan/London

Elad Lassry in conversation with Jörg Heiser

Jörg Heiser It's possible, if maybe a bit one-dimensional, to situate what you do in a canonical art-historic trajectory of the readymade and the appropriated image: from Duchamp, through Warhol and structuralist filmmaking, to the 1970s–80s Pictures Generation, and to the various "neo-conceptual" or "relational" approaches to exhibition framing and display that emerged in the 1990s.

What all these positions share is that they invest seemingly obscure or banal motifs – a snow shovel, a Kodak flower, a piece of found film stock, a series of commercial advertising photographs, or a prefabricated, computer-rendered anime character – with a sense of strangely heightened alertness, halfway between the uncanny and the comic, between gravity and irony, by "merely" choosing, isolating and displaying things, thus recontextualizing them dramatically.

The period in which your approach to the image and to object-making first emerged – starting in 2005, when you were studying at the University of Southern California – was also the time when (to state the obvious) a new type of image circulation started to become widely available, a digital folklore of the shared image that very much relies on a logic of the curiously inept and obscure becoming suddenly significant, simply by being selected and circulated. The imageboard website *4chan.org* was launched in 2003, and it was in this juvenile, anarchic realm that phenomena like the Anonymous movement and the very concept of the Internet meme – say, an endless series of funny pictures of cats, "lolcats" – first came about. I can imagine how you might cringe at the thought of your work being compared to digital image folklore and to Internet memes, given the meticulous attention you give to ideas of display, framing, and objecthood. All of which couldn't matter less to your typical meme-nerd, whose main aim is to simply let the image circulate in as viral and inflationary a manner as possible...

Elad Lassry This touches on a core issue, on something that is often misunderstood about the work, and in my opinion, about the general cultural analysis as well.
I don't cringe at all to hear you speaking of digital snapshots or phone-photographs in relation to my photographs, and work in general. The relationship is actually a close one. Perhaps in my mind, I'm always more curious about the photograph as a tool. And about the fact that the digital revolution has made that possibility more obvious. What I mean is that the photograph has always had the capacity to serve solely as a tool, but an emotional engagement made that capacity hard to embrace. You are describing the process through which the non-artist, the engaged citizen, became involved in moving pictures around. Using pictures to communicate, create a profile, establish an atmosphere. Aside from accessibility, I'm seeing a curious shift in the relation to the photo: the photo is being shifted online, well-crafted or not, and is being embraced as a free agent. Rarely, in this kind of activity, does the image have a relevant author.

JH What changed for your practice when it suddenly became possible for millions of people to make, archive, circulate, display and access any image imaginable instantly and globally?
How did that affect your own approach to finding and selecting, to isolating and putting on display?

EL What is interesting to me is that the suspension of the photographic image became the norm.

Pictures became tools for the masses. My photographs, like my practice in general, are certainly informed by that condition. I'd actually make the case that they would have not been made in the same way otherwise. I refused to address this condition in a literal manner, so I did not feel it was necessary for my photographs to *look* like what we think of as a digital photograph. However, they rely on heavy digital intervention – apparent to the trained eye, at least – while still referencing and participating in an analog conversation. It was important for me to claim a reference of inclusion, as I believe that these histories are extremely informed and affected by one another.

The capsulation of the image was meant to suggest the condensation of multiple images, perhaps not all that different from what you see on a sharing site. The seamless quality came about to distance myself from any authorship. To give a pan-quality to this specific archive that is made by me. Hence the "universal" scale: a certain democracy that the viewer can't escape in the singular solution of "the digital era" because it simultaneously claims an obvious relationship with traditional modes of photographic production. Yet not quite.

JH You have described yourself as a "very aggressive consumer of images..."

EL I'm not sure why I chose the word "aggressive" in describing my relationship to the consumption of images.
Seems a bit obnoxious in retrospect. Regardless, the fact that I "go through" many images a day is perhaps the reality behind me saying that. I look at many images, I *seek* them out. Which leads me to think I see more than what is made available for me out there.

JH "Going through" is a term that immediately makes sense to me. You go through many images, and as you do, they are not filtered so much as they are superimposed, or folded in on one

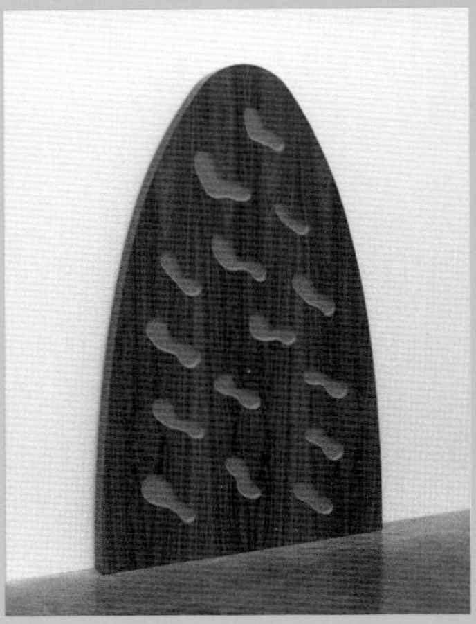

another, in your mind's eye – that's how I imagine it – to form a kind of intuitive continuum of the typically unlikely, with the individual surprising image wrought from it.

The strict seriality of many of your framed pictures having the same scale (and the frames being colored in a way that highlights or complements a dominant color in them) not only works within a conceptual logic inherent to the "body of work" or "series," but at the same time builds on a "pre-condition" in our minds shaped by encountering images that are similarly "framed," gridded, and formatted on computer screens or within the pages of printed matter. Yet that pre-condition is taken and pushed towards hybrid states between the flatness of the image and the physicality of its display, between strict seriality and the epiphanic occurrence of the singular image.

"Going through" is also a term that reminds me of sculptural modes. Of working through materials, objects, to form a constellation. Literally going through the doors/doorways that I saw in your show at White Cube in London in 2011 made me realize two things: sugarloaf-shaped as they are, with their rosewood finish and strangely comic-abstracted indentations, they are immediately reminiscent of a certain pop fantasy halfway between Tolkien's Shire of hobbits and the Flintstones, without fully being either; at the same time, they seem to pay homage to Richard Artschwager, and to a mode of handling material and its connotations that he established around 1964, as if unraveling the orthodoxy of Minimal art before it was even fully in existence.

Richard Artschwager sadly passed away on February 9, 2013. There is a question inherent in his work that some of your recent objects presented at PAC in Milan not only take up, but go further into. The question is to what extremes the tension between the thing as physically present and the thing as signifying cipher can be taken.

EL I began making objects hoping to elaborate on the potential of the picture and the moving picture to exist as objects.

Untitled (door 1), 2011
MDF, rosewood and paint
198.1 x 127 x 7.6 cm (78 x 50 x 3 in)
Photo: Ben Westoby
Courtesy White Cube, London/Hong Kong/São Paulo

I find that I work through a sort of "stretching;" by this I mean that I always get excited by the notion of pushing an idea to the point of potential absurdity, so that the photograph is being claimed as a sculpture. And there is this kind of playfulness: to put it backwards, why *isn't* the photograph a sculpture... The elaboration on that took the form of the cabinets, the stripped-down ornaments and the fixed installation of my films. All claimed a designated space that is "the picture." Perhaps the ornaments, which are essentially a sign-like sculpture, occupying the ratio of my photographs and installed in a relation of proximity, are the most extreme gesture. In them, I reduce the picture, with its Latin linguistic origin of also being a painting, a mark and so on, to a sign that is possibly compatible with the pictures adjacent to it. There was also a need to deal with the dimensions of the picture, specifically reflecting on the capacity for compression. With the objects, I make the volume of the compressed file come back to haunt us. Yet in such a ridiculously analog manner!

In 2010, Anthony Huberman invited me to do an exhibition with Richard Artschwager at the Museum of Contemporary Art St. Louis. Interestingly, the relationship had to do with my photographs. Huberman chose to focus on Artschwager's rubberized horsehair sculptures which were less known, and spoke about their connection to my photographs in the sense of having "fuzzy edges." I thought it was such a beautiful observation, and made such a great juxtaposition. Artschwager's sculptures were figures whose edges the viewer had to negotiate, due to the nature of the piled hair... And something similar happens within my photographs: it's often difficult to separate the foreground from the background, or the subject of the picture from the non-subject.

JH At PAC, the blue objects that look like two crossed slashes doubling as Christmas/Easter decorations (with the single balls dangling from them on ribbons) (*Untitled (Wall, Milan Blue)*, 2012) form a burlesque of the fetish object, presented austerely

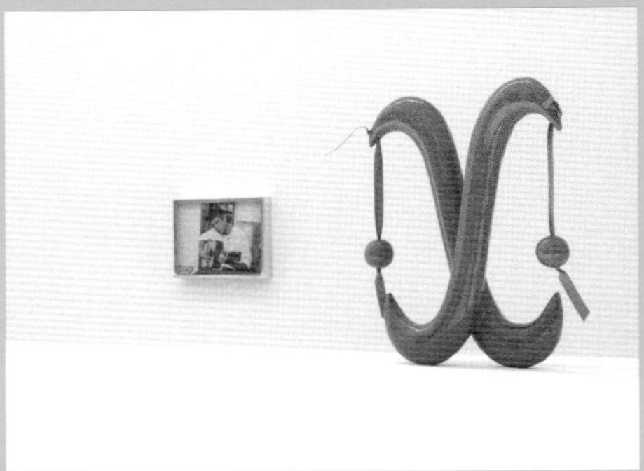

on a long white counter, just across from an equally long white wall with a series of images of Anthony Perkins strumming a guitar (another fetish burlesque: Norman Bates doubling as Elvis) (*Man (Guitar)*, 2012). You've mentioned Haim Steinbach as a major influence, and as in his work, that counter is a display mode that highlights the very idea of preciousness and elevation (of the commodity and how it is presented) as a curious, even funny thing we seem too often to take for granted. To quote good ol' Marx: "the form of wood, for instance, is altered, by making a table out of it. Yet, for all that, the table continues to be that common, every-day thing, wood. But as soon as it steps forth as a commodity, it is changed into something transcendent. It not only stands with its feet on the ground, but, in relation to all other commodities, it stands on its head, and evolves out of its wooden brain grotesque ideas, far more wonderful than 'table-turning' ever was." Your objects – are they uncanny siblings of real-life absurdist objects, taking cues from these "grotesque ideas"? What is it that brings them about – a process, as with the images, of "going through" many of them?

EL Once you establish that the picture is also an object, a sculpture, it's only fair to propose how the traditional sculpture can become a picture... Perhaps that's the perverse way I've always admired Steinbach's work. A way he might disagree with. I was thinking about how the "proposals" he made, the shelf pieces, introduced a set of problems where as a viewer I did not care much that the object on them was dimensional. It became a picture in my mind, mostly in the sense that it seemed to function on a level that was asking for quick reduction, only to unfold into something so much more complex. I love this condition in which sets of questions compete, pushing the viewer further.

JH The most obvious difference one could see between your films and the photographs and objects is that they explicitly indicate the passage of time.

Man (Guitar), 2012
Silver gelatin print, aluminum frame
21.6 x 26.6 x 3.8 cm (8.5 x 10.5 x 1.5 in each), triptych
Edition of 3, 2 AP
Courtesy Massimo De Carlo, Milan/London

Untitled (Wall, Milan Blue), 2012
Ceramic, ribbons
Dimensions variable
Courtesy Massimo De Carlo, Milan/London

But otherwise it's most notable how *close* your films still feel to the pictures and objects, and like them, their status as a medium feels infected by that of other media. They are like sculptures, like slowly tracked photographs, like documented performances (and as a result, your performances reportedly – I haven't experienced one in person – feel like films minus that documentative framing). And they are mute, which also supports this impression. They produce a certain nervousness (your word, I think) on the part of the viewer in terms of how to relate to this – what am I watching? what kind of slippery meaning am I eyeballing? – which is heightened by the fact that there is a certain bodily awkwardness or corniness (again, your word) at work that brings this nervousness close to some kind of hysteria. A hysteria, however, which feels sedated, as if locked in a diving bell. For example, a slow left-to-right pan of a zebra, from tail to head, followed by a slow left-to-right pan of a woman's face (*Zebra and Woman*, 2007), or equally stoically, mechanically/geometrically tracked movements of dancers in beige unitards, which culminate in them alternating between looking serious and smiling awkwardly in front of the cheerful painted abstraction that was tracked in the first part of the piece (*Untitled (Passacaglia)*, 2010).

What makes you want to do these films – is it their particular function in the exhibition, their interplay with the other elements in the show, their contribution to making the status of object and image and body-in-space feel all the more oxymoronic? Or is the emphasis rather on the idea of capturing what reveals itself in movement and through the explicit passage of time, like a slippage or excess, a sweaty forehead after too much exposure to a hot spotlight?

EL Recently, an art historian friend and I were talking about art, starting with an artist he was working with and then my own practice. He asked at some point: what is a good question about the work (to have the viewer arrive at)? We went back and forth; he suggested, to summarize, "What is a picture?" and I started laughing and said: "No... 'What *is* this?' is a good question." In a way, I truly like that place. A place of reconsidering something that my intellect would argue is very obvious. It would say "this is a picture of a dog, Elad." So, yes, taking a medium that is so heavily processed and turning it into "it," as if we were not sure what happens in the process of mediation.

So many funny statements are made these days about pictures... From the exhausting discussion of "the Internet" and "sharing pictures" to the non-stop academic symposia about how photography has advanced. Yet it's all gestural. The reason I insist on film is precisely because I don't think the problem with pictures lies solely in the iPhone, or digital file... Or the materiality of the print, and so on. In this case, I'm interested in the resistance that the medium brings with it. The potential for excruciating nostalgia. I personally am too young to remember a 16mm projector at home. But viewers do, and bring it up. I do remember home video. However, that's not the point. I have always made my films look as little as possible like the image you often get from 16mm film stock. They are shot at the highest f-stop, and I use lots of light to achieve a grain that is in visual contradiction with the lexicon of "experimental cinema". I credit Jack Goldstein with that strategy.

I am perhaps pleased that you recognize a sedated hysteria. That means I'm not completely alone with that one. I never heard that before. I also kind of dislike that. But it is there. Hence the sedation. There is a sort of hysteria that comes with pictures. It's embarrassing.

JH A very peculiar, odd experience which one otherwise may only get out of watching late nineteenth-century silent movie porn. Or particular Jacques Tati scenes, and maybe some of John Baldessari's films.

EL I absolutely agree that Tati is there as well, and possibly nineteenth-century porn. That all ties into this lack of fixed meaning, that slippery quality that filmic representation offers.

To me the films are a negation again. There is a picture that lingers, and maybe moves. There is a picture, in which its context, structure and history, its whole framework, are possibly competing with its presence.

JH As an aside, you've noted that I mention Richard Artschwager but not Ettore Sottsass, and that this was a whole different and long conversation.

On a more obvious level, there's a parallel between the two in regard to the use of laminated boxes, but on another level, Sottsass's work points to something that I meant to address, though admittedly not with him in mind: and that is the question of how all the media in your work that we've discussed so far – photography, sculpture, film, including fuzzy states in between – feed into what has variously been described as the "exhibition as medium."

A spatial arrangement experienced in time, a choreography of impulses, stimuli, statements, observations, states of "what *is* this..." Sottsass's *House Environment* of 1972 – with its series of containers holding household appliances – is a good example of a (designer's and architect's) approach to creating a "full environment" that however doesn't follow the integral or holistic idea of a *Gesamtkunstwerk*, but favors a modular, variable, fractured constellation. Just as Sottsass's environments seem to ask "How do we dwell, and could it be more exciting and intellectually stimulating?" your exhibitions – with their combination of various elements and media – seem to ask "How do we experience...?"

EL I mentioned Sottsass mostly to point out an interesting tendency that has to do with construction. Both he and Artschwager have been relevant to me. However, the mention of Sottsass, perhaps especially in relation to this Milanese exhibition, implies a direct source of tension that seems to take us directly to Memphis's postmodern practice via the modernist negotiations of the applied arts, i.e., the Viennese Secessionists and so on.

JH Could it ever be tempting to you to take your work, with all of that in mind, outside of the white cube, to “risk” it in surroundings that are impregnated with entertainment, usage, and/or the demands of other disciplines like theater or cinema? Lassry in a multiplex, an airport, a casino? Or do you prefer your work to remain in the gallery laboratory, with all its freedom and constraints?

EL Multiplexes, airports, casinos... Fascinating. I’d love for art to be there; however, it often feels like it’s in opposition. I like the fact that one visits a white-cube space precisely because of a need to see art. And I do like the focus and all the by-now familiar advantages of the white cube. Getting out is a bit of a pedagogy at the same time. The problem with public art often seems to be its institutionalization.

I see great works of art in the streets; they wouldn’t be defined as such, however, as they mostly seem to have been unintentional: they are accidents, or perhaps, ironically, born out of a lack of consideration. In West Hollywood there is an iron rack standing on a little grass lawn, where real estate agents drop off keys and rather biomorphically shaped add-on locks for apartments they show. It stands alone, decorated by a variety of bulbous pieces of metal (locks) that change every day. I see it as I walk my dogs in the evenings. Often when the LA sun is starting to set. It’s honestly up there with my top ten art experiences.

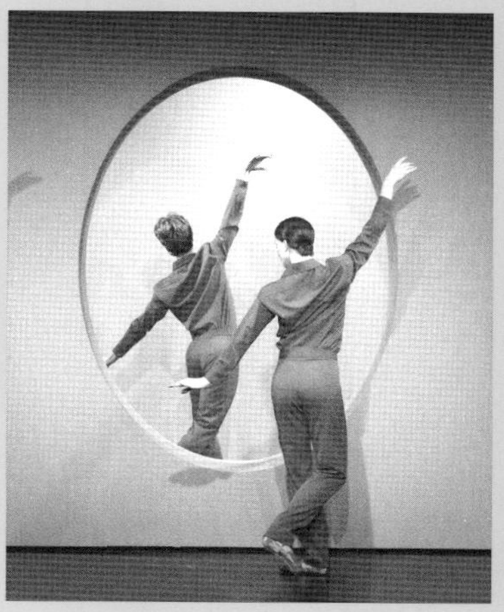

Untitled (Presence 2005) was performed March 2, 2012 at the Hayworth Theatre in Los Angeles.

ELAD LASSRY

Modern Life Is Noisy

by Aram Moshayedi

From the moment the sea of incandescent lights came up and the bright reflection of a polished, glimmering stage set filled the theater with its blinding glow in Elad Lassry's performance *Untitled (Presence)* (2012), the room was consumed by the sound of a low, barely audible hum. For a few brief seconds, before the dancers came onstage, the room was radiant and vibrating like a refrigerator on a hot summer day. The audience was left, for a moment, with this image – a piercing intensity comparable only to Chris Burden's reticent sculptural installation for the 1993 Whitney Biennial, *Fist of Light* (1993), which withheld the brute force of 1000 or more 500-watt quartz lights and was described by Burden as being potentially "so bright it turns black to white." Eventually, as eyes adjusted and the drone of electrical currents became familiar again in Lassry's performance, the pristine scene was interrupted by the rhythmic punctuation of the dancers' purposeful steps as they shuffled across the hardwood stage and the so-called silence was broken.

First presented at The Hayworth Theater in Los Angeles as *Untitled (Presence 2005)* and again on three occasions in a slightly modified form at The Kitchen in New York as *Untitled (Presence)*, the performance was an experience overwhelmed by sound.[1] The deafening silence that had, until then, only been disrupted by the occasional whirl of a 16 or 35mm projector in Lassry's earlier installations was brought to a fever pitch by the unmediated nature of a living, breathing event. Overcoming the ordered muteness that had subsumed his ear-

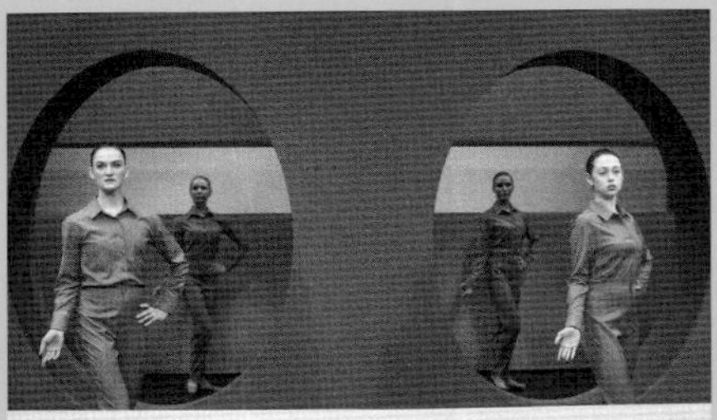

Untitled (Presence) was performed September 13-16, 2012 at The Kitchen in New York.

lier pictures, it seemed almost something of an aberration for Lassry to offer dancers in the flesh, whose bodies produce sounds that would otherwise have been edited out and sterilized by the inherently cold processes of representation. The artist's aspiration to achieve what he has described as a "perfect picture," up to this point, has produced an abundance of photographs and films that favor a decidedly visual noise over any actual recorded sounds. The overwhelming sense of visual discord produced by many of Lassry's photographs, though, might be best described as akin to such non-visual phenomena as tinnitus, or the uncontrollable perception of a high-frequency ringing in the ear when no sound is present. The annoyance of looking too hard or too closely at an image, or at one that is without fixity, produces the same result in the eye as it does the ear, and Lassry's tightly orchestrated *mise-en-scène* and a general overemphasis on the visual, in most cases, has shrouded his interest in the residual experiences of looking, in the echoes produced by pictures.

Curator Beatrix Ruf once wrote of Lassry's tendency to produce "visually seductive and irritating photographic and filmic works,"[2] while Douglas Crimp has referred to the artist's "productively irksome" performance at The Kitchen.[3] Often optically charged, bright and candy-colored, Lassry's pictures elicit responses of pleasure and displeasure such as these, perhaps a result of their tendencies toward dissonance, resonance, noise, and other sonic metaphors that have been used to describe them. This is not to say that the fetishistic polish, tacky satin, glistening smiles, designer objects, shapely breasts, insufferable cartoon characters, awkward patterns, plastic foodstuffs, chiseled bodies, stilted animals, perfect blondes, garish hues, double exposures, athletic actors, holiday schmaltz, grotesque tableware, exotic cats, and reflective surfaces aren't the source of this ambivalence. It is perhaps, instead, the muteness – like the blank stare on the face of a model with a fair white complexion and vacant blue eyes – that gets under our skin. With nondescript titles like *Man (Sky Blue)* (2012),

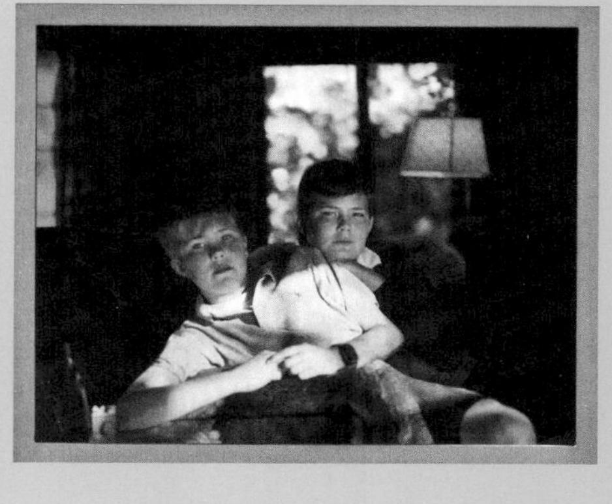

Boys, 2009
C-print, brass frame
20.3 x 25.4 x 3.8 cm (8 x 10 x 1.5 in)
Edition of 5
Courtesy Massimo De Carlo, Milan/London

Man 071 (2007), *Woman (Painting)* (2010), *Woman (Nude)* (2010), *Boys* (2009), *Girl (Green/Red)* (2011), and *Blonde Girl* (2009), it is apparent that the pictures on offer are not just representative of some infallible ideal, but are clearly from a world to which most of us will never belong, a world of generic perfection.

The world we inhabit, the one not conveyed by Lassry's pictures, is populated by bodies that sweat, that leak fluids and shit; bodies that decay, that have acne and scars and stringy hair; bodies that quiver, that shake; bodies that smell and that carry a stench; bodies with razor burn and saggy skin, misshapen genitals and droopy asses; and, most importantly, bodies that make noise, that grunt and groan and strain. When they are at their glossiest, Lassry's images keep forces like these at bay, and this is our greatest grievance: that we can never be as perfect as a picture or contain as much ghost-like elegance as a dancer on film. Taken one step further, it is the impossible silence of photographic images and of so-called "silent films" that makes this alienating experience all the more apparent. Collectively, we don't fit into the picture provided by the alienating quietness of most exhibitions, as our shoes scrape across a given concrete floor, our stomach grumbles, and our heavy breath gives away our age. The annoyance with ourselves is often most deafening when we are put on display by the performance of looking at an exhibition; and the scenes composed for Lassry's photographs and films, and by his architectural tableaux, are reminders of how inherently divorced we are from pictures, how difficult it is to penetrate and access picturehood despite the supposed visual literacy of our time.

An antidote to this is offered by the actors and dancers from the New York City Ballet, with whom Lassry has worked on a number of films including *Untitled (Passacaglia)* (2010), and who have been immunized by the camera that records the image of their bodies in space and divorces that image from the soundtrack of life as it is lived. Obscured by the filmic medium's function of making things available to vision, the bodies of the dancers in Lassry's films are best likened to ghosts, carrying none of the weight or messiness of the physical body. In *Untitled (Passacaglia)*, the dancers perform a series of movements based on Doris Humphrey's radical work of modern choreography, *Passacaglia* (1938), but their bodies are rendered mute, if not moot, by the camera's structural preoccupation with the splatter-coated set that frames their movements. In Lassry's clean and ordered field of vision, there is only room for a kind of visual dissonance provided by the film's backdrop and awkwardly cropped frame; there is a cleaning-up of any missteps or the unwanted sounds that otherwise accompany and express moving images. However, as seminal music theorist Douglas Kahn has argued when making the case for the cultural dominance of visuality over aurality, "Silence is all of the sound we don't intend. There is no such thing as absolute silence."[4]

A similar argument was perhaps also offered when art critic Sarah Lehrer-Graiwer succinctly asked after the performance of Lassry's *Untitled (Presence 2005)* in Los Angeles, "Can pictures – photographs – capture the pitter-patter sound of a ballerina tip-tip-tapping en pointe?"[5] Surely, an uncontrollable din is one of the many things evaded by photographic pictures, and perhaps one of the strongest forces that press against the conceptually "cinematic" and "pictorial" impulse behind Lassry's foray into live work. The edit-like lighting cues, shifting scenography, and overall transformation of the theater's black box into an aperture were key concepts behind Lassry's performances, with the pitter-patter, tip-tip-tapping as the only thing upsetting the pitch-perfect silence that consumed the audience's attention on most nights. For the most part, we behaved like an audience.[6] We sat in silence; the sounds came from the dancers' feet on stage, from the rolling casters as the set pieces were moved into position with each new scene; and hardly ever were we made aware of anything or anyone else.

It wasn't until the last performance of *Untitled (Presence)* in New York – when the persistent coughing and hacking of a man in The

Kitchen's adjacent bathroom permeated through much of the work's total 23 minutes – that the unbreakable sound was finally broken and our flaws reemerged and reinforced the distance between ourselves and the picture-perfect scene playing out before us. Members of the New York City Ballet and American Ballet Theater moved through their repertoire of dance phrases and composed the various pictorial vignettes by periodically moving the multicolored sets into a sequence of frame-like arrangements, through which audiences were provided a series of fragmented views. The sound of the dancers' staccato steps filled the cavernous black box, while the hollow floor acted like the head of a drum, and somewhere in the distance the sound of a man's last gasp was a reminder of our physicality. The sonic aspects of *Untitled (Presence)* overwhelmed the visual, and our bodies fell silent in response, but for these moments we uncontrollably shifted our weight from side to side. We felt embarrassed for the man and empathized with Lassry, who was seated among us.

However, the wretched sounds of the man coughing at The Kitchen provided an experience unlike any other from Lassry's earlier "dance" films, which have featured many of the same dancers in choreographies that similarly pull from the history of modern dance, and which have tended to be exhibited alongside Lassry's photographs at a one-to-one scale. The addition of any foreign sounds by viewers in the theater context, unlike galleries and museums for the most part, is an unwelcome disruption and annoyance: we are so often reminded to please silence our cell phones and respect our neighbors at the beginning of a theatrical presentation, yet hardly ever shushed in an exhibition. But on that night in New York there was a revelation, hacked up in the phlegm of a theatergoer's throat, something akin to a non-diegetic soundtrack for the pictures unfolding before our eyes. It was the sound of looking; it was the sound of our selves and of being disgusting.

1) *Untitled (Presence 2005)* was first presented as a 17-minute, 30-second performance in Los Angeles by David Kordansky Gallery at The Hayworth Theater on March 2, 2012, and again on three occasions at The Kitchen in New York in September of that same year as *Untitled (Presence)*, the duration of which was 23 minutes. The differences between the two performances included the increased length, as well as changes to the casting, set design, and overall choreography.

2) Beatrix Ruf, ed., "Introduction," *Elad Lassry* (Zurich: Kunsthalle Zürich and JRP Ringier, 2010), 71.

3) Douglas Crimp, "Being Framed," *Artforum International* 51 no. 5 (January 2013): 56.

4) Douglas Kahn, *Noise Water Meat: A History of Sound in the Arts* (Cambridge, Mass. and London: MIT Press, 1999), 163.

5) Sarah Lehrer-Graiwer, "Last Picture Show," *Artforum.com* (6 March 2012), http://artforum.com/diary/id=30458.

6) "Behave Like an Audience" is the name of a vinyl LP by the band Concert, published in 2013 by Sternberg Press, whose title was conceived by curator Mai Abu ElDahab.

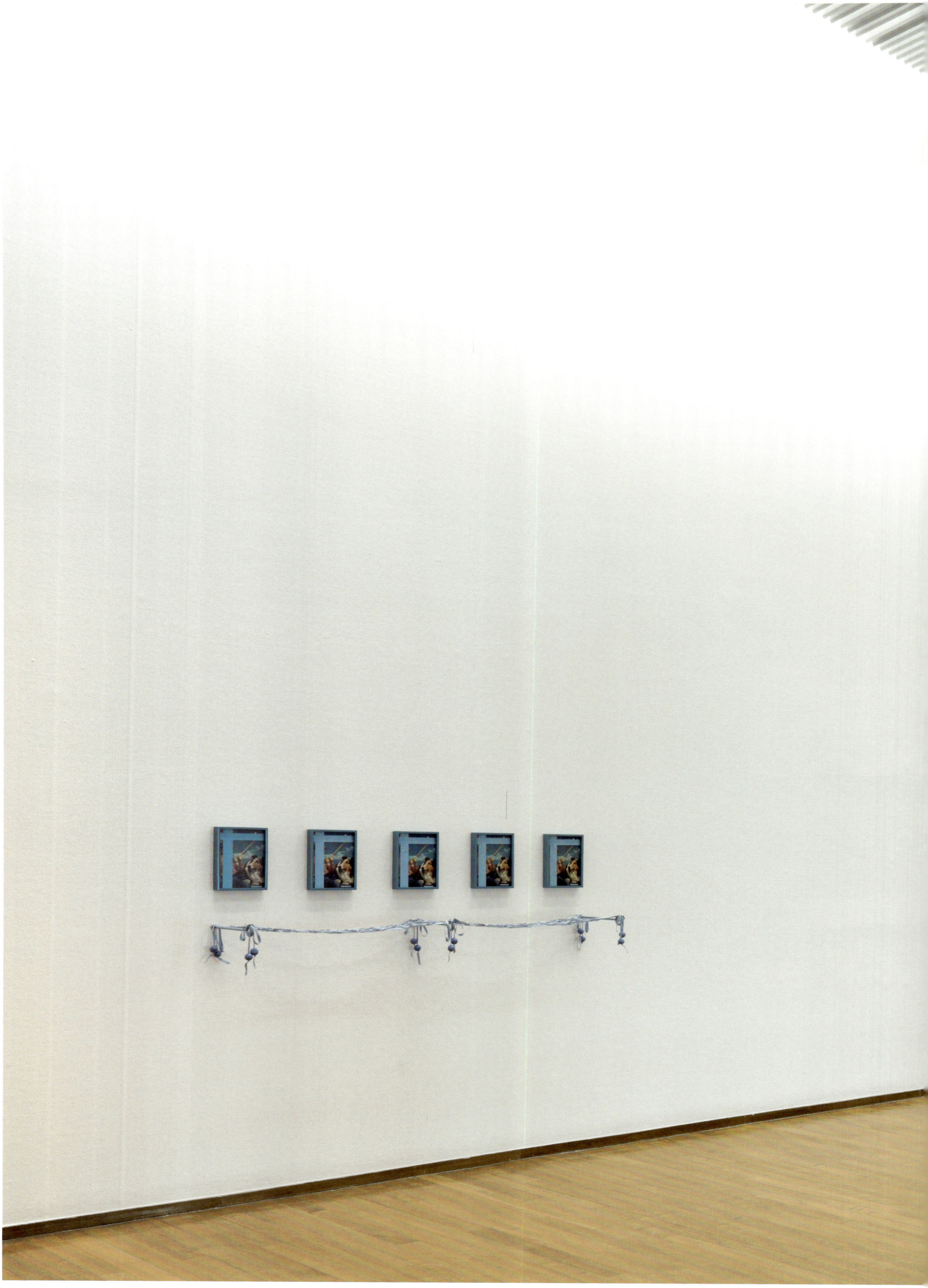

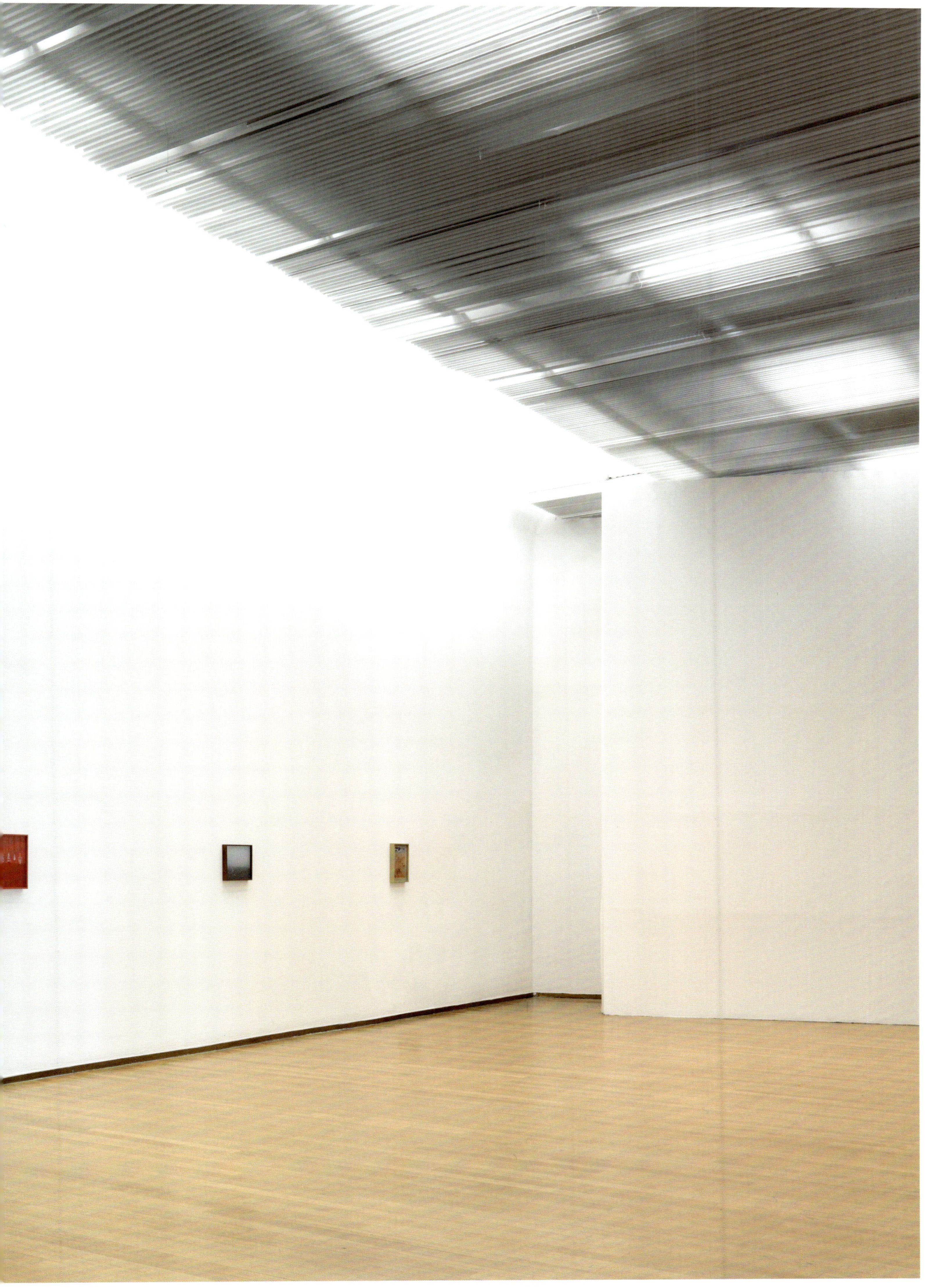

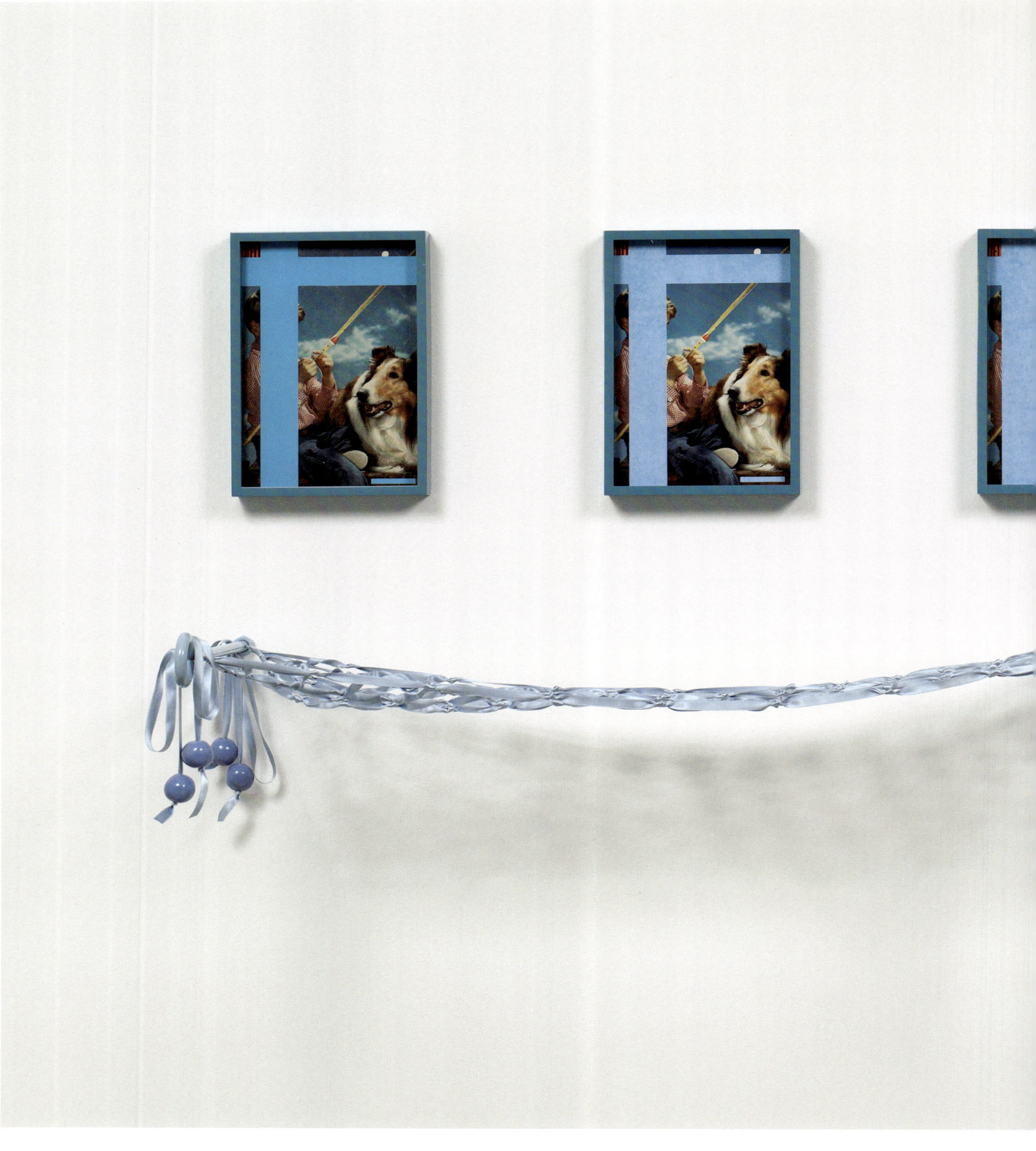

 Collie (Blue Bars), 2012

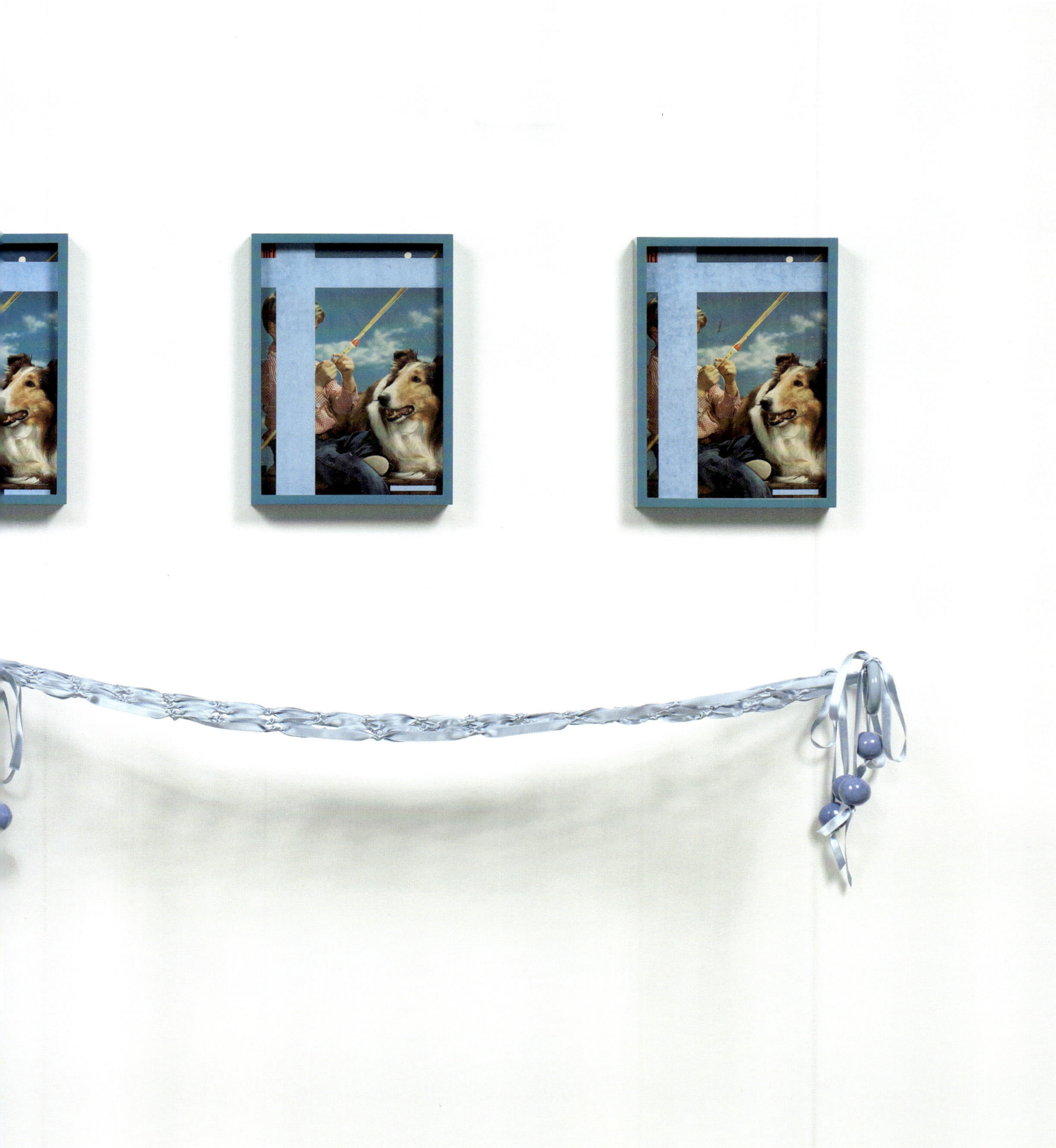

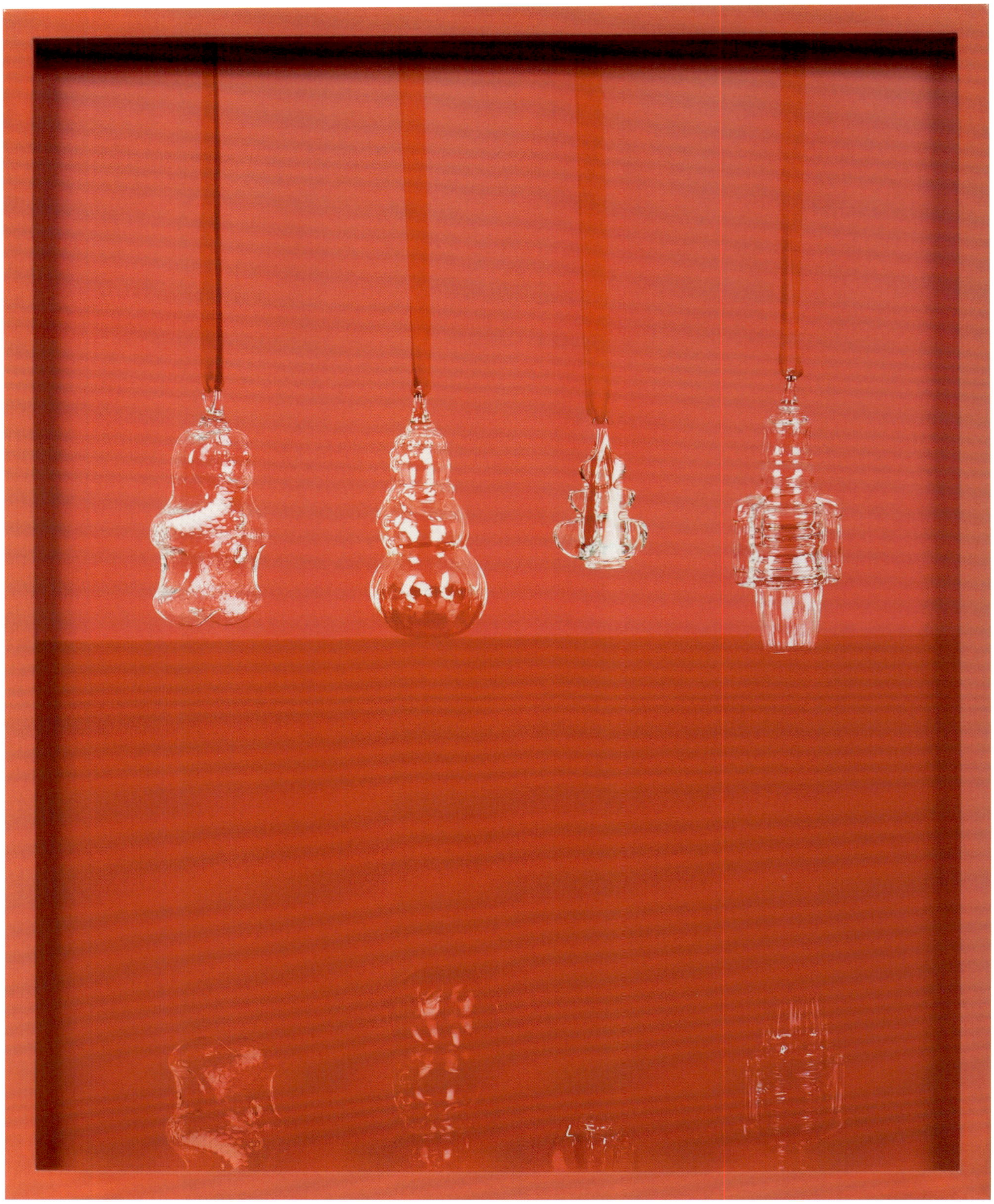

 Four Crystals (Red), 2011

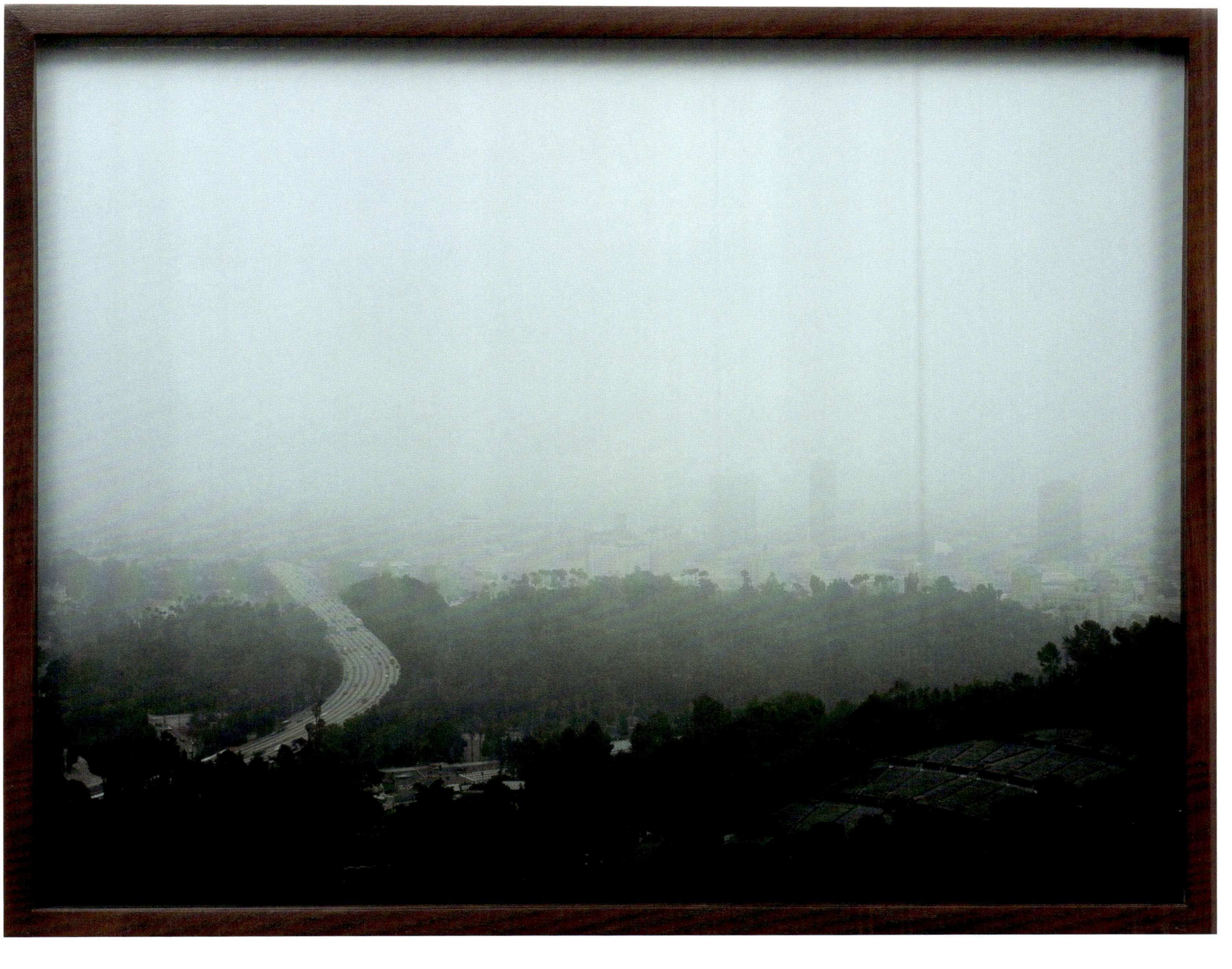

27 *Hollywood Bowl, Fog,* 2007

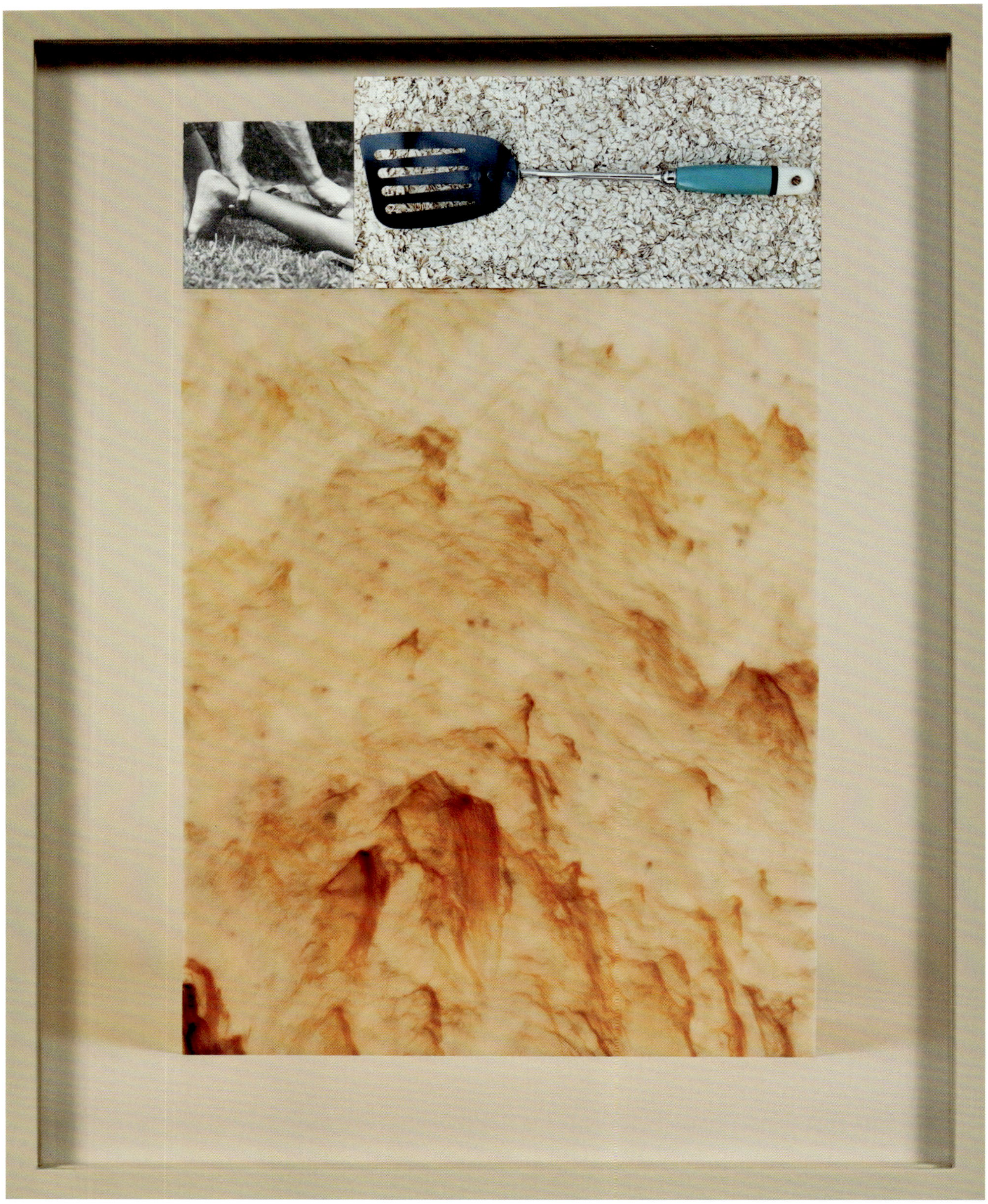

 Horn Structure (For Massage and Collectable), 2009

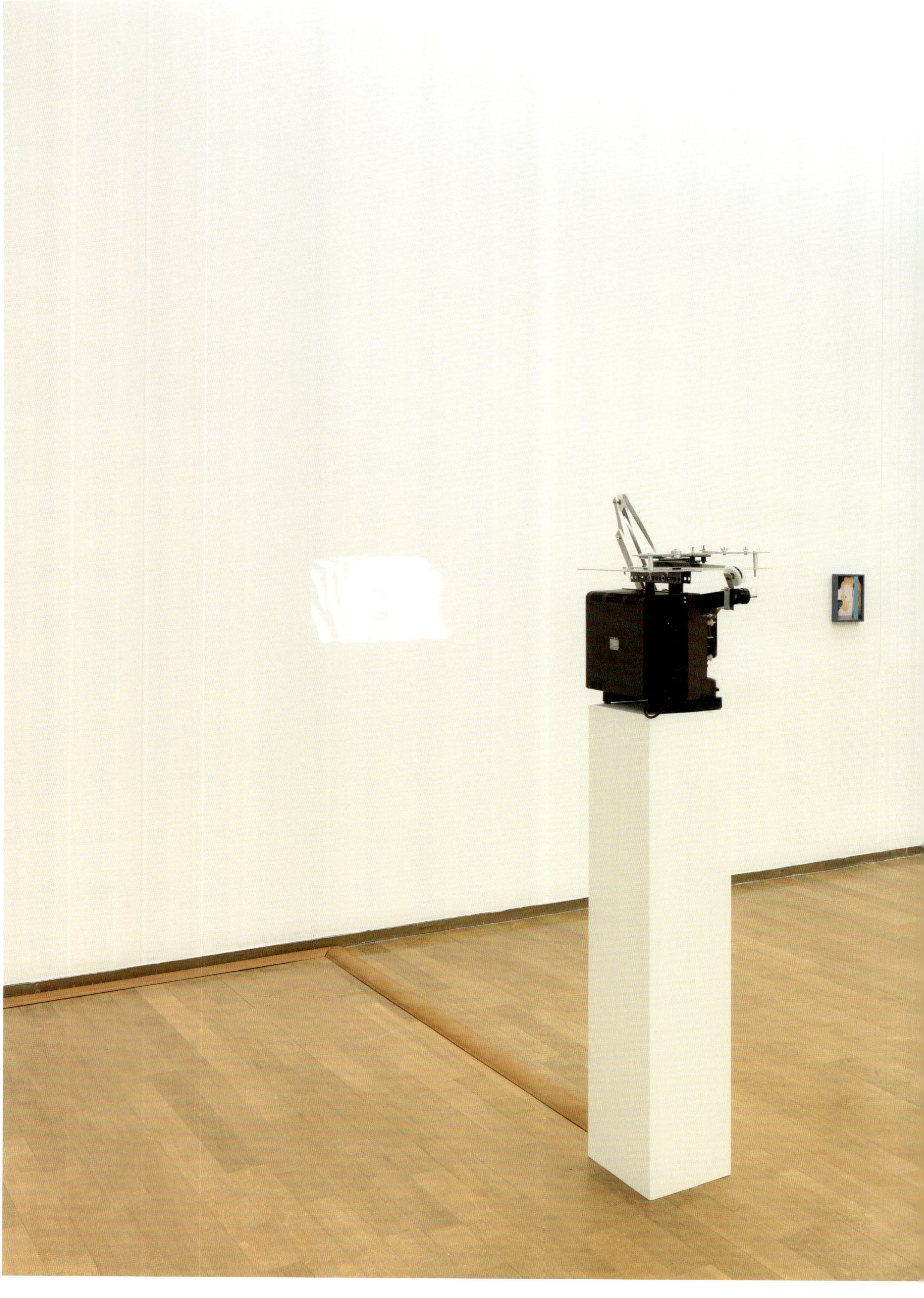

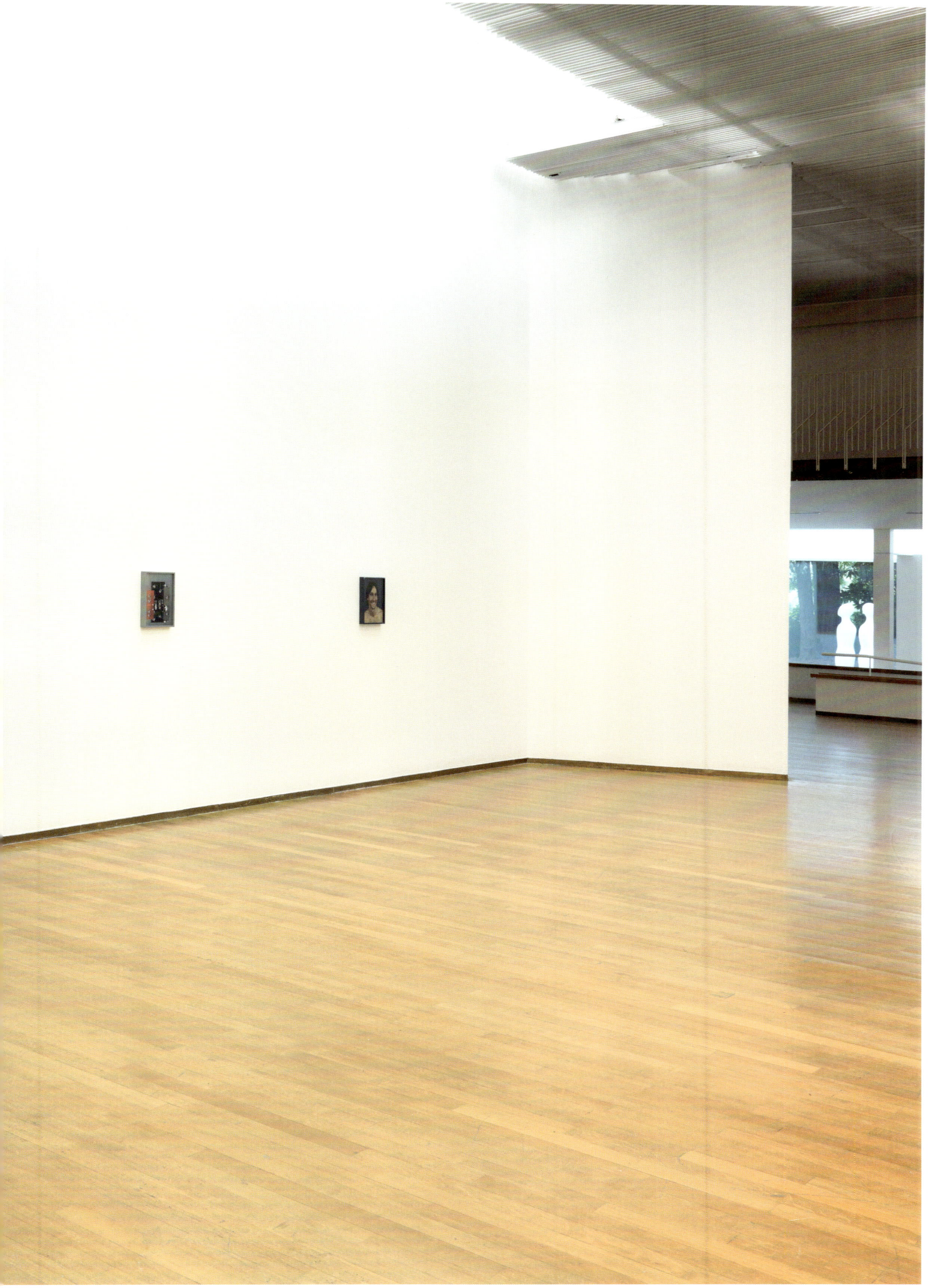

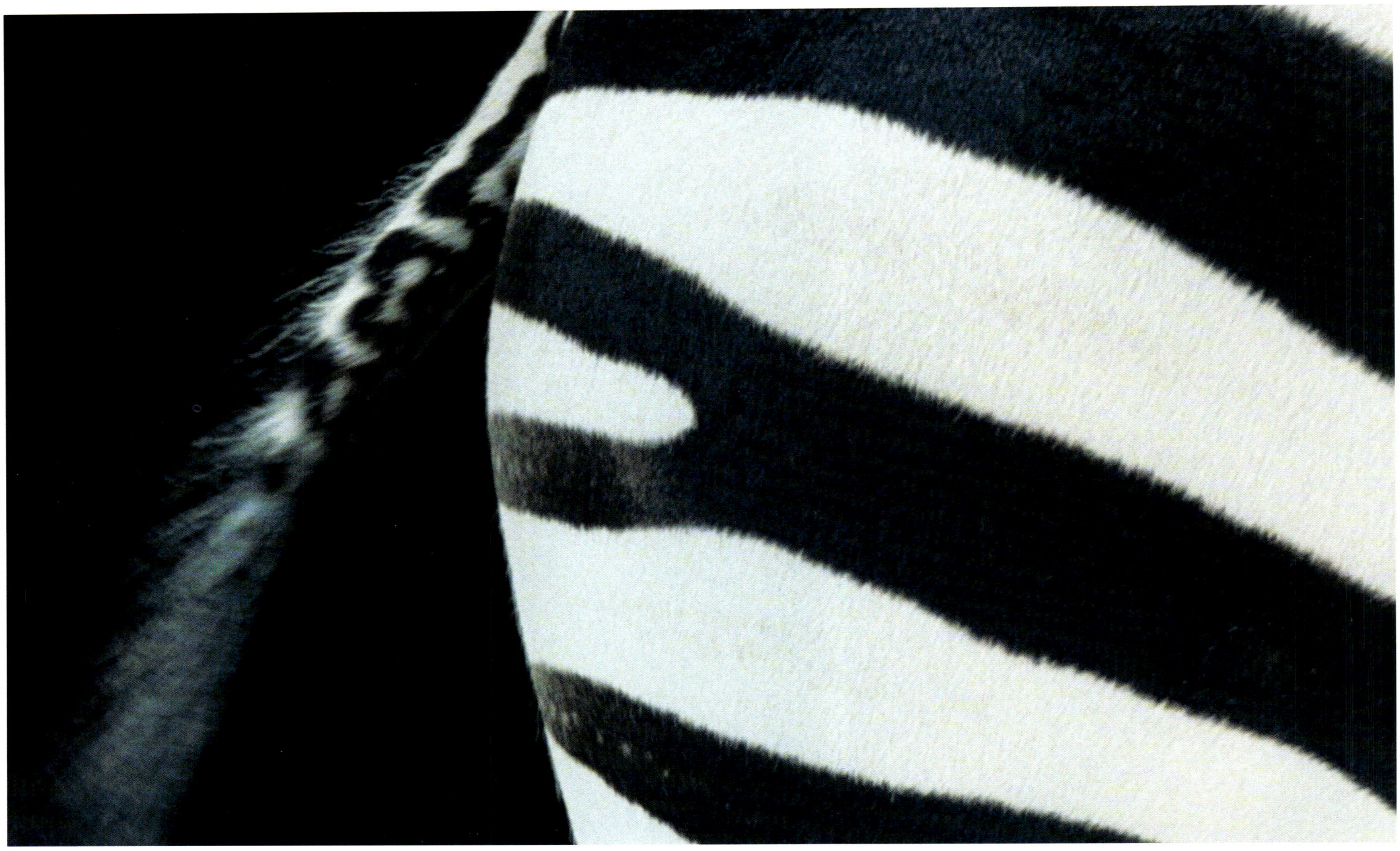

 Zebra and Woman, 2007

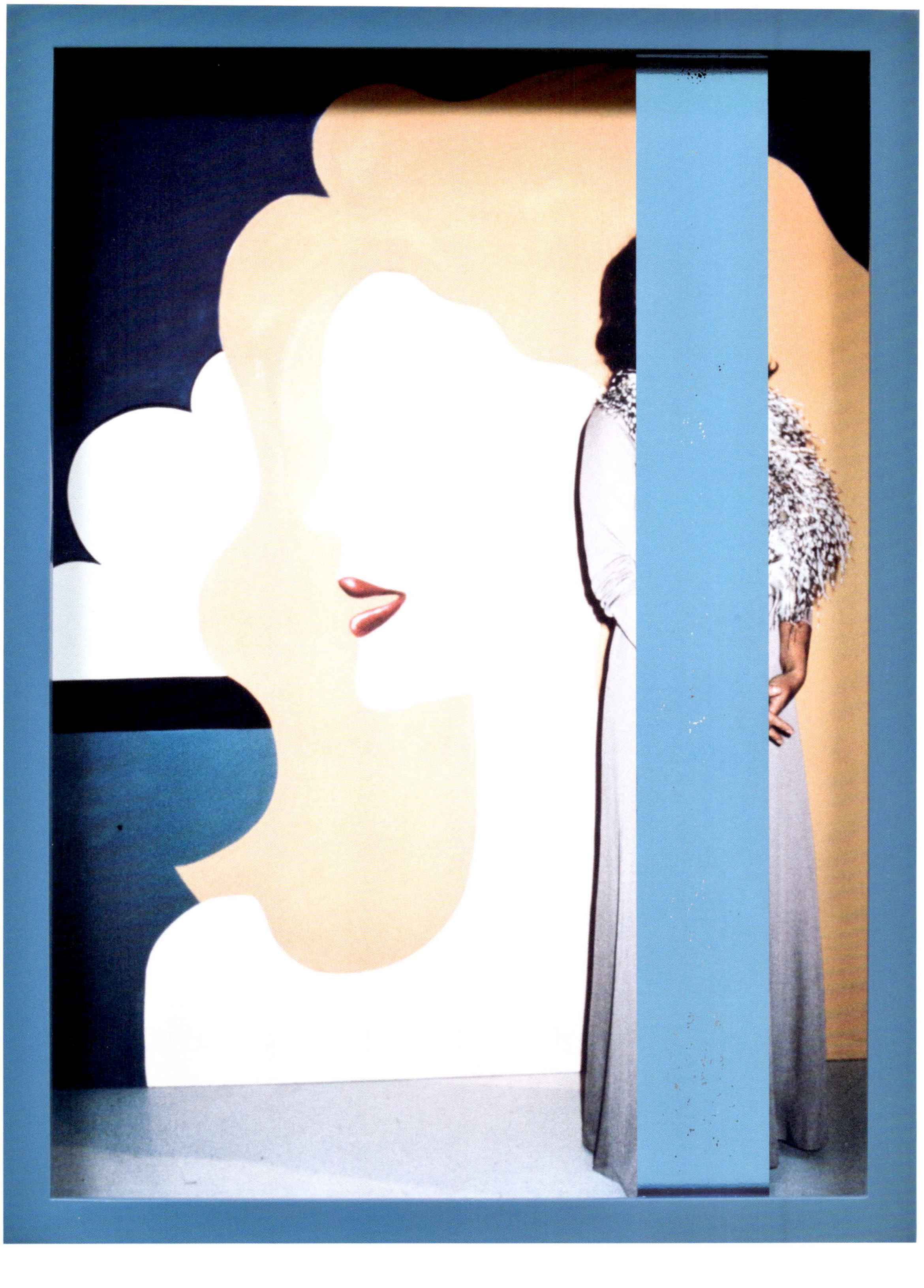

 Untitled (Blue Bar Mural), 2011

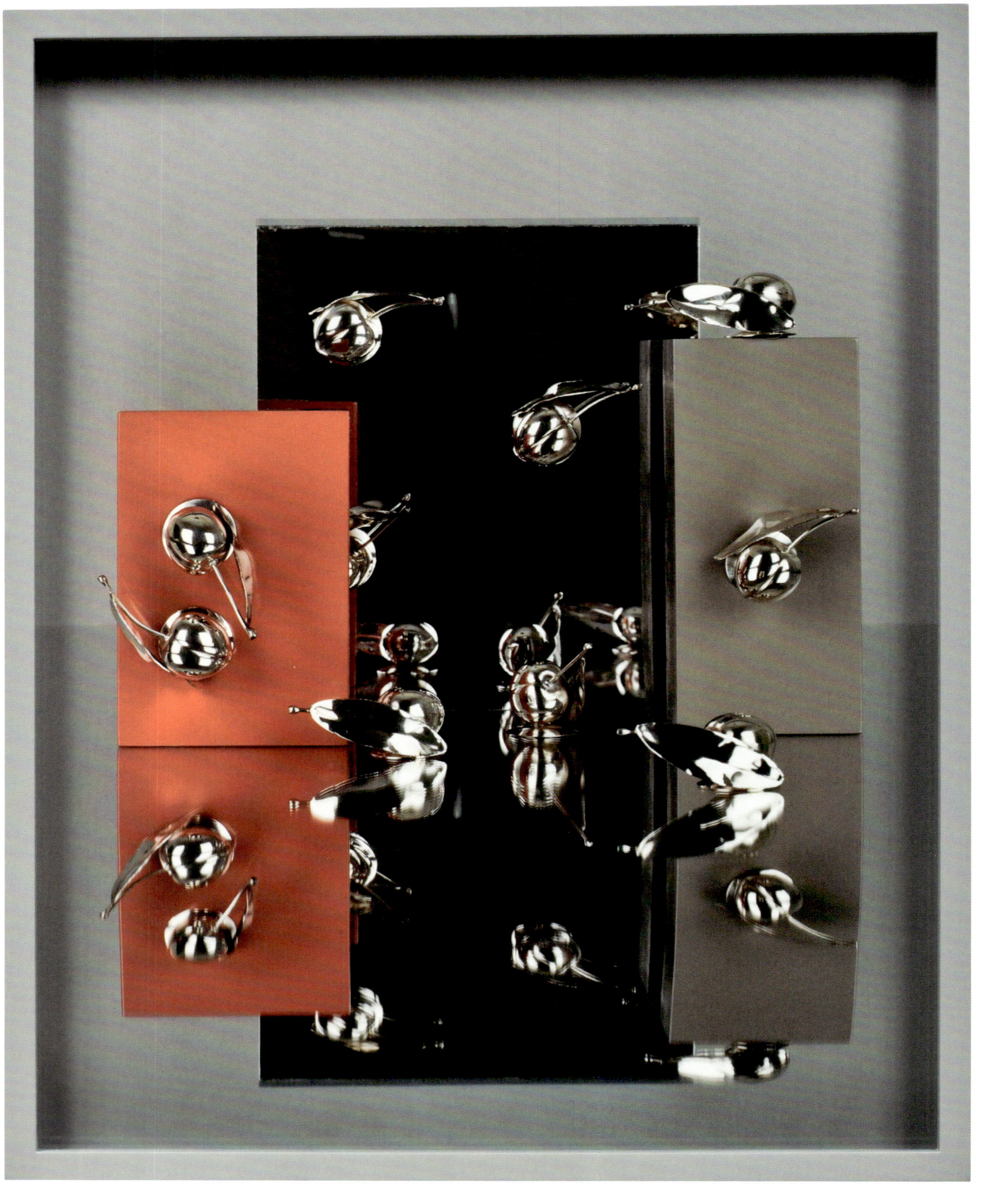

34 *Cherries (Silver)*, 2012

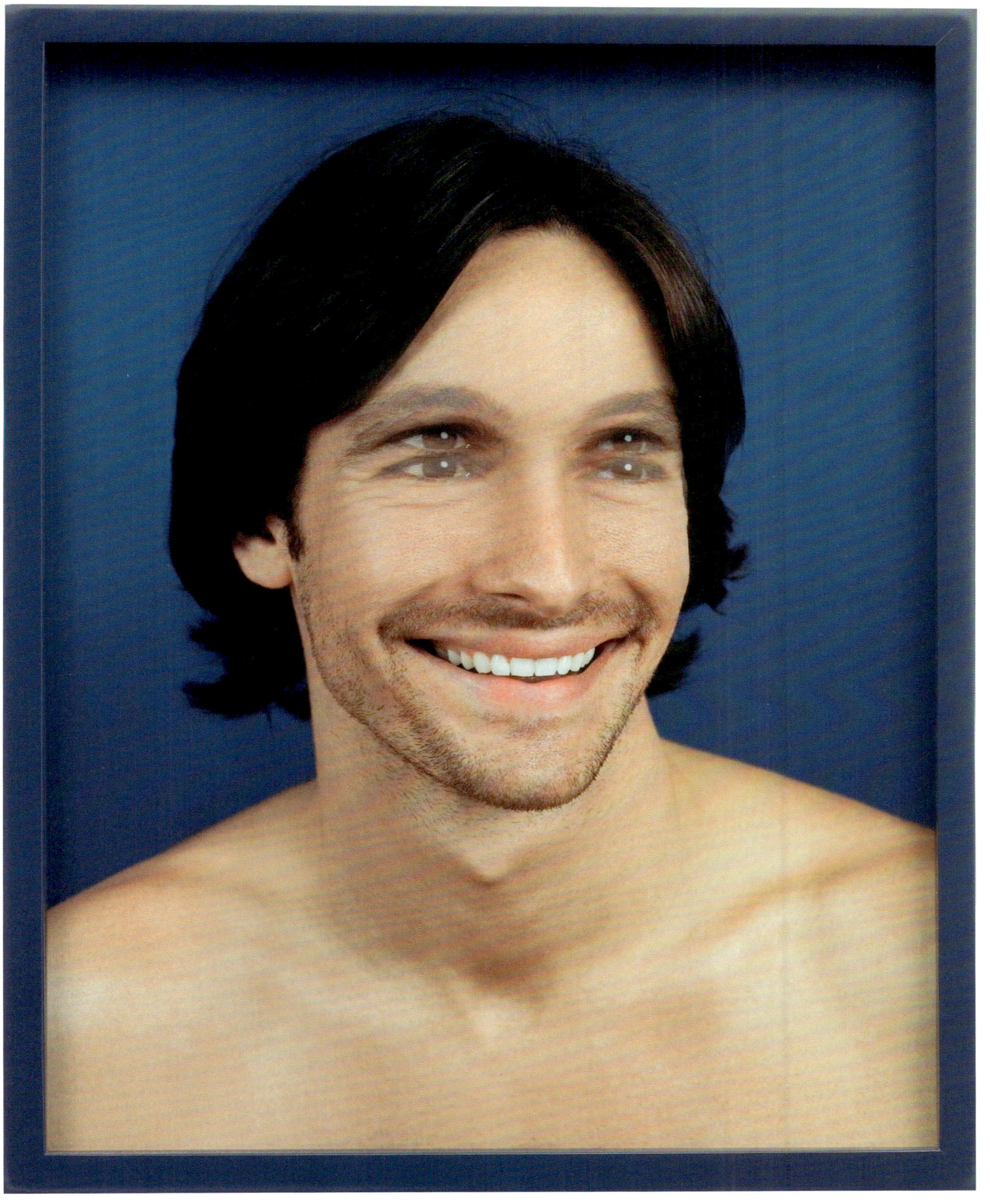

35 *Man 071*, 2007

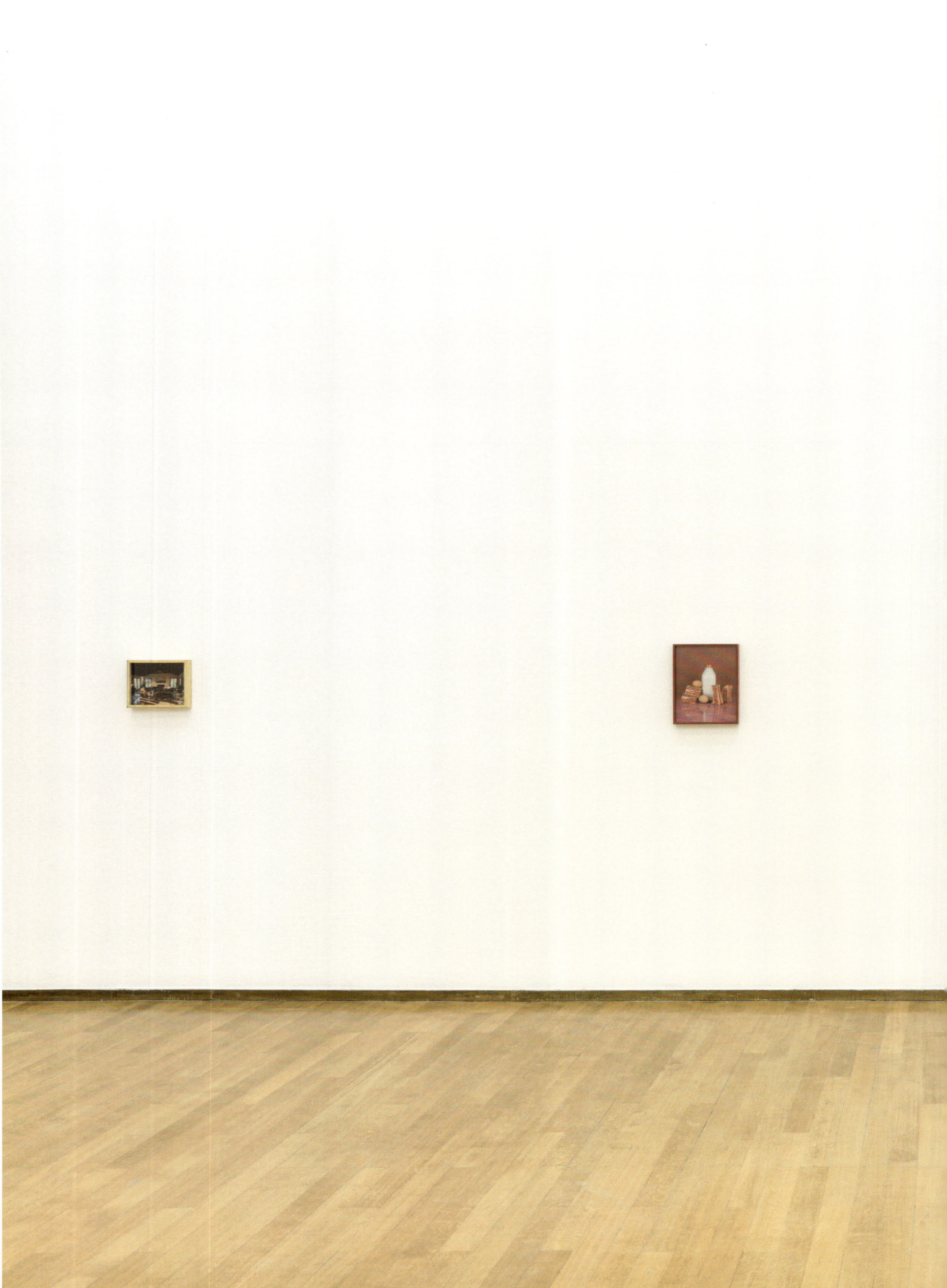

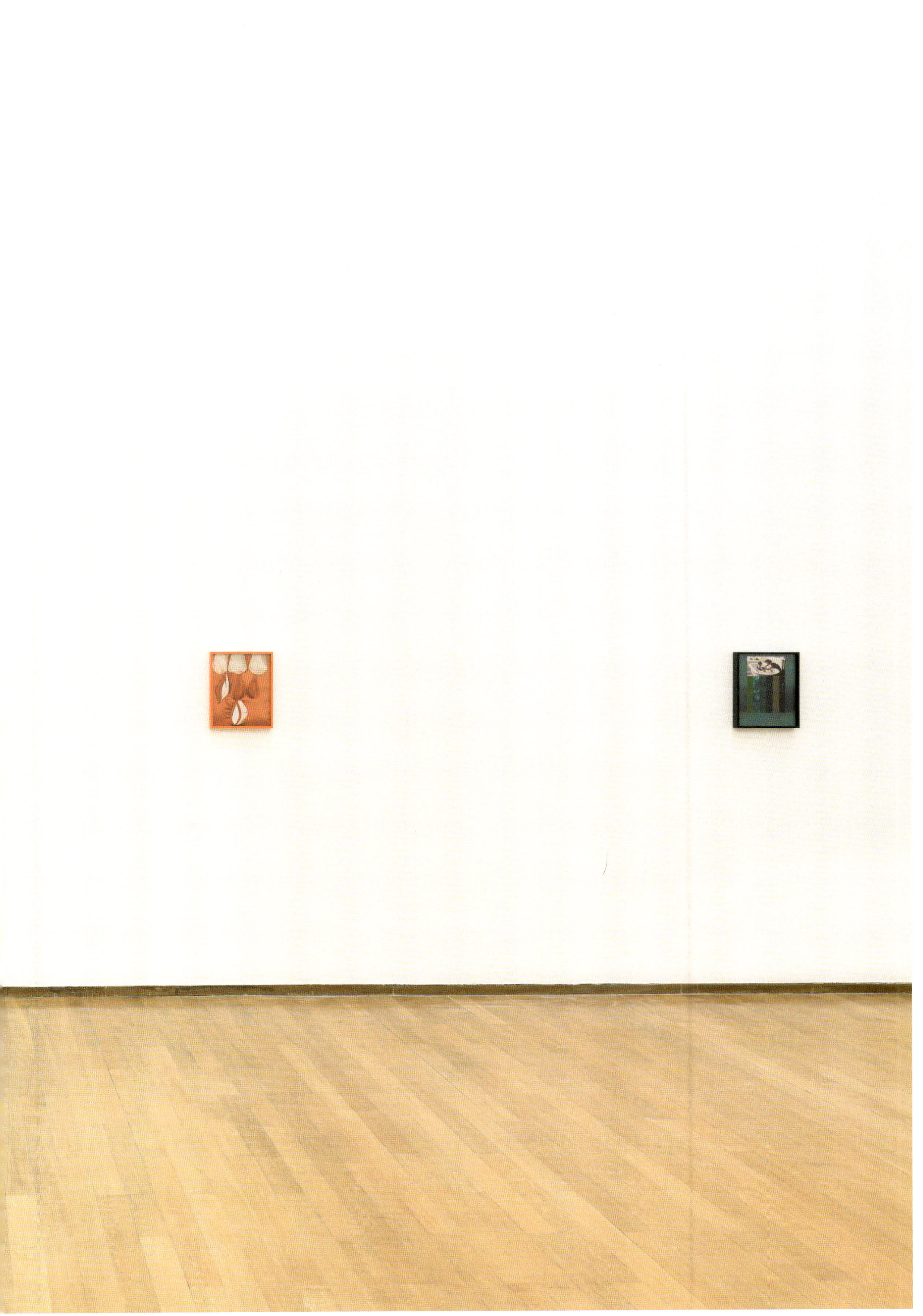

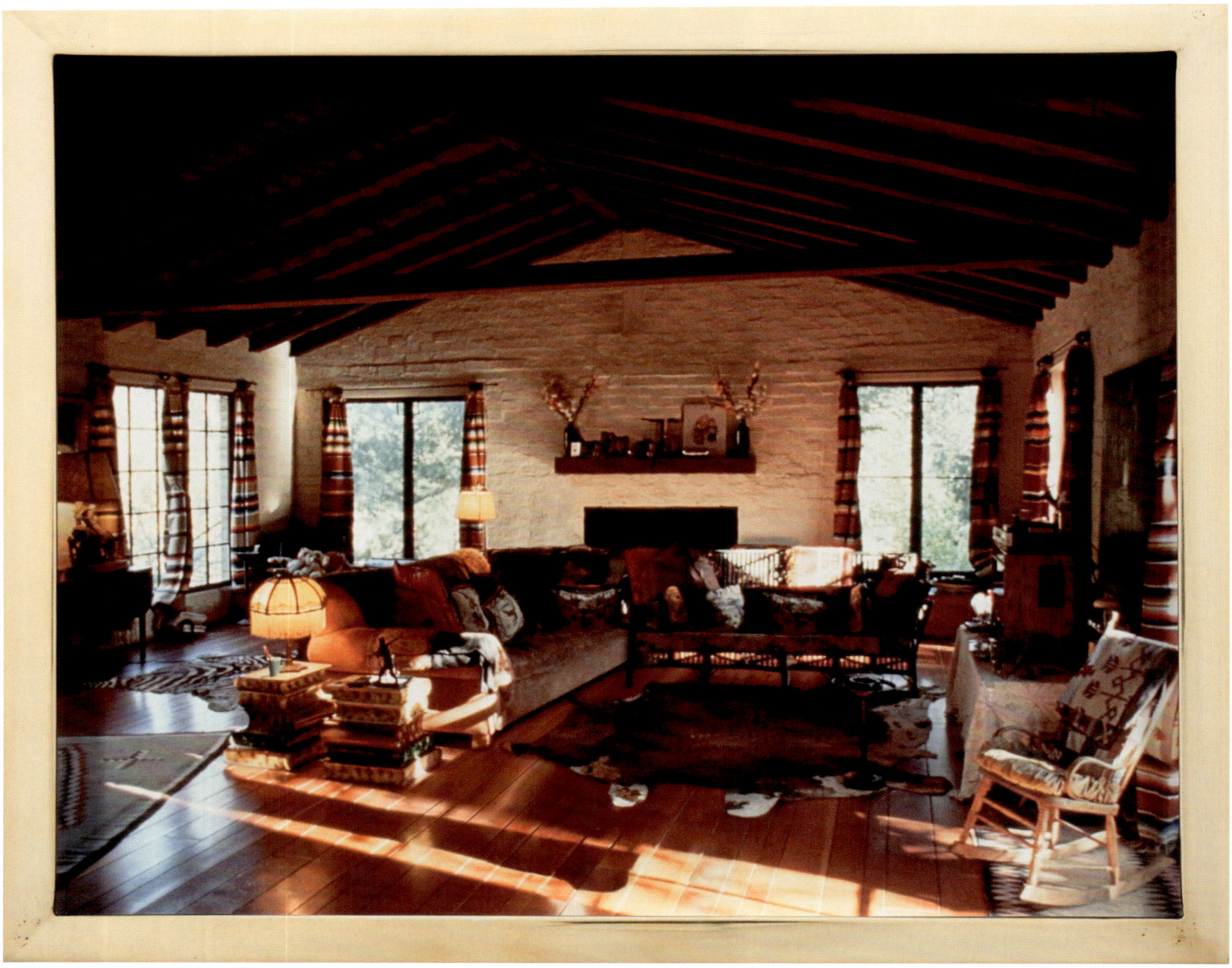

 Living Room, 2010

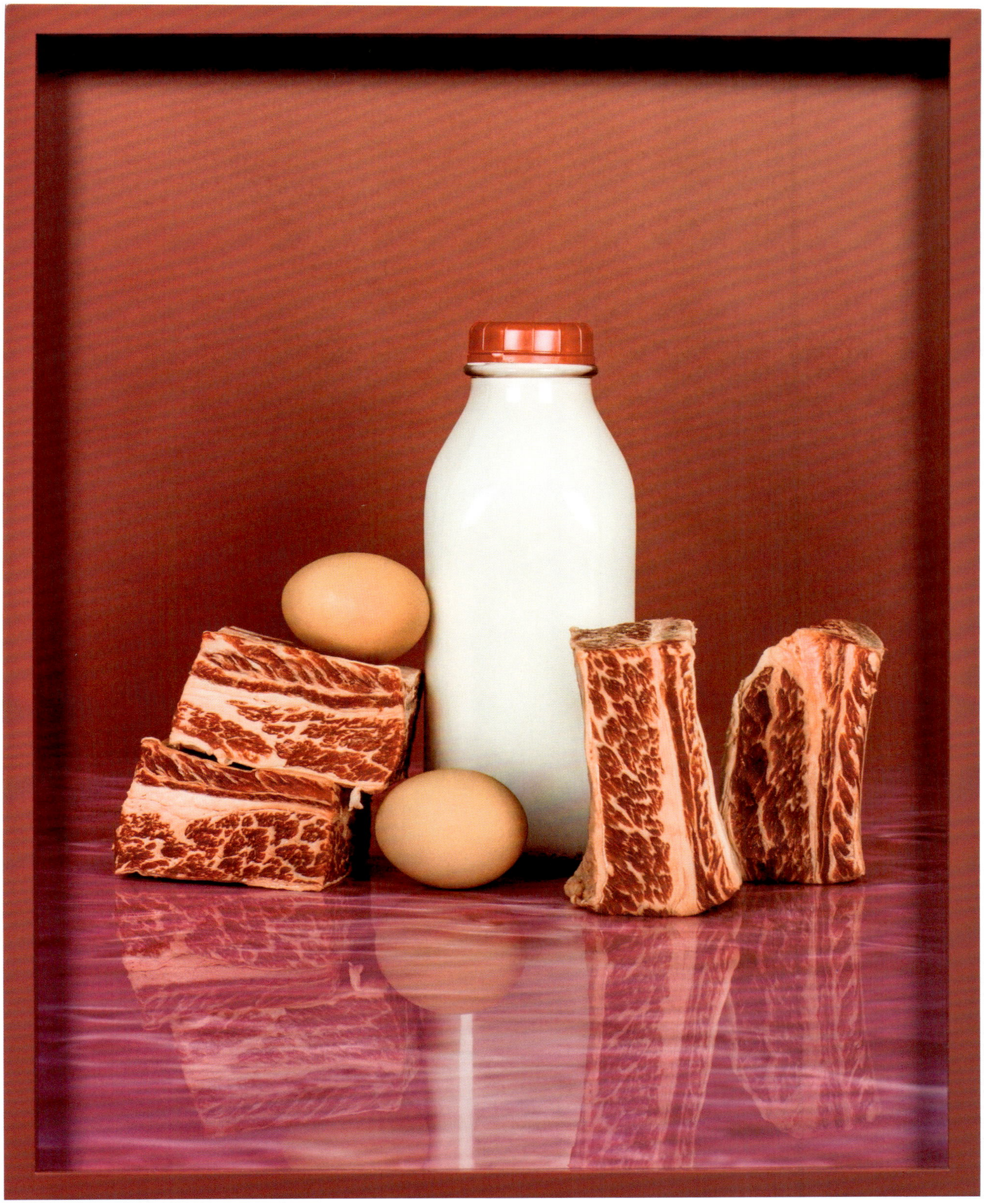

 Short Ribs, Eggs, 2012

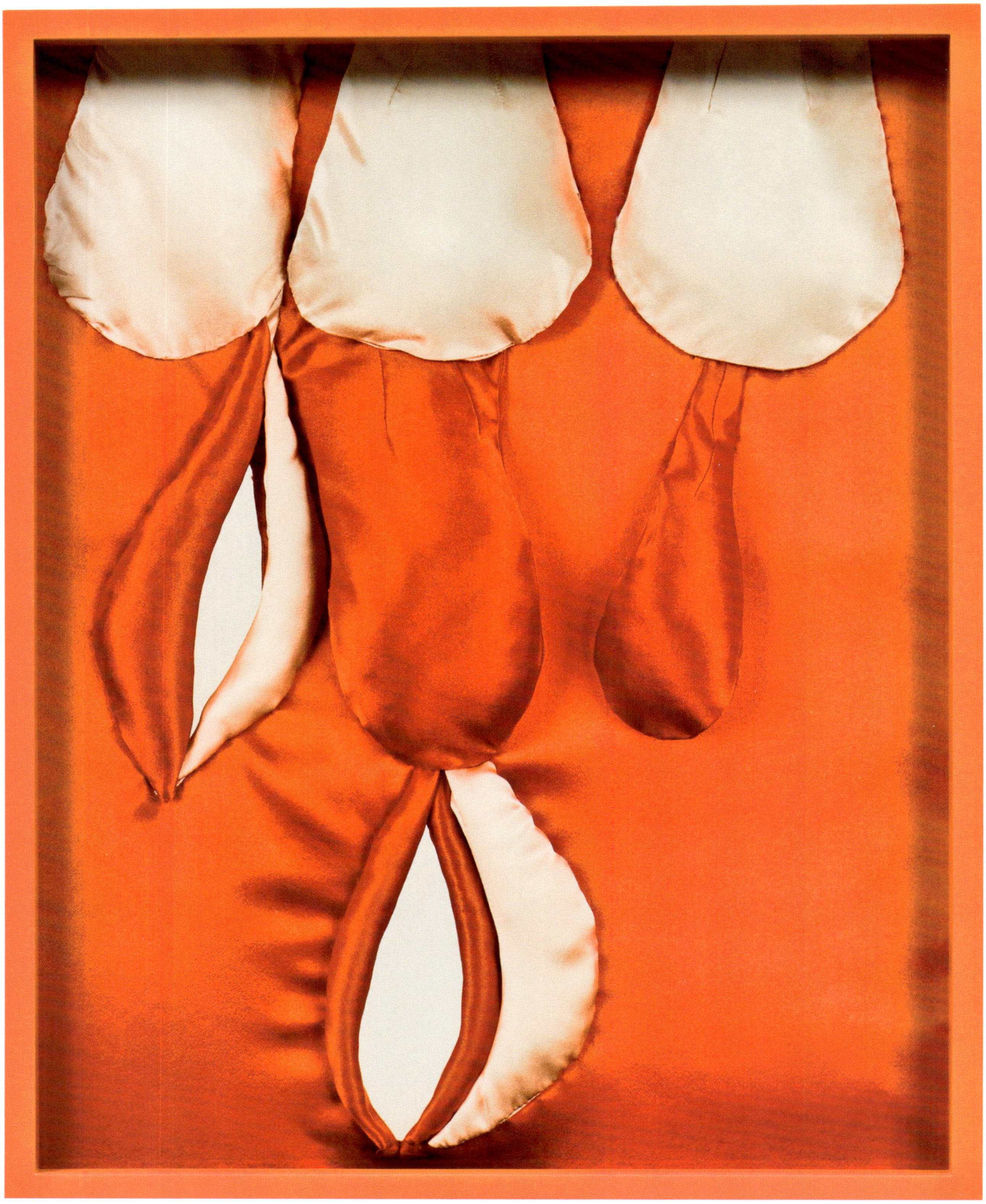

 Pillow 3, 2012

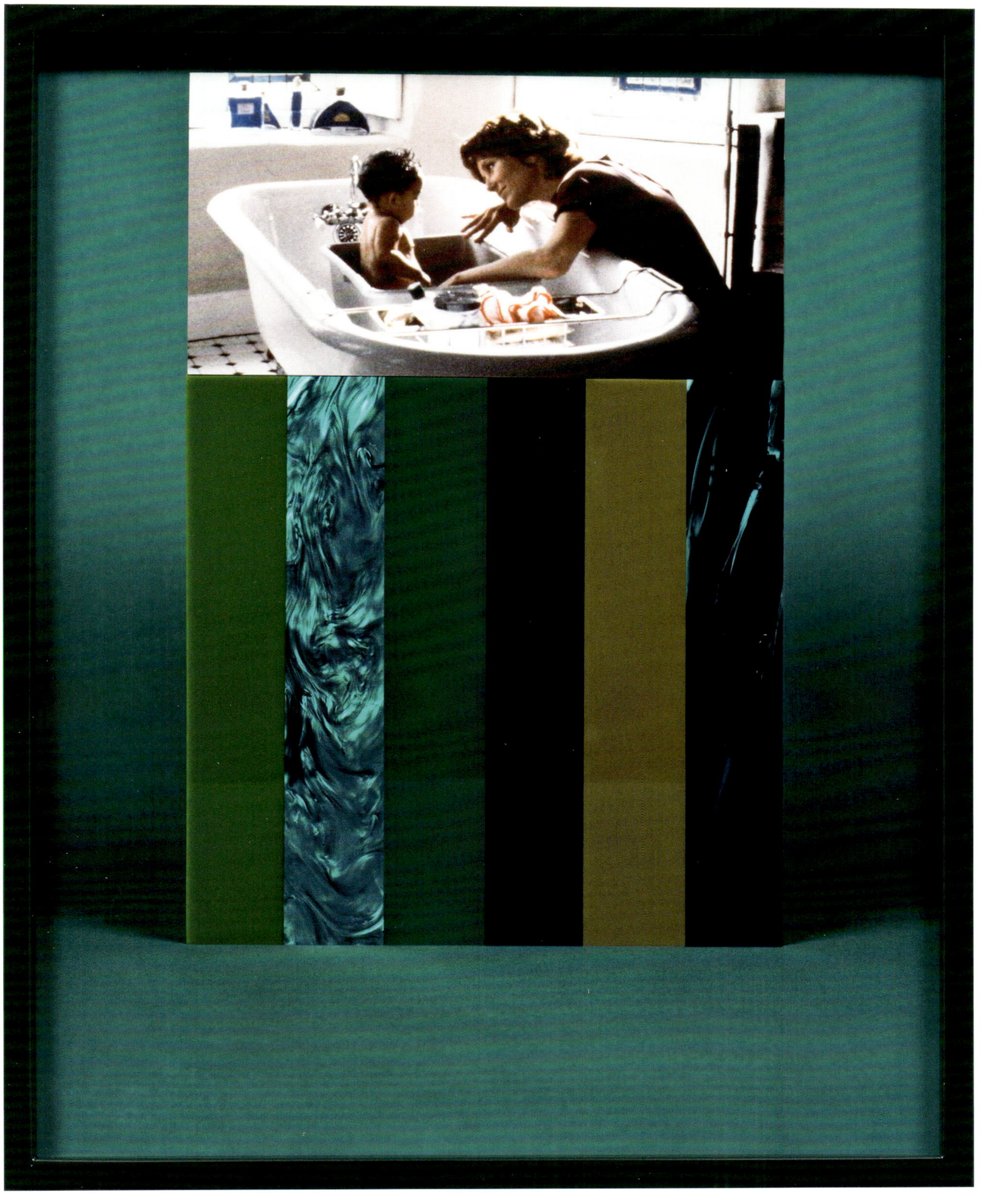

 Laminated Structure (For Bath), 2011

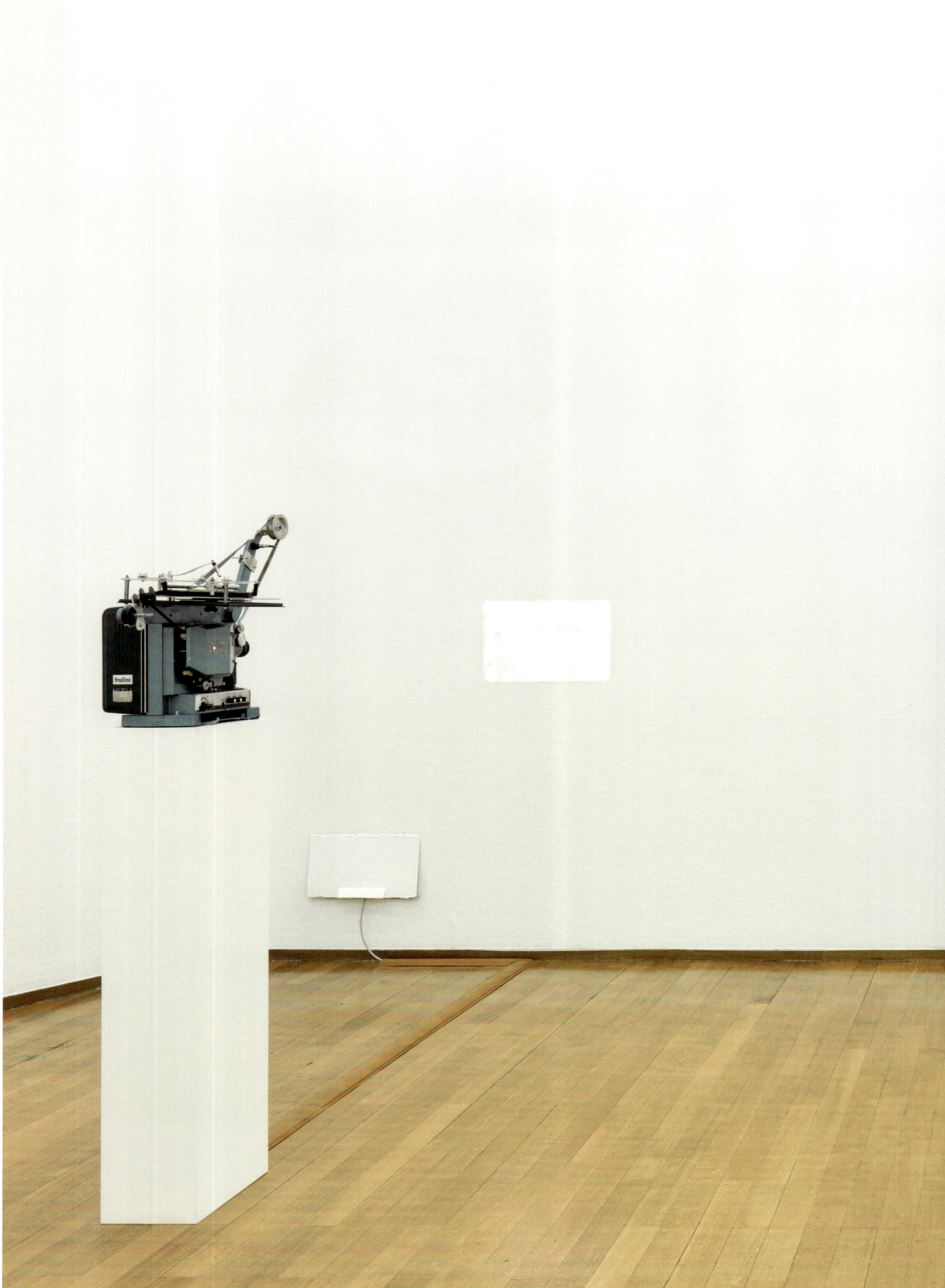
frollini

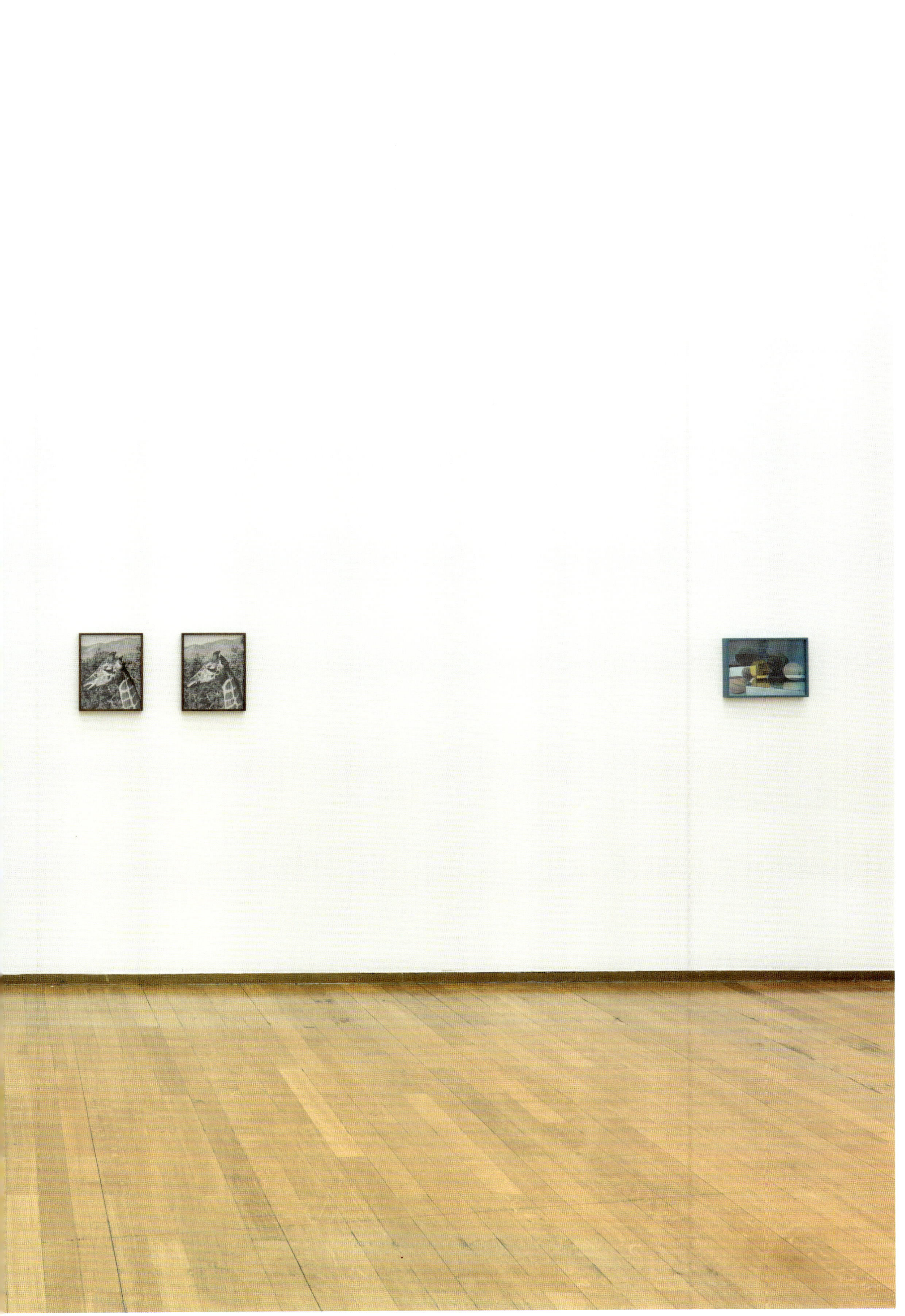

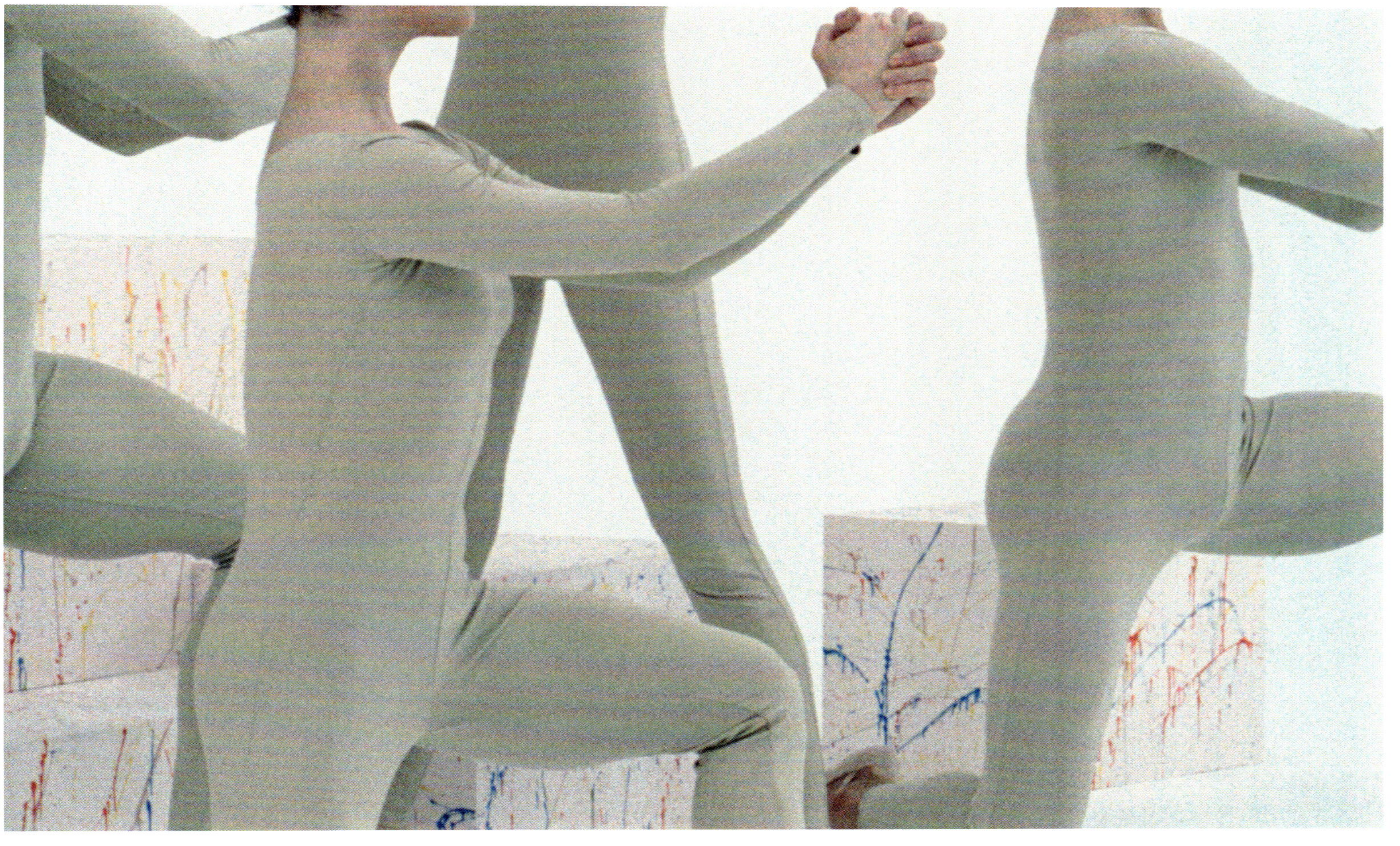

45 *Untitled (Passacaglia)*, 2010

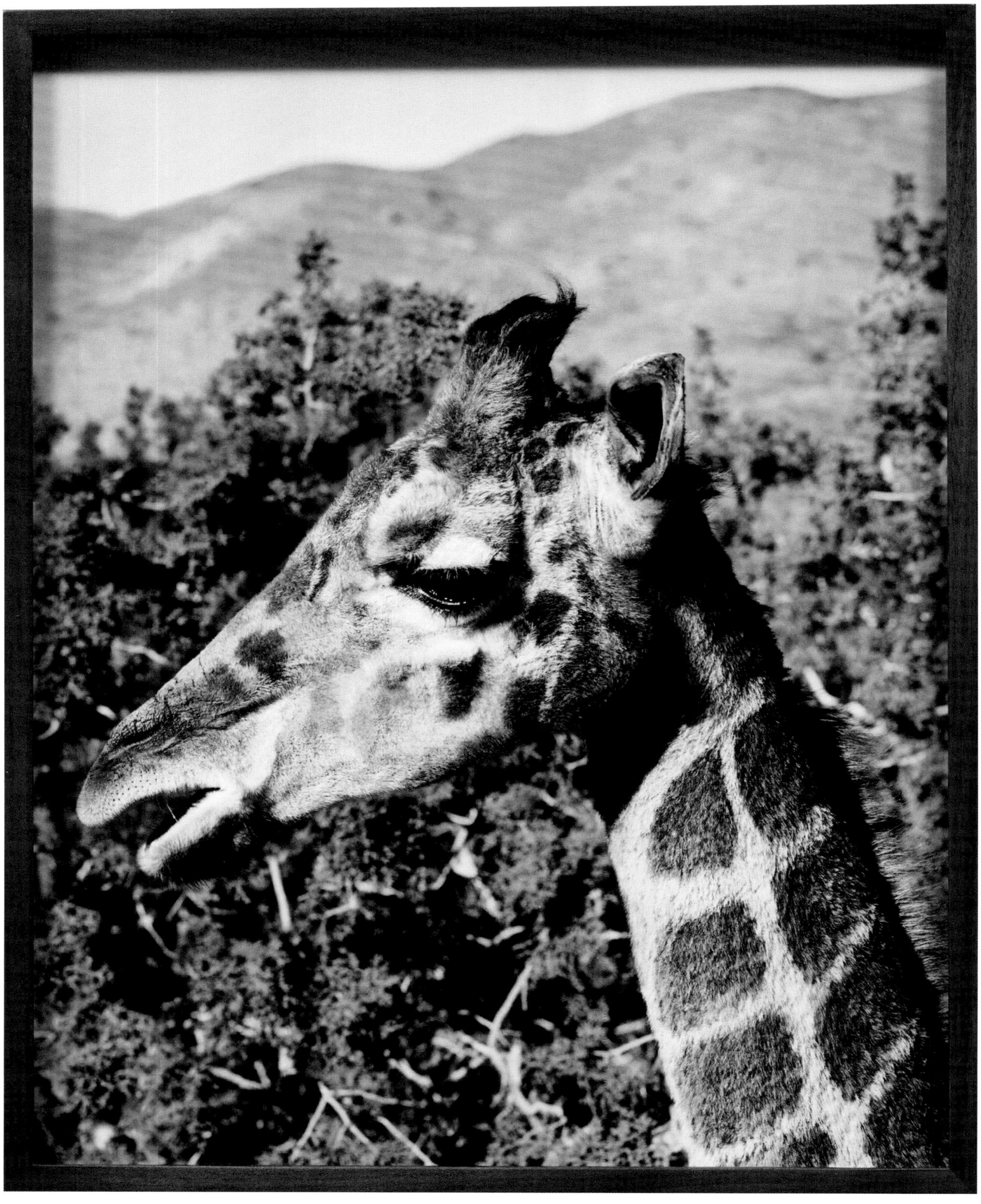

 Giraffe, 93040, 2011

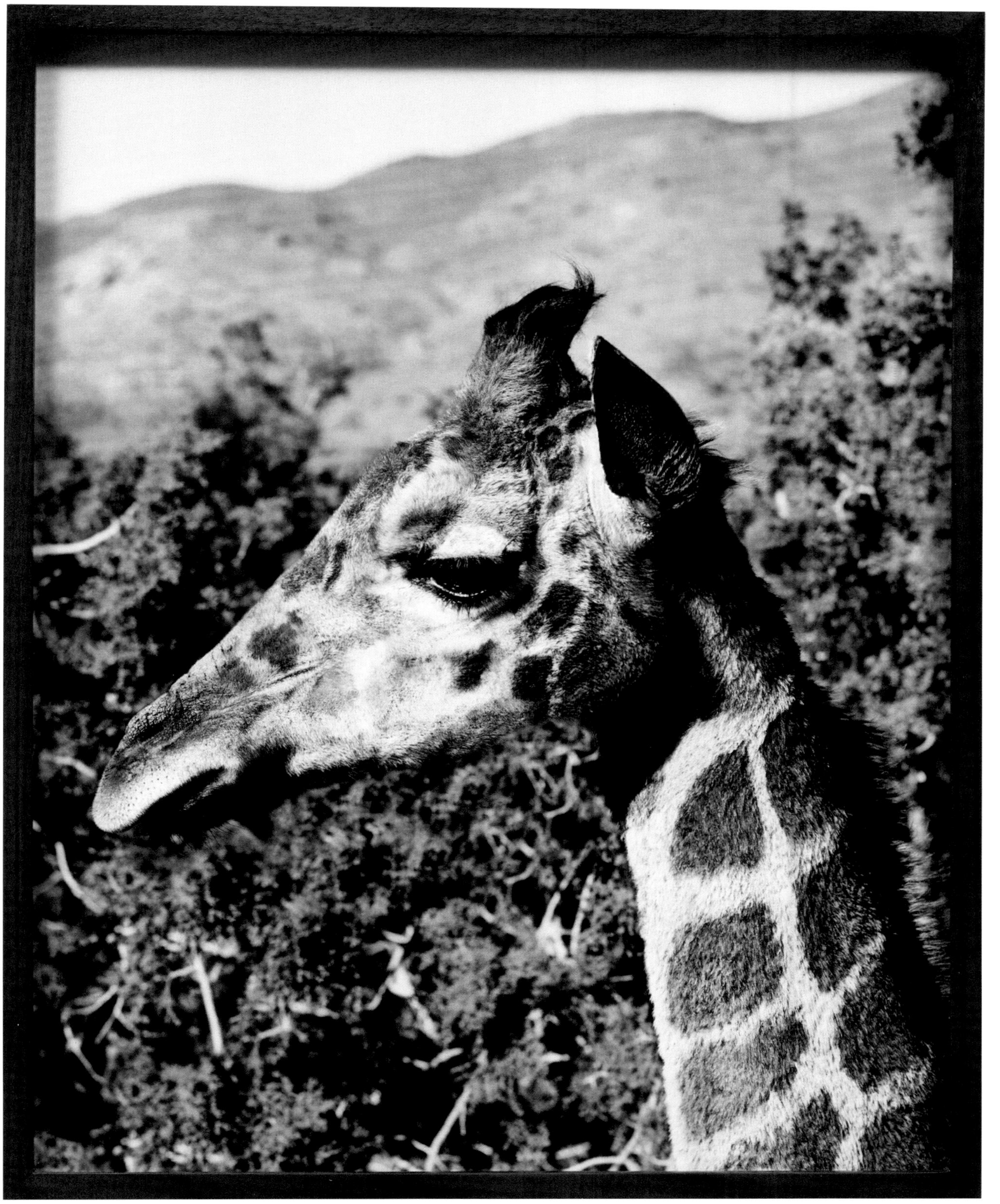

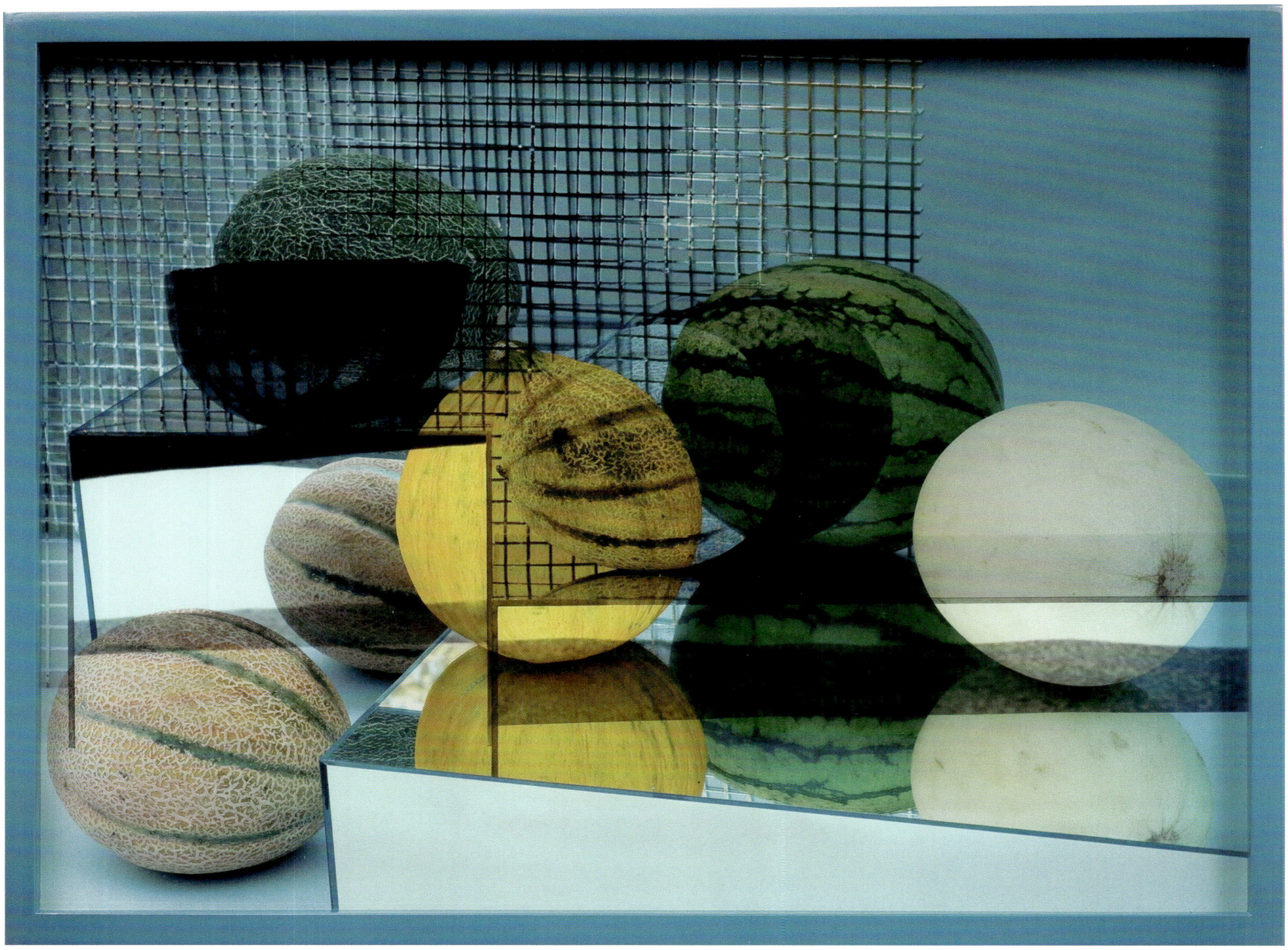

Circles and Squares (A Tasteful Organic Melons Arrangement) 2, 2007

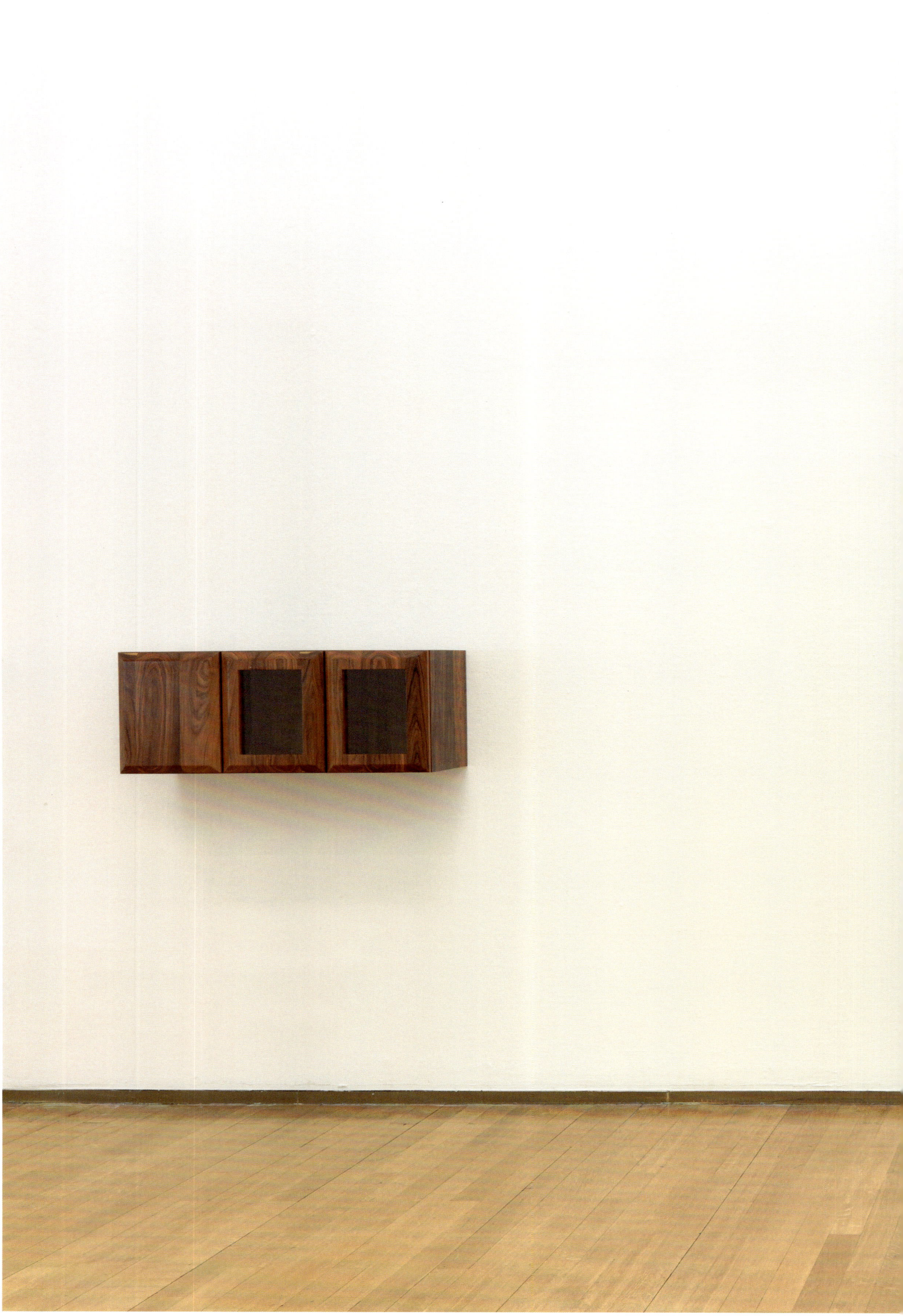

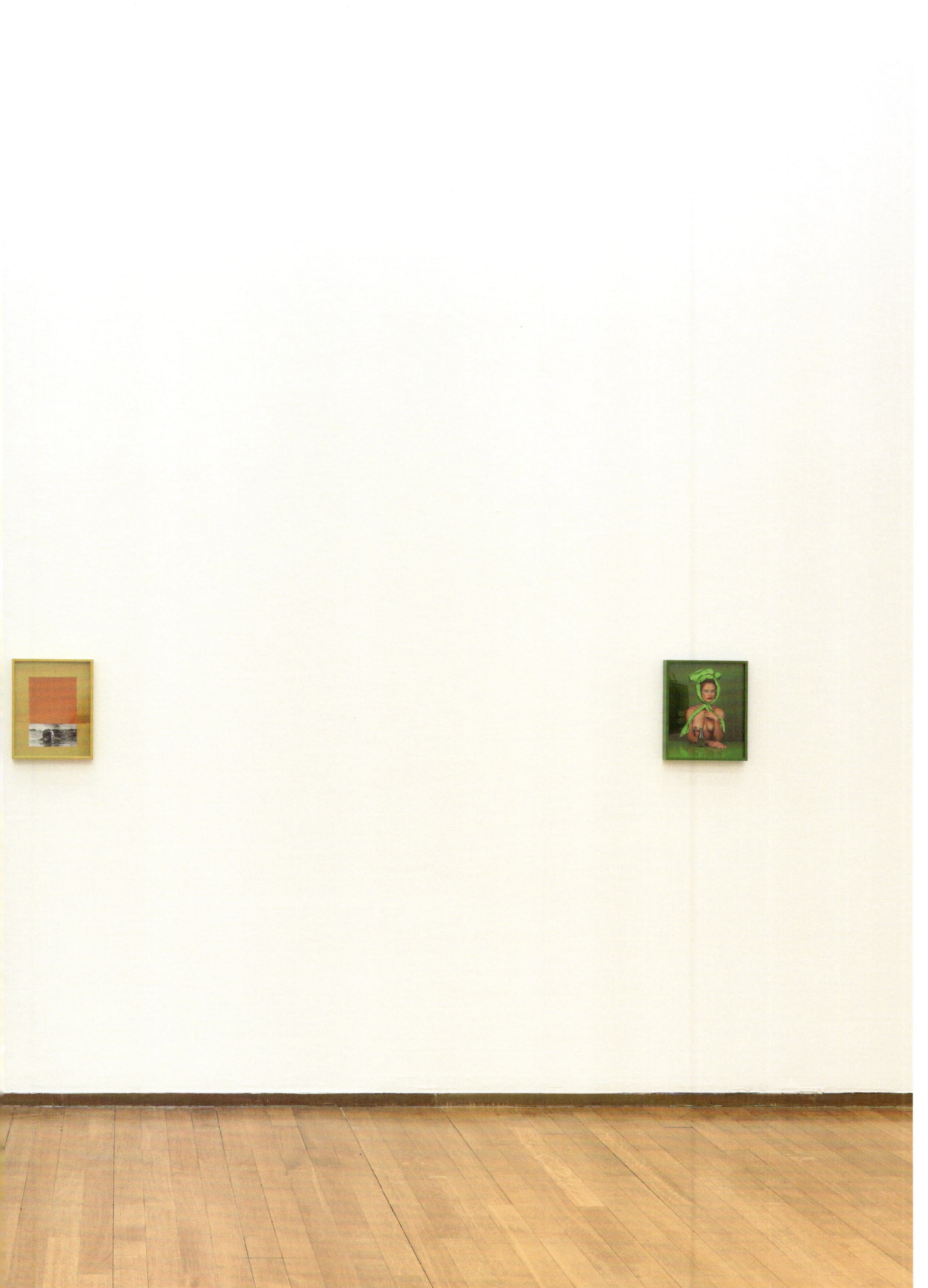

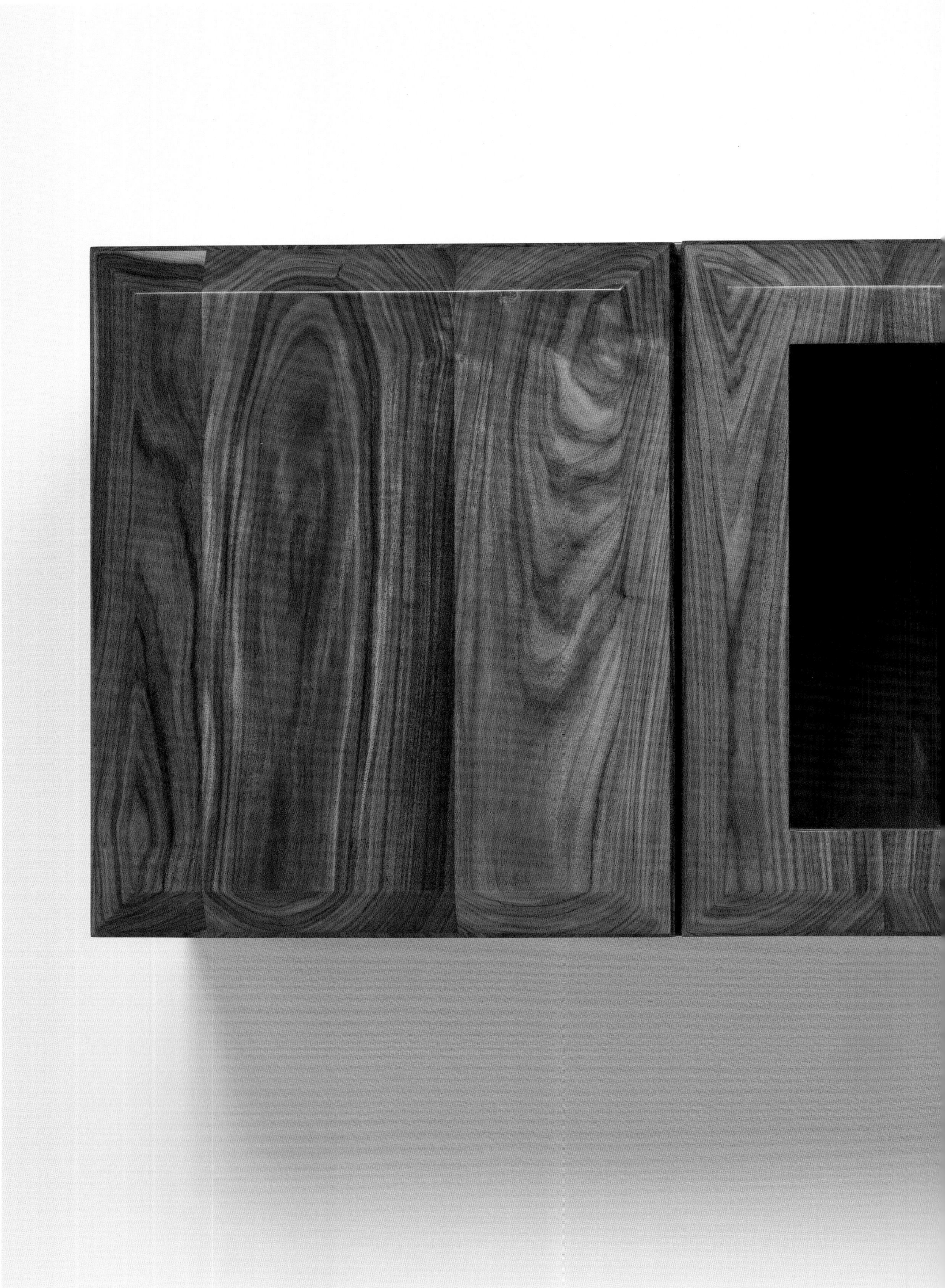

52 *Untitled (Rosewood Cabinet)*, 2012

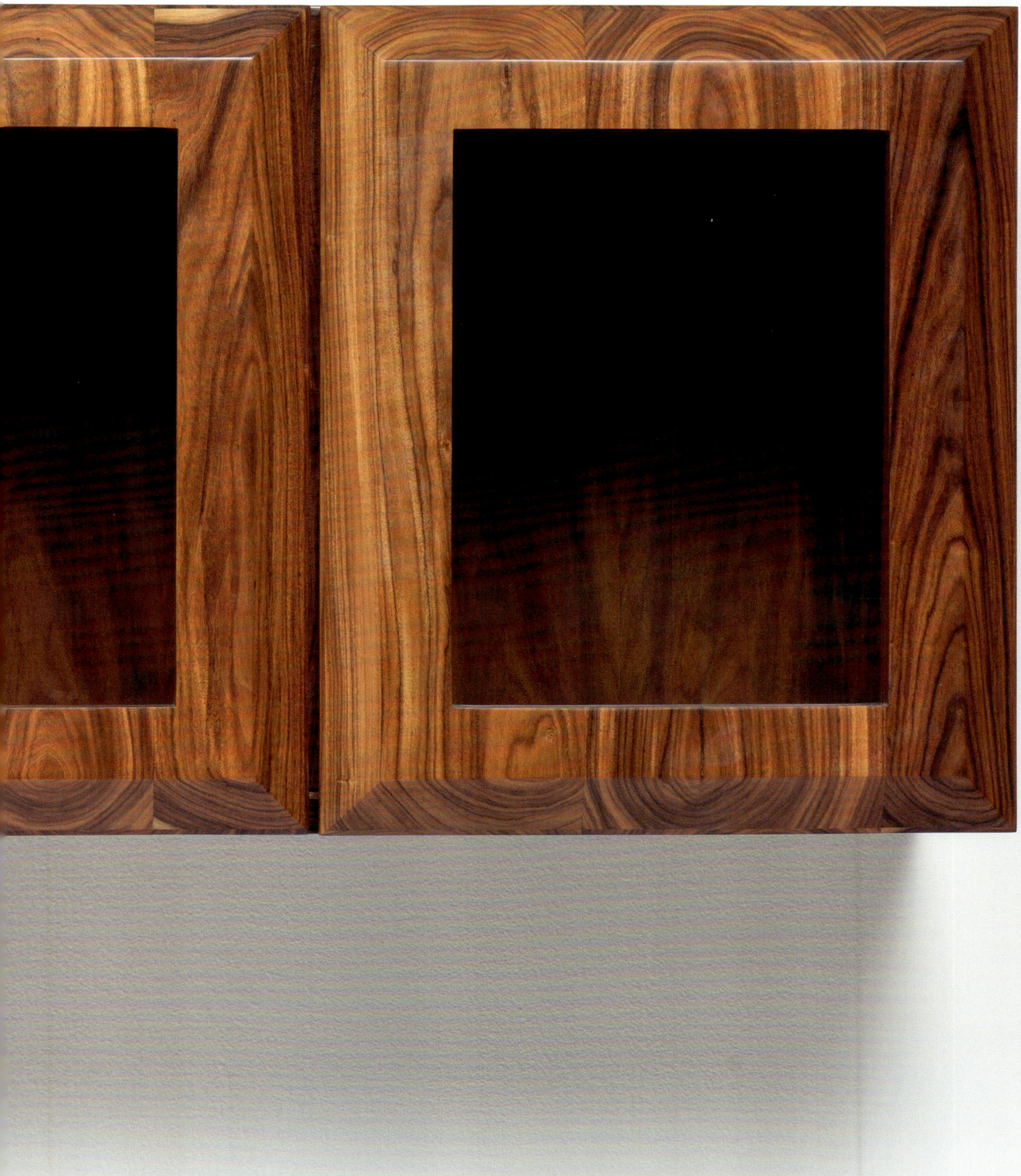

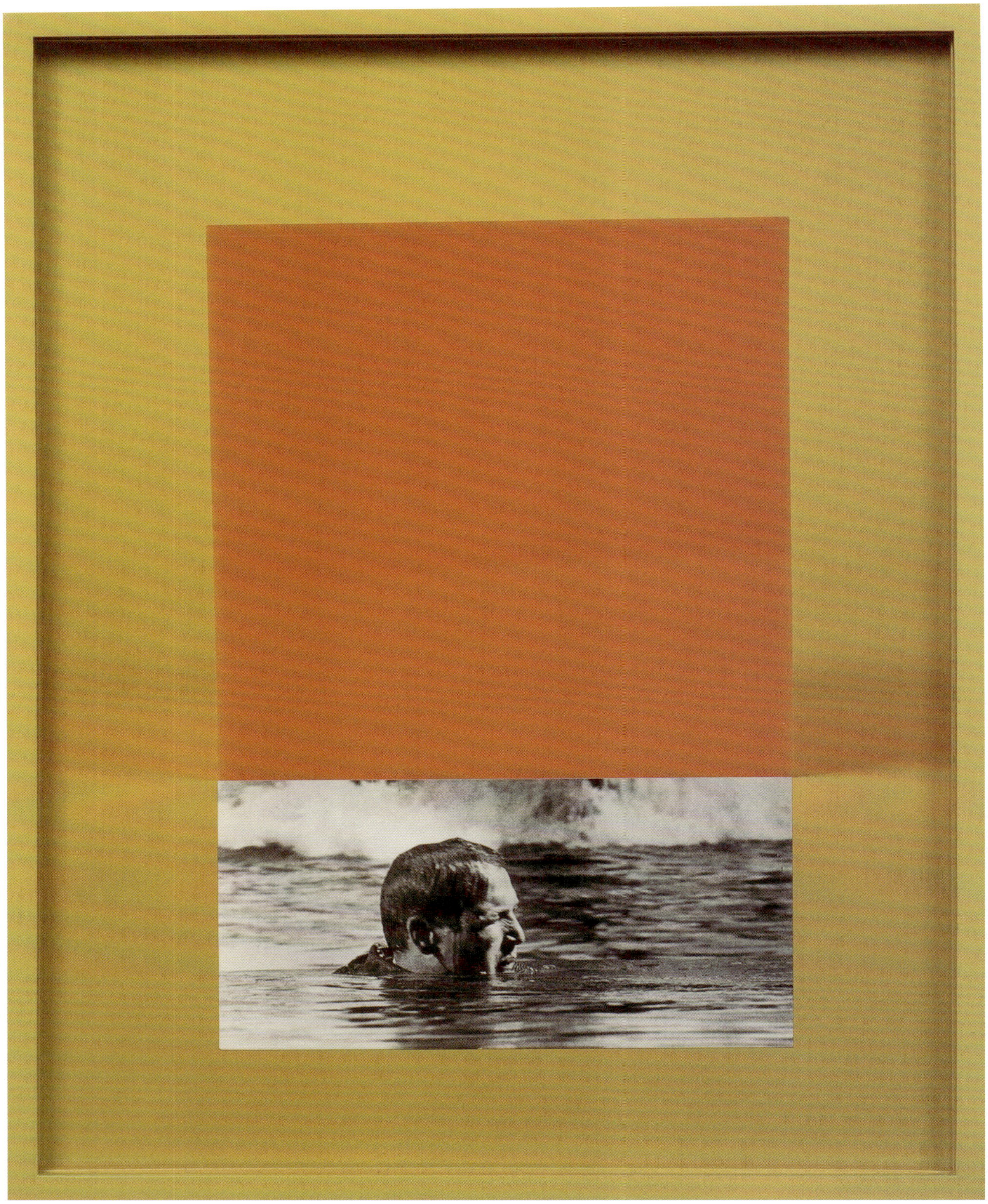

 Yellow Plinth, Wave Length, 2008

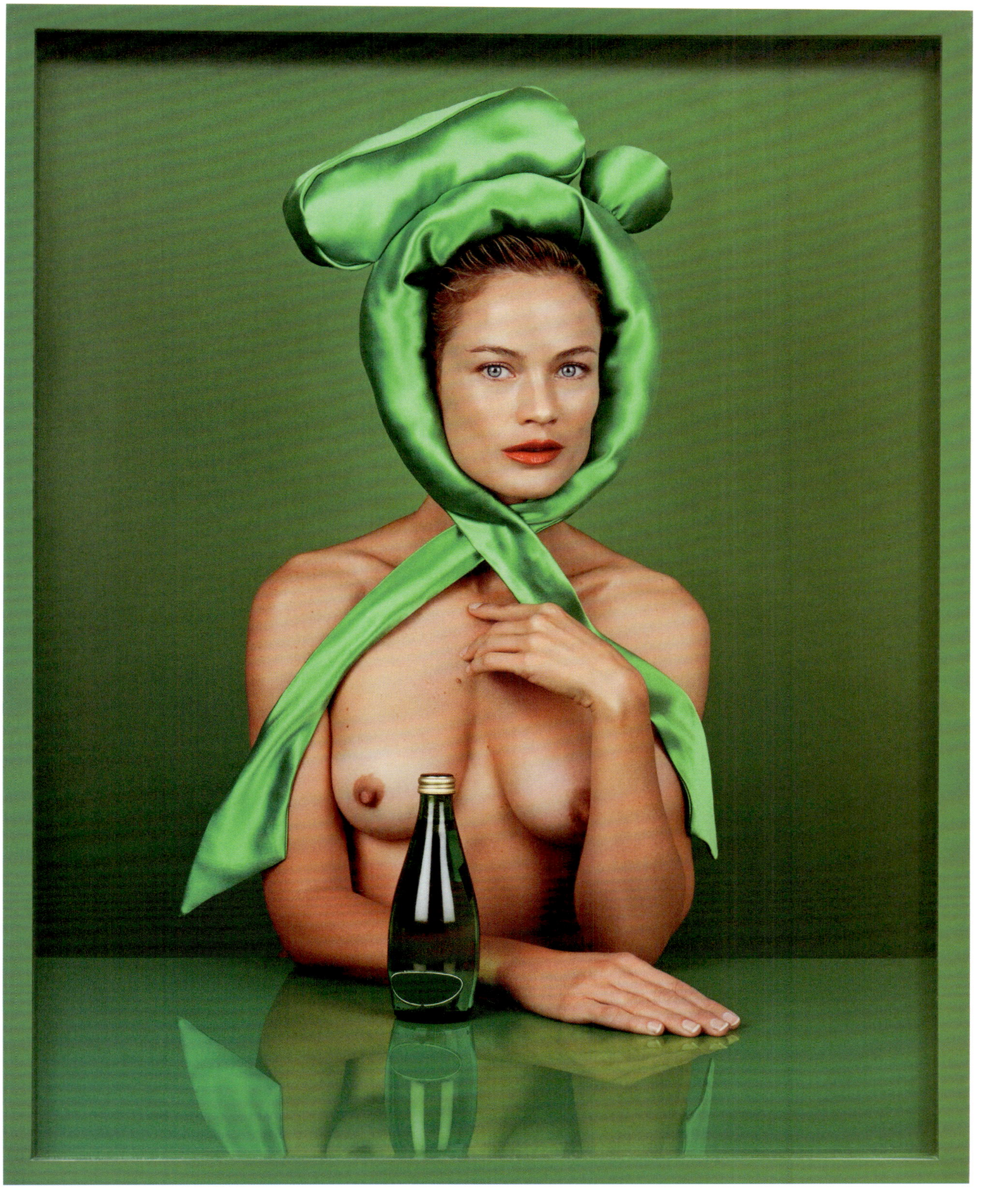

 Woman (Green Bow), 2012

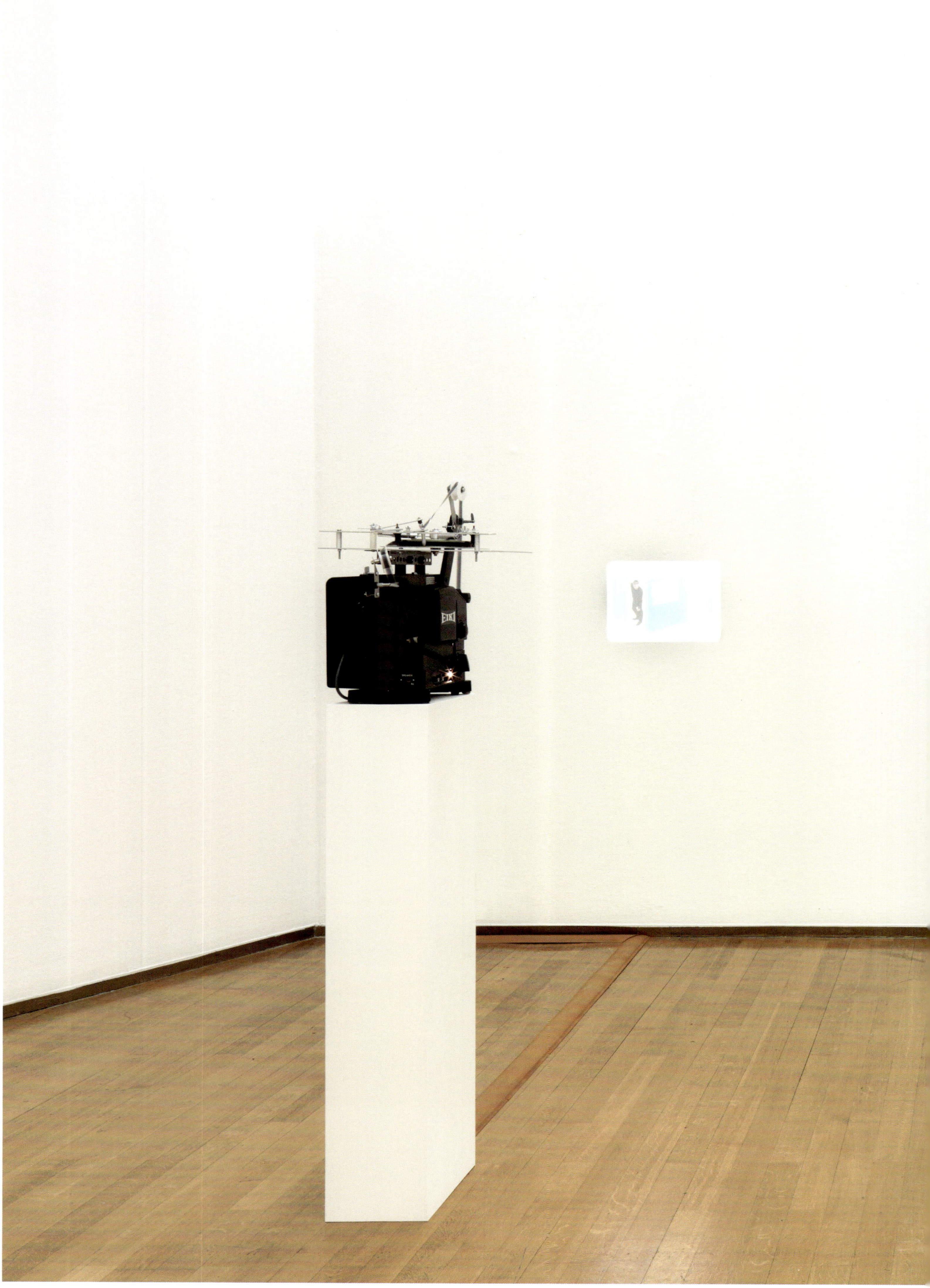

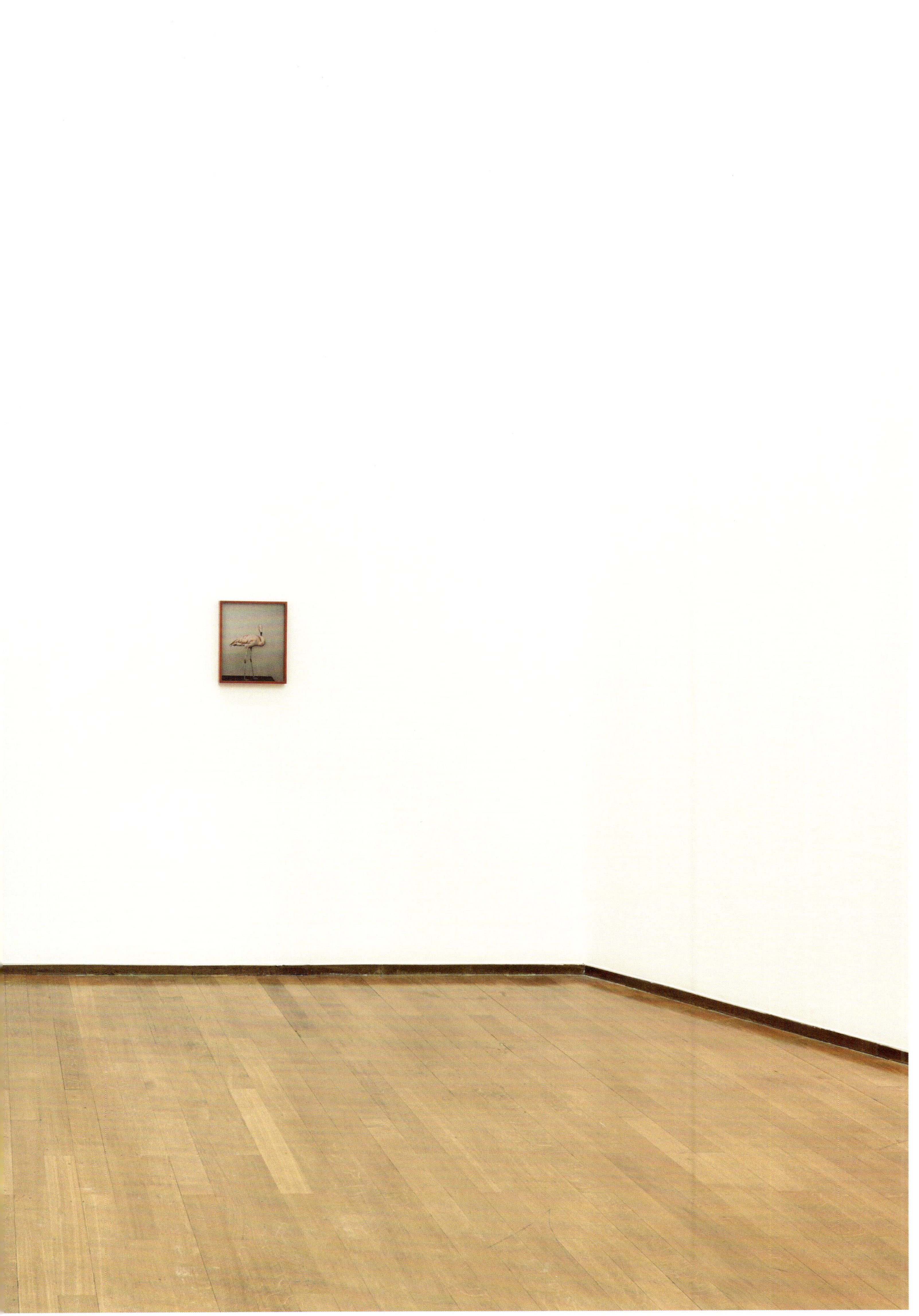

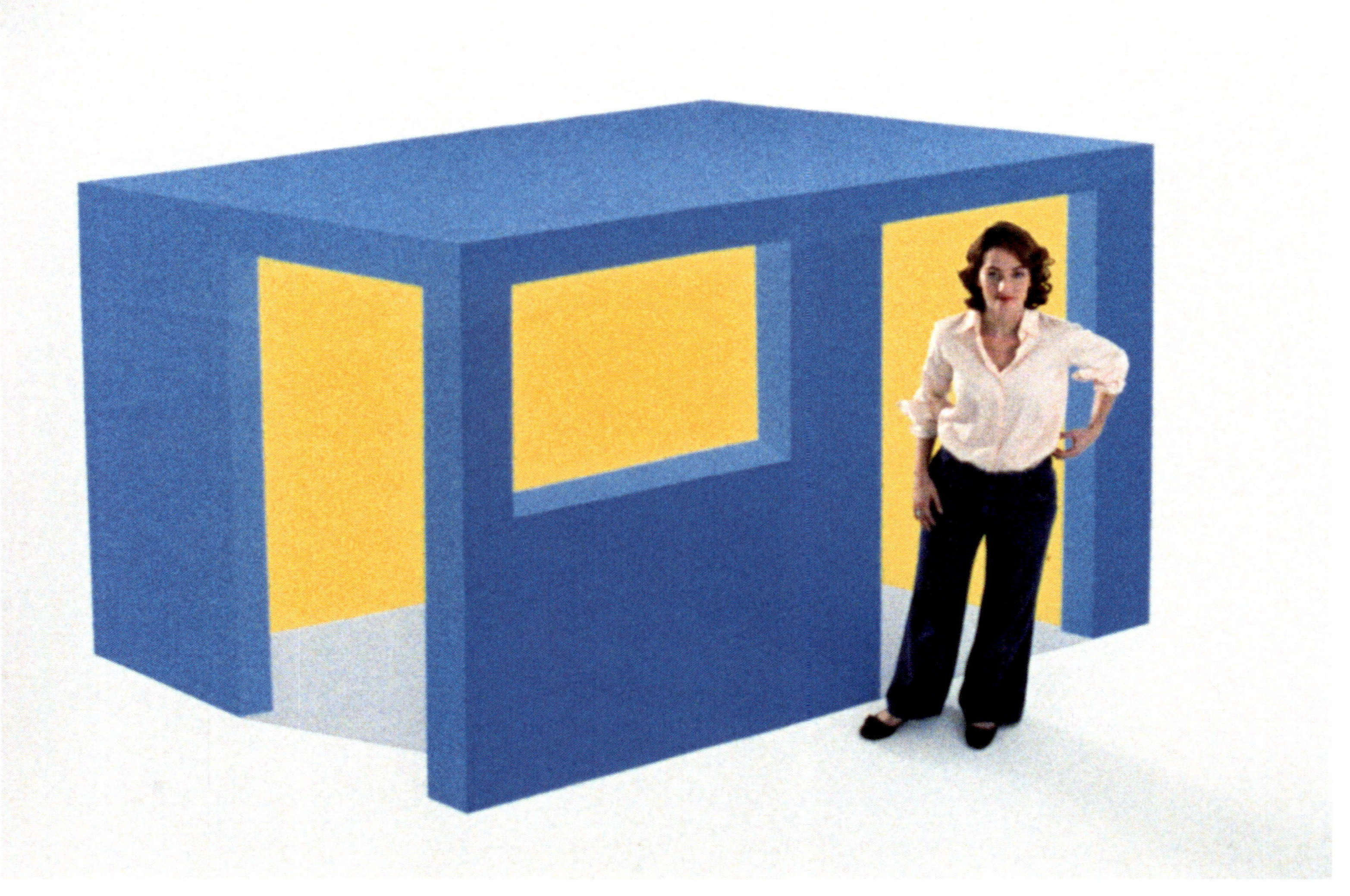

58 *Untitled*, 2008

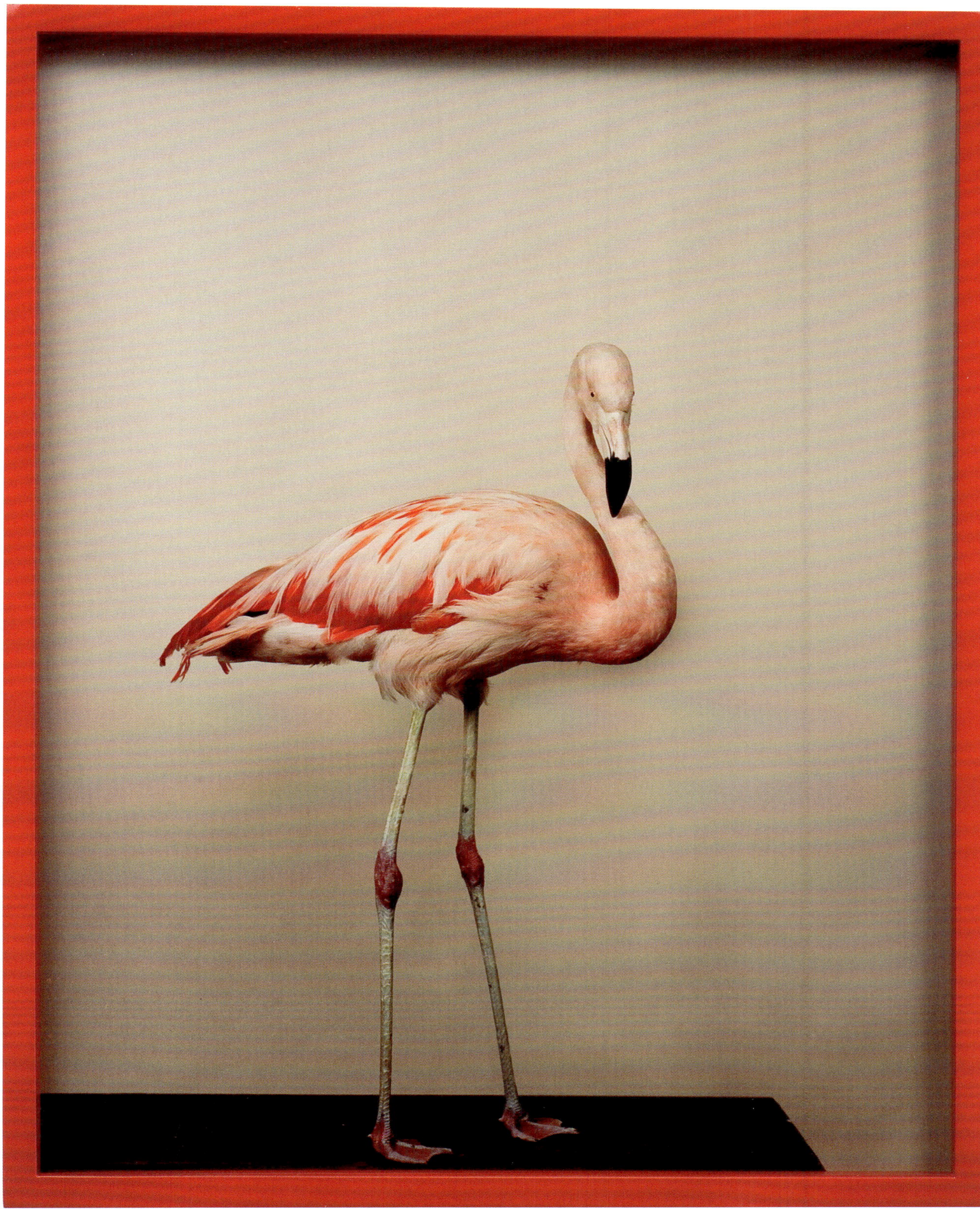

 Chilean Flamingo, 90028, 2007

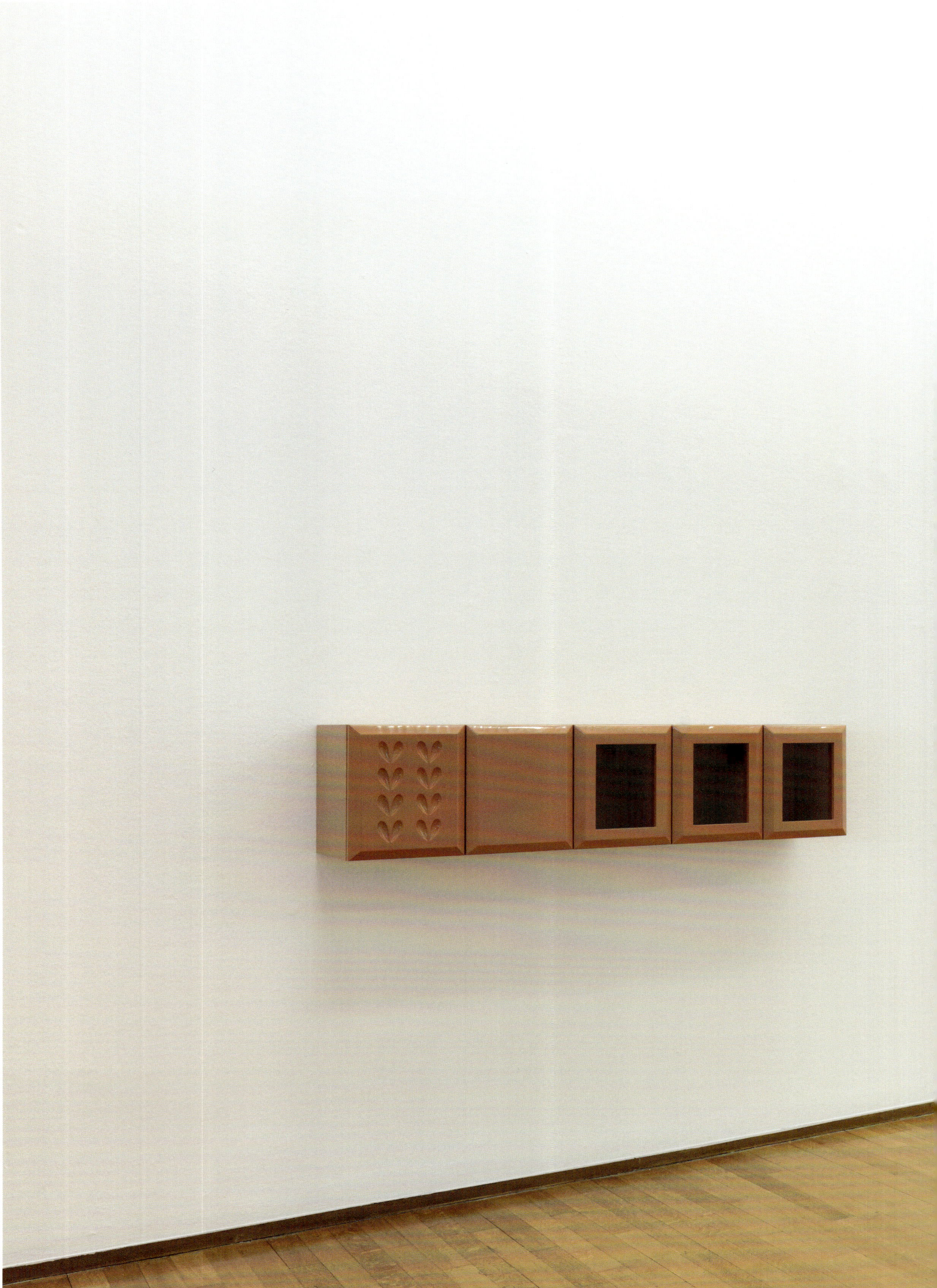

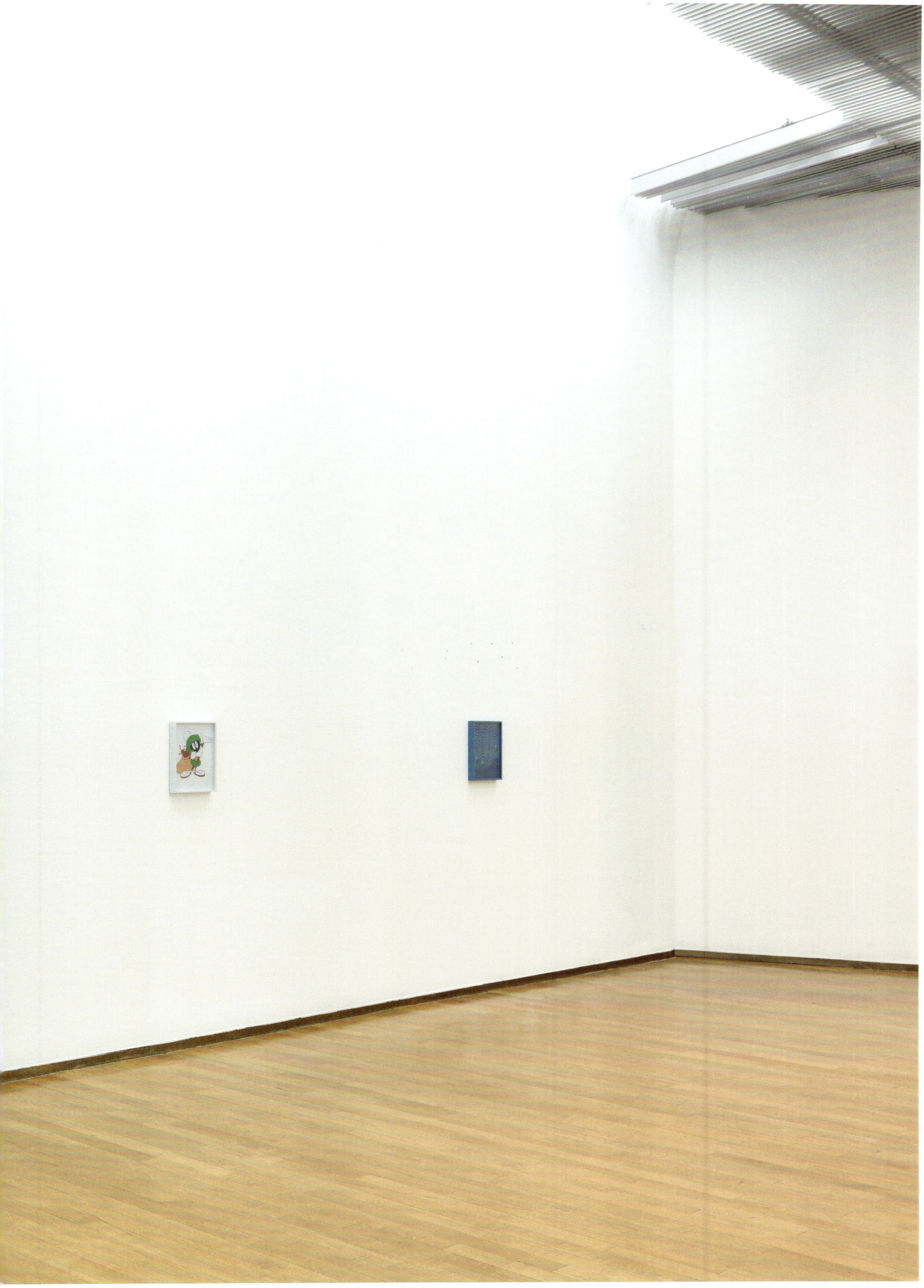

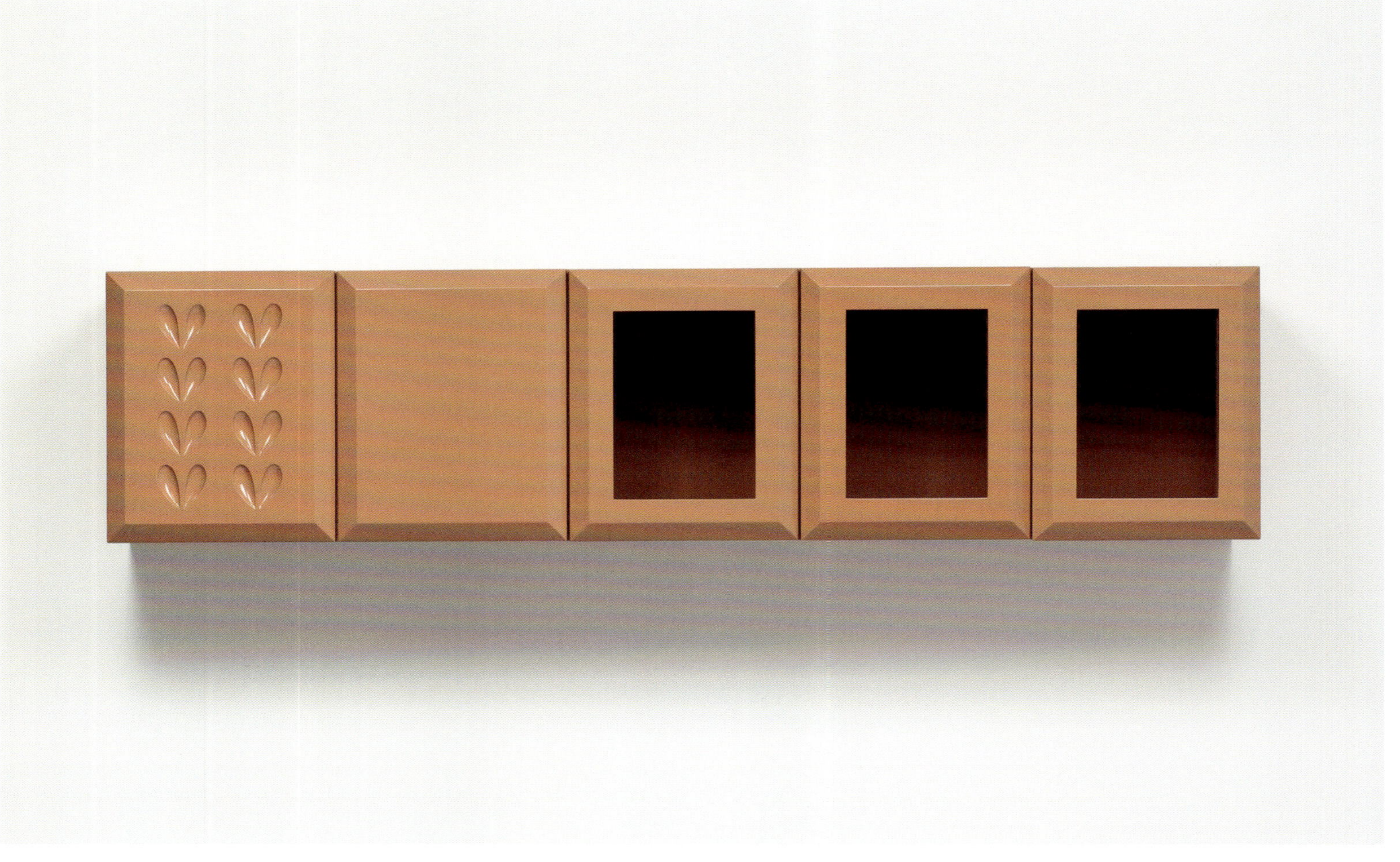

 Untitled (Light Brown Cabinet), 2012

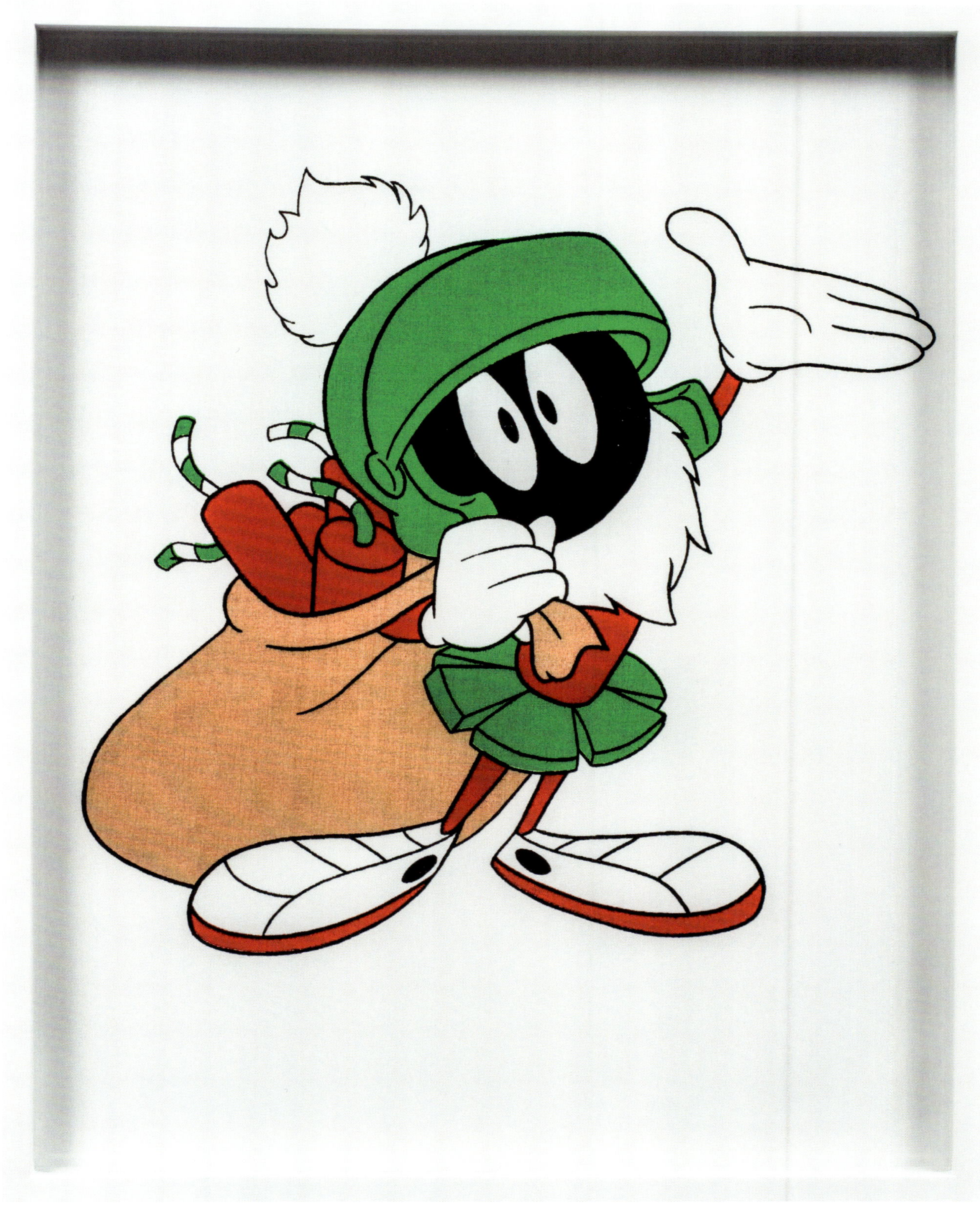

 Martian, 2012

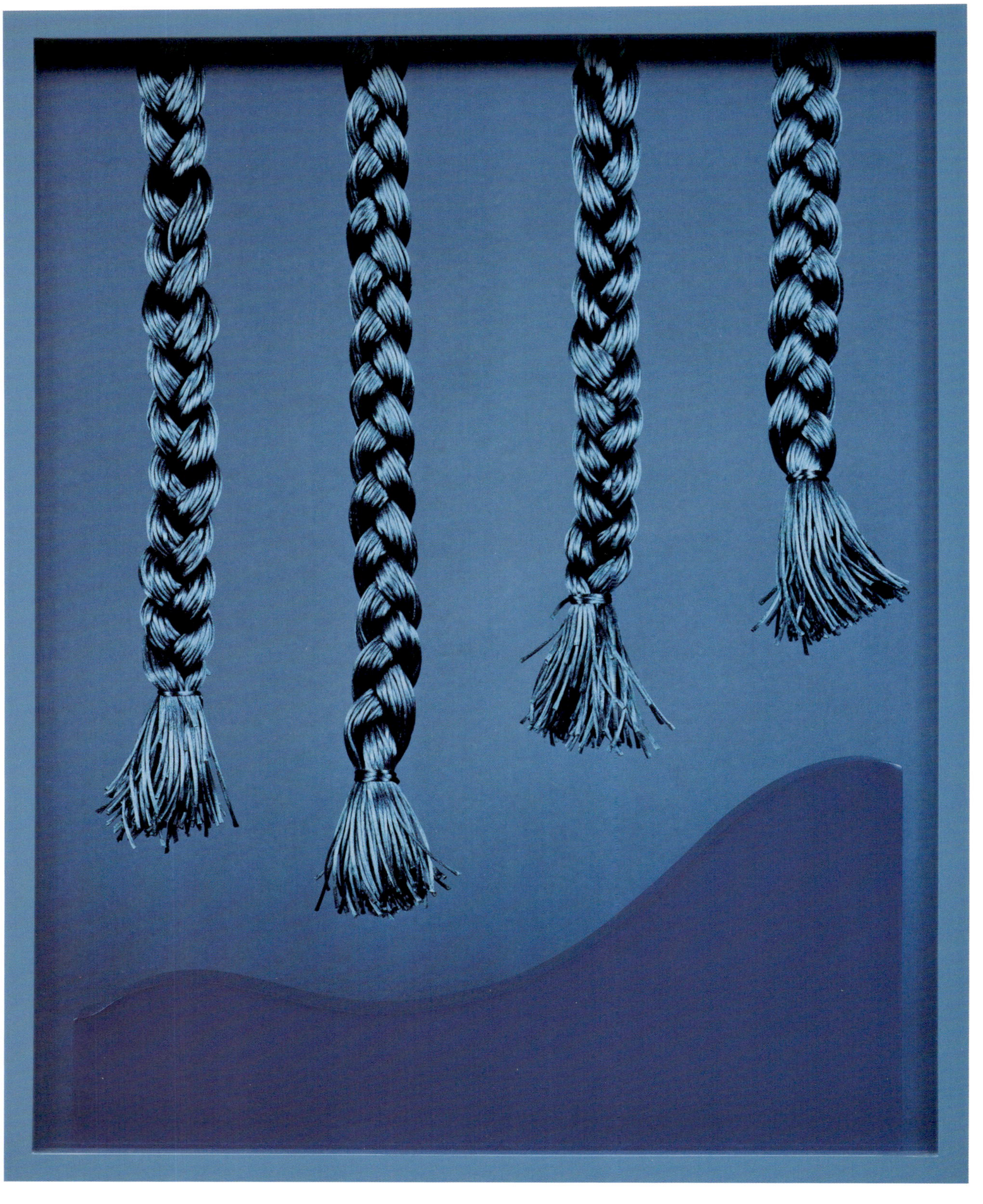

64 *Four Braids (Blue)*, 2012

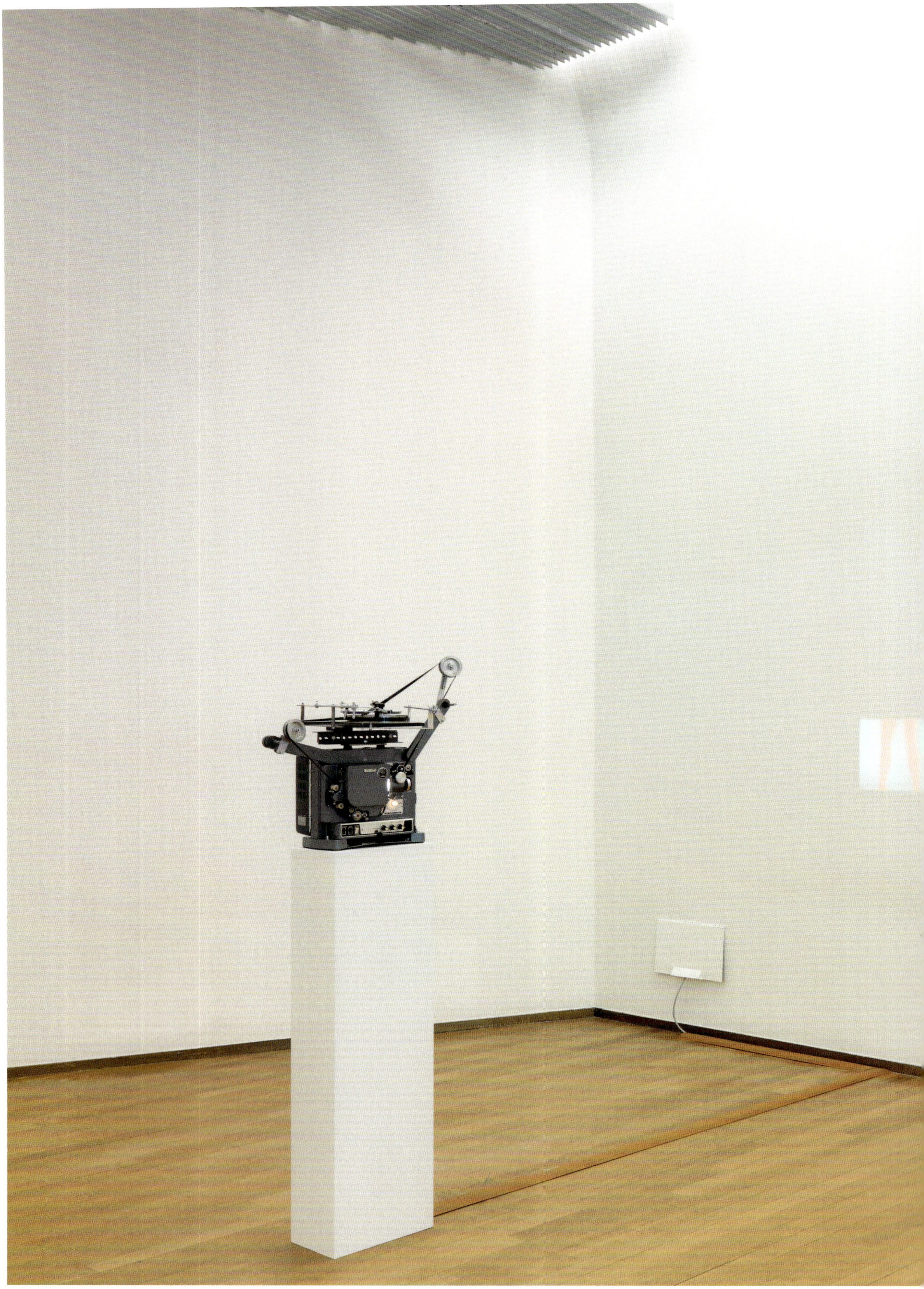

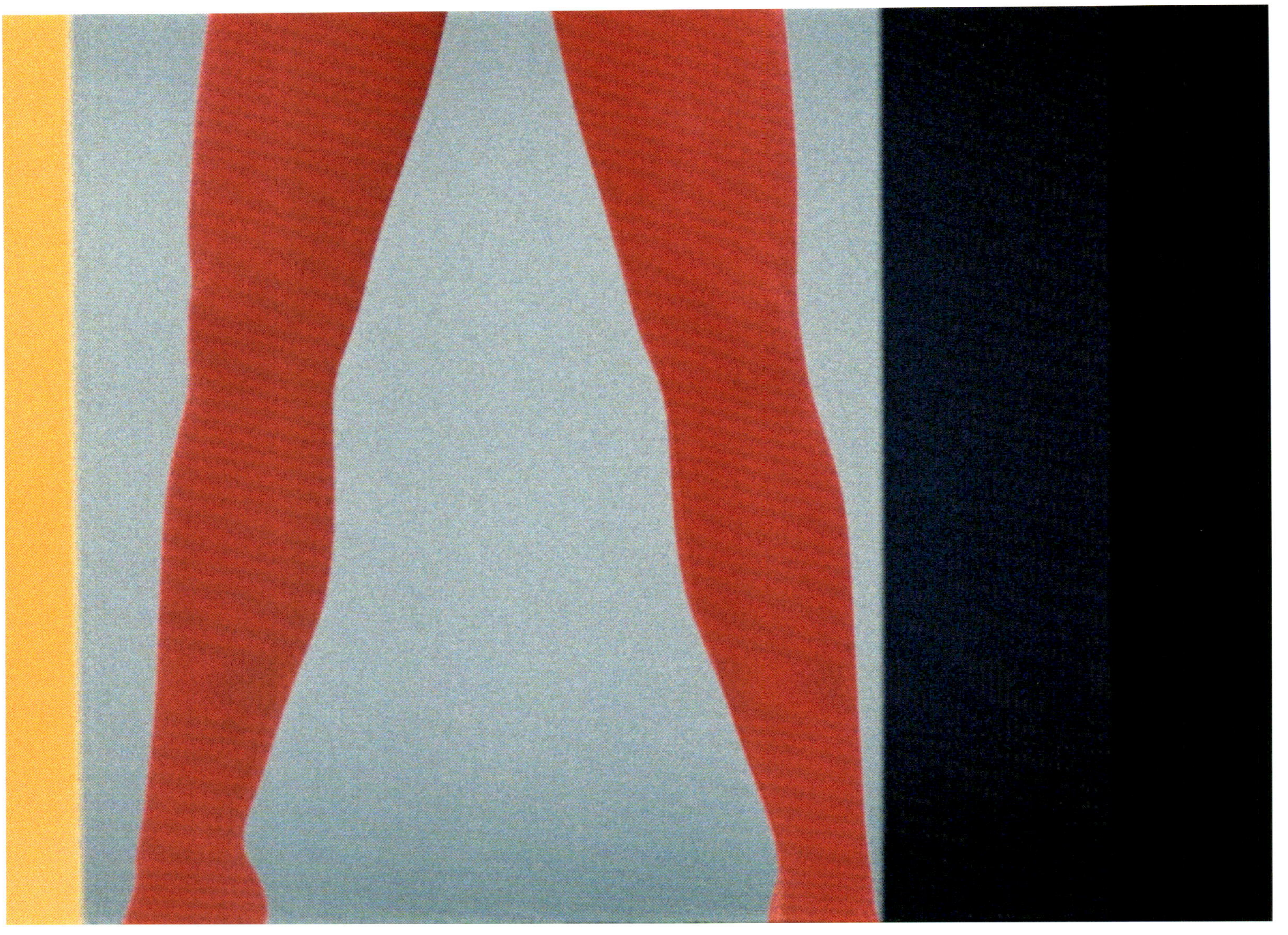

 Untitled, 2009

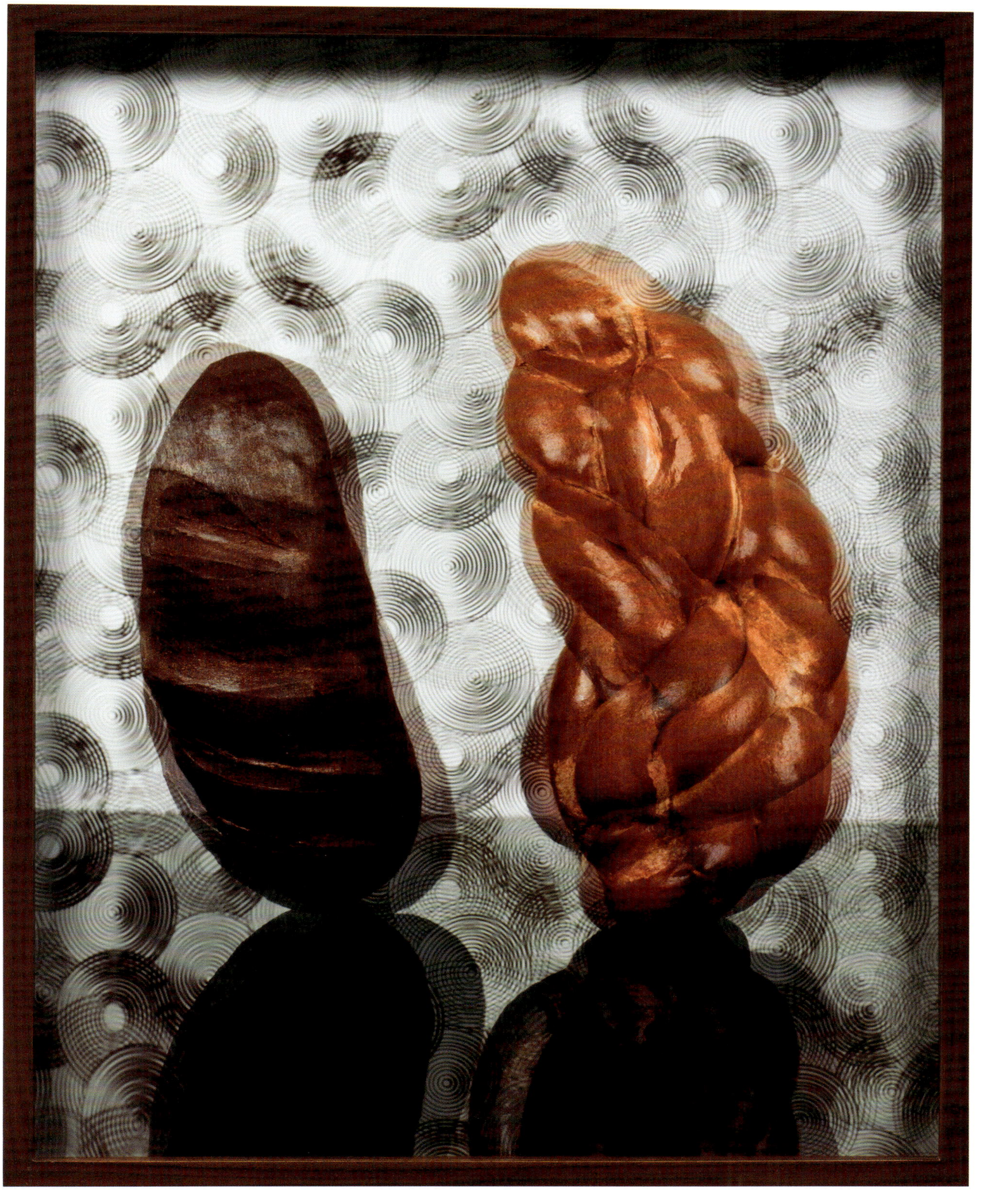

 Bread, 2012

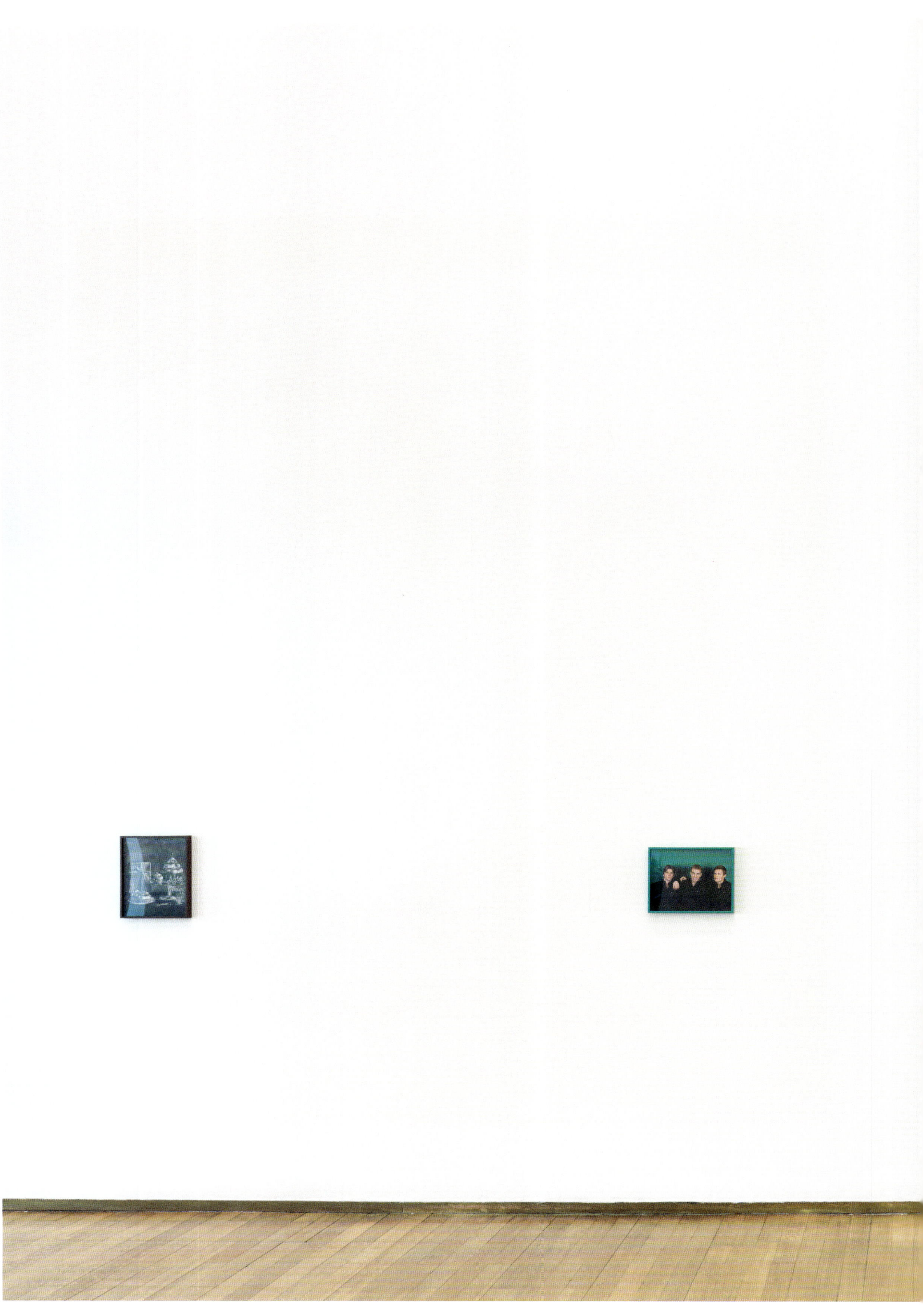

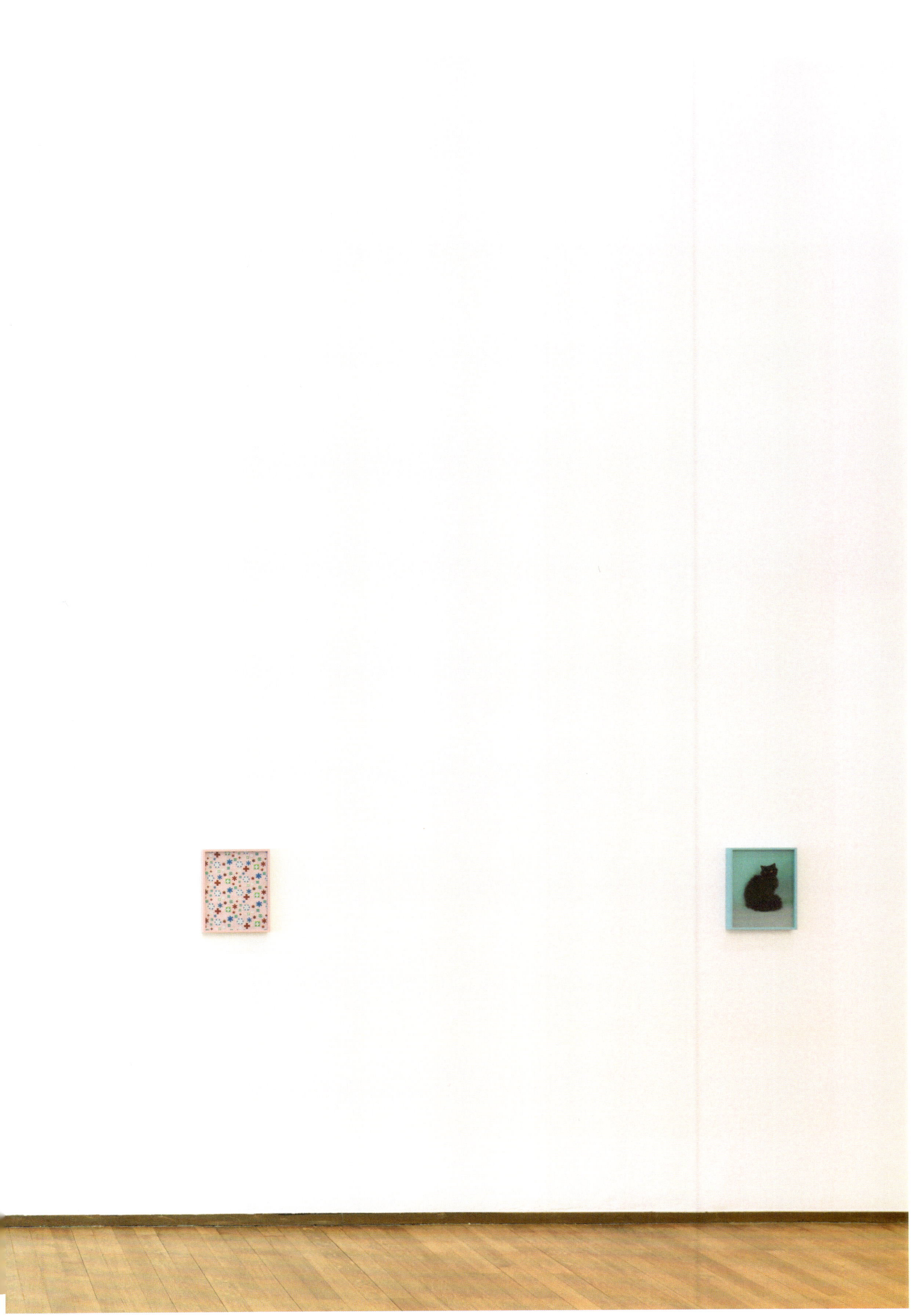

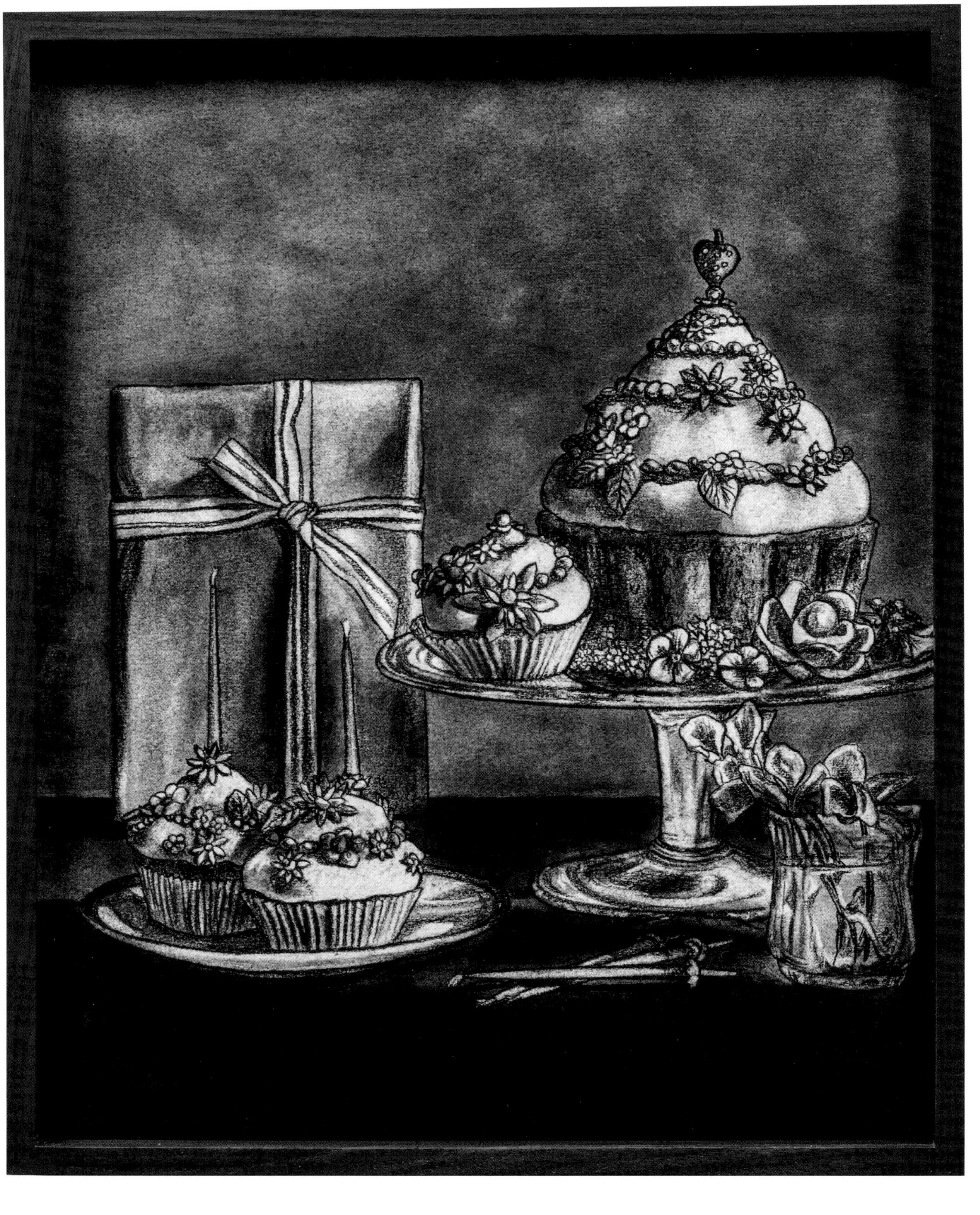

 Cupcakes, 2012

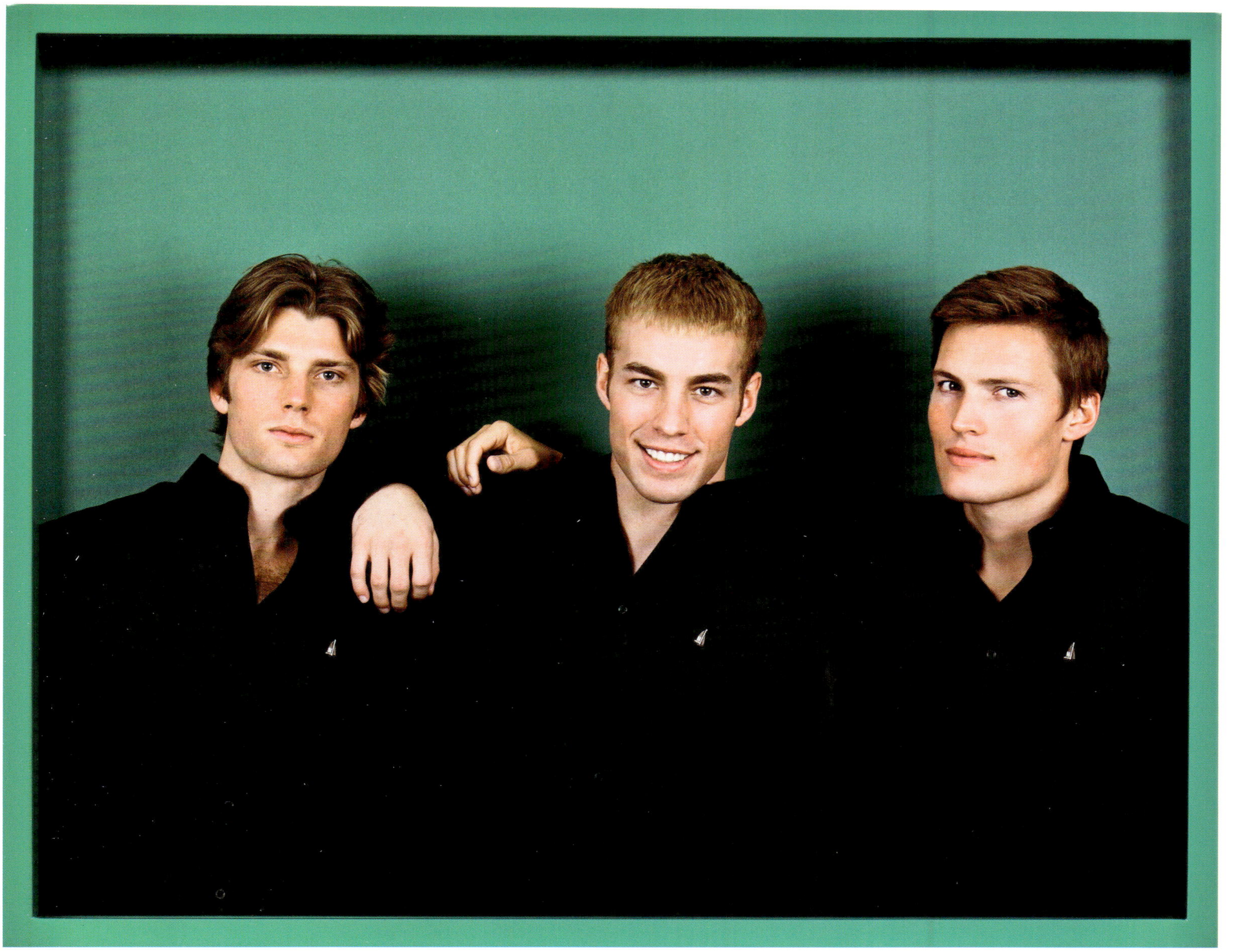

73 *Three Men*, 2011

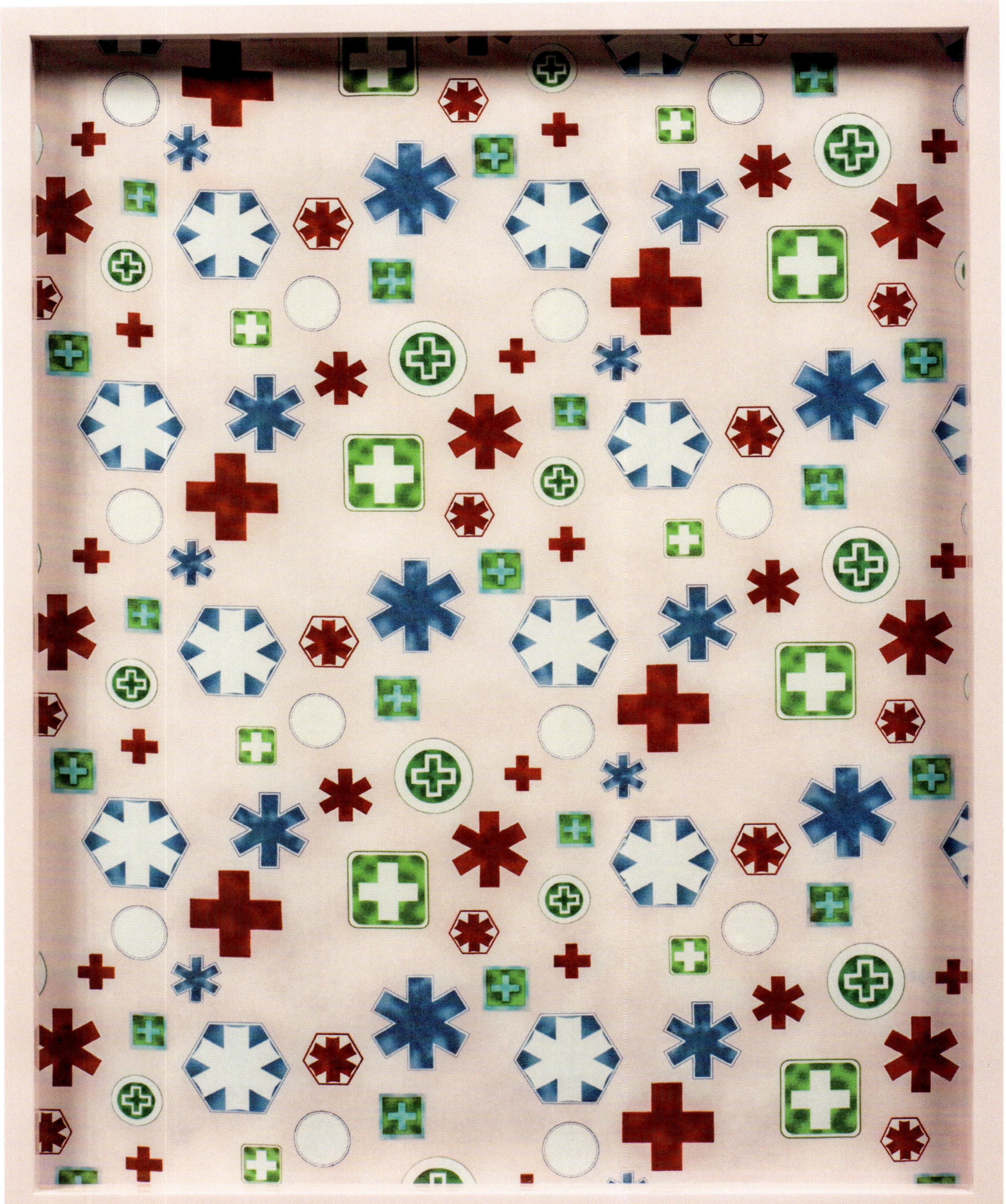

 Asterisks, 2012

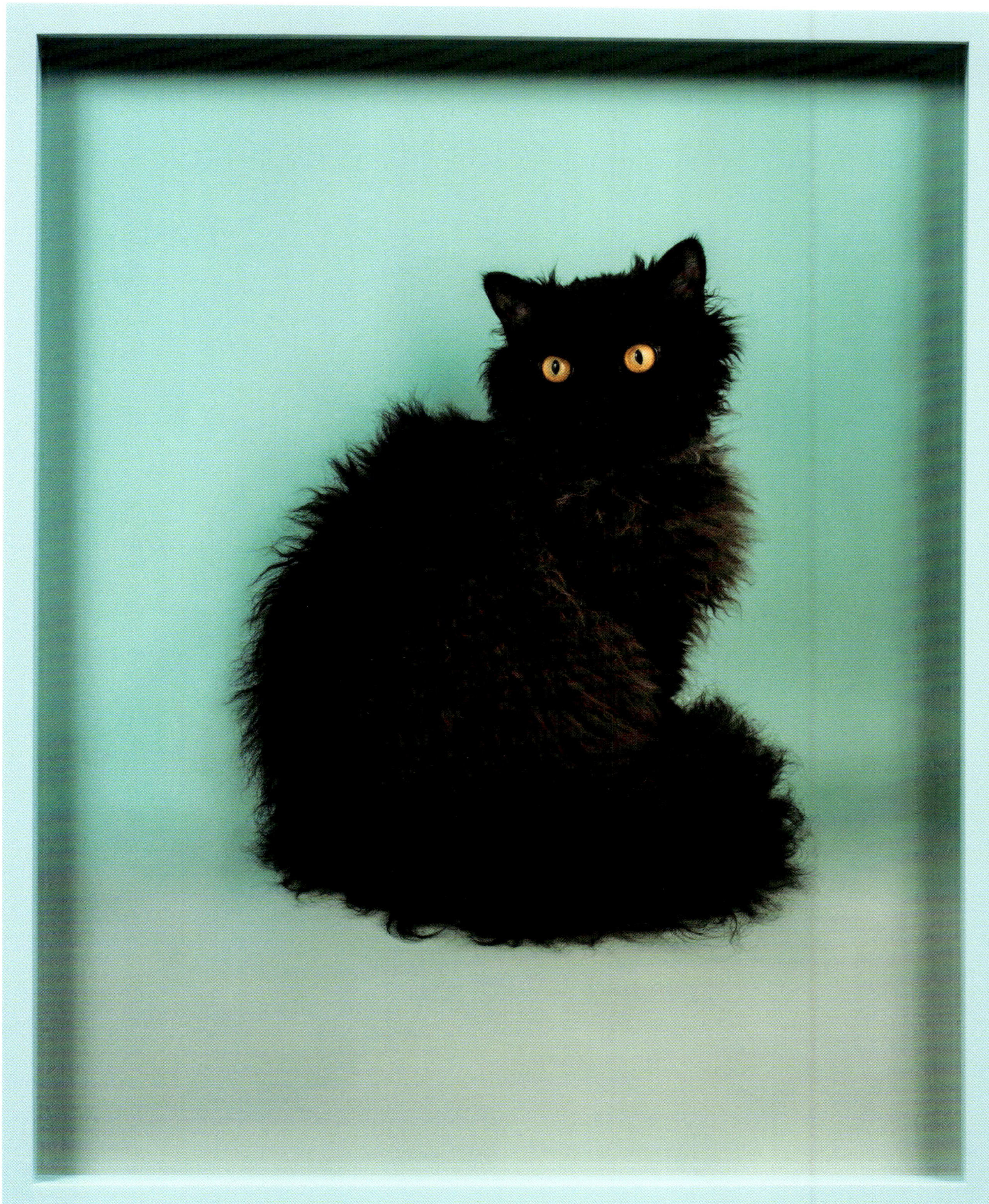

 Selkirk Rex, 2012

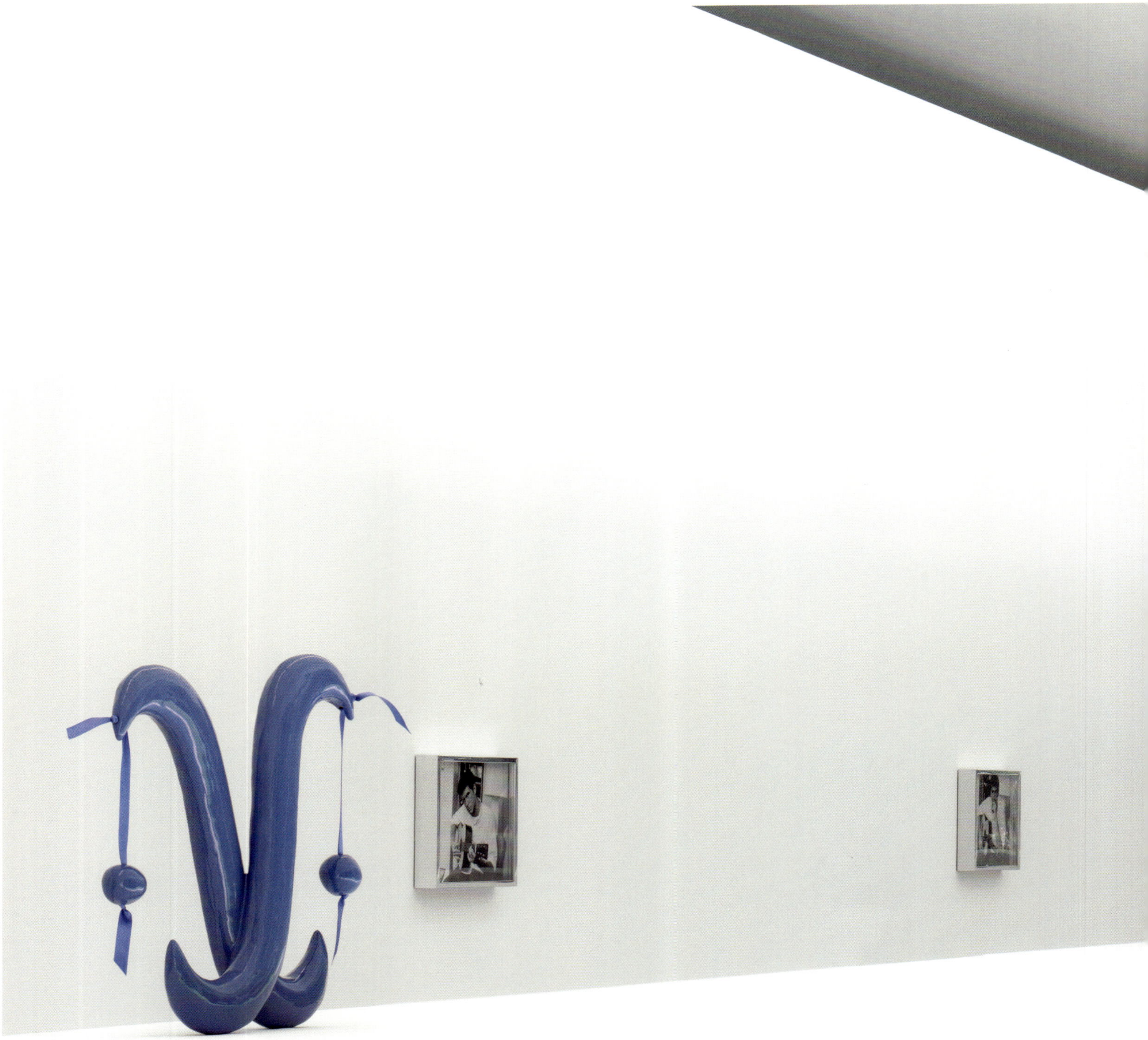

 Untitled (Wall, Milan Blue), 2012

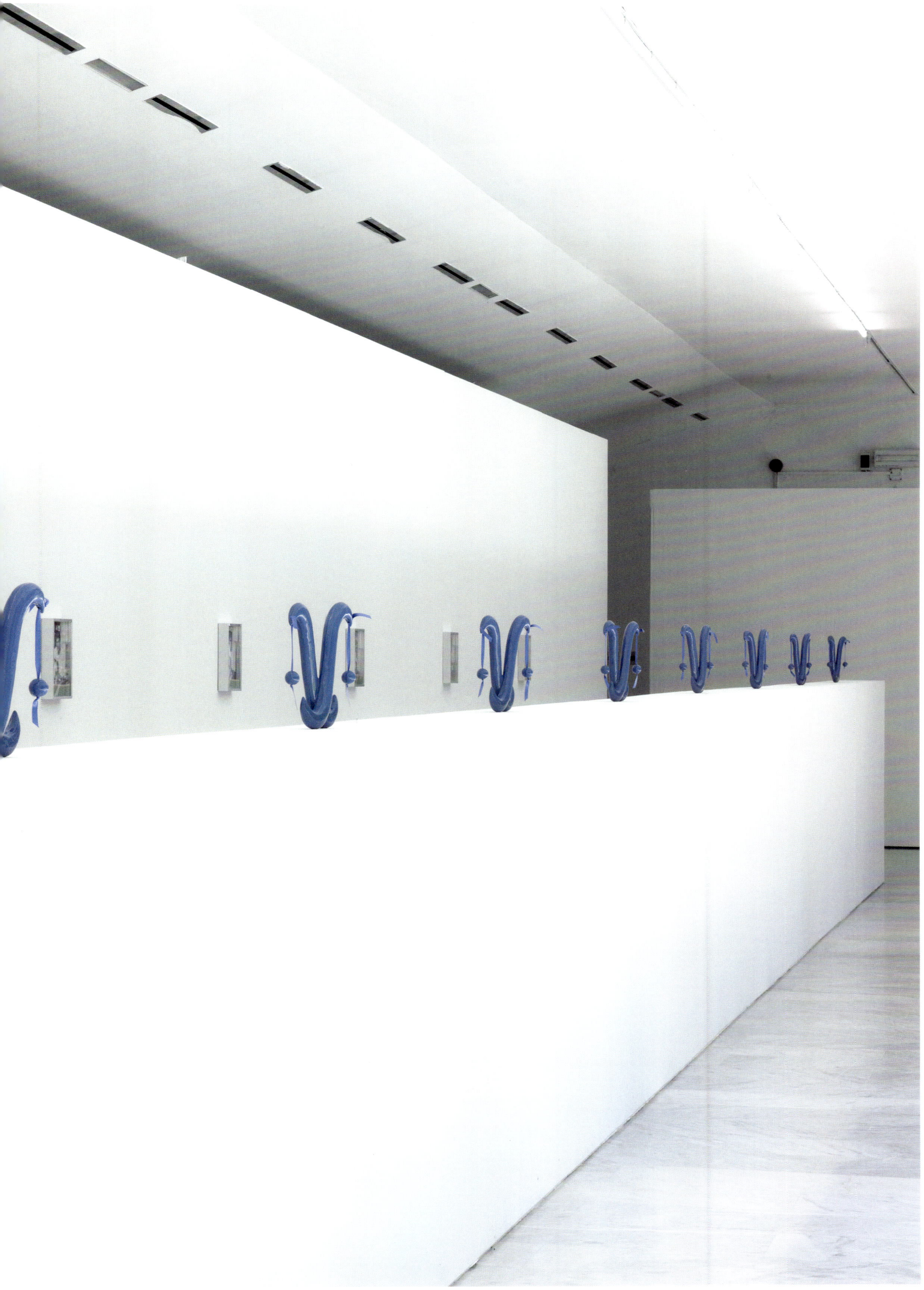

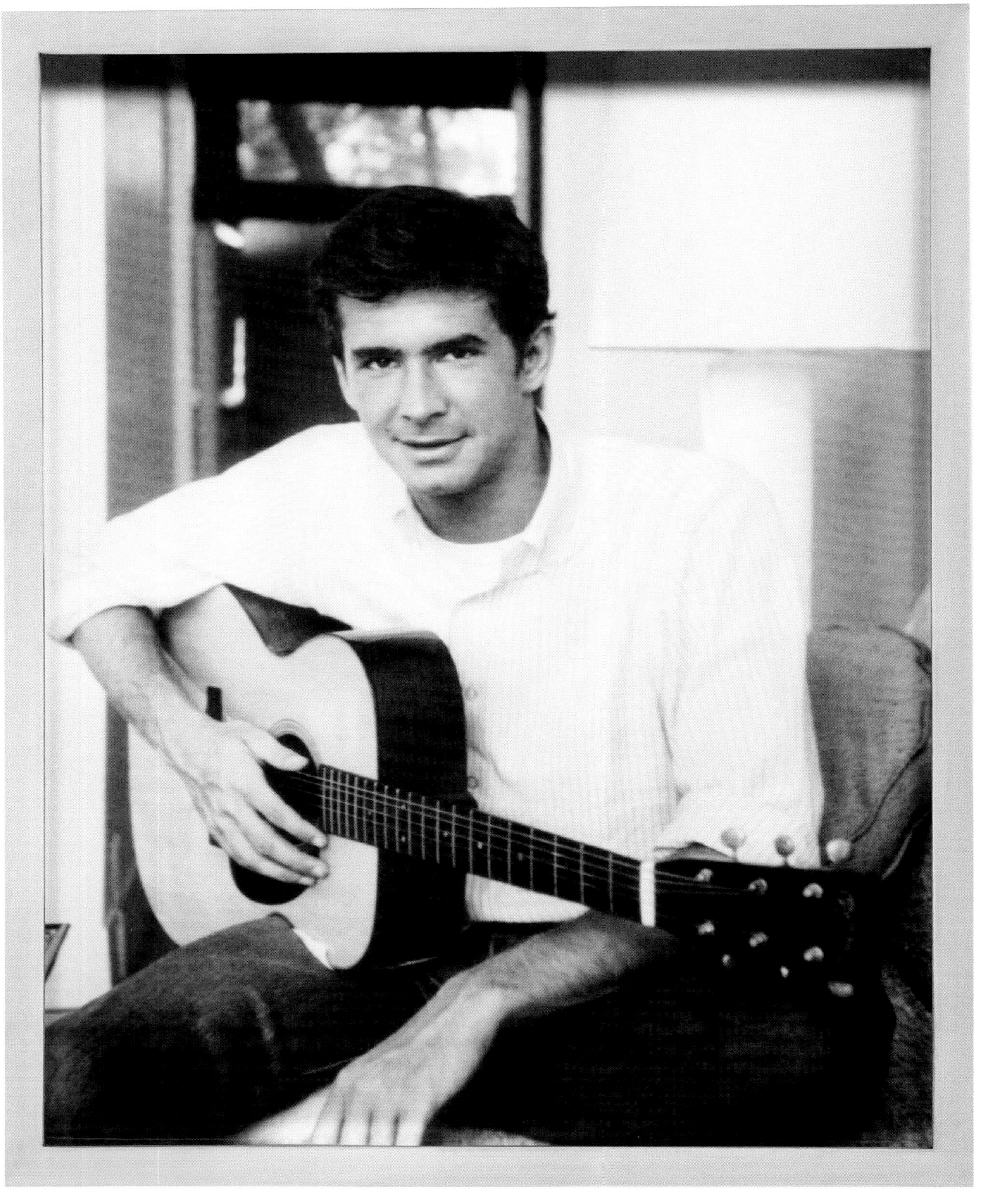

78 *Man (Guitar, Portrait)* (detail), 2012

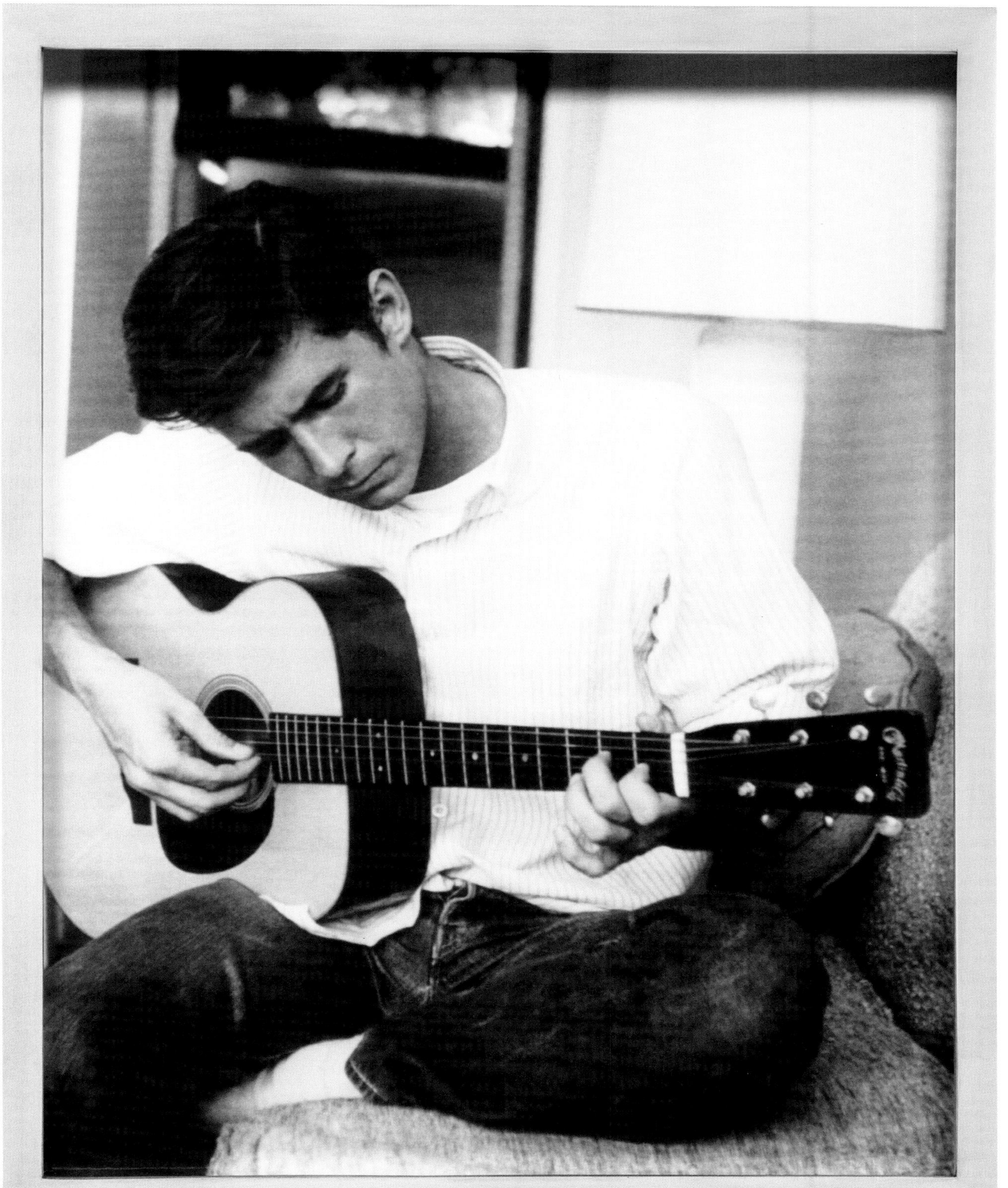

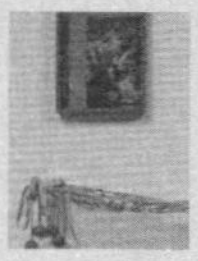

24.
Collie (Blue Bars), 2012
Foil on offset prints on paper, painted frames, painted metal poles, ceramic beads, ribbons
57.8 x 170.2 x 35.6 cm
(22.75 x 67 x 14 in)
Unique
Courtesy Massimo De Carlo, Milan/London

26.
Four Crystals (Red), 2011
C-print, painted frame
36.8 x 29.2 x 3.8 cm
(14.5 x 11.5 x 1.5 in)
Edition of 5, 2 AP
Courtesy Massimo De Carlo, Milan/London

27.
Hollywood Bowl, Fog, 2007
C-print, walnut frame
29.2 x 36.8 x 3.8 cm
(11.5 x 14.5 x 1.5 in)
Edition 5 of 5
Courtesy of the Hadley Martin Fisher Collection

28.
Horn Structure (For Massage and Collectable), 2009
Silver gelatin print and C-print on C-print, painted frame
36.8 x 29.2 x 3.8 cm
(14.5 x 11.5 x 1.5 in)
Unique
Courtesy Ringier Collection, Switzerland

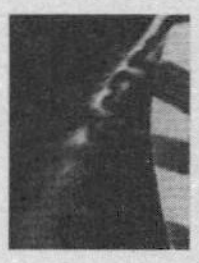

32.
Zebra and Woman, 2007
Super 16mm film, silent
5′ 21″
Courtesy Matt Mottahedan Collection

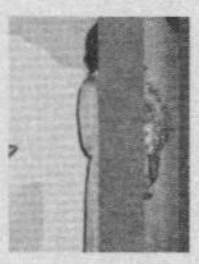

33.
Untitled (Blue Bar Mural), 2011
Foil on C-print, painted frame
25.7 x 19.1 x 3.8 cm
(10.125 x 7.5 x 1.5 in)
Unique
Courtesy Private Collection

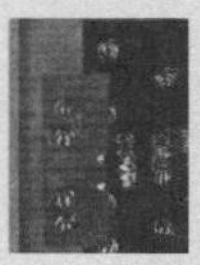

34.
Cherries (Silver), 2012
C-print, painted frame
36.8 x 29.2 x 3.8 cm
(14.5 x 11.5 x 1.5 in)
Edition of 5, 2 AP
Courtesy Massimo De Carlo, Milan/London

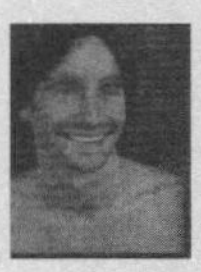

35.
Man 071, 2007
C-print, painted frame
36.8 x 29.2 x 3.8 cm
(14.5 x 11.5 x 1.5 in)
Edition 2 of 5
Courtesy Private Collection, Milan

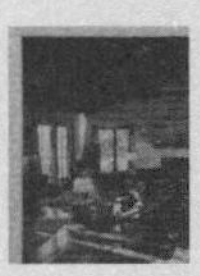

38.
Living Room, 2010
C-print, brass frame
20.3 x 25.4 x 3.8 cm
(8 x 10 x 1.5 in)
Edition 3 of 3
Courtesy Private Collection

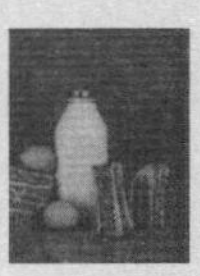

39.
Short Ribs, Eggs, 2012
C-print, painted frame
36.8 x 29.2 x 3.8 cm
(14.5 x 11.5 x 1.5 in)
Edition of 5, 2 AP
Courtesy Massimo De Carlo, Milan/London

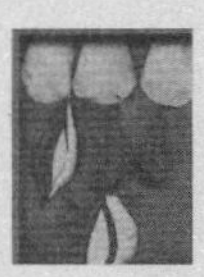

40.
Pillow 3, 2012
C-print, painted frame
36.8 x 29.2 x 3.8 cm
(14.5 x 11.5 x 1.5 in)
Edition of 5, 2 AP
Courtesy Massimo De Carlo, Milan/London

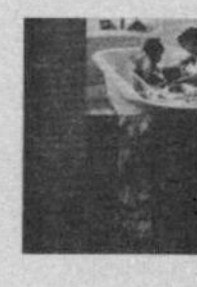

41.
Laminated Structure (For Bath), 2011
C-print on C-print, painted frame
36.8 x 29.2 x 3.8 cm
(14.5 x 11.5 x 1.5 in)
Unique
Courtesy Ringier Collection, Switzerland

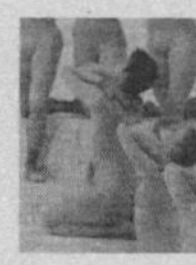

45.
Untitled (Passacaglia), 2010
Super 16mm film, silent
15′ 27″
Courtesy Massimo De Carlo, Milan/London

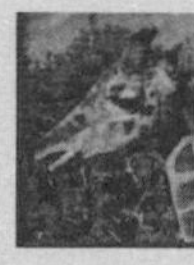

46.
Giraffe, 93040, 2011
Silver gelatin prints, walnut frames
36.8 x 29.2 x 3.8 cm
(14.5 x 11.5 x 1.5 in each)
Diptych
Edition 1 of 5
Courtesy Collection of Amy and Harris Schwalb

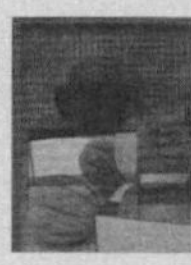

48.
Circles and Squares (A Tasteful Organic Melons Arrangement) 2, 2007
C-print, painted frame
29.2 x 36.8 x 3.8 cm
(11.5 x 14.5 x 1.5 in)
Edition 1 of 5
Courtesy Private Collection, Berlin

52.
Untitled (Rosewood Cabinet), 2012
Rosewood, tempered glass, varnish
43.2 x 106.7 x 35.6 cm
(17 x 42 x 14 in)
Unique
Courtesy Massimo De Carlo, Milan/London

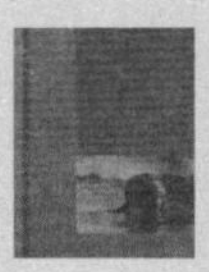

54.
Yellow Plinth, Wave Length, 2008
Silver gelatin print on C-print, painted frame
36.8 x 29.2 x 3.8 cm
(14.5 x 11.5 x 1.5 in)
Unique
Courtesy Collection of David Simkins

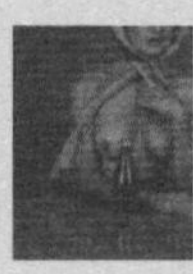

55.
Woman (Green Bow), 2012
C-print, painted frame
36.8 x 29.2 x 3.8 cm
(14.5 x 11.5 x 1.5 in)
Edition of 5, 2 AP
Courtesy Massimo De Carlo, Milan/London

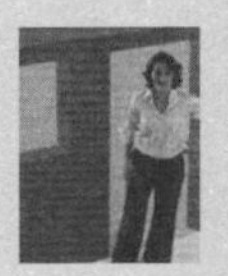

58.
Untitled, 2008
16mm color film, silent
9′ 20″
Courtesy Rennie Collection, Vancouver

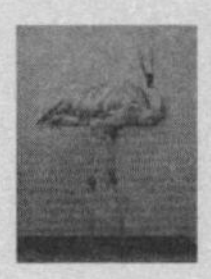

59.
Chilean Flamingo, 90028, 2007
C-print, painted frame
36.8 x 29.2 x 3.8 cm
(14.5 x 11.5 x 1.5 in)
Edition 5 of 5
Courtesy Private Collection, Berlin

62.
Untitled (Light Brown Cabinet), 2012
MDF, tempered glass, paint
43.2 x 177.8 x 35.6 cm
(17 x 70 x 14 in)
Unique
Courtesy Massimo De Carlo, Milan/London

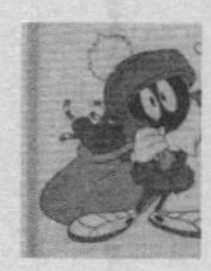

63.
Martian, 2012
C-print, painted frame
36.8 x 29.2 x 3.8 cm
(14.5 x 11.5 x 1.5 in)
Edition of 5, 2 AP
Courtesy Massimo De Carlo, Milan/London

64.
Four Braids (Blue), 2012
C-print, painted frame
36.8 x 29.2 x 3.8 cm
(14.5 x 11.5 x 1.5 in)
Edition of 5, 2 AP
Courtesy Massimo De Carlo, Milan/London

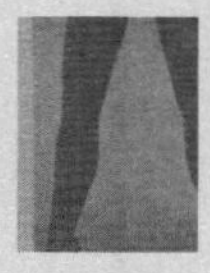

68.
Untitled, 2009
16mm film, silent
11′ 40″
Courtesy Rennie Collection, Vancouver

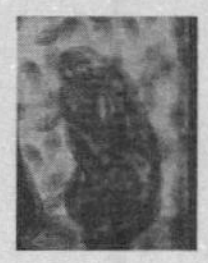

69.
Bread, 2012
C-print, walnut frame
36.8 x 29.2 x 3.8 cm
(14.5 x 11.5 x 1.5 in)
Edition of 5, 2 AP
Courtesy Massimo De Carlo, Milan/London

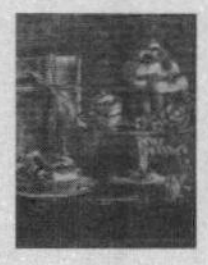

72.
Cupcakes, 2012
Charcoal drawing, walnut frame
36.8 x 29.2 x 3.8 cm
(14.5 x 11.5 x 1.5 in)
Unique
Courtesy Massimo De Carlo, Milan/London

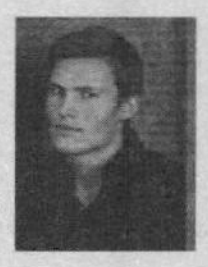

73.
Three Men, 2011
C-print, painted frame
29.2 x 36.8 x 3.8 cm
(11.5 x 14.5 x 1.5 in)
Edition 5 of 5
Courtesy of the artist and White Cube, London/Hong Kong/São Paulo

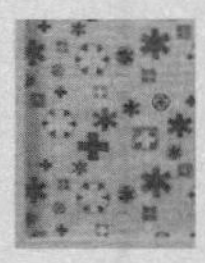

74.
Asterisks, 2012
C-print, painted frame
36.8 x 29.2 x 3.8 cm
(14.5 x 11.5 x 1.5 in)
Edition of 5, 2 AP
Courtesy Massimo De Carlo, Milan/London

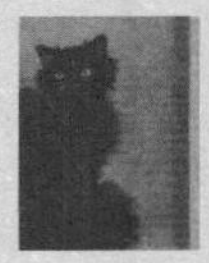

75.
Selkirk Rex, 2012
C-print, painted frame
36.8 x 29.2 x 3.8 cm
(14.5 x 11.5 x 1.5 in)
Edition of 5, 2 AP
Courtesy Massimo De Carlo, Milan/London

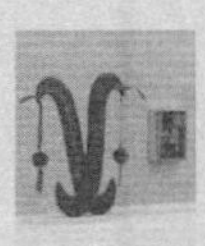

76.
Untitled (Wall, Milan Blue), 2012
Ceramic, ribbons
Dimensions variable
Unique
Courtesy Massimo De Carlo, Milan/London

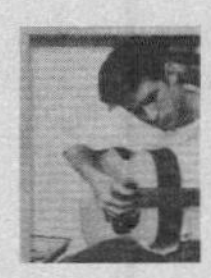

78.
Man (Guitar, Portrait) (detail), 2012
Silver gelatin prints, aluminum frames
21.6 x 26.7 x 3.8 cm
(8.5 x 10.5 x 1.5 in each)
Polyptych, 5 framed works
Edition of 3, 2 AP
Courtesy Massimo De Carlo, Milan/London

Milano
Comune di Milano
Cultura, Moda, Design

Mayor
Giuliano Pisapia

City Councillor for Culture
Filippo Del Corno

Cultural Director
Giulia Amato

PAC

Director
Domenico Piraina

Exhibition Coordination
Diego Sileo

Chiefs of Organization and Administration
Giovanni Bernardi, Simone Percacciolo

Organization
Giuliana Allievi, Luisella Angiari, Filomena Della Torre, Christina Schenk, Giulia Sonnante, Roberta Ziglioli

Administration
Roberta Crucitti, Valeria Giannelli, Laura Piermattei, Sonia Santagostino, Luisa Vitiello

Event Coordination
Anna Appratti

Chief of Technical Coordination
Paolo Arduini

Technical Coordination
Luciano Madeo, Lorenzo Monorchio, Andrea Passoni

Technical Coordination
Luciano Cantarutti

Communication and Promotion
Francesca La Placa, Antonietta Bucci

Press Office
Elena Conenna

Visual Communication
Dalia Gallico, Art Lab

Operational Assistance
Palma Di Giacomo, Giuseppe Premoli, Luciana Sacchi

Custodial Service
Corpo di Guardia Villa Reale/PAC

Exhibition

Curator
Alessandro Rabottini

Installation Project and Design
Muon

Graphic Design
Thomas Berloffa / auricchioberloffa.com

Transport
Piccin Arte Overdriveart

Insurance
Lloyd's, London

Conservation and Condition Report
Studio Formica

We are grateful to the following lenders
Mary Leigh Cherry and Tony de los Reyes, Massimo De Carlo, Hadley Martin Fisher, David Kordansky Gallery, Rennie Collection, Ringier Collection, Amy and Harris Schwalb, David Simkins, White Cube, and to all those who have asked to remain anonymous

Thanks also go to
Patrizia Bonadei, Enza Bruno, Paolo Daffara, Theresa Flemming, Cathrine Forni, Samantha Gregg, William Parks, Carola Quirico, Laura Ravelli, Christina Ruf, Simone Strummiello, Agnello Vitiello

PAC Sponsor
TOD'S

Teaching Activities
Marte
Marta Ferina & Co.

Sponsor for Teaching Activities
coop Lombardia

Imprint

Published by
Mousse Publishing

Editor
Alessandro Rabottini

Editorial Coordination
Antonio Scoccimarro, Ilaria Bombelli

With
Anna Peres

Texts
Jörg Heiser, Elad Lassry, Aram Moshayedi, Alessandro Rabottini

Editing and Proofreading
Johanna Bishop, Chiara Leoni

Translations
Stefano Asperti, Johanna Bishop

Design
Marco Fasolini, Fausto Giliberti, Andrea Novali, Francesco Valtolina

Typeface
Hercules Text, Neuzeit Book Heavy

Photo credits
All images courtesy of the artist, except for pp. 22, 24, 30, 36, 42, 50, 52, 56, 60, 62, 66, 70, 76, courtesy Alessandro Zambianchi

Color separation and print
Artigianelli, Brescia

This publication has been made possible thanks to
Massimo De Carlo, Milan/London

Further support was received from
303 Gallery, New York
David Kordansky Gallery, Los Angeles
Galerie Francesca Pia, Zurich
White Cube, London/Hong Kong/São Paulo